**JOHN W. TUKEY**
*Princeton University and
Bell Telephone Laboratories*

# Exploratory Data Analysis

**ADDISON-WESLEY PUBLISHING COMPANY**

*Reading, Massachusetts • Menlo Park, California*

*London • Amsterdam • Don Mills, Ontario • Sydney*

This book is in the

ADDISON-WESLEY SERIES IN

BEHAVIORAL SCIENCE: QUANTITATIVE METHODS

Consulting Editor:
FREDERICK MOSTELLER

The writing of this work, together with some of the research reported herein, was supported in part by the United States Government.

Copyright © 1977 by Addison-Wesley Publishing Company, Inc. Philippines copyright 1977 by Addison-Wesley Publishing Company, Inc.

All rights reserved. No part of this publication may be reproduced, stored in a retrieval system, or transmitted, in any form or by any means, electronic, mechanical, photocopying, recording, or otherwise, without the prior written permission of the publisher. Printed in the United States of America. Published simultaneously in Canada. Library of Congress Catalog Card No. 76-5080.

ISBN 0-201-07616-0

Dedicated to the memory of
CHARLIE WINSOR, biometrician, and
EDGAR ANDERSON, botanist,
data analysts both,
from whom the author learned much that
could not have been learned elsewhere

# Preface

This book is based on an important principle:

**It is important to understand what you CAN DO before you learn to measure how WELL you seem to have DONE it.**

Learning first what you can do will help you to work more easily and effectively.

This book is about exploratory data analysis, about looking at data to see what it seems to say. It concentrates on simple arithmetic and easy-to-draw pictures. It regards whatever appearances we have recognized as partial descriptions, and tries to look beneath them for new insights. Its concern is with appearance, not with confirmation.

### Examples, NOT case histories

The book does not exist to make the case that exploratory data analysis is useful. Rather it exists to expose its readers and users to a considerable variety of techniques for looking more effectively at one's data. The examples are not intended to be complete case histories. Rather they show isolated techniques in action on real data. The emphasis is on general techniques, rather than specific problems.

A basic problem about any body of data is to make it more easily and effectively handleable by minds--our minds, her mind, his mind. To this general end:

◇ anything that makes a simpler description possible makes the description more easily handleable.

◇ anything that looks below the previously described surface makes the description more effective.

So we shall always be glad (a) to simplify description and (b) to describe one layer deeper.

In particular:

◇ to be able to say that we looked one layer deeper, and found nothing, is a definite step forward--though not as far as to be able to say that we looked deeper and found thus-and-such.

◇ to be able to say that "if we change our point of view in the following way . . . things are simpler" is always a gain--though not quite as much as to be able to say that "if we don't bother to change our point of view (some other) things are equally simple".

Thus, for example, we regard learning that log pressure is almost a straight line in the negative reciprocal of absolute temperature as a real gain, as compared to saying that pressure increases with temperature at an evergrowing rate. Equally, we regard being able to say that a batch of values is roughly symmetrically distributed on a log scale as much better than to say that the raw values have a very skew distribution.

In rating ease of description, after almost any reasonable change of point of view, as very important, we are essentially asserting a belief in quantitative knowledge--a belief that most of the key questions in our world sooner or later demand answers to "by how much?" rather than merely to "in which direction?".

Consistent with this view, we believe, is a clear demand that pictures based on exploration of data should *force* their messages upon us. Pictures that emphasize what we already know--"security blankets" to reassure us--are frequently not worth the space they take. Pictures that have to be gone over with a reading glass to see the main point are wasteful of time and inadequate of effect. **The greatest value of a picture** is when it *forces* us to notice **what we never expected to see.**

We shall not try to say why specific techniques are the ones to use. Besides pressures of space and time, there are specific reasons for this. Many of the techniques are less than ten years old in their present form--some will improve noticeably. And where a technique is very good, it is not at all certain that we yet know why it is.

We have tried to use consistent techniques wherever this seemed reasonable, and have not worried where it didn't. Apparent consistency speeds learning and remembering, but ought not to be allowed to outweigh noticeable differences in performance.

In summary, then, we:

⋄ leave most interpretations of results to those who are experts in the subject-matter field involved.

⋄ present techniques, not case histories.

⋄ regard simple descriptions as good in themselves.

⋄ feel free to ask for changes in point of view in order to gain such simplicity.

⋄ demand impact from our pictures.

⋄ regard every description (always incomplete!) as something to be lifted off and looked under (mainly by using residuals).

⋄ regard consistency from one technique to another as desirable, not essential.

### Confirmation

The principles and procedures of what we call confirmatory data analysis are both widely used and one of the great intellectual products of our century.

In their simplest form, these principles and procedures look at a sample--and at what that sample has told us about the population from which it came--and assess the precision with which our inference from sample to population is made. We can no longer get along without confirmatory data anlysis. **But we need not start with it.**

The best way to **understand what CAN be done is no longer**--if it ever was--**to ask what things could,** in the current state of our skill techniques, **be confirmed** (positively or negatively). Even more understanding is *lost* if we consider each thing we can do to data *only* in terms of some set of very restrictive assumptions under which that thing is best possible--assumptions we *know we CANNOT check in practice*.

### Exploration AND confirmation

Once upon a time, statisticians only explored. Then they learned to confirm exactly--to confirm a few things exactly, each under very specific circumstances. As they emphasized exact confirmation, their techniques inevitably became less flexible. The connection of the most used techniques with past insights was weakened. Anything to which a confirmatory procedure was not explicitly attached was decried as "mere descriptive statistics", no matter how much we had learned from it.

Today, the flexibility of (approximate) confirmation by the jackknife makes it relatively easy to ask, for almost any clearly specified exploration, "How far is it confirmed?"

**Today, exploratory and confirmatory can--and should--proceed side by side.** This book, of course, considers only exploratory techniques, leaving confirmatory techniques to other accounts.

### Relation to the preliminary edition

The preliminary edition of *Exploratory Data Analysis* appeared in three volumes, represented the results of teaching and modifying three earlier versions, and had limited circulation. Complete restructuring and revision was followed by further major changes after the use of the structure and much of the material in an American Statistical Association short course. The present volume contains:

- ◇ those techniques from the first preliminary volume that seemed to deserve careful attention.
- ◇ a selection of techniques from the second preliminary volume.
- ◇ a few techniques from the third preliminary volume.
- ◇ some techniques (especially in chapters 7, 8, and 17) that did not apppear in the preliminary edition at all.

It is to be hoped that the preliminary edition will reappear in microfiche form.

### About the problems

The teacher needs to be careful about assigning problems. Not too many, please. They are likely to take longer than you think. The number supplied is to accommodate diversity of interest, not to keep everyone busy.

Besides the length of our problems, both teacher and student need to realize that many problems do not have a single "right answer". There can be many ways to approach a body of data. Not all are equally good. For some bodies of data this may be clear, but for others we may not be able to tell from a single body of data which approach is preferred. Even several bodies of data about very similar situations may not be enough to show which approach should be preferred. Accordingly, it will often be quite reasonable for different analysts to reach somewhat different analyses.

Yet more--to unlock the anlysis of a body of data, to find the good way or ways to approach it, may require a key, whose finding is a creative act. Not everyone can be expected to create the key to any one situation. And, to continue to paraphrase Barnum, no one can be expected to create a key to each situation he or she meets.

**To learn about data anlysis, it is right that each of us try many things that do not work**--that we tackle more problems than we make expert analyses of. We often learn less from an expertly done analysis than from one where, by not trying something, we missed--at least until we were told about it--an opportunity to learn more. Each teacher needs to recognize this in grading and commenting on problems.

### Precision

The teacher who heeds these words and admits that there need be *no one correct approach* may, I regret to contemplate, still want whatever is done to be digit-perfect. (Under such a requirement, the writer should still be able to pass the course, but it is not clear whether he would get an "A".) One does, from time to time, have to produce digit-perfect, carefully checked results, but forgiving techniques that are not too distributed by unusual data are also, usually, *little disturbed by SMALL arithmetic errors*. The techniques we discuss here have been chosen to be forgiving. It is to be hoped, then, that small arithmetic errors will take little off the problem's grades, leaving severe penalties for larger errors, either of arithmetic or of concept.

### Acknowledgments

It is a pleasure to acknowledge support, guidance, cooperation, and hard work. Both the Army Research Office (Durham), through a contract with Princeton University, and the Bell Telephone Laboratories have supported the writing financially. Charles P. Winsor taught the writer, during the 1940's, many things about data analysis that were not in the books. Although its formal beginning came after his death, this book owes much to S. S. Wilks, whose leadership for statistics in Princeton made possible the gaining of the insights on which it is based.

Careful reading of earlier versions by friends and colleagues, especially David Hoaglin and Leonard Steinberg, was most helpful, as were comments by those who have taught the course at various institutions. As noted above, student reaction led to many changes. Frederick Mosteller took his editorial responsibilities very seriously; the reader owes him thanks for many improvements. The arithmetic is much more nearly correct because of the work of Ms. Agelia Mellros.

Careful and skilled typing, principally by Mrs. Mary E. Bittrich and by Mrs. Elizabeth LaJeunesse Dutka (earlier versions), with significant contributions by Mrs. Glennis Cohen and Mrs. Eileen Olshewski, has been of vital importance. The cooperative attitude and judgment of the Addison-Wesley staff, particularly that of Roger Drumm (without whose long-continued encouragement this book might not be a reality), Mary Cafarella (production editor), Marshall Henrichs (designer), and Richard Morton (illustrator), were of great help.

*Princeton, New Jersey*                                                                              John W. Tukey
*Christmas, 1976*

# To the Student or Teacher

Everything illustrated or set as a problem can be done with pencil and paper. (If you have a hand-held calculator, fine.) The only tools the illustrator used (except for a few pictures that do not reflect the analysis of data) were a pen and a straightedge. Each of you could make pictures that are almost as nice if you tried moderately hard. (How to use graph paper effectively is discussed in Sections 2C and 5A.)

The first six chapters are the basic trunk from which everything grows. They belong in any version. What comes next?

It would be more conventional, and more useful for what people are used to doing, to go to chapters 10 and 11 on two-way plots, and, if time permitted, to their extensions in chapters 12 and 13.

I was, however, sufficiently impressed with the usefulness of smoothing (mainly from using it myself) to put chapter 7, and its applications to $(x,y)$ batches in chapters 8 and 9, next. Both students-on-their-own and classes may wish to experiment by following chapters 1 to 6 with either 10, 11, etc., or 7, etc.

Chapter 14 is of central importance, but it is not easy. I would have liked to put it earlier, but could not find a good enough excuse. (It has to follow chapter 7.)

Chapter 15 deals with the important question of fractions based upon counts. It could follow either chapters 1 to 6, chapters 10 and 11 (etc.), chapter 7 (etc.), or chapter 14. Where it fits depends on the kinds of data a particular group of students are likely to be interested in.

Chapter 17, which is based on chapter 16, deals with making sense out of ordinary "distributions" of numerical values into bins. Chapters 16 and 17 could follow chapters 1 to 6 directly, but it seems unlikely that they would deserve such priority. Chapter 18 deals with a special kind of distributional situation that is neither very common nor all that rare. Many teachers and many students-on-their-own may wish to put it on the back burner.

Chapters 19 and 20 are based on Chapter 17. They help tie the techniques used in that chapter back to what may be more familiar ground. For those who have already studied such standard distributions, these chapters will surely earn their keep. For others, their priority is quite low. These chapters are not likely to help anyone analyze data better; they may help in thinking about the results of chapter 17 analyses.

In summary, then, the basic group of chapters (1 to 6) can be followed by any or all of: smoothing (chapter 7ff), two-way analyses (chapters 10 and 11 and ff), essential points about $(x,y)$ batches (chapter 14), counted fractions (chapter 15), empirical distributions (chapters 16 and 17 and ff). After the basic group, chapters 7, 10–11, 14, and 15 are probably next most essential, unless chapter 17 (really chapter 16ff) is specially important to the particular group involved.

John Tukey

# Contents

| | | |
|---|---|---|
| **1** | **SCRATCHING DOWN NUMBERS (stem-and-leaf)** | **1** |
| | Comments about the index page | 2 |
| | 1A Quantitative detective work | 1 |
| | 1B Practical arithmetic | 3 |
| | 1C Scratching down numbers | 6 |
| | 1D Doing better with stem-and-leaf | 7 |
| | 1E Using the right number of stems | 11 |
| | 1F How to count by tallying | 16 |
| | 1G What does it mean to "feel what the data are like"? | 19 |
| | 1H How far have we come? | 20 |
| | 1K How to use stem-and-leaf to pick up additional information (optional technique) | 23 |
| | 1P Additional problems | 25 |
| **2** | **SCHEMATIC SUMMARIES (pictures and numbers)** | **27** |
| | 2A Extremes and median | 29 |
| | 2B Hinges and 5-number summaries | 32 |
| | 2C Box-and-whisker plots | 39 |
| | 2D Fences, and outside values | 43 |
| | 2E Schematic plots | 47 |
| | 2F Pros and cons; the Rayleigh example | 49 |
| | 2G Eighths, sixteenths, etc. | 53 |
| | 2H How far have we come? | 55 |
| **3** | **EASY RE-EXPRESSION** | **57** |
| | 3A Logarithms = logs | 59 |
| | 3B Quick logs | 61 |
| | 3C Comparisons of two batches | 64 |
| | 3D Quick roots and quick reciprocals | 69 |
| | 3E Looking quickly | 79 |
| | 3F Counted data | 83 |
| | 3G Relation among powers and logs (optional) | 86 |
| | 3H How far have we come? | 92 |
| | 3K How to think about logs (background) | 93 |
| | 3P Additional problems | 93 |
| **4** | **EFFECTIVE COMPARISON (including well-chosen expression)** | **97** |
| | 4A Alternative forms of display of summaries | 99 |
| | 4B Comparing several batches (continued) | 102 |
| | 4C A more extensive example | 105 |

|     |     |     |
| --- | --- | --- |
| 4D  | The meaning of comparison | 110 |
| 4E  | Adjustments, rough and exact | 110 |
| 4F  | Residuals | 113 |
| 4H  | How far have we come? | 115 |
| 4P  | Additional problems | 116 |

## 5  PLOTS OF RELATIONSHIP — 125

|     |     |     |
| --- | --- | --- |
| 5A  | How to plot $y$ against $x$ | 126 |
| 5B  | Looking at subtraction | 131 |
| 5C  | Subtracting straight lines | 135 |
| 5D  | Plotting the population of the U.S.A. | 141 |
| 5E  | Plotting the ratio of births to deaths | 148 |
| 5F  | Untilting defines "tilt" | 154 |
| 5H  | How far have we come? | 156 |
| 5P  | Additional problems | 157 |

## 6  STRAIGHTENING OUT PLOTS (using three points) — 169

|     |     |     |
| --- | --- | --- |
| 6A  | Looking at three points | 171 |
| 6B  | Re-expressing $y$ alone | 172 |
| 6C  | Re-expressing $x$ alone | 175 |
| 6D  | A braking example | 181 |
| 6E  | The vapor pressure of $H_2O$ | 187 |
| 6F  | Re-expressing the second variable | 191 |
| 6G  | Wise change of origin as a preliminary | 193 |
| 6H  | How far have we come? | 197 |
| 6P  | Additional problems | 199 |

## 7  SMOOTHING SEQUENCES — 205

|     |     |     |
| --- | --- | --- |
| 7A  | Medians of 3 | 210 |
| 7B  | Eye resmoothing | 214 |
| 7C  | Looking ahead | 216 |
| 7D  | Copying-on--and more, usually. | 221 |
| 7E  | Blurring the smooth--and setting the fences | 223 |
| 7F  | Splitting peaks and valleys | 227 |
| 7G  | Hanning | 231 |
| 7H  | How far have we come? | 235 |

## 7⁺  OPTIONAL SECTIONS FOR CHAPTER 7 — 237

|     |     |     |
| --- | --- | --- |
| 7I  | Breaking a smooth | 237 |
| 7J  | Choice of expression | 247 |
| 7K  | A two-section example | 259 |
| 7M  | How much more may we have learned? | 264 |

## 8  PARALLEL AND WANDERING SCHEMATIC PLOTS — 265

|     |     |     |
| --- | --- | --- |
| 8A  | Parallel schematic plots | 265 |
| 8B  | Smoothing the cross-medians | 274 |
| 8C  | Smoothing broken hinges | 276 |
| 8D  | Dealing with the two questions | 279 |

|   |     |                                                                       |     |
|---|-----|-----------------------------------------------------------------------|-----|
|   | 8E  | Wandering schematic plots                                             | 283 |
|   | 8F  | A more demanding example: Governor's salary and bank deposits         | 287 |
|   | 8G  | Further questions/analysis in the example                             | 298 |
|   | 8H  | How far have we come?                                                 | 306 |
|   | 8I  | The need to smooth both coordinates (optional)                        | 307 |
| **9**  | **DELINEATIONS OF BATCHES OF POINTS**                          | **309** |
|   | 9A  | E-traces and D-traces                                                 | 309 |
|   | 9B  | Simple delineation––Twin Rivers again                                 | 311 |
|   | 9C  | Reduced and schematic delineations                                    | 313 |
|   | 9D  | What our schematic plots and delineations have missed                 | 319 |
|   | 9E  | Three variables at once––or more                                      | 321 |
|   | 9H  | How far have we come?                                                 | 329 |
| **10** | **USING TWO-WAY ANALYSES**                                     | **331** |
|   | 10A | Two-way residuals; row-PLUS-column analysis                           | 332 |
|   | 10B | The row-PLUS-column fit                                               | 337 |
|   | 10C | Some points of technique                                              | 343 |
|   | 10D | Row-TIMES-column analysis                                             | 344 |
|   | 10E | Looking at row-PLUS-column fits and their residuals                   | 349 |
|   | 10F | Fitting one more constant                                             | 352 |
|   | 10G | Converting PLUS to TIMES; re-expression                               | 358 |
|   | 10H | How far have we come?                                                 | 360 |
| **11** | **MAKING TWO-WAY ANALYSES**                                    | **362** |
|   | 11A | Taking medians out                                                    | 363 |
|   | 11B | Alternative organizations of the arithmetic                           | 372 |
|   | 11C | Making the core of a two-way plot                                     | 374 |
|   | 11D | Going on with the residuals                                           | 378 |
|   | 11E | Coding residuals; condensing fits and residuals                       | 382 |
|   | 11F | We can combine!                                                       | 390 |
|   | 11G | Guidance for expression                                               | 396 |
|   | 11H | How far have we come?                                                 | 399 |
| **11⁺**| **OPTIONAL SECTIONS FOR CHAPTERS 10 AND 11**                   | **401** |
|   | 11I | Exploring beyond PLUS-one (extends Chapter 10)                        | 401 |
|   | 11J | Taking out any summary                                                | 404 |
|   | 11K | An example of re-expression––city killings                            | 408 |
|   | 11L | An unusual fit                                                        | 415 |
|   | 11M | How much more may we have learned?                                    | 419 |
| **12** | **ADVANCED FITS**                                              | **420** |
|   | 12A | PLUS-one fits                                                         | 421 |
|   | 12B | Pictures for "-PLUS-one" fits                                         | 424 |
|   | 12C | Making those pictures                                                 | 428 |
|   | 12D | Sometimes we can have parallel-line plots, still                      | 431 |
|   | 12E | More extended fits                                                    | 433 |
|   | 12F | Simplification is sometimes possible                                  | 438 |
|   | 12H | How far have we come?                                                 | 441 |

## 13 THREE-WAY FITS — 443
- 13A Three- and more-way analyses: Arrangement and tagging — 443
- 13B An analysis of the psychological example — 448
- 13C Making three-way analyses — 452
- 13D Three-way re-expression — 458
- 13E More about the example — 462
- 13H How far have we come? — 465

## 14 LOOKING IN TWO OR MORE WAYS AT BATCHES OF POINTS — 466
- 14A Coordinates and level traces — 467
- 14B Different middle traces for the same slices — 470
- 14C An explanation — 475
- 14D Changing the slicing coordinate — 476
- 14E What matters? — 481
- 14F Rematching and strength of relationship — 482
- 14H How far have we come? — 491
- 14I The ubiquity of medians (optional section) — 492

## 15 COUNTED FRACTIONS — 494
- 15A Started counts and counted fractions — 496
- 15B Three matched scales for counted fractions — 498
- 15C Quicker calculation — 502
- 15D Examples where careful expression clearly pays off — 508
- 15E Double folding--the 2 × 2 case — 513
- 15F Double folding--larger cases — 516
- 15G Easy froots and flogs with a slide rule (optional) — 520
- 15H How far have we come? — 522

## 16 BETTER SMOOTHING — 523
- 16A Reroughing — 523
- 16B Some examples — 526
- 16C If we want things still smoother — 531
- 16D Further possibilities — 534
- 16H How far have we come? — 542

## 17 COUNTS in BIN after BIN — 543
- 17A Root smooth and root rough· — 543
- 17B Counts of basic counts — 550
- 17C Fitting to smoothed roots — 555
- 17D Corn borers, wheat prices, and Student's simulations — 561
- 17E Bins of unequal width — 570
- 17F Double roots — 576
- 17G Cautionary examples — 582
- 17H How far have we come? — 587

## 18 PRODUCT-RATIO PLOTS — 588
- 18A Sizes and counts — 589
- 18B Product-ratio analysis — 594
- 18C Forcing the unusual to be noticed — 598

|      |                                                          |      |
|------|----------------------------------------------------------|------|
| 18D  | Comparisons between collections                          | 602  |
| 18E  | Looking at the smallest basic count                      | 604  |
| 18F  | When zeros are counted                                   | 605  |
| 18G  | Under the microscope                                     | 608  |
| 18H  | How far have we come?                                    | 612  |

## 19 SHAPES OF DISTRIBUTION — 614

|      |                                                          |      |
|------|----------------------------------------------------------|------|
| 19A  | Looking at shapes of distribution                        | 616  |
| 19B  | The Gaussian reference                                   | 623  |
| 19C  | Using letter values to look at shapes of distribution    | 626  |
| 19D  | Pushback technique (optional section)                    | 637  |
| 19H  | How far have we come?                                    | 644  |

## 20 MATHEMATICAL DISTRIBUTIONS — 646

|      |                                                                |      |
|------|----------------------------------------------------------------|------|
| 20A  | Binnings vs. distributions                                     | 648  |
| 20B  | Densities for distributions vs. densities for binnings         | 651  |
| 20C  | Tables and pictures comparing two sets of shapes of distributions | 654 |
| 20H  | How far have we come?                                          | 661  |

## 21 POSTSCRIPT — 662

|      |                                                          |      |
|------|----------------------------------------------------------|------|
| 21A  | Our relationship to the computer                         | 663  |
| 21B  | What has been omitted?                                   | 664  |
| 21C  | How should the past chapters look different?             | 665  |
| 21D  | What have we been introduced to?                         | 666  |

**GLOSSARY** — 667

**INDEX TO REFERENCE TABLES** — 677

**ALPHABETICAL INDEX** — 677

**FRONTPAPERS**
1. Break table for two-decimal logs
2. Break table for (square) roots
3. Main break table--digits of negative reciprocals

**REARPAPERS**
4. Pluralities, folded roots, folded logarithms
5. Values of $\log_e \sqrt{\text{count} + \frac{1}{6}}$
6. Values of $\sqrt{\text{count} + \frac{1}{6}}$

# Scratching down numbers (stem-and-leaf)

1

chapter index on next page

## 1A. Quantitative detective work

**Exploratory data analysis is detective work**--numerical detective work--or counting detective work--or graphical detective work.

A detective investigating a crime needs both tools and understanding. If he has no fingerprint powder, he will fail to find fingerprints on most surfaces. If he does not understand where the criminal is likely to have put his fingers, he will not look in the right places. Equally, the analyst of data needs both tools and understanding. It is the purpose of this book to provide some of each.

Time will keep us from learning about many tools--we shall try to look at a few of the most general and powerful among the simple ones. **We do not guarantee to introduce you to the "best" tools, particularly since we are not sure that there can be unique bests.**

Understanding has different limitations. As many detective stories have made clear, one needs quite different sorts of detailed understanding to detect criminals in London's slums, in a remote Welsh village, among Parisian aristocrats, in the cattle-raising west, or in the Australian outback. We do not expect a Scotland Yard officer to do well trailing cattle thieves, or a Texas ranger to be effective in the heart of Birmingham. Equally, very different detailed understandings are needed if we are to be highly effective in dealing with data concerning earthquakes, data concerning techniques of chemical manufacturing, data concerning the sizes and profits of firms in a service industry, data concerning human hearing, data concerning suicide rates, data concerning population growth, data concerning fossil dinosaurs, data concerning the genetics of fruit flies, or data concerning the latest exploits in molecular biology. A full introduction to data analysis in any one of these fields--or in any of many others--would take much more time than we have.

The Scotland Yard detective, however, would be far from useless in the wild west or the outback. He has certain general understandings of conventional detective work that will help him anywhere.

**In data analysis there are similar general understandings. We can hope to lead you to a few of them. We shall try.**

The processes of criminal justice are clearly divided between the search for the evidence--in Anglo-Saxon lands the responsibility of the police and other

# index for chapter 1

Comments about the index page   2
**1A. Quantitative detective work**   **1**
  review questions   3

**1B. Practical arithmetic**   **3**
  precision of arithmetic   3
  rounding   4
  cutting   4
  units and decimal points   4
  h for "and a half"   5
  * for place filler   5
  # for count   5
  review questions   6

**1C. Scratching down numbers**   **6**
  *batch*   6
  bold figures   6
  review questions   7

**1D. Doing better with stem-and-leaf**   **7**
  *stem-and-leaf displays*   8
  checking   10
  review questions   11

**1E. Using the right number of stems**   **11**
  stretched stem-and-leaf   11
  squeezed stem-and-leaf   12
  using mixed leaves   12
  review questions   16

**1F. How to count by tallying**   **16**
  used cars again   18
  review questions   19

**1G. What does it mean to "feel what the data are like"?**   **19**
  review questions   20

**1H. How far have we come?**   **20**
  what have we learned to do?   21
  where do we stand?   21

**1K. How to use stem-and-leaf to pick up additional information (optional technique)**   **23**
  parts of numbers   23

**1P. Additional Problems**   **24**

| EXHIBIT | | PAGE |
|---|---|---|
| | 1A | |
| | 1B | |
| | 1C | |
| 1★ | | 7 |
| | 1D | |
| 2★ | | 8 |
| 3 | | 9 |
| | 1E | |
| 4★ | | 12 |
| 5★ | | 13 |
| 6★ | | 14 |
| 7★ | | 15 |
| | 1F | |
| 8★ | | 17 |
| 9★ | | 18 |
| | 1G | |
| 10★ | | 20 |
| | 1H | |
| 11★ | | 22 |
| | 1K | |
| 12 | | 24 |
| | 1P | |
| 13★ | | 25 |
| 14★ | | 26 |
| 15★ | | 26 |

★ = includes problem(s)

**Comments about the Index Page**

The index page for this chapter--and for others--is *nonstandard*.
Two sorts of additional entries are given between the section titles:

◊ subtopics, printed in lightface type in the index, and centered in text.

◊ vital terms defined, printed in italics in the index when short, and displayed like formulas in text.

To the right of the index page we give pages for numbered "exhibits". This neutral word is used for tables, charts, graphs, plots, and the like. They are all numbered in a single sequence within each chapter, so they will be easy to find.
Their first main appearance will be identified by using bold numbers, as **14**, for *that* appearance and not for others.

investigative forces--and the evaluation of the evidence's strength--a matter for juries and judges. In data analysis a similar distinction is helpful. Exploratory data analysis is detective in character. Confirmatory data analysis is judicial or quasi-judicial in character. Only exploratory data analysis will be our subject here.

Unless the detective finds the clues, judge or jury has nothing to consider. **Unless exploratory data analysis uncovers indications, usually quantitative ones, there is likely to be nothing for confirmatory data analysis to consider.**

Experiments and certain planned inquiries provide some exceptions and partial exceptions to this rule. They do this because one line of data analysis was planned as part of the experiment or inquiry. **Even here, however, restricting one's self to the planned analysis--failing to accompany it with exploration--loses sight of the most interesting results too frequently to be comfortable.**

As all detective stories remind us, many of the circumstances surrounding a crime are accidental or misleading. Equally, many of the indications to be discerned in bodies of data are accidental or misleading. To accept all appearances as conclusive would be destructively foolish, either in crime detection or in data analysis. **To fail to collect all appearances because some--or even most--are only accidents would, however, be gross misfeasance deserving (and often receiving) appropriate punishment.**

> **Exploratory data analysis can never be the whole story, but nothing else can serve as the foundation stone--as the first step.**

We will be exploring numbers. We need to handle them easily and look at them effectively. Techniques for handling and looking--whether graphical, arithmetic, or intermediate--will be important. The simpler we can make these techniques, the better--so long as they work, and work well. When details make an important difference, they deserve--and will get--emphasis.

### review questions

What is exploratory data analysis? How is it related to confirmatory data analysis? How is preplanned analysis related to exploratory data analysis? Should we look only at appearances we are sure are correct?

## 1B. Practical arithmetic

A few arithmetic details deserve our attention before we go further.

### precision of arithmetic

We shall feel free to be inconsistent about keeping extra decimal places or rounding off. We shall try to use reasonable care and good judgment in view of the total behavior of the data involved. Thus, for example, we may do either of these:

$(1/3)(19 + 19 + 20) = 19.3$     (rest of data close by)
$(1/3)(19 + 19 + 20) = 19$      (rest of data farther away)

### rounding

When we do round, it will be to the nearest number, with ties broken by preferring even numbers. Thus:

$17.5 \rightarrow 18$
$18.5 \rightarrow 18$
$19.5 \rightarrow 20$
$20.5 \rightarrow 20$

As a result, we may have

$(1/2)(19 + 20) = 19.5$     (rest of data close by)
$(1/2)(19 + 20) = 20$       (rest of data farther away)
$(1/2)(20 + 21) = 20.5$     (rest of data close by)
$(1/2)(20 + 21) = 20$       (rest of data farther away)

### cutting

We often prefer to cut numbers by just dropping off the extra digits. This is somewhat quicker and easier, especially if one is not used to rounding.

It is hard to say just what is the proper balance between cutting and rounding. If more detail is easily available (say on the same page or an adjacent page), then surely one may as well cut. If the data is a matter of deep personal concern, then it is almost certain that one will round, if only for conscience's sake. We shall do both. We recommend to the reader that he or she practice both.

### units and decimal points

Numbers can be written down for storage--live or dead--or examination. We will almost always write them down to be looked at, either as part of the description of a computation or as part of results--final or intermediate.

When writing down a single number, keeping both raw units and decimal points often helps. To say that there are 200,000,000 people in the U.S., or that a sheet of plastic film is 0.00127 inches thick, is likely to convey about as much as we can.

Once we have a few numbers to look at, it usually helps us to move to easily understandable units that are more helpfully related to the numbers at hand. A population given as 201 millions--or 201.2--is likely to be seen more clearly than if it is given as 201,234,567. To give the thickness of a plastic film as 1.27 thousandths of an inch, instead of 0.00127 inches, is equally likely to be helpful.

Once we have more than a few numbers to look at, decimal points are a liability. They break up the appearance of the numbers, making it harder to make comparisons, especially among more than two numbers. The cure is to shift to units that are as convenient for the numbers at hand as possible. Instead of 201.2, 127.3, 63.4, ... millions of people, we do better to work with 2012, 1273, 634, ... hundreds of thousands. Instead of thicknesses of 1.27, 2.52, 0.62, 3.83, ... thousandths of an inch, we do better to work with 127, 252, 62, 384, ... hundredths of thousandths of an inch.

To be clear about what we are doing, it is often convenient to label parts (panels) of a table with the units involved. Thus, such headings as:

A) POPULATION--in hundreds of thousands

A) POPULATION--unit = 100,000 people

A) POPULATION--unit = $10^5$ people

B) FILM THICKNESS--unit = 0.00001 inch

B) FILM THICKNESS--unit = $10^{-5}$ inch

will occur fairly often.

### h for "and a half"

Even if the data is in whole numbers (integers), the numbers we calculate from it are often fractions. We try to keep our arithmetic simple. As a result, quite a few of these fractions are halves.

We would like to avoid 4.5, 13.5, and other distracting appearances of ".5". While not yet used extensively, the use of an "h", without space or decimal point, to stand for ".5" or "and a half" is very helpful. We will often use it (especially in later chapters).

When we want to count by halves, then, we may go 4, 4h, 5, 5h, 6, instead of 4, 4.5, 5, 5.5, 6, or 4.0, 4.5, 5.0, 5.5, 6.0.

### "*" for place filler

Digits in numbers often serve only as place-holders. A population reported as 201234567 is not likely to be accurately known to so many figures (though the figures given may accurately represent the number of people found and counted). We often gain by having a simple notation for a place filler. Here we shall use "*", always cutting the value. Thus 20123**** stands for a number between 201230000 and 201239999 that we do not choose to give as 20123(cut) tens of thousands.

### "#" for count

We shall often use "#" for "count" or "number" wherever its use seems helpful.

### review questions

Do we have to be consistent about keeping extra decimal places? How does rounding differ from cutting? How do we round halves? Why is a good choice of working units important? What is the effect of decimal points when handling numbers in bulk? What is meant by "137h"? By "137*"? By "#"?

### 1C. Scratching down numbers

Probably the simplest task that the reader is likely not to be either equipped or accustomed to do is to write down a batch of numbers in such a way as to give a general feel of "what they are like". Here, by a

### batch

of numbers we mean a set of similar values, obtained however they may have been. Simple examples might be (i) the weights of 21 students in a freshman class, (ii) the total winter snowfall at a selected place for each of the last 10 winters, (iii) the total sales this year of the 14 life-insurance agents who sold the most insurance last year, (iv) the number of electricity failures during the last decade in each of 11 high-income suburbs of New York City, (v) the numbers of mites found on each of 49 rats.

In each of these cases we are likely to desire a feel for the set of values as a whole. We may also want to copy them down, from whatever source we may have found them in, in a form that will be easy to use for almost any purpose. If one technique will serve both ends, so much the better. If two versions of the same technique are needed, we will use them.

Let us begin with the used-car advertisements in the *Sunday Standard-Times* of New Bedford, Massachusetts for 18 August 1968. Some 18 ads list three or more cars. The numbers listed are, in order: 6, 7, 7, 3, 5, 7, 3, 11, 16, 4, 17, 17, 6, 7, 9, 4, 7, 5. Just writing them down gives some idea of how many cars to an ad. Surely, however, we can do better. Exhibit **1** shows a way to write these numbers down that gives us more of a picture of them as a whole. In panel A each number is indicated by an "X". In panel B we gain a little by substituting the last digit for the X's, thus giving us something more to grasp with the flying eye, and providing a partial check on whether we put each mark in the correct line.

### bold figures

In the first exhibit of the first chapter we have already begun to use bold-face figures to distinguish one kind of entry from another. What is the user of pencil and paper supposed to do? Surely, if boldface helps in communication from writer to reader, something ought to be used to help in communication from scratcher-down to looker (who may be the same person at slightly different times).

Whenever we can do it, using **two colors** (or more than two colors in more complex situations) is likely to be the clearest method. (If the four-color BIC ballpoint is still available at a reasonable price, it provides an easy way to carry more than one color.) **Otherwise, the use of pen for bold and pencil for not bold works fairly well.**

At the last extreme, circling the numbers that would be bold is useful, though both more work and somewhat less effective.

### review questions

What is a batch? What is an example of a set of numbers that is not a batch? Why do we use bold figures? What choices have we in making a similar effect with pencil or pen? What are two ways of tallying numbers?

### 1D. Doing better with stem-and-leaf

The technique we have just been learning works well in some cases, but it is easy to find cases where it is not good enough. If we turn to the prices of second-hand Chevrolets (including Chevelles and Impalas) in the same 18 ads, we find the following (in dollars): 250, 150, 795, 895, 695, 1699, 1499, 1099, 1693, 1166, 688, 1333, 895, 1775, 895, 1895, 795. Running the eye over this string of numbers doesn't help too much. Neither would a direct application of the technique of the last section, since going from 150 to 1895 in single-dollar steps takes 1746 lines, of which only 17 would be marked.

exhibit **1** of chapter 1: cars/advertisement

**Tallying numbers of cars per advertisement in such a way as to give some feeling for the 18 numbers AS A WHOLE**

```
    A) CONVENTIONAL           B) DIGIT-REMINDER

 3 | x x                   3 | 33
 4 | x x                   4 | 44
 5 | x x                   5 | 55
 6 | x x                   6 | 66
 7 | x x x x x             7 | 77777
 8 |                       8 |
 9 | x                     9 | 9
10 |                      10 |
11 | x                    11 | 1
12 |                      12 |
13 |                      13 |
14 |                      14 |
15 |                      15 |
16 | x                    16 | 6
17 | x x                  17 | 77
```

C) PROBLEMS

1a) Collect a batch of data of interest to you.

1b) Display this data as in panel A.

1c) Do the same as in panel B.

We could, of course, drop off the last two digits, and keep only the complete hundreds of dollars. Cutting the numbers in this way would give: 2, 1, 7, 8, 6, 16, 14, 10, 16, 11, 6, 13, 8, 17, 8, 18, 7, which could easily be treated as in exhibit 1. We shall set this possibility aside for the moment, planning to come back to it shortly.

Let us return to panel B of exhibit 1. Exhibit **2** sets out the data when each line corresponds to two adjacent numbers of cars per ad--and when each corresponds to three adjacent numbers. Panels C and D of exhibit 2 show what happens when we condense 10 possible values on each line.

The panel-D version is quite efficient. One has even less to write down than if one were to write the given values down as is. Thus, 1∗ | 1677, for example, uses seven characters to represent 11, 16, 17, 17. The as-is form requires eight characters--or twelve if we count punctuation--rather than five--or seven.

The central idea of panel D is to give part of the information once for all at the start of each line and then give the rest of the information in the line, packed together as closely as makes sense. We shall call such displays

stem-and-leaf displays

Each line is a **stem**, each piece of information on a stem is a **leaf**. When, as

---

exhibit **2** of chapter 1: cars/advertisement

**The data and notation of panel B of exhibit 1 applied when each line includes 2, 3, or 10 possibilities**

(Note: Entries are final digits of observed numbers.)

A) 2 per LINE

```
 2-3 | 33
 4-5 | 4455
 6-7 | 6677777
 8-9 | 9
10-11| 1
12-13|
14-15|
16-17| 677
```

B) 3 per LINE

```
 3-5 | 334455
 6-8 | 6677777
 9-11| 91
12-14|
15-17| 677
```

C) 10 per LINE

```
 0-9 | 33445566777779
10-19| 1677
```

D) as C, ABBREVIATED--to be called SIMPLE STEM-and-LEAF

```
0∗ | 33445566777779
1∗ | 1677
```

E) PROBLEMS

2a) Display the data collected for (1a) as a simple stem-and-leaf display.

2b) Collect a batch of size at least 50, doing the collection in the form of a stem-and-leaf display.

here, the label for the stem is the first part of a number, to be followed by each leaf in turn, we call the label the **starting part**.

We like to use one-digit leaves, but--as we will see shortly--we can find the use of two- or more-digit leaves a good thing.

Any stem-and-leaf display is in terms of some unit. For our used-car prices, for example, $10 seems a convenient unit. Thus $250 becomes 25 and appears as 2∗|5.

What shall we do with $1099? We could cut to 109 or round to 110. Which should we do? This was discussed in the introduction to this chapter, but left open, as it will be here.

exhibit 3 of chapter 1: Chevrolet prices

**Three displays of 17 Chevrolet prices**

A) The PRICES--in dollars
   250, 150, 795, 895, 695, 1699, 1499, 1099, 1693, 1166, 688, 1333, 895, 1775, 895, 1895, 795.

B) The PRICES--cut to $10 units
   25, 15, 79, 89, 69, 169, 149, 109, 169, 116, 68, 133, 89, 177, 89, 189, 79.

| C) UNIT = $100[†] CHECK-OFF | D) UNIT = $10[†] STEM-and-LEAF one-digit leaf | E) UNIT = $1 STEM-and-LEAF two-digit leaf |
|---|---|---|
| 1 \| × | 1∗ \| 5 | 1∗∗ \| 50 |
| 2 \| × | 2 \| 5 | 2 \| 50 |
| 3 \| | 3 \| | 3 \| |
| 4 \| | 4 \| | 4 \| |
| 5 \| | 5∗ \| | 5∗∗ \| |
| 6 \| × × | 6 \| 98 | 6 \| 95, 88 |
| 7 \| × × | 7 \| 99 | 7 \| 95, 95 |
| 8 \| × × × | 8 \| 999 | 8 \| 95, 95, 95 |
| 9 \| | 9∗ \| | 9∗∗ \| |
| 10 \| × | 10 \| 9 | 10 \| 99 |
| 11 \| × | 11 \| 6 | 11 \| 66 |
| 12 \| | 12 \| | 12 \| |
| 13 \| × | 13∗ \| 3 | 13∗∗ \| 33 |
| 14 \| × | 14 \| 9 | 14 \| 99 |
| 15 \| | 15 \| | 15 \| |
| 16 \| × × | 16 \| 99 | 16 \| 99, 93 |
| 17 \| × | 17∗ \| 7 | 17∗∗ \| 75 |
| 18 \| × | 18 \| 9 | 18 \| 95 |

[†] Values cut, not rounded.

P) PROBLEM

3a) Explain, in words, the entries after each stem = 16.

Panel D of exhibit **3** shows the result of cutting our Chevrolet prices to a $10 unit and giving the results in standard stem-and-leaf form, namely with:

◇ all but the last digit remaining as the starting part, labelling the stem,

◇ the last remaining digit entered on that line as the leaf,

◇ enough starting parts followed by * to remind us, without unnecessary confusion, how many digits (here, one) are left for the leaf.

This display also shows the prices in $100 units à la panel A of exhibit **1**, and the full prices in a different stem-and-leaf display with two-digit leaves. Clearly:

◇ each of the three panels conveys a quite similar general picture of the 17 prices as a whole.

◇ the amount of detailed information increases as we go from panel C to panel D to panel E.

◇ the cost of this increase in information is negligible in going from panel C to panel D, but may not be negligible in going from panel D to panel E.

As a result, we see that we will do well to use the style of either panel D or panel E--or, often, both together--in such a situation as this.

### checking

One thing we regretfully learn about work with numbers is the need for checking. Late-caught errors make for painful repetition of steps we thought finished.

**Checking is inevitable; yet, if it is too extensive, we spend all our time getting the errors out of the checks. Our need is for enough checks but not too many.**

In scratching down data in a stem-and-leaf, we ought to make at least one check, to ask one question: Have we made as many entries as there were values to enter? If there are but few numbers in all, we can point at them in turn and count them up without writing anything down. For the present, though, in order to remind us of the need for checking, we will carry a "(check count)" column in parentheses. (An example will appear in the next exhibit.) To keep the calculation of such check columns from being an effort, we need to learn how to count the number of leaves on a stem rapidly and easily. Consider the 19 leaves that will appear on one stem of the next exhibit, namely

0121243122301214202

The recommended procedure is the following:

◇ place the left hand near the paper, with the forefinger pointing to the right.

⋄cover blocks of three digits at a time counting "3', "6", "9",... to one's self as this is done.

In this example, we would have, successively--the black figures are visible, the light ones concealed:

| | |
|---|---|
| 0121243122301214202 | "0" |
| 0121243122301214202 | "3" |
| 0121243122301214202 | "6" |
| 0121243122301214202 | "9" |
| 0121243122301214202 | "12" |
| 0121243122301214202 | "15" |
| 0121243122301214202 | "18" |

Clearly, "18" and "1" make 19.

If we find an error in the total count, or wish to check more carefully, we need to see whether we have the right leaves on the right stems. The easy way to do this is to go through the given values again, placing a dot over each leaf that is (either originally or after adjustment) correct. This process goes rapidly and seems quite effective.

### review questions

How can we enter different numbers in the same line of a tally? What is a stem-and-leaf display? What is a stem? A leaf? A starting part? How many digits in a leaf? How do we check the making of a stem-and-leaf display overall? In detail?

### 1E. Using the right number of stems

Sometimes splitting stem from leaf between adjacent digits does not seem satisfactory. One choice may involve so few stems that the values are too crowded for us to see them at all well, yet moving one more digit into the starting part makes so many stems that we are bothered by being too spread out. In such cases we may need to abandon the simple stem-and-leaf for a modification.

### stretched stem-and-leaf

One way out is to use two stems--two lines--for each starting part. In such a display, we use one line for leaves 0, 1, 2, 3, 4 and the other for leaves 5, 6, 7, 8, 9. The starting parts are repeated, with the * shown only for those that include leaves of 0 and 1. (Dots can be used to fill in.)

Exhibit **4** shows an example using the land areas of the 82 counties of Mississippi. In this instance, using the conventional stem-and-leaf with one-digit starting parts would have not worked badly (as the reader may easily verify). However, the two-line version gives us a better picture of how county areas behave in Mississippi. Note that a few counties have been identified. This both heightens realism and stirs the imagination.

## squeezed stem-and-leaf

Five stems--instead of one, two, or ten--can also be a good choice. Here we can use a scheme for tagging lines that makes it easy to avoid confusion, namely * for "zero" and "one", t for "two" and "three", f for "four" and "five", s for "six" and "seven", · for "eight" and "nine". (English digit names are surprisingly well adapted to a five-stem pattern.) Exhibit 5 shows a sample.

## using mixed leaves

Sometimes we use stem-and-leaf more as a way to collect numbers than as a way to look carefully at them. When we are doing this, we may be able--or

---

**exhibit 4 of chapter 1: county areas**

**The areas of the counties of Mississippi in a two-line stem-and-leaf display (stretched stem-and-leaf)**

A) SAMPLE DATA--in square miles

448, 405, 729, 724, 412, 917, 592, ...

B) Same ROUNDED--to 10's of square miles--(5's rounded to evens)

45, 40, 73, 72, 41, 92, 59, ...

C) The DISPLAY--in 10's of square miles

```
                                          (#)
3·|8                    Tate              (1)
4*|0121243121300214202                    (19)
4·|597886556569                           (12)
5*|142010                                 (6)
5·|977899958797                           (12)
6*|412441                                 (6)
6·|898598                                 (6)
7*|320341203                              (9)
7·|86657                                  (5)
8*|303                                    (3)
8·|8                    Hinds             (1)
9*|24                   Bolivar, Yazoo    (2)
                                          (82, √)
```

**Doubtful roundings (originally ending in 5) are made to nearest even number.**

P) PROBLEM

4a) Find another of data deserving a stretched stem-and-leaf display. Display it.

S) SOURCE

The World Almanac, 1966, page 370.

we may need--to do something special with values that trail out away from the bulk of the batch, as illustrated in exhibit **6**, including:

⋄ changing the division between starting part and leaf as seems convenient. (Here, and often, at multiples of ten; sometimes at other points.)

⋄ being careful to show this change **both** by a change in the number of *'s **and** by a blank line.

⋄ giving the batch in two forms, one for the quick look and one for storage for future use. (Sometimes we shall omit one or the other.)

Notice carefully, for example, what has happened to the second value in the original list. In the form for looking at, 1345 has become **1*** | 3, namely

exhibit **5** of chapter 1: later example

**A five-line stem-and-leaf display (squeezed stem-and-leaf)**

### A) FIVE-LINE VERSION

|    |         | (#)    |
|----|---------|--------|
| 1* | 1       | (1)    |
| t  | 2333    | (4)    |
| f  | 445555  | (6)    |
| s  | 66677   | (5)    |
| .  | 88      | (2)    |
| 2* | 0000011 | (7)    |
| t  | 23      | (2)    |
| f  | 445     | (3)    |
| s  | 6       | (1)    |
| .  | 9       | (1)    |
| 3* | 1       | (1)    |
| t  | 3       | (1)    |
| f  |         |        |
| s  |         |        |
| .  |         | (34√)  |

### B) The SIMPLE VERSION

|    |       | (#)   |
|----|-------|-------|
| 11 | 8     | (1)   |
| 12 | 0     | (1)   |
| 13 | 488   | (3)   |
| 14 | 08    | (2)   |
| 15 | 1266  | (4)   |
| 16 | 058   | (3)   |
| 17 | 08    | (2)   |
| 18 | 58    | (2)   |
| 19 |       | (·)   |
| 20 | 03688 | (5)   |
| 21 | 38    | (2)   |
| 22 | 1     | (1)   |
| 23 | 5     | (1)   |
| 24 | 05    | (2)   |
| 25 | 8     | (1)   |
| 26 | 3     | (1)   |
| 27 |       | ·     |
| 28 |       | ·     |
| 29 | 5     | (1)   |
| 30 |       | ·     |
| 31 | 2     | (1)   |
| 32 |       | ·     |
| 33 | 0     | (1)   |
| 34 |       | ·     |
|    |       | (34, √) |

**P) PROBLEM**

5a) Find another batch deserving a squeezed stem-and-leaf display. Display it.

**S) SOURCE**

Example in later chapter, related to exhibit 6, following.

exhibit 6/1: Scratching down numbers

**13\*\***, something between 1300 and 1399. In the form for storage, 1345 has become **1\*\*\***|345, namely 1345.

If we want to save detail for later use, there is no substitute for the "for storage" form. If we want to look at the pattern, there is no substitute for the

---

exhibit **6** of chapter 1: ultimate powers

**Ultimate power of hydroelectric plants of the Bureau of Reclamation (in megawatts)**

A) SAMPLE VALUES

30, 1345, 225, 900, 120, 162, . . .

B) TWO FORMS of STEM-and-LEAF DISPLAY

| (for looking at) | | (#) | (for storage) | |
|---|---|---|---|---|
| 1* | 65 | (2) | 1* | 6, 5 |
| 2 | 5244 | (4) | 2 | 5, 2, 4, 4 |
| 3 | 03626 | (5) | 3 | 0, 3, 6, 2, 6 |
| 4 | 508 | (3) | 4 | 5, 0, 8 |
| 5* | 0 | (1) | 5* | 0 |
| 6 | 0 | (1) | 6 | 0 |
| 7 | 51 | (2) | 7 | 5, 1 |
| 8 | | | 8 | |
| 9* | | | 9* | |
| 1** | 26350210 | (8) | 1** | 20, 62, 34, 50, 00, 20, 14, 08 |
| 2 | 285 | (3) | 2 | 25, 85, 50 |
| 3 | 7 | (1) | 3 | 79 |
| 4 | 2 | (1) | 4 | 24 |
| 5** | | | 5** | |
| 6 | | | 6 | |
| 7 | | | 7 | |
| 8 | | | 8 | |
| 9** | 0 | (1) | 9** | 00 |
| 1*** | 39 | (2) | 1*** | 345, 974 |
| 2 | | | 2 | |
| 3 | | | 3 | |
| | | (34√) | | |

C) NOTES

Source: The World Almanac, 1966, p. 263 (1967, p. 267).

Their source: U.S. Bureau of Reclamation.

Identification: If only one form had been used, leaving more space, some of the large and small plants would have been identified on the display, including Boysen 15, Elephant Butte 16, . . . , San Luis 424 Glenn Canyon 900, Hoover 1345, Grand Coulee 1974.

P) PROBLEM

6a) Find another set of data worthy of being shown in both of these ways together. Show it.

"for looking at" form. (In special circumstances, we may want to use compromise forms.)

A useful approach is illustrated by the righthand side of panel B in exhibit 6. Notice the use of a comma between each pair of leaves to make clear the leaf length. If we want to look harder at such a display, we can revert to single-digit leaves, as on the left of panel B.

This type of stem-and-leaf is very effective in accumulating information about values spread from many quite small to a few quite large. Often, as in this example, the numbers tend to pile up on--and just after--each "1"--an example of what is sometimes called the "abnormal law of large numbers".

exhibit **7** of chapter 1: arbitrary example

**Mixed-leaf stem-and-leaf applied to 25 values, some trailing**

### A) THE VALUES

5, −52, −27, −83, 8, −14, −122, −110, 112, 58, −119, 33, 18, −52, −19, 12, −82, 14, 25, −182, −40, 64, −56, 5, 13.

### B) The STEM-and-LEAF for STORAGE

```
                              (#)
1** | 12,                     (1)
0** | 64,                     (1)

 5* | 8                       (1)
  4 |
  3 | 3                       (1)
  2 | 5                       (1)
 1* | 2, 4, 3, 8              (4)
  0 | 5, 8, 5                 (3)
 −0 |
−1* | 4, 9                    (2)
 −2 | 7                       (1)
 −3 |
 −4 | 0,                      (1)
−5* | 2, 6, 2                 (3)

−0** | 83, 82                 (2)
−1** | 22, 10, 19, 82         (4)
                             (25, √)
```

### P) PROBLEMS

7a) Make a "for looking at" stem-and-leaf of the same data.

7b) Make a stem-and-leaf of the same data using only one size of leaves.

7c) Criticize the display mode in (7b). How could it be improved?

7d) Improve it.

When this is so, we see a triangular appearance for each "1" to "9"--as if the leaves had washed against the blank lines--an appearance which covers up most of whatever interesting things we might see. (We will learn how to deal with this before long.)

It often makes things even simpler, however, to feel free to jump to longer leaves in the middle of a digit-length, as illustrated in exhibit **7**, where one-digit leaves are used from −59 to +59, and two-digit leaves begin at −60 and +60. This enables us to use 16 lines where 31 (+11 to +0 and −0 to −18) would have been required if all starting parts had used the same number of *'s.

Notice in this exhibit:

◇ the use of stem-and-leaf for both positive and negative values--and the consequent need for both a "+0" stem and a "−0" stem.

### review questions

How many stems is it natural to use for one starting part? How do we choose? What can we gain? Do we need to be consistent in our starting parts within one display? What is at stake? What form of stem-and-leaf is intended for storage? What form for looking-at? How free can our choices be? How many "0" starting parts can there be?

### 1F. How to count by tallying

The basis of stem-and-leaf technique, entering an additional digit--or digits--to mark each value, works well for batches of limited size. Once we have much more than 20 leaves on a stem, however, we are likely to feel cramped--and our stems begin to be hard to count. We ought to be able to escape to some other way of handling such information, whenever the other way gives us enough detail.

The fast methods involve one pencil (or pen) stroke per item. One method counts by fives in this style:

$$/ \quad // \quad /// \quad //// \quad \cancel{////}$$

This has been widely used. The writer finds it treacherous, especially when he tries to go fast. (It is too easy for him to do

$$\cancel{//} \quad \text{or} \quad \cancel{/////}$$

for this approach to give satisfactory performance.)

The recommended scheme uses first dots, then box lines, then crossed lines to make a final character for 10. Thus:

| | | |
|---|---|---|
| 4 | is | ∷ |
| 8 | is | ☐ |
| 10 | is | ⊠ |

The order in which the four dots are placed around the square is unimportant, as is the order in which the four sides of the square are filled in, and the order in which the two diagonals are crossed.

Once anyone is used to this scheme, the occurrence of errors is much less than for the other approach. The closure of the four dots completing the outline of the square is clear and definite; there is no tendency to go on to lines after three points--or to add a fifth point. The same applies to the completion of the sides of the square. (A few novices report a tendency to move on to the next character before crossing the diagonals. Their eights--□--are, however, distinctive enough from tens--⊠--to be picked up once the tallying has been completed and the result is being converted into digits.

Exhibit **8** shows a variety of simple examples.

**exhibit 8 of chapter 1: arbitrary examples**

**Tallying by tens variously illustrated**

| A) TALLY | COUNT |
|---|---|
| . | 1 |
| ∶ or .. | 2 |
| ∴ or ∶. (etc.) | 3 |
| ∷ | 4 |
| ⊡ or \|∷ (etc.) | 5 |
| ⌈∷ or └∷ (etc.) | 6 |
| ⊓∷ or ⊏∷ (etc.) | 7 |
| □ | 8 |
| ⊠ or ⊠ | 9 |
| ⊠ | 10 |
| ⊠ ⊠ ⊓ | 27 |
| ⊠ ⊠ ⊠ ⊠ . | 42 |

**B) A SIMPLE EXAMPLE**

```
 0 | ⊠ ⊠ □
 1 | ⊠ ⊠ ∶
 2 | ⊠ ⊠
 3 | ⊠ ⌈
 4 | ⊠ .
 5 | ⊠ ⊠ ∶·
 6 | ⊠
 7 | ⊓
 8 | |∶
 9 | ∷
10 | ·
```

**C) A TWO-WAY EXAMPLE**

|  | Freshmen | Sophs | Juniors | Seniors |
|---|---|---|---|---|
| Boys | ⊠ ⊠ ⊠ ⊠ .. | ⊠ ⊠ ∶ | ⊠ ⌈ | ⊠ · |
| Girls | ⊠ ⊠ ⌈ | ⊠ ⊠ ⊠ ∷ | ⊠ ∶· | ⊠ ⊠ ∶ |

**P) PROBLEM**

8a) Collect a batch of size at least 100 by tallying by tens, and display as in B and A.

## exhibit 9/1: Scratching down numbers

### used cars again

For exhibit **9** we return to the *Sunday Standard-Times* and its used-car ads. This display gives--in both tallied and digit form--the number of cars of each model year advertised:

⋄ in ads for one or two cars

⋄ in ads for three or more cars.

We can get a quick grasp of appearances easier and faster from the tallies-by-ten than from the same counts expressed as numbers. (It may even be worthwhile, on occasion, to turn digit-given counts into tallies-by-ten to look at them.)

The general impressions easily gained from this display include:

⋄ a tendency for small ads (one or two cars) to offer somewhat older cars.

⋄ a question as to why '55 appears to be popular.

exhibit **9** of chapter 1: auto years

**Model years of used cars advertised (comparison of numbers with tallies-by-tens)**

| A) ADS for 1 or 2 CARS | | B) ADS for 3 or MORE CARS | |
|---|---|---|---|
| Number | Tally | Number | Tally |
| 1 | '54 \| . | — | '54 \| |
| 2 | '55 \| . . | 1 | '55 \| . |
| — | '56 \| | — | '56 \| |
| 1 | '57 \| . | 1 | '57 \| . |
| 2 | '58 \| . . | 1 | '58 \| . |
| 5 | '59 \| ∴ | 1 | '59 \| . |
| 5 | '60 \| ∴ | 4 | '60 \| ∷ |
| 2 | '61 \| . . | 10 | '61 \| ⊠ |
| 11 | '62 \| ⊠ . | 14 | '62 \| ⊠ ∷ |
| 12 | '63 \| ⊠ ∴ | 17 | '63 \| ⊠ ⊓ |
| 10 | '64 \| ⊠ | 25 | '64 \| ⊠ ⊠ ∴ |
| 14 | '65 \| ⊠ ∷ | 34 | '65 \| ⊠ ⊠ ⊠ ∷ |
| 12 | '66 \| ⊠ . . | 27 | '66 \| ⊠ ⊠ ⊓ |
| 13 | '67 \| ⊠ ∴ | 12 | '67 \| ⊠ . . |
| 3 | '68 \| ∴ | 4 | '68 \| ∷ |

**P) PROBLEM**

9a) Collect two related batches of at least 80 values each, and display as above.

**S) SOURCE**

Sunday Standard-Times, New Bedford, Mass. for 18 August 1968 (page 51, Fourth Section).

Each of us can provide or invent reasons for the first of these. As for the second, reexamination of the ads shows that two of the three 55's are "classic Thunderbirds".

While tally-by-tens is an effective technique, it does not tell us as much as stem-and-leaf. It should be almost exclusively used as an overflow procedure. Indeed, overflow need not occur as often as we might think, for it is often very helpful to accumulate information in parts--parts which we are then free to combine or keep apart. Exhibit 9 shows collection in two parts--had we gone further and collected in three or four parts, it would have been reasonable--so far as size of counts go--to collect each part using stem-and-leaf. If there were further information to collect, it would be not only reasonable but desirable to do this.

#### review questions

What trouble does tallying by fives get into? What are the steps in tallying by tens? Can it pay to turn numerical counts into tallies by 10's? Why? Can it pay to collect tallied information in parts? What do we gain? What tends to shift our choice between stem-and-leaf and tallying-by-tens?

### 1G. What does it mean to "feel what the data are like"?

We have been looking at well-behaved batches of numbers. True, there was something to pick up about 1955 used cars, but generally we saw little that was striking in any of our stem-and-leaf displays.

Exhibit **10** shows the height of the highest point in each of the 50 states. Clearly the states fall into three groups, which turn out to be:

◇ Alaska.

◇ the Rocky Mountain States, California to Washington, and Hawaii.

◇ all others.

The things that are frequently thrust upon us by stem-and-leaf displays are:

◇ appearances of separation into groups.

◇ unsymmetric trailing off, going farther in one direction than another.

◇ appearance of unexpectedly popular (or unpopular) values.

◇ about where the values are "centered".

◇ about how widely the values spread.

The stem-and-leaf itself guides us to the first three. A little help with the last two may be useful, as we shall shortly see.

Notice that, where apparent "breaks" occur, it may be useful to identify values near the breaks as well as at the ends.

## review questions

What do stem-and-leaf displays frequently thrust upon our attention? Would tallyings-by-ten do the same?

### 1H. How far have we come?

This first chapter has been devoted to (i) a discussion of what exploratory data analysis is all about, and (ii) an introduction to the simple tools that allow us to scratch down information about batches of numbers in ways that:

exhibit **10** of chapter 1: state heights

**The heights of the highest points in each state**

A) STEM-and-LEAF--unit 100 feet

```
                                            (#)
 0* | 43588       Del, Fla, La, Miss, RI    (5)
  1 | 237886                                 (6)
  2 | 484030                                 (6)
  3 | 45526                                  (5)
 4* | 80149                                  (5)
  5 | 34307                                  (5)
  6 | 376                                    (3)
  7 | 2           S. Dak                     (1)
 8* | 8           Texas                      (1)
  9 |
 10 |
 11 | 2           Oregon                     (1)
12* | 768                                    (3)
 13 | 81258                                  (5)
 14 | 544         Calif, Colo, Wash          (3)
 15 |
16* |
 17 |
 18 |
 19 |
20* | 3           Alaska                     (1)
                                           (50, √)
```

P) PROBLEMS

10a) Make a similar display for the lowest point in each of the 50 states.

10b) Make the comments suggested by the results of (10a).

S) SOURCE

The World Almanac, 1966, page 269.

Their source: National Geographic Society.

◊give us a feeling of how each batch seems to behave as a whole

◊or help to collect and store the values in a form that is convenient for later use.

Sometimes we use different forms for storage and for insight.

### what have we learned to do ?

Exploratory data analysis is detective work--in the purest sense--finding and revealing the clues.

Confirmatory data analysis, which we do not discuss in this book, goes further, assessing the strengths of the evidence.

The techniques for dealing with batches of numbers we have met are all closely related. All of them display the numbers before our eyes by putting down the most informative mark that we have thought of and that will be effective. (Too much detail or too much sprawl can detract from what we see.) Each technique deserves to be checked at least as far as the total count. (Overdotting makes checking of stem-and-leaf displays easy.)

When we make a stem-and-leaf display, we have a choice among one-digit leaves, two-digit leaves, etc., and mixed leaves. We make this partly for convenience, partly in view of our purposes, partly for necessity. Sometimes we make two different styles at the same time.

When we have many values to collect or display, and can give up further details, tallying-by-tens offers us real advantages.

Among the more important things that detailed displays of batches can show us are separation, asymmetry, irregularities, centering, and width.

### where do we stand ?

As this chapter has begun to illustrate, one thing the data analyst has to learn is how to expose himself to what his data are willing--or even anxious-- to tell him. Finding clues requires looking--in the right places and with the right magnifying glass.

Most of us know something about looking at a single number. It may not be easy to draw all the implications from the fact that there were 46,930 deaths from motor vehicle accidents in the U.S.A. in 1964, but the difficulty is not in understanding the one number.

The same is often true of comparing a few firmly based numbers, like these 1962–64 average death rates per hundred million person miles:

| | |
|---|---|
| ◊trains | 0.09 |
| ◊planes | 0.16 |
| ◊buses | 0.17 |
| ◊cars (turnpike) | 1.2 |
| ◊cars (general) | 2.3 |

exhibit **11**/1: Scratching down numbers

exhibit **11** of chapter 1: auto makes

**Collection of model-year and name information for advertised used cars**

**A)** The NAME CODE

| | |
|---|---|
| Bonneville | b |
| Buick | B |
| Cadillac | A |
| Chevelle | h |
| Chevy II | 2 |
| Chevrolet | C |
| Chrysler | H |
| Classic | 1 |
| Comet | 3 |
| Corvair | 4 |
| Dodge | D |
| Eldorado | E |
| Ford | F |
| Futura | f |
| Galaxie | g |
| Grand Prix | G |
| Impala | I |
| Jaguar | J |
| Lincoln | L |
| Mercury | M |
| Mustang | m |
| Oldsmobile | O |
| Plymouth | p |
| Pontiac | P |
| Rambler | r |
| Simca | 5 |
| Studebaker | s |
| Sunbeam | S |
| Tempest | t |
| T-bird | T |
| Valiant | V |
| Volkswagon | v |

**B)** The COLLECTION

| | |
|---|---|
| '55 | T |
| '56 | |
| '57 | E |
| '58 | C |
| '59 | I |
| '60 | AIAv |
| '61 | CItHvFFFHT |
| '62 | 4CIP2O3CTIIP2 |
| '63 | IIFrBFrCrLOICCbP |
| '64 | Pg1BDD4CIlrBl3PshCAOCM |
| '65 | 5H4C4OCpDCFCCCbBICSCPOIDCCBACC2CCCr |
| '66 | ftvh1Cv3B4hFlvACOhmPJFFCp3 |
| '67 | VPPFGmvPmCAG |
| '68 | HpDp |

**C)** The COLLECTION ORGANIZED

| | |
|---|---|
| ·55 | T |
| '56 | |
| '57 | E |
| '58 | C |
| '59 | I |
| '60 | AAIv |
| '61 | CFFFHHItTv |
| '62 | CCIIIOPPT2234 |
| '63 | bBCCCFFIIILOPrrr |
| '64 | ABBCCCDDghIIIMOPPrs134 |
| '65 | CCCCCCCCCCCCCCC |
| '65 | AbBBDDFHIIOOpPrS245 |
| '66 | ABCCCfFFFhhhIJmOpPtvvv1334 |
| '67 | ACFGGmmPPPvV |
| '68 | DHpp |

**P)** PROBLEM

11a) Find another set of data that deserves this type of display. Display it.

**S)** SOURCE

Ads for three or more cars, page 51, Fourth Section, Sunday Standard-Times, New Bedford, Mass. for 18 August 1968.

**(1H) 1K: picking up additional information (optional technique)**   23

Again we are likely to have difficulty with the implications, but not with understanding the numbers themselves.

The simplest case where we need to learn to grasp the numbers more firmly or more rapidly comes when we face a batch of similar numbers. Hence the techniques of this chapter and those that follow.

(The end of the section called "How far have we come?" is the formal close of this chapter--and of each chapter that follows. Any further sections are--in one sense or another--optional. Some of them may introduce convenient techniques, but these techniques are not essential to our main aims or practices.)

### 1K. How to use stem-and-leaf to pick up additional information (optional technique)

The basic principle--stem to locate where we write it down; leaf to say more--need not be confined to the sort of examples we have so far seen. Let us return to the *Sunday Standard-Times* again, and the 18 ads for three or more cars. Let us now use the stem-and-leaf technique to collect combined information on the model year and name of the cars offered. Exhibit **11** shows the technique and the results. Some aspects of the coding scheme are worth noting:

◊ we have used suggestive letters where available, but have not hesitated to step out to numbers when convenient letters are not available.

◊ we have distinguished capital and small letters whenever their handwritten forms are clearly distinct, but have not tried to use both otherwise.

Panel C of this exhibit shows the result of organizing the entries alphabetically for each model year (and providing two lines for '65). This makes quite clear a tendency for certain name-year combinations to appear frequently. The most obvious example is the fifteen '65 Chevrolets, but others can be found.

#### pairs of numbers

In the last examples, we were really storing pairs of the form

(number, make)

Rather clearly we can go on to store pairs of the form

(1st number, 2nd number)

Exhibit **12** does this for logarithmic expressions of bank deposits (1st number) and governor's salary (2nd number) for which the identified data and source are given in exhibit 21 of chapter 7. Here the stem 10 is followed by leaves (4, 574) and (6, 556), thus specifying the pairs

$$(104, 574) \quad \text{and} \quad (106, 556).$$

exhibit **12** of chapter 1

**Using a stem-and-leaf display for 50 pairs of numbers (see exhibit 21 of chapter 8 for identification)**

```
−3 | (7, 505)
−2 |
−1 | (5, 301)
−0 | (2, 398), (0, 398)
 0 | (6, 243), (9, 301)
 1 | (2, 255), (3, 255), (4, 525), (5, 366), (6, 544)
 2 | (3, 342), (7, 477), (7, 301)
 3 | (0, 398), (6, 398)
 4 | (2, 477), (2, 0), (3, 398), (6, 439), (9, 255)
 5 | (0, 653), (9, 301)
 6 | (0, 455), (3, 398), (4, 301), (7, 477), (9, 398)
 7 | (4, 398), (6, 453), (6, 477), (9, 267)
 8 | (0, 512), (0, 544), (1, 628), (4, 477)
 9 | (3, 398), (6, 439), (7, 544), (8, 398)
10 | (4, 574), (6, 556)
11 |
12 | (0, 544), (8, 602)
13 | (0, 544), (0, 602), (7, 602)
14 | (7, 653)
15 | (0, 653)
16 | (6, 644)
17 |
18 |
19 |
20 | (9, 699)
```

## 1P. Additional problems

exhibit **13** of chapter 1: data and problem

**Viscosity at a temperature of 190°C and a low rate of shear**

A) DATA

| Run (for) each sample | Viscosity in 100,000's of poises | | |
|---|---|---|---|
| | Sample I | Sample II | Sample III |
| 1  | .384 | .661 | 3.54 |
| 2  | .376 | .671 | 3.66 |
| 3  | .376 | .688 | 3.42 |
| 4  | .371 | .644 | 4.10 |
| 5  | .385 | .668 | 4.09 |
| 6  | .377 | .648 | 3.77 |
| 7  | .365 | .706 | 4.17 |
| 8  | .384 | .715 | 3.91 |
| 9  | .365 | .647 | 4.61 |
| 10 | .384 | .682 | 3.87 |
| 11 | .378 | .692 | |
| 12 |      | .729 | |

P) PROBLEM

13a) In 1963, McGlanery and Harban gave the values in panel A, showing how well they could measure the viscosity of liquids with a device called a capillary rheometer. Make appropriate stem-and-leaf displays for each of the three samples; comment on the appearance of each.

S) SOURCE

R. M. McGlanery and A. A. Harban 1963. Two instruments for measuring the low-shear viscosity of polymer melts. *Materials Research and Standards* 3: 1003–1007. Table 2 on page 1004.

exhibit **14** of chapter 1: data and problem

**Radial velocities of stars near α Persei**

A) The RADIAL VELOCITIES--in kilometers per second

| | | | | | | |
|---|---|---|---|---|---|---|
| +50.0 | −2.4  | +1.0  | −0.5  | +2.2  | −11.8 | −11.9 |
| −36.1 | +10.0 | +3.7  | −15.9 | −4.1  | −19.1 | −16.2 |
| −7.0  | +7.0  | −9.5  | +2.0  | −3.0  | −22.1 | −10.8 |
| +2.2  | 0.0   | −8.4  | +5.0  | +3.2  | −9.1  | −6.0  |
| −2.8  | −1.0  | +0.5  | +4.4  | −4.4  | −17.2 | −23.9 |
| +3.0  | −0.7  | +2.2  | +1.6  | −0.3  | +12.8 | −8.2  |
| +24.7 | +15.9 | +18.0 | +6.0  | +14.5 | −10.5 | −13.6 |
| +4.8  | +9.0  | −17.5 | +5.9  | −18.4 | −17.2 | −4.4  |
| | | | | | | +15.9 |
| | | | | | | −25.7 |

P) PROBLEM

14a) In 1958, Heckmann and Lübeck gave the radial velocities in panel A for stars near the star Alpha Persei. Make an appropriate stem-and-leaf. Comment on its appearance.

S) SOURCE

O. Heckmann and K. Lübeck 1958. "Das Farben-Helligkeits-Diagramen des Bewegungshaufens um Alpha Persei". *Zeitschrift für Astrophysik* **45**: 243–263. Figure 2 on page 248 is based on Tables 2 and 3 on pages 247 and 248.


exhibit **15** of chapter 1: data and problem

**Association constant of the first nitrate complex ion of plutonium (IV), measured in perchloric–nitric acid solutions with 2.00-M hydrogen ion concentration**

A) RESULTS OF REPEATED EXPERIMENTS

1.97, 2.13, 2.02, 1.55, 2.83, 2.92, 2.40, 3.24, 3.21, 3.88, 3.85, 3.89, 2.84, 3.08, 4.17, 2.64, 3.67, 3.06, 2.70, 1.91, 2.34, 2.49, 2.76, 3.50

P) PROBLEM

15a) In 1949, Hindman gave the numbers in panel A, which refer to the stability of a complex ion combining 4-valent plutonium ions with nitrate ions. Make an appropriate stem-and-leaf diagram. Comment on its appearance.

S) SOURCE

J. C. Hindman 1949. Complex ions of plutonium. The nitrate complex ions of plutonium (IV). Pages 388–404 of *The Transuranium Elements*; edited by G. T. Seaborg, J. J. Katz, and W. M. Manning, McGraw-Hill (volume 14B of the Plutonium Project Record). Table 1 on page 390.

# Easy summaries-- numerical and graphical    2

chapter index on next page
___

We have learned to scratch down batches. Stem-and-leaf displays let us do this easily and rapidly. Even more importantly, they let us look at the general pattern of the batch.

When this pattern is distinctive, we can easily say so in words. Exhibit 10 of chapter 1, for example, shows a pattern of three quite distinct groupings. (This does not happen very often.) The patterns of exhibits 4 and 5 of that chapter, for another, stop much more abruptly toward smaller values than toward larger ones. (This happens all too often.)

We would like to be able to summarize the most frequently occurring characteristics of the pattern of a batch. It will be good to do this in terms of a few numbers that are easily understood, and to agree on what these numbers are, what they are called, how they are marked, how they are routinely written down, and how they are usefully and easily shown graphically!

**To do this well, we must limit our objectives.** It would be wrong to expect a standard summary to reveal the unusual--no matter how important the unusual may be when it occurs. Things like the separation of the pattern into groupings--illustrated by exhibit 10 of chapter 1--ought not to show in an easy summary of routine form. On the one hand, such a separation deserves words. On the other, trying to show such occasional happenings would complicate our procedures unduly and confuse the summaries for all the many batches where nothing of the sort happens.

We ought to depend on stem-and-leaf displays--or other visual devices for details--to show us the unexpected. Standard summaries--whether numerical or visual--should only be able to reveal the more or less expected.

**Summaries can be very useful, but they are not the details.** So long as the detail is not so great as to confuse us hopelessly, there will be no substitute for having the full detail where we can look at it, set out in as clear a way as we can easily manage. Summaries are necessary for vast amounts of data--and often convenient for smaller amounts. They are not supposed to--and cannot be expected to--replace the corresponding details. Often, of course, the details will add little, but it is important to prepare for the occasions when they add much.

### review questions

Why would anyone want to summarize a batch? How completely can we expect to do this? How broad can our objectives be? What is the reasonable

# index for chapter 2

**General comment** 27
review questions 27

**2A. Extremes and median** 29
*extremes* 29
*median* 29
*counting-in* 29
*rank* 29
*rank down* 30
*rank up* 30
*depth* 30
*interpolated ranks* 31
review questions 31

**2B. Hinges and 5-number summaries** 32
*hinges* 33
*5-number summary* 33
*letter-value display* 33
*examples* 33
*further examples* 37
review questions 38

**2C. Box-and-whisker plots** 39
*rules* 39
basic ideas of plotting 42
tracing paper 42
scale values 43
plotting without graph paper 43
review questions 43

**2D. Fences, and outside values** 43
*H-spread* 44
*step* 44
*inner fences* 44
*outer fences* 44

*adjacent values* 44
*outside values* 44
*far out values* 44
*fenced letter display* 44
*ranges* 46
*range* 46
*trimeans* 46
*trimean* 46
review questions 47

**2E. Schematic plots** 47
*rules for same* 47
review questions 49

**2F. Pros and cons; the Rayleigh example** 49
a better use 50
aim reiterated 51
choice of plots 52
review questions 53

**2G. Eighths, sixteenths, etc.** 53
(almost optional here; used in later chapters)
*eighths* 53
*sixteenths* 53
*E,D,C, etc.* 53
*7-number summaries* 53
*9-number summaries* 53
*letter values* 54
review questions 54

**2H. How far have we come?** 55
what have we learned to do? 55
where do we stand? 56

| EXHIBIT | PAGE |
|---|---|
| **2A** | |
| 1 | 30 |
| 2★ | 32 |
| **2B** | |
| 3★ | 34 |
| 4★ | 38 |
| **2C** | |
| 5★ | 40 |
| 6★ | 41 |
| **2D** | |
| 7★ | 44 |

| EXHIBIT | PAGE |
|---|---|
| **2E** | |
| 8★ | 48 |
| **2F** | |
| 9 | 49 |
| 10 | 50 |
| 11 | 51 |
| 12 | 52 |
| **2G** | |
| 13★ | 54 |
| **2H** | |

★ = includes problem(s)

purpose of standard summaries? What ought more detailed presentations do for us? How frequently ought they be important?

## 2A. Extremes and median

If we are to select a few easily-found numbers to tell something about a batch as a whole, we certainly want:

◇ the extremes--the highest and lowest values--which we will mark "1" (for their depth).

◇ a middle value.

We would like these values to be easy to find and write down, whether the total count of the batch is 8, or 82, or 208.

Once we have the batch in stem-and-leaf form, we have its values almost in order. Once we have the check count, we can count up or down a stem at a time. The easy values to find are those that are fixed by counting.

If the batch has an even number of values, there are two in the middle by count. If there are 50 in all, halfway from 1 to 50 is 25.5, so that the 25th and 26th are the middle pair.

If the count is an odd number, there is a single value in the middle by count. If there are 13 in all, halfway from 1 to 13 is 7, so the 7th is it.

The median is defined to be

$$\text{Median} = \begin{cases} \text{the single middle value} \\ \text{the mean of the two middle values} \end{cases}$$

We shall mark it "M".

For exhibit 10 of Chapter 1, we find these values

```
1 | 203
M |  46    (hundreds of feet)
1 |   3
```

Once we have a stem-and-leaf display, the extremes are very easy. The total count is 50. The 25th and 26th values are 44 and 48, so the median is

$$46 = \tfrac{1}{2}(44 + 48)$$

### counting-in

To get an easy number in the middle of the batch--assuming that the batch is in stem-and-leaf form--we count in from either end, starting by counting the extreme value "1". This procedure assigns a

**rank**

to each number in the batch.

## exhibit 1/2: Easy summaries--numerical and graphical

We can start from either extreme. In such phrases as "he was first in his class", it seems natural to start by ranking the highest value as "1", to

### rank down

Many times, by contrast, we like to have the rank increase as the value increases, to

### rank up

This calls for ranking the lowest value "1".

The least rank--either up or down--that we can give to an observation is its

### depth

The depth of an extreme is always 1. Exhibit **1** shows an example.

---

**exhibit 1 of chapter 2: Chevrolet prices**

**The two directions of ranking illustrated (data of exhibit 3 of chapter 1)**

| Upward rank | Prices | Downward rank | Depth |
|---|---|---|---|
| 1  | 150  | 17 | 1 |
| 2  | 250  | 16 | 2 |
| 3  | 688  | 15 | 3 |
| 4  | 695  | 14 | 4 |
| 5  | 795  | 13 | 5 |
| 6  | 795  | 12 | 6 |
| 7  | 895  | 11 | 7 |
| 8  | 895  | 10 | 8 |
| 9  | 895  | 9  | 9 |
| 10 | 1099 | 8  | 8 |
| 11 | 1166 | 7  | 7 |
| 12 | 1333 | 6  | 6 |
| 13 | 1499 | 5  | 5 |
| 14 | 1693 | 4  | 4 |
| 15 | 1699 | 3  | 3 |
| 16 | 1775 | 2  | 2 |
| 17 | 1895 | 1  | 1 |

Remark:

upward rank + downward rank = 1 + total count

for each and every entry. This is a general property.

### P) PROBLEM

1a) Find a small set of numbers that interests you, put them in order, and write them out with both sets of ranks.

Whichever way we rank a batch of 17, the median has rank 9, where $9 = \frac{1}{2}(1 + \text{total count})$

In a batch of six, say for example,

$$7$$
$$12$$
$$18$$
$$20$$
$$27$$
$$54$$

the median is the mean of the two center values, here

$$\tfrac{1}{2}(18 + 20) = 19$$

The ranks of these values are 3 and 4, no matter which end we rank from. Half-way between rank 3 and rank 4 should come rank 3h. (Recall that we use "h" for "and a half".)

Again, 3h is $\frac{1}{2}(1 + \text{total count})$. The same sort of thing happens for batches of any even count. If we do the natural thing, and use half-integer

### interpolated ranks

for the midvalues half-way between the values to which we have given adjacent integer ranks, we will have the same expression

$$\text{depth or rank of median} = \frac{1 + \text{count}}{2}$$

for batches of all counts, odd or even. Exhibit **2** shows some examples.

To deal with medians, we need ranks or depths in halves and the corresponding interpolated values. The same ideas can, as we shall soon see, make it easy to describe other useful quantities. We can--and very occasionally do--use ranks or depths in quarters (possibly even more closely), but all such can be considered quite exceptional.

### review questions

What is an extreme? How do we mark it? How is a median defined? Where in the batch is it? What is an interpolated rank? What is a rank? What is the difference between ranking up and ranking down? What is a depth? How do we find the value corresponding to an interpolated rank? What is an interpolated depth? What formula gives the interpolated depth of the median? Its interpolated rank? Is there any value whose interpolated rank up is the same as its interpolated rank down?

## 2B. Hinges and 5-number summaries

We have grasped 3 numbers to define the ends and the middle of a batch. These are useful, but are rarely enough.

The median can be found by counting half-way from one extreme to the other. If we are going to add two more numbers to form a 5-number summary, it is natural to find them by counting half-way from each extreme to the median.

exhibit **2** of chapter 2: Chevrolet prices

**Examples of interpolated ranks**

**A)** In the DATA of EXHIBIT 1

| Interpolated rank | Value and its genesis |
|---|---|
| 1h | 200 = $\frac{1}{2}$(150 + 250) |
| 2h | 469 = $\frac{1}{2}$(250 + 688) |
| 3h | 691.5 = $\frac{1}{2}$(688 + 695) |
| 4h | 745 = $\frac{1}{2}$(695 + 795) |
| 5h | 795 = $\frac{1}{2}$(795 + 795) |

(and so on)

**B)** In the BATCH of 6

| Ranks | | Interpolated | |
|---|---|---|---|
| Rank | Number | Number | Rank |
| 1 | 7 | | |
| | | 9.5 | 1h |
| 2 | 12 | | |
| | | 15 | 2h |
| 3 | 18 | | |
| | | 19 | 3h |
| 4 | 20 | | |
| | | 23.5 | 4h |
| 5 | 27 | | |
| | | 40.5 | 5h |
| 6 | 54 | | |

**P) PROBLEMS**

2a) Find the values corresponding to upward ranks of 10, 10h, 11, 11h, and 12 in exhibit 1.

2b) Find the values corresponding to downward ranks of 10, 10h, 11, 11h, and 12 in exhibit 1.

2c) Find the two values corresponding to each of depths 4, 4h, and 5 in exhibit 1.

## 2B: hinges and 5-number summaries

Since we can think of finding first the median and then these new values as a folding process, it is natural to call the new values

### hinges.

If we have 9 values in all, the 5th from either end will be the median, since $\frac{1}{2}(1 + 9) = 5$. Since $\frac{1}{2}(1 + 5) = 3$, the third from either end will be a hinge. If we have 13 values, the 7th will be the median—and the 4th from each end a hinge. In folded form, a particular set of 13 values appears as follows:

```
    -3.2                1.5                 9.8
       -1.7         1.2     1.8          6.4
          -0.4   0.3             2.4   4.3
              0.1                 3.0
```

The five summary numbers are, in order, $-3.2$, $0.1$, $1.5$, $3.0$, and $9.8$, one at each folding point.

We usually symbolize the 5 numbers (extremes, hinges, median) that make up a

### 5-number summary

by a simple summary scheme like this:

```
#13
M7 |      1.5      |
H4 |  0.1     3.0  |
 1 | -3.2     9.8  |
```

where we have shown the count (marked #), the depth of the median (marked M), the depth of the hinges (marked H), and the depth of the extremes (always 1; needs no other mark) to the left of an inverted-U enclosure, and have written the corresponding values in the enclosure. We put the median in the middle, the upper values on one side, and the lower values on the other (which is which does not matter, and can change from time to time and from person to person). Such a scheme we call a

### letter-value display.

(We will slowly learn about more and more detailed letter-value displays.)

Since the depth of the extremes is ONE, the depth of the median is half of the count plus ONE. When the depth of the median is a half-integer, we first discard this half and then add ONE, before halving. Thus for a batch of 20, $\frac{1}{2}(1 + 20) = 10h$, and $\frac{1}{2}(1 + 10h)$ is taken as $\frac{1}{2}(1 + 10) = 5h$.

### examples

Exhibit 3 shows three examples of the calculation of hinges and 5-number summaries. Panel A is in folded form. Panel B starts from a modified stem-and-leaf where (a) the leaves are divided into columns and (b) rows (stems) have been divided in such ways as to make the folded structure clearer.

**34  exhibit 3(A-B)/2: Easy summaries--numerical and graphical**

exhibit **3** of chapter 2: varied data

**Hinges illustrated**

A) The 17 AUTO PRICES of EXHIBIT 1--in folded form

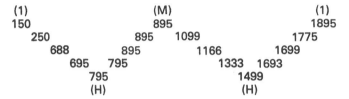

```
(1)                       (M)                        (1)
150                       895                       1895
   250               895       1099             1775
      688        895              1166      1699
         695  795                   1333  1693
            795                        1499
           (H)                         (H)
```

[17 prices, 1HMH1: 150, 795, 895, 1499, 1895 dollars]

B) The 34 ULTIMATE POWERS of EXHIBIT 6 of CHAPTER 1--folded stem-and-leaf; note irregular splitting of stems and letter display

|       | Extremes and median | Inter- mediate values | Hinges |
|-------|---------|-------------|--------|
| 1*    | 5       |             |        |
| 1·    |         | 6           |        |
| 1·    |         |             |        |
| 2*    |         | 2445        |        |
| 3*    |         | 02          |        |
| 3·    |         |             | 3      |
| 3·    |         | 66          |        |
| 4*    |         | 058         |        |
| 5*    |         | 0           |        |
| 6*    |         | 0           |        |
| 7*    | 1       |             |        |
| 7·    | 5       |             |        |
| 8*    |         |             |        |
| 9*    |         |             |        |
| 10    |         | 08          |        |
| 11    |         | 4           |        |
| 12    |         | 00          |        |
| 13*   |         | 4           |        |
| 1**   |         | 50          | 62     |
| 2     |         | 25, 50, 85  |        |
| 3**   |         | 79          |        |
| 0***  |         | 424         |        |
| 0     |         | 900         |        |
| 1***  |         | 345         |        |
| 1     | 974     |             |        |

```
          34  ult. powers
        M17h |   73
         H9  | 33    162
          1  | 15   1974
```

Panel C shows what will be the conventional stem-and-leaf display, one that includes cumulative counts from each end. In this example we have put the cumulative counts on the half-lines, where they really belong. In routine practice, we let the cumulative counts fall outward into the last lines included in them, as will be illustrated in Panel E. Panel D shows how we can count in, leaning on the cumulative counts. For the median, for example, we know from panel C that the cumulative count through value 54 is 38, and that rank 39 corresponds to 55, and so on. Once we have followed this pattern a few times,

exhibit **3** of chapter 2 (continued)

C) The 82 AREAS in exhibit 4 of chapter 1 with CUMULATIVE COUNTS ADDED--on the half line

|  |  | (#) |  |
|---|---|---|---|
| 3· | 8 | (1) |  |
|  |  |  | 1 |
| 4* | 0000111112222223344 | (19) |  |
|  |  |  | 20 |
| 4· | 555566678899 | (12) |  |
|  |  |  | 32 |
| 5* | 001124 | (6) |  |
|  |  |  | 38 |
| 5· | 577778899999 | (12) | √ 82 = 38 + 12 + 32 |
|  |  |  | 32 |
| 6* | 112444 | (6) |  |
|  |  |  | 26 |
| 6· | 588899 | (6) |  |
|  |  |  | 20 |
| 7* | 001223334 | (9) |  |
|  |  |  | 11 |
| 7· | 56678 | (5) |  |
|  |  |  | 6 |
| 8* | 033 | (3) |  |
|  |  |  | 3 |
| 8· | 8 | (1) |  |
|  |  |  | 2 |
| 9* | 24 | (2) |  |

Note that 82 = 38 + 12 + 32 is the count check.

D) CALCULATIONS for PANEL C

| first steps | counting-in ||| result |
|---|---|---|---|---|
|  | (near 41h) | (near 21↓) | (near 21↑) |  |
| 82 | 38  54 | 20  44 | 23  68 | 82 |
| M41h | 39  55 | 21  45 | 22  69 | M41h   57 |
| H21 | 40  57 | 22  45 | 21  69 | H21 \|45  69\| |
| 1 \|38  94\| | 41  57 | 23  45 | 20  70 | 1 \|38  94\| |
|  | 42  57 |  |  |  |

→

we will find it easy to do just by pointing at digits in the stem-and-leaf, writing down only the answers.

Though it is somewhat concealed, the tendency of the data in panel C to trail out further toward high values can be seen in the summary of panel D, whose heart is

```
| 57      |
| 45  69  |
| 38  94  |
```

exhibit **3** of chapter 2 (continued)

E) CUMULATIVE COUNTS in VARIOUS FORMS

| As scratched down | | For record | |
|---|---|---|---|
| (right of display) | | (left of stub) | |
| (1) | ⟨1 | 1 | 3· |
| (19) | ⟨20 | 20 | 4* |
| (12) | ⟨32 | 32 | 4· |
| (6) | ⟨38 | 38 | 5* |
| (12) | 12 | 82 √ (12) — | 5· |
| (6) | ⟨32 | 32 | 6* |
| (6) | ⟨26 | 26 | 6· |
| (9) | ⟨20 | 20 | 7* |
| (5) | ⟨11 | 11 | 7· |
| (3) | ⟨6 | 6 | 8* |
| (1) | ⟨3 | 3 | 8· |
| (2) | ⟨2 | 2 | 9* |

P) PROBLEMS

**Imitate** Panel A for:

3a) The 17 Chevrolet prices of exhibit 3 of chapter 1.

3b) The 25 values of exhibit 7 of chapter 1.

3c) A batch of 12 to 25 values of interest to you.

**Imitate** Panel B for

3d) A batch of 25 to 50 values of interest to you.

3e) The 50 heights of exhibit 10 of chapter 1.

**Imitate** Panel C for

3f) The 58 radial velocities in exhibit 13 of chapter 1.

3g) The 21 measurements of an association constant in exhibit 14 of chapter 1.

3h) A batch of 100 or more values of interest to you.

**Find** the 5-number summaries for

3i) The 93 model-years of panel A of exhibit 9 of chapter 1.

3j) The 141 model-years of panel B of exhibit 9 of chapter 1.

3k) A batch of size at least 100 that interests you.

The hinges, 45 and 69, are each 12 away from the median, thus being quite symmetrical, to be sure. The extremes, however, are respectively, 7 and 25 units outside the hinges, which shows a considerably longer trail-off toward higher values.

Panel E shows two forms of cumulative counts. First, the scratch-down version--placed where we have been putting the stem counts, and distinguished by a row of ⟨'s. (Note the omission of ⟨ for the line containing the median. If the median falls between lines, we draw a horizontal line, as in exhibit 16 of chapter 3.) Then the more formal version, used in most exhibits, but often avoided in paper and pencil work. This version goes to the left of the stem-heads. The lines separating off the line with the median--and the giving of that line's individual count--are important.

### further examples

Exhibit **4** shows two further examples.

The summary of panel A with its

```
|   46      |
| 20   112  |
|  3   203  |
```

could also be interpreted as just another trail-off toward higher values. As we know, this batch is clearly separated into 3 groupings. As we have said, the sort of summary we are making should not detect this separation. Clearly, in this example, it does not.

The populations of the 50 states are shown in panel B. Note that dealing with mixed leaves is not a complication. This batch shows a more common pattern, where extremes, hinges, and median come quite close to telling the tale. The second version of a five-number summary below the stem-and-leaf-- the one converted to millions--summarizes the situation quite well. We could--probably equally well--have rounded the values further to a box

```
|    2.5    |
| 0.9   4.3 |
| 0.2   17  |
```

which would still remind us that:

◊ 50 states are considered.

◊ the smallest state was about 200,000 population in 1960.

◊ about half the states fell between 0.9 million and 4.3 million--with about a quarter below 0.9 and a quarter above 4.3.

◊ about half the states were smaller and about half larger than 2.5 million.

◊ the largest was about 17 million.

## exhibit 4 (A)/2: Easy summaries

These reminders really give us a fairly good feel for this whole batch of 50 populations.

The populations of the states clearly trail off much farther above the middle than below. In this instance, this keeps us from condensing this summary still further, and is something we will need to be able to remedy.

### review questions

What is a hinge? How is its depth found? What approximation is almost always made? What values make up a 5-number summary? What are two ways to write them down as a summary scheme? How are total count and depths shown in a summary scheme? What is the natural order to find these depths?

### exhibit 4 of chapter 2: various

**Further examples of hinges and 5-number summaries**

A) The 50 HEIGHTS of EXHIBIT 10 of CHAPTER 1--with cumulative counts in the easy places

```
  5      0* | 34588      Fla, Del, La, Miss, Rl
 11      1  | 236788
 17      2  | 003448                  count check is
 22      3  | 24556                   22 + 5 + 23 = 50 ✓
 √5      4* | 01489
 23      5  | 03347
 18      6  | 367
 15      7  | 2          S.Dak
 14      8* | 8          Texas
  ·      9  |
  ·     10  |
 13     11  | 2          Oregon
 12     12* | 678        Idaho, Ariz, Mont
  9     13  | 12588      Nev, N.Mex, Utah, Wyo, Hawaii
  4     14  | 445        Wash, Colo, Calif
  ·     15  |
  ·     16* |
  ·     17  |
  ·     18  |
  ·     19  |
  1     20* | 3          Alaska
```

| 50 heights in hundreds of feet | | | 50 heights in thousands of feet | | |
|---|---|---|---|---|---|
| M25h | 46 | | M25h | 4.6 | |
| H13 | 20 | 112 | H13 | 2.0 | 11.2 |
| 1 | 3 | 203 | 1 | 0.3 | 20.3 |

Note: As you get more used to simple one-digit stem-and-leafs, you may want to use a single * above the vertical line in place of all the *'s and spaces just to the left of that line. (This lets you snuggle the stems up to the line.) Wholly optional.

What do we include in a stem-and-leaf to make counting-in easier? What kinds of behavior do the two panels of exhibit 4 illustrate?

### 2C. Box-and-whisker plots

We always want to look at our results. We usually can. In particular, we want to look at 5-number summaries. Exhibit 5 shows an easy way to do just this. We draw a long, thinnish box that stretches from hinge to hinge, crossing

exhibit **4** of chapter 2 (continued)

**B)** The 1960 POPULATIONS of the 50 STATES--unit = 10,000 people (rounded)

```
↓        1*|
2        2 |3,9                    Alaska, Nevada
4        3 |9,3                    Vermont, Wyoming
5        4 |5,                     Delaware
         5*|
11       6 |3,7,7,1,3,8
         7 |6,
13       8 |9,
15       9*|7,5

 21      1**|30,79,74,41,77,86
√(7)     2  |54,76,18,18,33,38,85
 22      3  |27,94,04,26,10,41,57,97,95
13       4  |95,66,32,56            Massachusetts
 9       5**|15,
 8       6  |07,                    New Jersey
 7       7  |82,                    Michigan
         8  |
 6       9**|71,58                  Ohio, Texas

 4       1***|572,008,678,132       Calif, Ill, NY, Pa
 ↑       2   |
```

50 populations in tens of thousands

| M25h | 246 | |
|---|---|---|
| H13 | 89 | 432 |
| 1 | 23 | 1678 |

50 populations in millions

| M25h | 2.5 | |
|---|---|---|
| H13 | 0.89 | 4.3 |
| 1 | 0.23 | 17 |

**P) PROBLEM**

4a) Find another set of data deserving such treatment. Give the treatment and plot the display.

**S) SOURCE**
The World Almanac, 1966, p. 325.
Their source: U.S. Bureau of the Census

it with a bar at the median. Then we draw a "whisker" from each end of the box to the corresponding extreme.

This process shows us the five-number summaries quite clearly, so clearly as to give us a clear idea of (some of) what we may have been missing. There is, inevitably, more empty space in a box-and-whisker plot than in a listing of a 5-number summary. There is more space for identification. We can at least identify the extreme values, and might do well to identify a few more.

exhibit **5** of chapter 2: various heights

**Box-and-whisker plots of 5-number summaries**

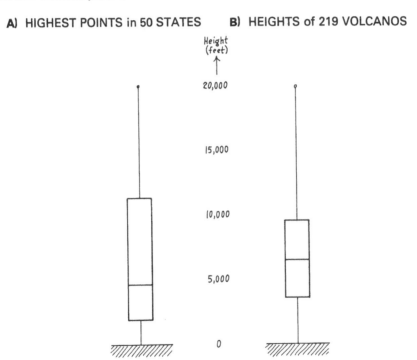

A) HIGHEST POINTS in 50 STATES     B) HEIGHTS of 219 VOLCANOS

P) PROBLEMS

Imitate these plots for the data of:

5a) Panel A of exhibit 3.

5b) Panel B of exhibit 3.

5c) Panel C of exhibit 3.

5d) Panel B of exhibit 4.

5e) A set of data of interest to yourself.

Exhibit **6** shows the same examples as in exhibit 5, with some identification added. Clearly, we learn more from such displays.

In exhibit 6 we have stopped the whiskers at the innermost identified values. This is recommended.

exhibit **6** of chapter 2: various heights

**Box-and-whisker plots with end values identified**

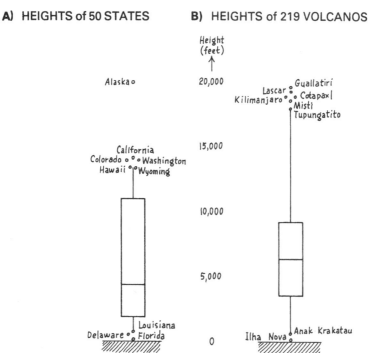

**P) PROBLEMS**

Imitate these plots for the data of:

6a) Panel A of exhibit 3.

6b) Panel B of exhibit 3.

6c) Panel C of exhibit 3.

6d) Panel B of exhibit 4.

6e) A set of data of interest to yourself.

6f) Make a similar plot for the heights of some of the highest mountains in the U.S.

6g) Make a similar plot for the heights of some of the highest mountains in the world.

### basic ideas of plotting

With box-and-whisker plots we are beginning our plotting education. The first thing we need to do is separate, in our mind, what it takes to make plotting easy from what it takes to make plotting effective. The lines ruled on graph paper help to make plotting easy, but they do not make plotting effective for seeing what is going on--instead they get in the way of seeing what we ought to see. (If we are using a graph as a substitute for a table--as a way to look up values of a function--things are likely to be different. This, however, is something we almost never need to do in exploratory data analysis.)

### tracing paper

If we want to see what our plots ought to tell us, there is no substitute for the use of tracing paper (or acetate). If we slip a well-printed sheet of graph paper just below the top sheet of a pad of tracing paper, we can plot on that top sheet of tracing paper almost as easily as if it were itself ruled. Then, when we have the points plotted, some boundary or reference lines drawn, and a few scale points ticked, we can take away the graph sheet and look at the points undisturbed by a grid. We often gain noticeably in insight by doing this. (And we have had to pay for a sheet of tracing paper rather than for a sheet of graph paper.) Alternatively, we can plot on a sheet of graph paper and then trace the result.

In doing either of these we gain very much by having:

◊ well-printed graph paper, which means (a) good quality lines, (b) every 10th line heavy, (c) every 5th line medium heavy. ("Five and dime" graph paper or quadrille-ruled paper will make our work much harder. For more detailed discussion, see section 5A, below.)

◊ tracing paper that erases cleanly and easily (quality equal to that of the Clearprint Paper Co. of San Francisco is worth the difference).

◊ a transparent plastic rule or triangle.

An alternative that:

◊ can be even more effective,

◊ is no more expensive,

◊ takes a little more trouble to prepare for,

replaces the tracing paper by the thin sheets of transparent plastic (acetate) made for use in overhead projectors. Two cautions are important:

1. You can only use markers specially made for the purpose. (A number of manufacturers make satisfactory products, but it may not be easy to find any of them. The writer prefers the temporary--wash-off--styles to those that cannot be erased.)

2. It is important to keep one's fingers off the plastic until the picture is completed. (A piece of thin graph paper, placed upside down, works very well as a hand shield.)

### scale values

We ought to put as many scale values on the graph paper preliminary as will help us make the plot easily. On the tracing paper final, however, we ought not show more than three or four numbers along a scale. More clutters up the picture and distracts the eye from what it ought to see. (Scales for dates are sometimes an exception. It can matter whether an appearance came in 1929 or 1928, in 1776 or 1775.)

People are used to scales on the left and below. So be it--for the picture, perhaps. When one is plotting the points, however, it is much more convenient to put the horizontal scale ABOVE the plot, where you do not have to move your hand to see it. (It would be rational to plot from detailed scales above and left, and to produce a final picture with a few scale points shown below and right; but such rationality is usually not worth the possibility of occasional confusion.)

### plotting without graph paper

We almost always want to look at numbers. We do not always have graph paper at hand. **There is no excuse for failing to plot and look.**

We usually have ruled paper at hand. For emergency graph paper, take out one sheet of ruled paper, turn it on its side, and place it beneath another sheet of ruled paper. If these two sheets have a light-colored backing--often provided by the rest of the pad or notebook--the vertical lines on the lower sheet are almost certain to show through well enough, combining with the horizontal lines on the top sheet to form a grid on which plotting is reasonably easy. (The first step in this sort of plotting is to mark--by ticks or unobtrusive dots--enough information on the top sheet to make it easy to get the lower sheet back to its original position after it slips.)

With this technique, one can make useful, if not decorative, plots almost anywhere.

### review questions

What is a box-and-whisker plot? What do its parts show forth? What rules does it obey about showing values individually? About identifying values? What must we separate in our minds about plotting? What are the essentials of convenient, effective plotting? How can we, in an emergency, plot without graph paper?

## 2D. Fences, and outside values

Hinges are for our convenience. They can--and will--serve various purposes for us. Their role in 5-number summaries is only the beginning.

When we look at some batches of values, we see certain values as apparently straying out far beyond the others. In other batches straying is not so obvious, but our suspicions are alerted. It is convenient to have a rule of thumb that picks out certain values as "outside" or "far out". To do this, we set up appropriate "fences" and use "outside" and "far out" accordingly.

A useful rule of thumb runs as follows:

⋄ "H-spread" = difference between values of hinges.

⋄ "step" = 1.5 times H-spread.

⋄ "inner fences" are 1 step outside hinges.

⋄ "outer fences" are 2 steps outside hinges (and thus 1 step outside of inner fences).

⋄ the value at each end closest to, but still inside, the inner fence is "adjacent".

⋄ values between an inner fence and its neighboring outer fence are "outside".

⋄ values beyond outer fences are "far out".

Exhibit **7** shows some examples. Notice the practice of writing the value of the H-spread just outside the box (under a horizontal "eave"), and then putting the value of the step in a "penthouse" on the second part of the display, which contains the fences and is written after the main letter display, either below it or to its right. Panels C to F show a convenient standard form, combining summary scheme, fences, and identification of outside values. We can call it a

fenced letter display

exhibit **7** of chapter 2: various examples

**Examples of calculation of fences and identification of adjacent, outside, and far outside values (based on exhibits 3 and 4)**

**A)** For panel A of exhibit 3

17 Chev. prices

```
M9 |    895    |
H5 | 795  1499 | 704
 1 | 150  1895 |
       | 1056 |
    f| -261  2555 |
      | xxx   xxx | out
    F| -1317 3611 |
      | xxx   xxx | far
```

adj: 150, 1895

Note that, here:

704 = 1499 − 795
1056 = (1.5)(704)
−261 = 795 − 1056
2555 = 1499 + 1056
−1317 = −261 − 1056
3611 = 2555 + 1056

➡

exhibit **7** of chapter 2 (continued)

**B)** For panel B of exhibit 3--see exhibit 6 of chapter 1 for details

34 ultimate powers

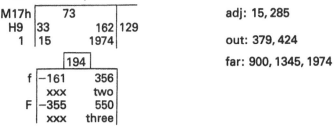

adj: 15, 285

out: 379, 424

far: 900, 1345, 1974

Note that, here: 121h = 156 − 34h and 182 = 3(121h)/2

**C)** REARRANGED VERSION of (B), with IDENTIFICATION

34 ultimate powers

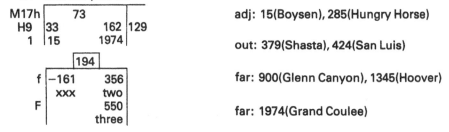

adj: 15(Boysen), 285(Hungry Horse)

out: 379(Shasta), 424(San Luis)

far: 900(Glenn Canyon), 1345(Hoover)

far: 1974(Grand Coulee)

**D)** For panel C of exhibit 3--see panel D of exhibit 3 also

82 areas

adj: 38(Tate), 94(Yazoo)

**E)** For panel A of exhibit 4--hundreds of feet

50 heights

M25h | 46
H13 | 20    112 | 92
 1  |  3    203 | 200
     | 138 |
   f |−118   250
     | xxx   xxx
   F |

adj: 3(Florida), 203(Alaska)

➔

## ranges

One thing that a box-and-whisker plot brings to eye and mind is the spread from one extreme to the other, called the

## range

of the batch. For the various panels of exhibit 7 we have:

(for patch A)  range = 1895 − 150 = 1745 dollars.
(for panel B)  range = 1974 − 15 = 1959 megawatts.
(for panel C)  range = 74 − 38 = 36 tens of sq. miles.
(for panel D)  range = 203 − 3 = 200 hundreds of feet.
(for panel E)  range = 1678 − 23 = 1655 tens of thousands.

If we were to quote only one measure of spread, we usually do better to quote the H-spread, unless we are specifically concerned with either extremes or stray values. Giving both the range and the H-spread is often useful, however. Since the extremes are in the 5-number summary, finding the range is easy. We recommend writing it down rather often. Clearly the place for it in a letter display is on the line for extremes, just to the right of the box--as in panels E and F.

## trimeans

Another thing that box-and-whisker plots convey to us is an impression of location or centering that combines both median and hinges. The arithmetic that comes closest to matching this impression is probably the

## trimean

defined to be

$$\frac{\text{lower hinge} + 2(\text{median}) + \text{upper hinge}}{4}$$

exhibit **7** of chapter 2 (continued)

**F)** For panel B of exhibit 4--10,000's of people

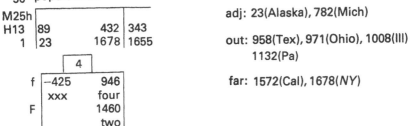

adj: 23(Alaska), 782(Mich)

out: 958(Tex), 971(Ohio), 1008(Ill)
     1132(Pa)

far: 1572(Cal), 1678(NY)

**P) PROBLEMS**

7a/b/c/d/e/f/g/h/i/j/k) Make a fenced letter display for the batch described in the correspondingly lettered problem of exhibit 3.

Sometimes we like to use trimeans instead of medians to give a somewhat more useful assessment of location or centering. We can use the trimean almost any place where the median is used. Readers should feel free to do so.

### review questions

Do hinges have more than one purpose? Vaguely, what is a stray value? How might we identify stray values? What is an H-spread? A step? An inner fence? An outer fence? Which values are outside? Which far outside? Which adjacent? What is a standard way of combining a summary scheme with fences and identification of outside values? What is a range? How useful is it? Can it replace an H-spread? Can it supplement an H-spread?

## 2E. Schematic plots

Sometimes it is worthwhile to have some rules about which values deserve to appear individually in a box-and-whisker plot. The simplest convenient set of rules appear to be these:

⋄ outside, and far out values should appear separately.

⋄ whiskers should be dashed, ending with dashed crossbars at the adjacent values.

⋄ far out values should be marked impressively and identified in capital letters

⋄ outside and adjacent values should all be identified--in small letters--unless identification would confuse the picture (this rarely happens unless there are more than six at a single end). Identification of adjacent values may be less obtrusive.

⋄ abbreviated identifications are welcome whenever likely to be understood.

We expect to be consistent about the whiskers: always using dashes for **schematic plots** according to these rules, always using solid lines for free-form box-and-whisker plots that do not (necessarily) conform to these rules.

For an example where these rules are well illustrated, we can turn to the areas of the 83 counties of Michigan (the stem-and-leaf will be presented as part of exhibit 4 of chapter 3). The fenced letter-value display is:

```
      83 areas
    M42 ┌────57────┐
   H21h │ 54h   78h│ 24
      1 │ 32    184│
        └───┌─36─┐─┘
          f │18h  114h│    adj: 32(Benzie), 111(Gogebic)
            │none four│    out: 118(Delta), 120(Iron)
          F └─    150h┘    out: 120(Schoolcraft), 132(Ontonagon)
                  two      far: 158(CHIPPEWA), 184(MARQUETTE)
```

**exhibit 8/2: Easy summaries**

The corresponding schematic plot appears as in exhibit **8**. Notice the treatment of a tie.

exhibit **8** of chapter 2: county areas

**Michigan counties, area in mi²**

A) SCHEMATIC PLOT

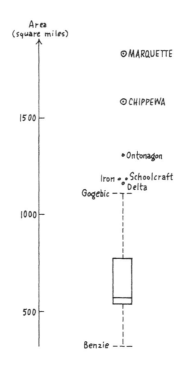

P) PROBLEMS

8a) Do you think the areas of counties in most states would include detached values? Why or why not?

8b) Make a suitably-identified schematic plot of the county areas for the state in which you were born (or of corresponding subdivisions of the country in which you were born).

8c) What can be said about the counties in Michigan whose areas are identified in panel A? (Use map.)

8d) What display would your answer to (8c) lead you to propose?

8e) Make that display.

### review questions

What is a schematic plot? What points are plotted separately? What points are identified? How do the whiskers show whether or not a box-and-whisker plot is a schematic plot? What is a fenced letter display? What rules are used in writing one down?

### 2F. Pros and cons; the Rayleigh example

We need now to look harder at the advantages and disadvantages of schematic plots. Let us begin with an example that shows both their weakness and their strength. During the winter of 1893-94, Lord Rayleigh was investigating the density of nitrogen from various sources. He had previously found indications of a discrepancy between the densities of nitrogen produced by removing the oxygen from air and nitrogen produced by decomposition of a chemical compound. The 1893-94 results established this difference with great definiteness, and led him to investigate further the composition of air chemically freed of oxygen. This led to the discovery of argon, a new gaseous element. Rayleigh's observations are summarized in exhibit **9**.

exhibit **9** of chapter 2: gas densities

**Rayleigh's weights of a standard volume of "nitrogen"**

**A) DATA**

| Date | Source of "nitrogen" | | Weight |
|---|---|---|---|
| | Origin | Purifying agent | |
| 29 Nov. '93 | NO | Hot iron | 2.30143 |
| 5 Dec. '93 | " | " | 2.29816 |
| 6 Dec. '93 | " | " | 2.30182 |
| 8 Dec. '93 | " | " | 2.29890 |
| 12 Dec. '93 | Air | " | 2.31017 |
| 14 Dec. '93 | " | " | 2.30986 |
| 19 Dec. '93 | " | " | 2.31010 |
| 22 Dec. '93 | " | " | 2.31001 |
| 26 Dec. '93 | $N_2O$ | " | 2.29889 |
| 28 Dec. '93 | " | " | 2.29940 |
| 9 Jan. '94 | $NH_4NO_2$ | " | 2.29849 |
| 13 Jan. '94 | " | " | 2.29889 |
| 27 Jan. '94 | Air | Ferrous hydrate | 2.31024 |
| 30 Jan. '94 | " | " | 2.31030 |
| 1 Feb. '94 | " | " | 2.31028 |

**S) SOURCE**

Lord Rayleigh, 1894. "On an anomaly encountered in determinations of the density of nitrogen gas," *Proc. Roy. Soc.* (Lond.) 55, 340-344 (also in his *Scientific Papers*, Vol. 4, pp. 104-108).

Exhibit **10** shows dot and schematic plots for Rayleigh's 15 weights of nitrogen, treated as a single batch. Here the main issue is the violent separation into two quite isolated subgroups. This is made quite clear by the individual values of the dot plot--and almost completely covered up by the schematic plot. (Only **almost,** because the experienced viewer--finding the whiskers so short, in comparison with box length--is likely to become suspicious that he should see more detail.)

Clearly we cannot rely on schematic plots to call our attention to structure near the center of the batch; we must use some other approach when this is needed. (Either dot plots or stem-and-leaf displays will serve.)

**a better use**

Exhibit **11** uses the schematic plots for one of the purposes for which they are best fitted: comparison of two or more batches. In it, the two batches of

exhibit **10** of chapter 2: gas densities

**Rayleigh's 15 weights of "nitrogen"**

A) DOT and SCHEMATIC PLOTS

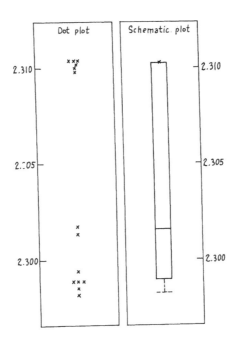

Rayleigh's weights, one for "nitrogen" from air, the other for "nitrogen" from other sources, are set out and compared. Clear to every eye are:

⋄ the relatively huge difference between the batches.

⋄ a slight tendency of the "other sources" weights to straggle upward-- noticeable, but far from offering any basis for explaining their difference from the air values.

⋄ a possible greater variability of the "other sources" weights, over and beyond this straggling.

Here schematic plots have done everything we could ask.

### aim reiterated

The Rayleigh data lets us make an old point once more. **Aims can differ, and plots should follow aims.** We can try to show either **general behavior** or **detail.** We should almost always choose one or the other.

Many books on graphical presentation stress getting the proper impression of relative size across to the reader. This may be sound for most articles in the

exhibit **11** of chapter 2: gas densities

**Rayleigh's two batches of weights of "nitrogen"**

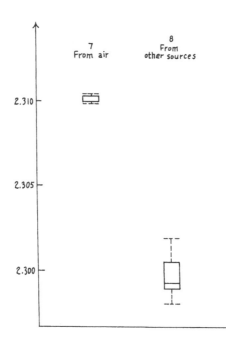

**52** exhibit 12/2: Easy summaries

daily newspaper, where making small differences more visible by starting the vertical scale far from zero is usually thought of as a "journalistic" or "political" trick.

We need to remember that doing just this is a vital scientific tool. If we had started the vertical scale at zero and used bar graphs, Rayleigh's results--so clear in exhibit 11--would have appeared as in exhibit **12.** Whatever exhibit 12 may look like, it is neither evidence for a new chemical element nor a basis for a Nobel prize.

From the point of view of freight charges for compressed gases, exhibit 12's impression of "negligible difference" gives a useful impression, but it takes exhibit 11 to make clear the scientific issue. **There is often no substitute for the detective's microscope--or for the enlarging graph.**

### choice of plots

We are now well equipped. Schematic plots will be the standard form, but we will feel free to use other plots instead whenever we feel that this will help.

exhibit **12** of chapter 2: gas densities

**Rayleigh's weights again**

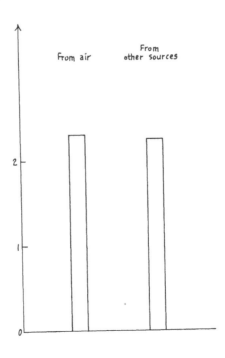

Clearly, we can use schematic plots in comparing three, four, or any reasonable number of batches. (When the number of batches becomes large, we will probably want to squeeze our bars down to solid lines, and thin our dotted lines out.)

As a visual approach to most comparisons especially when supplemented by more detailed looks at the residuals, these plots work very well, often providing very effective comparisons.

Comparing more than a few is something that stem-and-leaf cannot do at all well. A display involving more than two--or perhaps three--stem-and-leaf displays tends to leave us with no clear impression.

The key advantages of schematic plots are, then:

◊ focusing attention where it is most likely to be useful.

◊ reducing "busyness" to a point where many batches can be compared.

### review questions

What was Lord Rayleigh up to? What did he discover? Does exhibit 9 show that he discovered anything? Does the schematic plot of exhibit 10 hint at unusualness? If so, how? Does exhibit 11 show that something was discovered? What do books on graphical presentation often insist upon? Dare we pay attention to what they say? What does exhibit 12 show? What does it fail to show?

## 2G. Eighths, sixteenths, etc. (almost optional here; used in later chapters)

We sometimes want to go beyond the 5-number summary. Especially in larger batches, there can be advantages to pinpointing one or more values between the hinges and the extremes. A simple and useful approach is to continue halving in terms of depths (naming the results eighths, sixteenths, etc., since the hinges and median divide the batch into approximate quarters). This leads to:

$$\text{Depth of eighths} = \frac{1 + \text{depth of hinges}}{2}$$

$$\text{Depth of sixteenths} = \frac{1 + \text{depth of eighths}}{2}$$

and so on, where, as for hinges, we drop any h before dividing by 2. It seems natural to use E for eighths and then go on to "D", "C", "B", "A", "Z", "Y", ... as may be needed. The corresponding spreads will be used in chapter 19.

Exhibit **13** shows some examples of 7-number and 9-number summaries, adding first E's and then also D's.

When we need a name for M, H's, E's, etc., we call them

**letter values**

often including the extremes under this name also.

### review questions

What is an eighth? A sixteenth? A thirty-second? How do we mark them? What formulas help us find them? What approximation is usually needed? What is an E-spread? What is a D-spread? What values appear in a 7-number summary? In a 9-number summary? In an 11-number summary? How do the corresponding summary schemes appear? What is an extended schematic summary? How is it extended?

exhibit **13** of chapter 2: varied examples

### Examples of 7- and 9-number summaries

A) The 1960 POPULATIONS of the 50 STATES--unit = 10,000 people; stem-and-leaf at exhibit 4, panel B

```
                          50 populations
(Seven-number summary)  M25h | 246
                        H13  | 89    432 | 343 (H-spread)
                        E7   | 63    782 | 719 (E-spread)
                        1    | 23   1678 | 1655 (range)

                          50 populations
(Nine-number summary)   M25h | 246
                        H13  | 89    432 | 343 (H-spread)
                        E7   | 63    782 | 719 (E-spread)
                        D4   | 39   1008 | 969 (D-spread)
                        1    | 23   1678 | 1655 (range)
```

B) The 82 COUNTY AREAS of MISSISSIPPI--in 10's of square miles; stem-and-leaf at exhibit 3, panel C

```
     82 areas                      82 areas
M41h |    57           |     M41h |    57
H21  | 45   69 | 24    |     H21  | 45   69 | 24
E11  | 42   75 | 33    |     E11  | 42   75 | 33
1    | 38   94 | 56    |     D6   | 41   80 | 39
                              1   | 38   94 | 56
```

P) PROBLEMS

(13a) Imitate these schemes for the 50 heights of panel A of exhibit 4.

13b) Extend panel A to an 11-number summary.

13c) Extend panel B to first an 11-number summary and then a 13-number summary.

## 2H. How far have we come?

This chapter has been concerned with the simplest summaries for batches of numbers, as written down in numbers or as displayed graphically. Our central themes have been (1) summarization by picking out five numbers from among the values--and from among the means of pairs of adjacent values--that will tell us about the general behavior of the batch, and (2) plotting and identification of individual values that may be unusual. In summary, we have focused on extremes, hinges, and median, and have introduced both schematic summaries and box-and-whisker plots to show their values and relationships.

### what have we learned to do?

Ranks are defined by counting in, from either one end or the other, of a list of the values in order. (We usually do this on a stem-and-leaf display.) The lesser of the two ranks for a value is its depth. Half-integer ranks or depths (like rank 4h) refer to the mean of the values with adjacent ranks or depths (ranks 4 and 5 for rank 4h).

The median is half-way through the ordered list. Its two ranks (and its depth) are the same--half of the total count PLUS one. The depth of each hinge is half of one more than the integer part of the depth of the median. (For instance, median at 8h, integer part 8, one more 9, hinge at 4h.)

The difference between the extremes is the **range;** that between the hinges is the **H-spread.**

Extremes, hinges, and median make up a 5-number summary. When written down schematically, we give, outside an enclosure, the total count and the depths of median, hinges, and extremes, giving the values in the enclosure. We also give far out, outside, and adjacent values, often with identification. Often we add the corresponding spread to the right of the enclosure (under an eave).

Making even as simple a picture as a box-and-whisker plot deserves care and technique. Use of tracing paper (or acetate) and proper labeling are important.

The **step**--taken from hinge to first fence and from first fence to second fence--is one and one-half times the H-spread, which is the difference between the hinges. Values outside the second fences are **far out.** Values outside the first fences are **outside.** Values nearest to--but still inside--the first fences are **adjacent.**

Box-and-whisker plots show the same information more pictorially (sticking to our conventions matters). A schematic plot is a box-and-whisker plot that displays separately and identifies all outside values (all far out values with extra emphasis), whose whiskers are dashed and stop at the adjacent values.

We can include 8ths, 16ths, etc., in our schematic summaries when we wish. We use E, D, C, ... as tags and stick to the rule

$$\text{new depth} = \tfrac{1}{2}(1 + \text{integer part of old depth}).$$

### where do we stand ?

**We have not looked at our results until we have displayed them effectively.** In this chapter we have learned how to display summaries of batches effectively. Now we owe it to ourselves to make such displays routinely.

**Except when learning the numerical part of a new technique, no problem of exploratory data analysis is "solved" without something to look at.** A stem-and-leaf display is something to look at; so is a letter-value display, a schematic plot, or a more general box-and-whisker plot. Words that are both sufficiently descriptive and clearly informing can be something to look at--but it may not be possible to find words to express what a display or plot may show.

# Easy re-expression    3

chapter index on next page

We have learned how to collect the values of a batch--in most cases the natural collection scheme lets us look easily and effectively at the "shape" of the numbers as a whole. There are, however, rather frequent exceptions, as exhibit 6 of chapter 1 and panel B of exhibit 4 of chapter 2 illustrate. It is time to learn how to deal with these exceptions.

Only one broad approach seems natural. If the way the numbers were gathered or collected does not make them easy to grasp, we should change them--preserving as much information as we can use--into a form where we can grasp them more easily, both personally and with our displays and our arithmetic.

Data for us usually means numbers--whether given in digits, by tallies, or on a graph. Three broad classes of numbers are important enough to be worth both names and some discussion:

⋄ amounts and counts can never be less than zero and could be arbitrarily large. Heights, powers, areas, distances, numbers of deaths or of people, all fall here. The simplest indicator of whether re-expression is likely to help us look at a single batch is the ratio of the largest value to the smallest value. If this is small, close to 1, re-expression cannot change the appearance of the batch seriously. If this ratio is quite large, say 100 or more, we are almost sure to need re-expression, just to see what is going on. (More about counts when we reach section 3F.)

⋄ balances, where both positive and negative values occur. Profit-and-loss, actual minus predicted, and observed-minus fitted are all natural examples. One can usually think of this sort of data as the difference of two amounts or counts. Re-expression of the balance rarely helps, but re-expression of the amounts or counts before the subtraction sometimes helps a lot.

⋄ counted fractions and percentages, where the reasonable values are hemmed in on both sides. Re-expression is often very helpful, though the techniques are more special. (For the present we will avoid such cases. We will eventually come to dealing with counted fractions and percentages in chapter 15.)

⋄ grades and other ordered versions--including A, B, C, ..., E and -, +, + +, + + +, + + + + as examples--often respond well to slightly more complex techniques. (We may get to them in a later volume.)

# /3: Easy re-expression

review questions  59

**3A. Logarithms = logs**  59
review questions  61

**3B. Quick logs**  61
*break tables*  62
review questions  64

**3C. Comparisons of two batches**  64
five-figure logs may not be enough  67
review questions  69

**3D. Quick roots and quick reciprocals**  69
quick roots  69
quick reciprocals  70
*use negative reciprocals*  70
a volcano example  70
looking easier  75
reciprocal times  76
review questions  77

**3E. Looking quickly**  79
combining evidence across batches  83
review questions  83

**3F. Counted data**  83
review questions  84

**3G. Relation among powers and logs (optional)**  86
*trivial re-expressions*  88
quick reciprocal roots  89
quick squares  89
review questions  89

**3H. How far have we come?**  92

**3K. How to think about logs (background)**  93

**3P. Additional problems**  93

| EXHIBIT | PAGE |
|---|---|
| 3A | |
| 1 | 60 |
| 3B | |
| 2 | 62 |
| 3★ | 63 |
| 3C | |
| 4★ | 65 |
| 5★ | 67 |
| 6 | 68 |
| 3D | |
| 7 | 71 |
| 8 | 72 |
| 9 | 73 |
| 10 | 74 |
| 11 | 75 |
| 12 | 76 |
| 13 | 78 |
| 3E | |
| 14★ | 80 |
| 15★ | 82 |
| 3F | |
| 16★ | 85 |
| 3G | |
| 17★ | 86 |
| 18 | 90 |
| 19 | 91 |
| 3H | |
| 3K | |
| 3P | |
| 20★ | 94 |
| 21★ | 95 |
| 22★ | 96 |

**review questions**

What is an amount? A balance? A count? A counted fraction? What is the most urgent reason for re-expression? What suggests that re-expression will probably be needed?

## 3A. Logarithms = logs

What kinds of re-expression we will most frequently find helpful will depend upon the habits of our fellow men, on the ways men habitually choose to write down the numbers that come to us. Civilized "people" evolved from bears, for example, might have different habits than those evolved from apes. If so, they would be likely to find other forms of re-expression most useful.

As we have noticed, numbers that are always positive and that are not too closely alike are likely to gain from some form of re-expression. The most likely form is taking logarithms. (The next most likely form is taking square roots, to which we will come in turn.)

Logarithms--or as we shall regularly say, logs--can be made either simple or mysterious. We shall try to keep them simple.

What do you need to know about logs? In the first instance,

⋄ how to find good enough logs fast and easily (see next section).

⋄ that equal differences in logs correspond to equal ratios of raw values. (This means that wherever you find people using products or ratios--even in such things as price indexes--using logs--thus converting products to sums, and ratios to differences--is likely to help.)

Some readers may like a review of the few simple things they need to know about logs; this they can find in section 3K. There are different kinds of logs, but this makes little difference to us here. Of any two kinds, each must be a constant multiple of the other, so that no more care is needed than is needed in keeping feet and inches straight. We will thus routinely write merely "log"-- meaning the kind of log that makes arithmetic easiest, logs to the base 10, also called common logarithms. (Should we need any other kind, we would make a special note.)

The relation of logs to various powers, particularly fractional ones, can also be of interest. (We will come to this in section 3G.)

So much for generalities. Let us take up the example of exhibit 6 of chapter [1]. If we look back to that display, we see a strong tendency for the leaves to be washed into the corners where the leading digits are 1 or 2. What happens if we look at the logs of the same values? Exhibit **1** shows the result.

Notice that extremes, hinges, and median are given both in logs and in megawatts. If often helps to display both re-expressed and raw values when we are working with re-expressed numbers.

This stem-and-leaf display lets us get a rather good feel for the ultimate sizes of the Bureau of Reclamation's power plants. We see, perhaps more easily in the compressed form, a tendency to spread out to high values that is greater than the tendency to spread to low values. The statement that only plants of 15 megawatts or more are included now begins to be related to something we can see. Clearly, the stem-and-leaf in logs is more effective than the stem-and-leaf in raw megawatts. Getting the values where we can look at them better has meant getting them much more nearly symmetrical.

exhibit **1** of chapter 3: ultimate powers

**Ultimate power of 34 hydroelectric plants of the Bureau of Reclamation (logs found from powers in megawatts)**

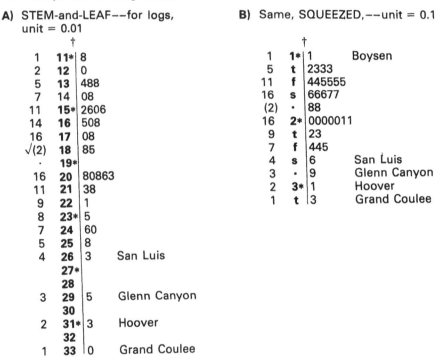

A) STEM-and-LEAF—for logs, unit = 0.01

B) Same, SQUEEZED,—unit = 0.1

† = does not include plants smaller than 15 megawatts (118 for panel A, 11 for panel B)

C) LETTER-VALUE DISPLAYS—logs and raw

34 pwrs (log megawatts)

| M17h | 1.86 | |
| H 9 | 1.52 | 2.21 .69 |
| 1 | 1.18 | 3.30 2.12 |

34 pwrs (megawatts)

| M17h | 73 | |
| H 9 | 33 | 162 129 |
| 1 | 15 | 1974 1959 |

S) SOURCES

**The World Almanac, 1966, page 263.**
Their source: U.S. Bureau of Reclamation.

Raw version: Exhibit 6 of chapter 1. (Also appears in exhibit 5 of chapter 1, and exhibit 3 of chapter 2.)

### review questions

What is the single most needed form of re-expression? What is the minimum that we should know about logs? How many letter-value displays are likely to help in summarizing numbers re-expressed to be looked at?

## 3B. Quick logs

We have just seen an example where two-decimal logs were quite precise enough. Indeed, as the righthand side of exhibit 1 shows, we can do quite well in this example with one-decimal values. This is but one example of many. Though there are exceptions, in most situations where re-expression in logs really helps, two-decimal logs help as much as many-decimal ones.

Since we can find two-decimal logs easily, as we shall soon see, our sailing instructions are clear:

⋄ plan to use two-decimal logs first always.

⋄ in rare cases where we need more precision, go back and start again with many-decimal logs.

**Overall, we will do much less work this way.**

To get few-decimal logs easily, special kinds of tables are very helpful. They make looking up logs just a matter of finding our place and reading the answer. (No interpolation!)

For one-decimal logs, we can use the first two columns of the following little table, whose use is easily learned. (It gives only the fractional part of the log; you have to find the integer part as usual.)

| First nonzero digits of $x$ | First decimal of log $x$ | Break |
|---|---|---|
| 8913 to 1122 | .0 | .0   8913 |
| 1123 to 1412 | .1 | .1   1122 |
| 1413 to 1778 | .2 |       1413 |
| 1779 to 2238 | .3 | .2   1778 |
| 2239 to 2818 | .4 | .3   2239 |
| 2819 to 3548 | .5 | .4   2818 |
| 3549 to 4466 | .6 | .5   3549 |
| 4467 to 5623 | .7 | .6   4466 |
| 5624 to 7079 | .8 | .7   5624 |
| 7080 to 8912 | .9 | .8   7079 |
|  |  | .9   8913 |

If we use the standard code--"when in doubt, go to an even answer"--we can compress this table quite a lot further. We have only to use the last two columns, entering between the breaks. (We go to an even answer when we come in exactly at the break.)

## exhibit 2/3: Easy re-expression

Tables that work like this are often called critical tables. We will try to be more specific and call them

**break tables.**

Exhibit 2 uses two break tables to make finding two-decimal logs easy.

**exhibit 2 of chapter 3: reference table**

**Break table for two-decimal logs**

**A) MAIN BREAK TABLE**

| Break | log | Break | log | Break | log | Break | log | Break | log |
|---|---|---|---|---|---|---|---|---|---|
| 9886 | .00 | 1567 | .20 | 2483 | .40 | 3936 | .60 | 6237 | .80 |
| 1012 | .01 | 1603 | .21 | 2541 | .41 | 4027 | .61 | 6383 | .81 |
| 1035 | .02 | 1641 | .22 | 2600 | .42 | 4121 | .62 | 6531 | .82 |
| 1059 | .03 | 1679 | .23 | 2661 | .43 | 4217 | .63 | 6683 | .83 |
| 1084 | .04 | 1718 | .24 | 2723 | .44 | 4315 | .64 | 6839 | .84 |
| 1109 | .05 | 1758 | .25 | 2786 | .45 | 4416 | .65 | 6998 | .85 |
| 1135 | .06 | 1799 | .26 | 2851 | .46 | 4519 | .66 | 7161 | .86 |
| 1161 | .07 | 1841 | .27 | 2917 | .47 | 4624 | .67 | 7328 | .87 |
| 1189 | .08 | 1884 | .28 | 2985 | .48 | 4732 | .68 | 7499 | .88 |
| 1216 | .09 | 1928 | .29 | 3055 | .49 | 4842 | .69 | 7674 | .89 |
| 1245 | .10 | 1972 | .30 | 3126 | .50 | 4955 | .70 | 7852 | .90 |
| 1274 | .11 | 2018 | .31 | 3199 | .51 | 5070 | .71 | 8035 | .91 |
| 1303 | .12 | 2065 | .32 | 3273 | .52 | 5188 | .72 | 8222 | .92 |
| 1334 | .13 | 2113 | .33 | 3350 | .53 | 5309 | .73 | 8414 | .93 |
| 1365 | .14 | 2163 | .34 | 3428 | .54 | 5433 | .74 | 8610 | .94 |
| 1396 | .15 | 2213 | .35 | 3508 | .55 | 5559 | .75 | 8810 | .95 |
| 1429 | .16 | 2265 | .36 | 3589 | .56 | 5689 | .76 | 9016 | .96 |
| 1462 | .17 | 2317 | .37 | 3673 | .57 | 5821 | .77 | 9226 | .97 |
| 1496 | .18 | 2371 | .38 | 3758 | .58 | 5957 | .78 | 9441 | .98 |
| 1531 | .19 | 2427 | .39 | 3846 | .59 | 6095 | .79 | 9661 | .99 |
| 1567 |  | 2483 |  | 3936 |  | 6237 |  | 9886 |  |

When in doubt use an even answer; thus, 1462 gives .16 and 1496 gives .18.

**B) SETTING DECIMAL POINTS**

| | | | |
|---|---|---|---|
| 1 | +0 | −1 | 1 |
| 10 | +1 | −2 | 0.1 |
| 100 | +2 | −3 | 0.01 |
| 1000 | +3 | −4 | 0.001 |
| 10,000 | +4 | −5 | 0.0001 |
| 100,000 | +5 | −6 | 0.00001 |
| 1,000,000 |  |  | 0.000001 |

**C) EXAMPLES**

| Number | B  A | log number |
|---|---|---|
| log 137.2 | 2 + .14 = | 2.14 |
| log 0.03694 | −2 + .57 = | −1.43 |
| log 0.896 | −1 + .95 = | −0.05 |
| log 174,321 | +5 + .24 = | 5.24 |

Let us give an example of how this works in practice. Exhibit **3** gives the ultimate power in megawatts of the hydroelectric stations and power plants of the Corps of Engineers, first as raw values and then as (unsorted) logs. Let us trace a few examples. 14 comes between 1396 and 1429, so its two-decimal log ends in .15. Similarly, 18 comes between 1799 and 1841, so its two-decimal log ends in .26. Note the entry--15, 26,--of these logs in the first line of panel B.

The gain from going to logs is shown in the stem-and-leaf for logs in Panel C. Although a little long and gangling--something we will soon learn to cure---it does give us a reasonable picture of how these ultimate powers are spread out. (Panel A gave good storage but little help in visualization.)

exhibit **3** of chapter 3: ultimate powers

**Re-expression of the ultimate power in megawatts of 62 hydroelectric stations and power plants of the Corps of Engineers**

A) STEM-and-LEAF

| | | |
|---|---|---|
| 2 | 1* | 4,8 |
| 4 | 2 | 8,6 |
| 9 | 3 | 6,0,0,4,0 |
| 11 | 4 | 3,5 |
| 13 | 5* | 4,2 |
| 17 | 6 | 0,8,8,1 |
| 21 | 7 | 5,6,0,6 |
| 22 | 8 | 6 |
| 24 | 9* | 0,6 |
| 30 | 10 | 0,0,0,0,0,0 |
| √(4) | 11 | 0,2,8,0 |
| 28 | 12 | 4 |
| 27 | 13* | 0,5,5,0 |
| 23 | 1** | 75,65,40 |
| 20 | 2 | 50,80,04,00,70 |
| 15 | 3 | 40,20,30 |
| 12 | 4 | 68,00 |
| 10 | 5** | 18,40,95 |
| 7 | 6 | 00 |
| | 7 | |
| 6 | 8 | 10,10 |
| | 9** | |
| 4 | 1*** | 728,400,743 |
| 1 | 2 | 700 |

B) RE-EXPRESSED in LOGS (unit = 0.01)

| | | |
|---|---|---|
| 1** | 15,26 | Philpott, St. Mary's |
| ·· | 45,42 | J. Percy Priest, Narrows |
| ·· | 56,48,48,53,48 | |
| ·· | 63,65 | |
| ·· | 73,72 | |
| ·· | 78,83,83,79 | |
| ·· | 88,88,85,88 | |
| ·· | 93, | |
| 1·· | 95,98 | |
| 2** | 00,00,00,00,00,00 | |
| ·· | 04,05,07,04 | |
| ·· | 09 | |
| ·· | 11,13,13,11 | |
| ·· | 24,22,15 | |
| ·· | 40,45,31,30,43 | |
| ·· | 53,51,52 | |
| ·· | 67,60 | |
| ·· | 71,73,77 | |
| ·· | 78 | Dworshak |
| ·· | | |
| ·· | 91,91 | Little Goose, Lwr Monumental |
| ·· | | |
| 3** | 24,15,24 | Cheatham, McNary, The Dalles |
| 3·· | 43 | John Day |

62 powers (megawatts)

| M31h | | 110 | |
|---|---|---|---|
| h16 | 68 | 280 | 212 |
| 1 | 14 | 2700 | 2686 |

62 powers (log megawatts)

| M31h | | 2.04 | |
|---|---|---|---|
| H16 | 1.83 | 2.45 | .62 |
| 1 | 1.15 | 3.43 | 2.28 |

➡

### review questions

What is a break table? How do we use exhibit 2? Did re-expression help in exhibit 3? Why?

### 3C. Comparisons of two batches

One notable use of stem-and-leaf displays is for comparison of two batches. Use of logs can also help here. Exhibit **4** compares the land areas of

exhibit **3** of chapter 3 (continued)

C) STEM-and-LEAF for LOGS--in 0.01

```
11*| 5         (Philpott)
12 | 6         (St. Mary's)
13 |
14 | 25888     (Narrows, J. Percy Priest, --, --, --,)
15*| 36
16 | 35
17 | 2389
18 | 335888
19*| 358
20 | 00000044579
21 | 11335
22 | 24
23*| 01
24 | 035
25 | 123
26 | 07
27*| 1378
28 |
29 | 11        (Little Goose, Lower Monumental)
30 |
31*| 5         (McNary)
32 | 44        (Cheatham, The Dalles)
33 |
34 | 3         (John Day)
```

P) PROBLEMS

3a) Sort out the log values into a conventional stem-and-leaf.

3b) Do you like the result? How would you improve it?

3c) Carry out the improvement.

S) SOURCE

**The World Almanac, 1966, and Book of Facts, page 265.**

Their source: Corps of Engineers, U.S. Army.

the 83 counties of Mississippi with those of the 82 counties of Michigan. We see that county areas in Michigan:

◇ tail out much more toward large areas.

◇ are strikingly concentrated between 545 and 595 square miles (26 out of 83).

Panel B shows another way of displaying the same information. Some will prefer the one, others the other.

Let us look to see what the effect is of going over to logs. Exhibit **5** shows the same comparison for logs. On the whole, the comparison is clearer:

◇ setting aside the 23 Michigan counties with logs of 1.75 or 1.76, the rest of the Michigan counties appear as a rather dinosaur-shaped pattern-- running from, say, 1.6 (blunt end) to 2.2 (tail)--while the pattern of the Mississippi counties appears rather rectangular--running from near 1.6 to near 1.9.

exhibit **4** of chapter 3: county areas

**The land areas of the counties of Michigan and Mississippi**

**A)** SIDE-by-SIDE STEM-and-LEAF--unit 10 square miles

| | Michigan, 83 counties | | Mississippi, 82 counties |
|---|---|---|---|
| 3* | 2 | 3* | |
| 3· | 75 | 3· | 8 |
| 4* | 75 | 4* | 0121243121301214202 |
| 4· | 8596689 | 4· | 597886556569 |
| 5* | 1042043214 | 5* | 142010 |
| 5· | 75877677685776776666678666 | 5· | 977899958797 |
| 6* | 4011 | 6* | 412441 |
| 6· | 865 | 6· | 898598 |
| 7* | 120142 | 7* | 320341203 |
| 7· | 65 | 7· | 86657 |
| 8* | 3212 | 8* | 303 |
| 8· | 668 | 8· | 8 |
| 9* | 101 | 9* | 24 |
| 9· | 6 | 9· | |
| 10* | 313 | 10* | |
| 10· | | 10· | |
| 11* | 1 | 11* | |
| 11· | 8 | 11· | |
| 12* | 00 | 12* | |
| 12· | | 12· | |
| 13* | 2 | 13* | |
| 13· | | 13· | |
| 1** | 58,84 | | |

→

**exhibit 4 of chapter 3 (continued)**

**B) BACK-to-BACK STEM-and-LEAF--leaves sorted, as well**

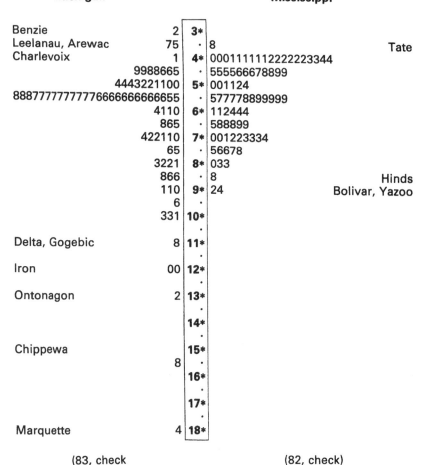

```
            Michigan                              Mississippi

Benzie                         2 | 3*|
Leelanau, Arewac              75 |  .| 8                     Tate
Charlevoix                     1 | 4*| 000111111222223344
                            9988665 |  .| 555566678899
                       4443221100 | 5*| 001124
          888777777777766666666666655 |  .| 577778899999
                            4110 | 6*| 112444
                             865 |  .| 588899
                          422110 | 7*| 001223334
                              65 |  .| 56678
                            3221 | 8*| 033
                             866 |  .| 8                    Hinds
                             110 | 9*| 24             Bolivar, Yazoo
                               6 |  .|
                             331 |10*|
                                 |   .|
Delta, Gogebic                 8 |11*|
                                 |   .|
Iron                          00 |12*|
                                 |   .|
Ontonagon                      2 |13*|
                                 |   .|
                                 |14*|
                                 |   .|
Chippewa                         |15*|
                               8 |   .|
                                 |16*|
                                 |   .|
                                 |17*|
                                 |   .|
Marquette                      4 |18*|

        (83, check                                (82, check)
```

**P) PROBLEMS**

4a) Imitate panel A with some data of interest to you.

4b) Imitate panel B with same.

**S) SOURCE**
The World Almanac, 1966, pages 369-370.

Here it might seem that the gain from re-expression came mainly from improving the picture for the separate stem-and-leaf pictures. Actually, however, the spreads are much more similar for logs, and the shape difference between states is more easily described in the log display. Thus the comparison is considerably improved.

Making stem-and-leafs "easier to look at" has so far mainly involved making patterns more symmetrical--sometimes the one and only pattern, sometimes some or most of the patterns. We might as well plan to try to make stem-and-leafs as symmetrical as we can do easily. This will do us no harm if we do not take doing it too seriously, and its byproducts will often reward us.

### five-figure logs may not be enough

A different sort of example comes by turning to what was known in 1927 about the atomic weights of the chemical elements. The values in exhibit **6** are the results found by different investigators in the years 1922 to 1925, which were the basis for the German 1927 standard values for the atomic weights of

exhibit **5** of chapter 3: county areas

**Logs of areas of counties of Michigan and Mississippi (from exhibit 4)**

A) BACK-to-BACK STEM-and-LEAF--unit = 0.01 in log

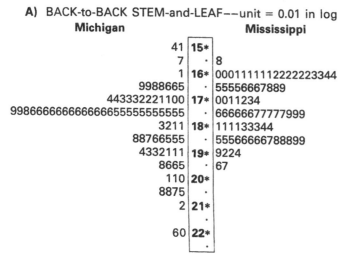

|  Michigan | | Mississippi |
|---:|:---:|:---|
| 41 | 15* | |
| 7 | · | 8 |
| 1 | 16* | 0001111112222223344 |
| 9988665 | · | 55556667889 |
| 443332221100 | 17* | 0011234 |
| 99866666666666655555555555 | · | 66666677777999 |
| 3211 | 18* | 111133344 |
| 88766555 | · | 55566666788899 |
| 4332111 | 19* | 9224 |
| 8665 | · | 67 |
| 110 | 20* | |
| 8875 | · | |
| 2 | 21* | |
|  | · | |
| 60 | 22* | |
|  | · | |

Identifications on panel B of exhibit 4.

B) PROBLEMS

5a) Imitate panel A of exhibit 4 with the log ultimate powers of exhibits 1 and 3.

5b) Imitate panel A above with same.

exhibit **6** of chapter 3: atomic weights

**Atomic weights of hydrogen and bromine regarded as best available in 1927**

A) RAW ATOMIC WEIGHTS

| Hydrogen | Bromine |
|---|---|
| 1.00779 | 79.916 |
| 1.00781 | 79.920 |
| 1.00777 | 79.920 |
| 1.00782 | 79.927 |
| 1.00775 | 79.912 |
| 1.00766 | 79.926 |
| 1.00779 | 79.915 |
| 1.00769 | 79.915 |
| 1.00783 | 79.918 |
|  | 79.909 |

B) STEM-and-LEAF for RAW VALUES --
unit = 0.001, values cut

| Hydrogen | Bromine |
|---|---|
| 100* | 777777777 | 7990* | 9 |
|  | 7991 | 62558 |
|  | 7992* | 0067 |

C) LOG ATOMIC WEIGHTS --
six decimal values

| Hydrogen | Bromine |
|---|---|
| .003370 | 1.902634 |
| 379 | 656 |
| 361 | 656 |
| 383 | 694 |
| 353 | 612 |
| 314 | 688 |
| 370 | 628 |
| 327 | 628 |
| 387 | 645 |
|  | 596 |

D) COMPARATIVE STEM-and-LEAF --
log values, unit = .000001

| Hydrogen |  | Bromine |  |
|---|---|---|---|
| 00331* | 4 | 190259* | 6 |
| 00332 | 7 | 190260 |  |
| 00333 |   | 190261 | 2 |
| 00334 |   | 190262* | 88 |
| 00335* | 3 | 190263* | 4 |
| 00336 | 1 | 190264 | 5 |
| 00337 | 090 | 190265 | 66 |
| 00338* | 37 | 190266 |  |
|  |  | 190267* |  |
|  |  | 190268 | 8 |
|  |  | 190269 | 4 |

E) NOTES

1. The comparison of panel B shows us very little.
2. In panel D the bulk of the two batches is about equally spread out, and we see clearly the tendency of some (i.e., one or two) individual values to stray away from most others.

S) SOURCE

W. A. Roth and K. Scheel, 1927. Erster Ergänzungsband (1st supplement to the 5th edition of) Landolf-Bornstein's Physikalische-Chemische Tabellen.

hydrogen and bromine. Clearly, any attempt to compare by stem-and-leaves with equal units fails for the raw values.

When we turn to logs, we find that five-decimal logs are not good enough, so we use six-decimal values. Panel D, which compares log atomic weights in units of the sixth decimal, shows that stem-and-leaf pictures with a common unit can do well in comparing these batches, if applied to the right expression. Moreover, they can reveal the spread and straying characteristics of the original batches.

Here, the gain from re-expression comes from matching the stem-and-leaf displays better to each other. As we will see in the next chapter, this tends to be more important than gaining symmetry.

That log expression works well for atomic weights should not surprise us. For a given sort of amount, the best measurement accuracy--at a given date --is likely to be characterized by fluctuations and errors that are, roughly, a given fraction of that amount. We usually speak of having about so many significant figures. To have about a certain number of decimals in the log is usually closer to the truth, because this is closer to making "smallest change/value" a constant fraction.

### review questions

Are stem-and-leaf displays useful for comparison? Can logs help? Always? Why/why not? Was exhibit 5 better than exhibit 4? Why/why not? How many decimal places should we use in logs to start with? May we find we need more? Fewer? Why should we expect logs to do fairly well with atomic weights? Name two sorts of advantages that we may be able to gain by re-expression? Which is the more important? Which gains its value mainly from byproducts?

### 3D. Quick roots and quick reciprocals

#### quick roots

We are also likely to want to find quick square roots. Here we can still do well with a break table, although the situation is not quite as nice, since we can't reuse a square-root table until $x$ has changed by a factor of 100--not 10, as for logs--as we see from the following simple example:

$$\sqrt{2} = 1.42$$
$$\sqrt{20} = 4.47$$
$$\sqrt{200} = 14.2$$

We could get two-figure accuracy in our roots--we shall drop the "square-" except where it is really needed--with a table of the same form as exhibit 3, but we would not be likely to like it when we had it. Going from

$$\sqrt{100} = 10 \quad \text{to} \quad \sqrt{121} = 11$$

and from

$$\sqrt{10000} = 100 \quad \text{to} \quad \sqrt{12100} = 110$$

is a big step. Going from

$$\sqrt{96.04} = 9.8 \quad \text{to} \quad \sqrt{98.01} = 9.9$$

and from

$$\sqrt{9604} = 98 \quad \text{to} \quad \sqrt{9801} = 99$$

is a much smaller step--only about one-tenth the size, however you figure it. These would be the smallest steps in the answers from an exactly two-figure table of square roots.

We can do better by taking smaller steps near 11 and larger steps near 98. Exhibit 7 is a break table that does just this.

This table will be put to use in the next example.

### quick reciprocals

From time to time, we find re-expressing numbers by reciprocals useful. When we do this, it is often convenient to preserve order, to have the more positive raw value correspond to the more positive reciprocal. This doesn't happen automatically--in fact, quite the reverse. Thus, 3 is greater than 2, while 1/3--the reciprocal of 3--is less than 1/2--the reciprocal of 2.

The way out is simple:

## Use negative reciprocals

These work well, since $-1/3$ is more positive than $-1/2$. Since we are likely to start with numbers like 57, it is often convenient to work with $-1000/\text{number}$.

Exhibit **8** gives a break table for negative reciprocals. Except for the minus sign, no new feature. To look up $-1000/43.7$, we note that 437 comes between 4347 and 4425, so that the nonzero figures of $-1000/43.7$ are $-228$ (to this accuracy). Fixing the decimal point, using panel A, gives us $-22.8$ for $-1000/43.7$.

### a volcano example

Examples where steam-and-leaf displays of roots clearly give us the best grasp are harder to find than examples where this happens for logs--the latter seem to turn up all over.

Exhibit **9** sets out the heights of those 219 volcanoes that the National Geographic Society and the 1966 *World Almanac* found interesting enough to list. Clearly we have more trailing off to large heights than to small ones.

exhibit **7** of chapter 3: reference table

**Break table for (square) roots**

### A) EXAMPLES

Start by dividing number into periods of two digits, decimal point falling between periods. Thus 124.2 is 1 24 2, but 1242 is 12 42. Like wise 0.00654 is 00 65 4 or 65 4.

| Number | Periods | from B | from C | Number |
|---|---|---|---|---|
| 124.2 | 1 24 2 | ab. | 112 | 11.2 |
| 1242 | 12 42 | ab. | 35 | 35. |
| .00654 | 00 65 4 | .0x | 80 | .080 |

### B) BREAK TABLES to SET DECIMAL POINT -- enter between bold figures, leave with light figures

| | | |
|---|---|---|
| a | .x | **1** .01 |
| ab. | .0x | .**00** 01 |
| abc. | .00x | .**00 00** 01 |
| abcd. | .000x | .**00 00 00** 01 |

### C) MAIN BREAK TABLE -- in and out as in panel B

| Break | Root | Break | Root | Break | Root | Break | Root | Break | Root |
|---|---|---|---|---|---|---|---|---|---|
| 98 01 | | 2 49 64 | | 5 66 | | 15 60 | | 35 40 | |
| 1 02 01 | 100 | 2 62 44 | 160 | 5 86 | 240 | 16 40 | 40 | 37 40 | 60 |
| 1 06 09 | 102 | 2 75 56 | 164 | 6 20 | 246 | 17 22 | 41 | 39 69 | 62 |
| 1 10 25 | 104 | 2 89 00 | 168 | 6 50 | 252 | 18 06 | 42 | 42 25 | 64 |
| 1 14 49 | 106 | 3 02 76 | 172 | 6 81 | 258 | 18 92 | 43 | 44 69 | 66 |
| 1 18 81 | 108 | 3 16 84 | 176 | 7 13 | 264 | 19 80 | 44 | 47 61 | 68 |
| 1 23 21 | 110 | 3 31 24 | 180 | 7 45 | 270 | 20 70 | 45 | 50 41 | 70 |
| 1 27 69 | 112 | 3 45 96 | 184 | 7 78 | 276 | 21 62 | 46 | 53 29 | 72 |
| 1 32 25 | 114 | 3 61 00 | 188 | 8 12 | 282 | 22 56 | 47 | 56 25 | 74 |
| 1 36 89 | 116 | 3 76 36 | 192 | 8 47 | 288 | 23 52 | 48 | 59 79 | 76 |
| 1 41 61 | 118 | 3 92 04 | 196 | 8 82 | 294 | 24 50 | 49 | 62 41 | 78 |
| 1 48 84 | 120 | 4 08 04 | 200 | 9 30 | 30 | 25 50 | 50 | 65 61 | 80 |
| 1 58 76 | 124 | 4 24 36 | 204 | 9 92 | 31 | 26 52 | 51 | 68 89 | 82 |
| 1 69 00 | 128 | 4 41 00 | 208 | 10 56 | 32 | 27 56 | 52 | 72 25 | 84 |
| 1 79 56 | 132 | 4 57 96 | 212 | 11 22 | 33 | 28 62 | 53 | 75 69 | 86 |
| 1 90 44 | 136 | 4 75 24 | 216 | 11 90 | 34 | 29 70 | 54 | 79 21 | 88 |
| 2 01 64 | 140 | 4 92 84 | 220 | 12 60 | 35 | 30 80 | 55 | 82 81 | 90 |
| 2 13 16 | 144 | 5 10 76 | 224 | 13 32 | 36 | 31 92 | 56 | 86 49 | 92 |
| 2 25 00 | 148 | 5 29 00 | 228 | 14 06 | 37 | 33 06 | 57 | 90 25 | 94 |
| 2 37 16 | 152 | 5 47 56 | 232 | 14 82 | 38 | 34 22 | 58 | 94 09 | 96 |
| 2 49 64 | 156 | 5 66 44 | 236 | 15 60 | 39 | 35 40 | 59 | 98 01 | 98 |

See text, page 75, for example of use.

exhibit **8** of chapter 3: reference table

**Break table for (negative) reciprocals** (using −1000/number)

A) BREAK TABLES for SETTING DECIMAL POINT

| Break | Start | Start | Break |
|---|---|---|---|
| 1000 | .x | a. | 1000 |
| 10,000 | .0x | ab. | 100 |
| 100,000 | .00x | abc. | 10 |
| 1,000,000 | .000x | abcd. | 1. |
| 10,000,000 | .0000x | abcde. | 0.1 |
| 100,000,000 | .0000x | abcde. | 0.01 |

Examples

| Number | A | B | −1000/number |
|---|---|---|---|
| 124.2 | a. | −80 | −8.0 |
| .04739 | abcde. | −212 | −212**. |
| 1242. | .x | −80 | −.80 |

B) MAIN BREAK TABLE—digits of negative reciprocal

| Break | Value | Break | Value | Break | Value | Break | Value | Break | Value |
|---|---|---|---|---|---|---|---|---|---|
| 990 | −100 | 1639 | −60 | 2469 | −40 | 4115 | −240 | 617 | −160 |
| 1010 | −98 | 1681 | −59 | 2532 | −39 | 4202 | −236 | 633 | −156 |
| 1030 | −96 | 1709 | −58 | 2597 | −38 | 4274 | −232 | 649 | −152 |
| 1053 | −94 | 1739 | −57 | 2667 | −37 | 4347 | −228 | 666 | −148 |
| 1075 | −92 | 1770 | −56 | 2740 | −36 | 4425 | −224 | 685 | −144 |
| 1099 | −90 | 1802 | −55 | 2816 | −35 | 4504 | −220 | 704 | −140 |
| 1124 | −88 | 1835 | −54 | 2899 | −34 | 4587 | −216 | 725 | −136 |
| 1149 | −86 | 1869 | −53 | 2985 | −33 | 4672 | −212 | 746 | −132 |
| 1176 | −84 | 1905 | −52 | 3077 | −32 | 4762 | −208 | 769 | −128 |
| 1205 | −82 | 1942 | −51 | 3175 | −31 | 4854 | −204 | 793 | −124 |
| 1235 | −80 | 1980 | −50 | 3287 | −30 | 4950 | −200 | 820 | −120 |
| 1266 | −78 | 2020 | −49 | 3367 | −294 | 505 | −196 | 840 | −118 |
| 1299 | −76 | 2062 | −48 | 3448 | −288 | 515 | −192 | 855 | −116 |
| 1333 | −74 | 2105 | −47 | 3509 | −282 | 526 | −188 | 870 | −114 |
| 1370 | −72 | 2151 | −46 | 3584 | −276 | 538 | −184 | 885 | −112 |
| 1408 | −70 | 2198 | −45 | 3663 | −270 | 549 | −180 | 901 | −110 |
| 1449 | −68 | 2247 | −44 | 3745 | −264 | 562 | −176 | 917 | −108 |
| 1493 | −66 | 2299 | −43 | 3831 | −258 | 575 | −172 | 935 | −106 |
| 1538 | −64 | 2353 | −42 | 3922 | −252 | 588 | −168 | 952 | −104 |
| 1587 | −62 | 2410 | −41 | 4016 | −246 | 602 | −164 | 971 | −102 |
| 1639 | | 2469 | | 4115 | | 617 | | 990 | |

See text, page 78, for example of use.

exhibit **9** of chapter 3: volcano heights

**Heights of 219 of the world's volcanoes**

A) STEM-and-LEAF--unit =100 feet (rounded)

```
    8    0  | 99766562
   18    1  | 9761009630
   40   2* | 6998776654442221109850
   58    3  | 8766554120995511426
   80    4  | 9998844331929433361107
  103   5* | 9766666655442221009773 1
 √(18)   6  | 898665441077761065
   98    7  | 98855431100652108073
   78    8  | 653322122937
   66   9* | 377655421000493
   51·  10  | 0984433165212
   38   11  | 4963201631
   28   12  | 45421164
   20  13* | 47830
   15   14  | 00
   13·  15  | 676
   10   16  | 52
    8  17* | 92
    6   18  | 5
    5   19  | 39730
```

B) VARIOUS SUMMARIES--identifications at exhibit 6 of chapter 2

219 hghts (hunds. of feet)

```
M110  |      65      |
H  55h| 37h    95h   | 58
E  28 | 24    121    | 97
    1 |  2    199    |197
```

```
         | 87  |
f | -49h      182h |   adj:  2 and 179
  |  xxx      six  |   out:  185, 190,
F |           269h |   out:  193, 193
  |           xxx  |   out:  197, 199,
```

S) SOURCE
**The World Almanac, 1966, pp. 282–283.**
Their source: National Geographic Society.

74    exhibit 10/3: Easy re-expression

exhibit **10** of chapter 3: volcano heights

**Logarithmic heights of volcanoes (from exhibits 9 and 2)**

A) STEM-and-LEAF--units 0.01 in log height in hundreds of feet

```
   1      3*|  0
    .     4 |
    .     5 |
    .     6 |
   5      7* |8808
   6      8  |5
   8      9  |55
  12     10  |4000
  13     11* |1
  18     12  |83080
  28     13  |8884442200
  43     14  |266533220650989
  58     15* |876644319944316
  80     16  |9998844331929433361107
 107     17  |7655555544332221007662198988
 √(33)   18  |343222111333221998876555588655956
  79     19* |03322111115247998887665557
  52     20  |004322110321016865545655999889
  22     21  |00344215599
  11     22  |0215479998
   1     23* |0
```

B) VARIOUS SUMMARIES--some identifications at exhibit 6 of chapter 2

```
  219  logs (in 0.01) of hghts.†
M 110 |      181
H  55h| 157h      198 | 40h
E  28 | 138       208 | 70
    1 | 30        230 | 200
```

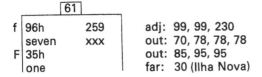

```
f  96h        259        adj: 99, 99, 230
   seven      xxx        out: 70, 78, 78, 78
F  35h                   out: 85, 95, 95
   one                   far: 30 (Ilha Nova)
```

(† **heights in 100's of feet**)

```
  219  logs (hghts in 100's of feet)
M 110 |      1.81
H  55h| 1.58      1.98
    1 | 0.30      2.30

  219  logs (hghts in feet)
M 110 |      3.81
H  55h| 3.58      3.98
    1 | 2.30      4.30
```

Note: 70 (Anak Krakatau), 78 (Surtsey), 78 (Matthew I.), 78 (Fonualei), 85 (Tavurvur), 90 (Niuafo'ou), 95 (Didicas), 99 (Taal), 99 (Guguan), 230 (Kilimanjaro).

Exhibit **10** shows the result of taking logarithms. We now have more trailing toward low logs than we have toward high logs. Clearly, we want something in between raw values and logs. Square roots are just the thing.

Exhibit **11** shows the stem-and-leaf for the roots. Now we have a nice symmetric picture, one that fits in well with a median of nearly 6500 feet and extremes near 200 and 19,900. If one wanted to compare interesting volcanoes with other interesting mountains, or to compare the interesting volcanoes of the northern and southern hemispheres--in each case as to height--we would do well to look at square roots of heights.

### looking easier

Rather than peering at the three stem-and-leafs of exhibits 9, 10, and 11, we can do better by looking at the corresponding schematic plots. Exhibit **12** shows them to us. To many of us the message is even clearer.

**exhibit 11** of chapter 3: volcano heights

**Square roots of heights of 219 volcanoes (from exhibits 9 and 7, latter rounded)**

A) STEM-and-LEAF--units 0.1 of roots of heights in hundreds of feet (equivalently, roots of height in feet)

```
   1      1* | 4
   6      2  | 65525
  13      3  | 0032262
  28      4  | 410409997776655
  51      5* | 144322110430998675996877
  73      6  | 2000220666644666684448
 107      7  | 00000006644444444422220066662288888
 √(34)    8  | 242220002222208886666444486444848886
  78      9* | 22220000424688888666644466
  52     10  | 004422220220006866 4686
  29     11  | 0222000226886488
  13     12  | 44488
   8     13* | 2266
   4     14  | 0000
```

B) LETTER-VALUE DISPLAY

```
  219    √hghts (in feet)

  M 110  |    80    |
  H  55h | 61    98 | 37
  E  28  | 49   110 | 61
      1  | 14   140 |126
```

adj: 14 (Ilha Nova), 140 (Kilimanjaro)

out: none

far: none

### reciprocal times

It often seems natural to measure times for things to happen. How long does it take a rat--in training to learn a maze--to run through the maze (and get his reward)? How long does it take individual fish to stop swimming when exposed to a given concentration of poison? How long after the earthquake did this house fall down? A major difficulty with using times is the rat that never finishes the maze, the fish that never stops swimming, the house that never falls down.

Clearly, the rat who never finishes the maze is much more like the slow rat who takes 1000 seconds than either is like the very fast rat who runs in four seconds. Indeed, the 1000-second rat is very much like the 800-second rat, much more so than the 204-second rat is like the four-second rat.

exhibit **12** of chapter 3: volcano heights

**Heights of 219 volcanoes, using three different expressions (schematic plots; for further identification see exhibit 6 of chapter 2)**

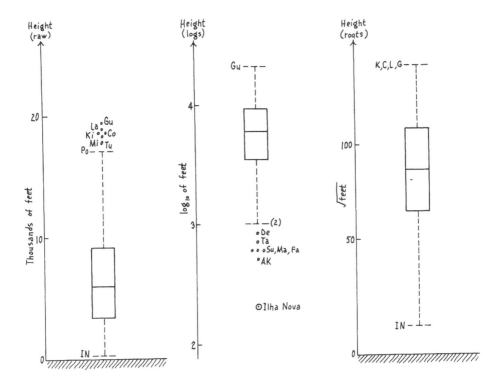

> The visible difficulty with "never" cases is a warning of a concealed difficulty with "very long time" cases.

We ought not to be using times in such cases; rather we should be using the reciprocal of time. For the rats we have mentioned as illustrations, and some others, we get the following sorts of results:

| Time | −1000/time | 1000/time |
|---|---|---|
| "never" | −0 | 0 |
| 1000 sec. | −1 | 1 |
| 800 sec. | −1.2 | 1.2 |
| 204 sec. | −49 | 49 |
| 10 sec. | −100 | 100 |
| 6 sec. | −167 | 167 |
| 5 sec. | −200 | 200 |
| 4 sec. | −250 | 250 |

If we want to continue to talk about slowness, the negative reciprocals are much better than the time. If we are willing to shift to fastness instead, the 1000/time values without the minus signs work very well indeed. (They are, of course, speeds or rates.)

Exhibit 13 shows examples from a rat-running study by Hull. Panel B shows some long times for the first pair of rats and an overall tendency for the second pair of rats to take less time. The pattern of spread is somewhat uncertain. Panel D shows the stem-and-leaf patterns for the reciprocal times. Three of the four rats now show a roughly symmetric scatter of about-the same size, while the scatter for "Rat 2" is reasonably symmetrical, though somewhat more spread out. The general pattern--and the comparison--seems clearer for the reciprocal times.

If we had stuck to panel B, we would have thought that, so far as spread of results go, those for rats 0 and 1 were unusually spread out, while those for rat 3 were unusually compressed. Looking at panel D, we get a quite different impression--one that is more likely to be correct. In these terms, only rat 2 looks unusually spread out (because of his two very quick trips). **How we look at numbers can matter.**

Once in a long while, data will come to us with both "never" and "zero" as raw values. In such cases, we can usually replace "zero" by "half the smallest nonzero value that would have been written down if observed." We should, however, complain about the data collection.

### review questions

What are the most common re-expressions besides logs? How do we use exhibit 7? Exhibit 8? What re-expression seemed to work best for volcano heights? Can schematic plots help us pick a re-expression? What does difficulty

**exhibit 13/3: Easy re-expression**

with values that are either possibly or actually infinite suggest about large values? What is the simplest way to avoid such difficulties? What is the most likely re-expression for rats' running times? Would you expect this to help with horse-racing times?

exhibit **13** of chapter 3: rat running

**Times and reciprocal times for rats on second day of training (from start through section 2 of a runway)**

|  | Rat 0 | Rat 1 | Rat 2 | Rat 3 |
|---|---|---|---|---|
| A) TIMES -- in 0.1 second | 76 | 119 | 108 | 56 |
|  | 127 | 186 | 39 | 70 |
|  | 261 | 93 | 65 | 81 |
|  | 137 | 224 | 29 | 57 |
|  | 74 | 128 | 59 | 46 |

B) STEM-and-LEAF displays for TIMES

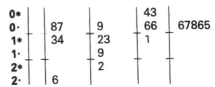

| | Rat 0 | Rat 1 | Rat 2 | Rat 3 |
|---|---|---|---|---|
| C) RECIPROCAL TIMES -- in 1000/seconds | 132 | 84 | 92 | 180 |
|  | 78 | 54 | 258 | 144 |
|  | 38 | 108 | 152 | 124 |
|  | 72 | 45 | 340 | 176 |
|  | 136 | 78 | 168 | 216 |

D) STEM-and-LEAF displays for RECIPROCAL TIMES

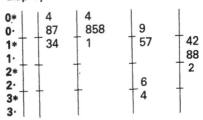

S) SOURCE
Clark L. Hull, 1934, "The rat's speed-of-locomotion gradient in the approach to food." *J. of Comparative Psychology* **17**: 393–422. (Data from table 3 on page 401.)

### 3E. Looking quickly

Making three stem-and-leaf displays seems to be a hard way to find out what re-expression makes the batch reasonably symmetrical. Let us find an easier way.

The sort of re-expressions we ordinarily make--including roots, logs, and negative reciprocals--have the simple property of preserving order. Thus, for example, because 57 is larger than 43, $\log 57 = 1.76$ is larger than $\log 43 = 1.63$.

**Preserving order necessarily preserves ranks and depths.** Thus, for example, median, hinges, eighths, ..., and extremes of log values are logs of the corresponding summaries of raw values; and the same is true for roots or negative reciprocals.

We have already noticed (in chapter 2) that the spacing of summaries throws light on the symmetry of the batch. How can we use such summaries most easily to judge symmetry? Probably by defining midsummaries, including a midhinge midH, a mideighth midE, and a midextreme mid1--often elsewhere called midrange--by

$$\text{Midhinge} = \frac{\text{lower hinge} + \text{upper hinge}}{2}$$

$$\text{Mideighth} = \frac{\text{lower eighth} + \text{upper eighth}}{2}$$

.
.
.

$$\text{Midextreme} = \frac{\text{lower extreme} + \text{upper extreme}}{2}$$

We then have only to look at the sequence

$$\text{median, midhinge, midextreme}$$

or at a longer sequence, such as

$$\text{median, midhinge, mideighth, midextreme}$$

and see if there is an indication of a trend.

To do this for a particular re-expression requires only re-expressing the corresponding summaries and then finding the midsummaries. Exhibit **14** does this for the volcano example. Whether we look at 3 midsummaries (from a 5-number summary), or 4 midsummaries (from a 7-number summary), or 5 midsummaries (from a 9-number summary), the answer is clear and strong: "use roots". (Only one stem-and-leaf--that for raw heights--was needed.) Panel D shows how midsummaries can be neatly attached to our "letter-value displays."

Several comments are in order:

◊ since the midextreme can be greatly affected by a single stray value, we are often reluctant to trust it as a base for picking out a re-expression.

◊ accordingly, starting from a 7-number or 9-number summary is often worthwhile.

◊ we often work with batches smaller than 219. When we do, we cannot expect as clear signals as we have found for this example.

◊ working with small batches, it is often worthwhile, when it is possible, to combine evidence from several batches.

exhibit **14** of chapter 3: volcano heights

**Use of midsummaries to pick a re-expression for volcano heights**

**A) WORKING with the VALUES of a 5-NUMBER (MH1) SUMMARY**

|     |      | (raw) | (root) | (log)   | Negative reciprocal |
|-----|------|-------|--------|---------|---------------------|
| 219 | 1    | 199   | 140    | 230     | −5                  |
|     | H    | 95h   | 98     | 198     | −10                 |
| 110 | M    | 65    | 80     | 181     | −15                 |
| 55  | H    | 37h   | 62†    | 157     | −26                 |
| 1   | 1    | 2     | 14     | 30      | −500                |
|     | M    | 65    | 80     | 181     | −15                 |
|     | midH | 66h   | 80h    | 177h    | −18                 |
|     | mid1 | 100h  | 77     | 130     | −152                |
|     | (trend) | (up) | (small) | (down) | (DOWN)           |

† Re-expression of 37h is 62; mean of re-expressions of 37 and 38 is (60 + 62)/2 = 61.

**B) WORKING with the VALUES of a 7-NUMBER SUMMARY--added values bold**

|     |      | (raw) | (root) | (log) | Negative reciprocal |
|-----|------|-------|--------|-------|---------------------|
| 219 | 1    | 199   | 140    | 230   | −5                  |
|     | E    | **121** | **110** | **208** | **−8**          |
|     | H    | 95h   | 98     | 198   | −10                 |
| 110 | M    | 65    | 80     | 181   | −15                 |
| 55  | H    | 37h   | 62     | 157   | −26                 |
| 28  | E    | **24** | **49** | **138** | **−42**           |
| 1   | 1    | 2     | 14     | 30    | −500                |
|     | M    | 65    | 80     | 181   | −15                 |
|     | midH | 66h   | 80     | 177h  | −18                 |
|     | midE | **72h** | **79** | **173** | **−25**           |
|     | mid1 | 100h  | 77     | 130   | −252                |
|     | (trend) | (up) | (small) | (down) | (DOWN)         |

➡

exhibit **14** of chapter 3 (continued)

C) WORKING with the VALUES of a 9-NUMBER SUMMARY--added values bold

|  | (raw) | (root) | (log) | Negative reciprocal |
|---|---|---|---|---|
| 1 | 199 | 141 | 230 | −5 |
| D 14h | 140 | 118 | 215 | −7 |
| E 28 | 121 | 110 | 208 | −8 |
| H 55 | 95h | 98 | 198 | −10 |
| M110 | 65 | 80 | 181 | −15 |
| H 55 | 37h | 62 | 157 | −26 |
| E 28 | 24 | 49 | 138 | −42 |
| D 14h | 16 | 40 | 120 | −62 |
| 1 | 2 | 14 | 30 | −500 |
| **M** | 65 | 80 | 181 | −15 |
| **midH** | 68h | 80 | 177h | −15 |
| **midE** | 72h | 79h | 173 | −25 |
| **midD** | **78** | **79** | **167h** | **−34** |
| **mid1** | 100h | 77 | 130 | −252 |
| (trend) | (up) | (close) | (down) | (DOWN) |

D) EXTENDED LETTER-VALUE DISPLAYS--of two alternative lengths

| 219 | (mid) | **raw** |  | (spr) | (mid) | **root** |  | (spr) | (mid) | **log** |  | (spr) |
|---|---|---|---|---|---|---|---|---|---|---|---|---|
| M110 | 65 | 65 |  |  | 80 | 80 |  |  | 181 | 181 |  |  |
| H 55h | 66h | 37h | 95h | 58 | 80 | 62 | 98 | 36 | 177h | 157 | 198 | 41 |
| 1 | 100h | 2 | 197 | 195 | 77 | 14 | 140 | 126 | 130 | 30 | 230 | 200 |

|  |  | raw |  |  |  | root |  |  |  | log |  |  |
|---|---|---|---|---|---|---|---|---|---|---|---|---|
| 219 | (mid) |  |  | (spr) | (mid) |  |  | (spr) | (mid) |  |  | (spr) |
| M110 | 65 | 65 |  |  | 80 | 80 |  |  | 181 | 181 |  |  |
| H 55h | 66h | 37h | 95h | 58 | 80 | 62 | 98 | 36 | 177h | 157 | 198 | 41 |
| E 28 | 72h | 24 | 121 | 97 | 79h | 49 | 110 | 61 | 173 | 138 | 208 | 70 |
| D 14h | 78 | 16 | 140 | 124 | 79 | 40 | 118 | 78 | 167h | 120 | 215 | 95 |
| 1 | 100h | 2 | 197 | 195 | 77 | 14 | 140 | 126 | 130 | 30 | 230 | 200 |

P) PROBLEMS

14a) Make a set of extended letter displays (MHE1) intermediate between the two sets of panel D.

14b) Extend the analysis of the volcano data to mid-summaries to an 11-number (MHEDC1) summary.

14c/d/e) Examine the areas of the counties of Michigan--data in exhibit 4--for a preferred re-expression using 5/7/9-number summaries.

14f/g/h) Do the same for Mississippi (same source) using 5/7/9-number summaries.

14i) Return to the log values of exhibit 1 and find midsummaries. What further re-expression is needed for symmetry?

14j) How much sense do you think this further re-expression makes?

**exhibit 15 of chapter 3: warp breaks**

**Tippett's 54 counts of warp breaks (9 in each of 6 conditions)**

A) DATA

| | |
|---|---|
| AL | 26,30,54,25,70,52,51,26,67 |
| AM | 18,21,29,17,12,18,35,30,36 |
| AH | 36,21,24,18,10,43,28,15,26 |
| BL | 27,14,29,19,29,31,41,20,44 |
| BM | 42,26,19,16,39,28,21,39,29 |
| BH | 20,21,24,17,13,15,15,16,28 |

B) SUMMARIES and MIDSUMMARIES, RAW--in counts

| | 1 | E | H | M | H | E | 1 | M | midH | midE | mid1 |
|---|---|---|---|---|---|---|---|---|---|---|---|
| AL | 25 | 26 | 26 | 51 | 54 | 67 | 70 | 51 | 40 | 46h | 47h |
| AM | 12 | 17 | 18 | 21 | 30 | 35 | 36 | 21 | 24 | 26 | 24 |
| AH | 10 | 15 | 18 | 24 | 28 | 36 | 43 | 24 | 23 | 25h | 26h |
| BL | 14 | 19 | 20 | 29 | 31 | 41 | 44 | 29 | 25h | 30 | 29 |
| BM | 16 | 19 | 21 | 28 | 39 | 39 | 42 | 28 | 30 | 29 | 29 |
| BH | 131 | 15 | 15 | 17 | 21 | 24 | 28 | 17 | 18 | 19h | 20h |
| | | | | | sum of all batches | | | 170 | 160h | 176h | 176h |
| | | | | | sum except AL | | | 119 | 120h | 130 | 129 |

(Except for the sum for all of either M or midH, increasing outwards)

C) SUMMARIES and MIDSUMMARIES, ROOTS--in 0.1's

| | 1 | E | H | M | H | E | 1 | M | midH | midE | mid1 |
|---|---|---|---|---|---|---|---|---|---|---|---|
| AL | 50 | 51 | 51 | 72 | 74 | 82 | 84 | 72 | 62h | 66h | 67 |
| AM | 35 | 41 | 42 | 46 | 55 | 59 | 60 | 46 | 48h | 50 | 47h |
| AH | 32 | 39 | 42 | 49 | 53 | 60 | 66 | 49 | 47h | 49h | 49 |
| BL | 37 | 44 | 45 | 54 | 56 | 64 | 66 | 51h | 50h | 54 | 51 |
| BM | 40 | 44 | 46 | 53 | 62 | 62 | 64 | 53 | 54 | 53 | 52 |
| BH | 36 | 39 | 39 | 41 | 46 | 49 | 53 | 41 | 42h | 44 | 44h |
| | | | | | sum of all batches | | | 315 | 305h | 317h | 311h |
| | | | | | sum except AL | | | 243 | 243 | 250h | 244h |

(reasonably level)

P) PROBLEMS

15a/b/c/d/e/f) Make stem-and-leafs and check the calculations of summaries for: AL/AM/AH/ /BL/BM/BH.

15g) Imitate panel C for log counts.

15h) Find a set of counts involving several batches and make a similar computation.

S) SOURCE

L. H. C. Tippett (1950). **Technological Applications of Statistics.** New York: John Wiley and Sons. London: Williams and Norgate. Page 106.

### combining evidence across batches

Exhibit **15** presents counts, used by Tippett, of the number of warp breaks during a fixed amount of weaving for each of six types of warp--AL, AM, AH, BL, BM, BH--differing in kind of cotton and tightness of twist. The results for AL are a little peculiar (E and H both 26, M 51), so we look at both the sums over all six batches and sums over all five except AL. For raw counts, while there is some irregularity involving either the sum of medians or the sum of midH's when we look at all six, there is little doubt that the sum of midsummaries increases outward. For square roots of counts, it is clear that the sums are relatively constant.

If we want symmetrical spreading of our warp counts, we are thus told to use roots.

### review questions

What is easy and natural to use in looking for symmetry--or lack of it--for a batch? What is a midhinge? A mideighth? A midextreme? How do we label each? Why do the sorts of summaries and midsummaries we use save us work? How clearly do midsummaries point to a particular re-expression for volcano heights? How do we fit midsummaries into letter-value displays? Should we believe midextremes? Implicitly? Not at all? How can we combine midsummaries across batches?

## 3F. Counted data

Most of our examples have so far involved measured data. We have slighted counted data (except for our last example, Tippett's warp counts).

(Some readers feel quite happy re-expressing heights, areas, megawatts, amounts of money, and the like, but develop a curious, almost unmanageable conscience when someone suggests re-expressing the number of times something occurred. All that listening to such a "conscience" can do is to make it more likely for them to **miss what they could have found**.)

When we have to deal, in later chapters, with special cases, such as

◇ rather small counts (especially when zero appears),

◇ how many out of a given number survived, smiled, were satisfied, etc.,

◇ counts arranged in special patterns,

there will be some refinements we can wisely use in re-expression.

For the present, though, so long as we deal with reasonably large counts, we may as well plan to re-express batches of counts **just as freely, and for all the same reasons,** as batches of amounts. The re-expressions most frequently useful for counts are logs and (square) roots.

In fact, it is rather hard to find a set of counts that are better analyzed as raw counts than as root counts. (Some sets of counts are all so nearly alike as to make their expression unimportant, but we may as well go at least as far as root counts with them.)

Exhibit **16** shows in logs the 1960 population of the 50 states, shown in raw counts in panel B of exhibit 4 of chapter 2. Clearly, it takes re-expression for us to look at these values. We do particularly well with the squeezed stem-and-leaf of panel B.

Exhibit **17** shows a still more extreme example--the numbers of deaths for selected causes in 1964 (omitting "all other diseases"). What appears to have led to the selection of these "causes" is a curious mixture of popular interest (acute poliomyelitis), ease of diagnosis (motor vehicle accidents), and importance (arteriosclerotic heart disease). Again, we see stem-and-leafs both for raw values and for logs. Panel B provides extra detail about the very smallest values. The condensed stem-and-leaf for logs in panel C gives us our first clear picture. We see:

⋄ a pronounced tendency to trail off toward low values.

⋄ a weak tendency to trail off toward high values.

⋄ a tendency toward asymmetry, even when these extreme trailing values are eliminated, the lower values being stretched out farther.

Even if we disregard the extreme trailing values, the use of square roots deserves a trial. This is done in panel D. We find that, even in the compressed form, we still do not get too clear a picture of all these values as a batch. The tendency of the three highest values--for "malignant neoplasm of digestive organs and peritoneum", "vascular lesions affecting central nervous system", and "arteriosclerotic heart disease"--to appear disconnected from the main body of causes is now more marked.

We are thus led to think that this particular collection of causes of death includes:

⋄ a number of unusually small (by comparison with the others) counts, mainly associated with infectious diseases.

⋄ a general mass of medium-sized counts.

⋄ three unusually large counts, which seem rather like "catch-alls".

By re-expressing our counts, we have learned something about even this unpromising set of data.

**review questions**

Should we be doubtful about re-expressing counts? Which re-expressions are most frequently useful for counts? What did we learn about causes of death? How much did re-expression help?

exhibit **16** of chapter 3: state populations

**The 1960 populations of the 50 states (raw values in tens of thousands as in panel B of exhibit 4 of chapter 2)**

A) LOGS--unit 0.01　　　B) SAME, SQUEEZED--unit = 0.1
　(for storage)　　　　　　(for viewing)

```
   1    13*| 6                 1*|
   2    14 | 6             1     t| 3           Alaska
   4    15 | 29            4     f| 455
   5    16 | 5             6     s| 67
   6    17*| 9            15     1·| 888888999
  12    18 | 003338       17     2*| 11         Ariz, Nebr
  15    19 | 589         √25     t| 22223333
        20 |              25     f| 444445555
  17    21*| 15           16     s| 666666677
  21    22 | 4557          7     2·| 899
  25    23 | 4478          4     3*| 00         Ill, Pa
 √25    24 | 04589         2     t| 22          Cal, NY
  20    25*| 1135
  16    26 | 0004679
   9    27 | 18
   7    28 | 9
   6    29*| 89
   4    30 | 05
        31 |
   2    32 | 02
        33*|
```

C) VARIOUS SUMMARIES of 50 STATE POPULATIONS

| | in millions | | | in logs of 10,000's | | | in logs of 1's | |
|---|---|---|---|---|---|---|---|---|
| M25h | 2.5 | 2.5 | | 2.39 | 2.39 | | 4.39 | 4.39 |
| H13 | 2.6 | 0.89 | 4.3　3.4 | 2.29h | 1.95 | 2.64 .69 | 4.29h 3.95 | 4.64 .69 |
| 1 | 8.6 | 0.23 | 17 | 2.29 | 1.36 | 3.22 | 4.29　3.36 | 5.22 |

P) PROBLEMS

16A) Imitate panels A and B for the 1970 populations of the 50 states. Does an apparent gap between smaller populations and larger populations continue?

16b) Imitate panels A and B for

　　　　　　log 1970 population − log 1960 population

　Compare your results for those for the separate years.

16c) Go back to earlier censuses, and continue the comparison.

S) SOURCE
The World Almanac, 1966, page 325.
Their source: U.S. Bureau of the Census

## 3G. Relation among powers and logs (optional)

We have said that logs and square roots are more likely to be useful ways of re-expression than any others. We have seen--by way of the heights of interesting volcanoes--that the square root does seem to fit in halfway between raw values and logs. This is enough to take us through most situations.

As preparation for the future, though, we may wish to look at a wider variety of powers, and to learn in more detail how logs fit in among the powers.

exhibit **17** of chapter 3: deaths by cause

**Deaths for 59 selected causes, 1964 (total 1,798,051, less "all other diseases 54,000")**

### A) SMALL COUNTS

| Raw | Log  | Cause |
|-----|------|-------|
| 17  | 1.23 | Polio |
| 42  | 1.62 | Diphtheria |
| 93  | 1.97 | Whooping cough (WC below) |
| 95  | 1.98 | Scarlet fever and strep throat (SFST below) |

### B) RAW VALUES--in 100's

```
0* | 7,3,8,4,2
1* | 7,8,3,1
2  | 6,5
3  | 5,8
4  | 9,4,6,4
5* | 9,4
6  | 5
7  | 6
8  | 2
9* | 9,9,8

1**  | 35,35,67,59,22,11,10
2    | 62,77,57,34,32,03,52,53,06
3    | 28,23,72,07
4    | 92,00,69
5**  | 32,74,78,69

0*** | 932,
1*** | 982,
2
3
4*** | 454,
```

### C) LOGS--in 0.1's

```
     .  |
     s  | 7              Polio
     f  |
     t  | 3              Diph
  -0*   | 00             WC, SFST
   0*   |
     t  | 3              Abortion
     f  | 4              Dysentery
     s  | 6              Measles
     .  | 89
   1*   | 01
     t  | 22
     f  | 4455
     s  | 666677
     .  | 8899
   2*   | 00000011
     t  | 23333
     f  | 444444555
     s  | 6667777
     .  | 9†
   3*   |
     t  | 3†
     f  |
     s  | 7†
     .  |
```

† See text for identification

exhibit **17** of chapter 3 (continued)

**D) ROOTS--of counts (cut)**

```
0** | 001111
  t | 22233
  f | 44555
  s | 6666777
  · | 889999
 1* | 00111
  t | 22
  f | 4455
  s | 666667
  · | 889
 2* | 01
  t | 23
  f | 444
  s |
 3* |
  t | 1†
  f |
  s |
  · |
 4* |
  t |
  f | 5†
  s |
  · |
0** | 7†
```

† See text for identification

**P) PROBLEMS**

17A) Collect (as in panel B) similar data for deaths in 1969.

17b) Imitate panel C.

17c) Imitate panel D.

17d) Find and display changes in logs from 1964 to 1969.

17e) Find display changes in roots from 1964 to 1969.

17f) Discuss the results of (17d) and (17e).

**S) SOURCE**

The World Almanac, 1966, page 299

Their source: National Center for Health Statistics

We are now trying to think--and learn--about re-expression in general, not just re-expression for making better stem-and-leaf displays. Thus we should be fairly general, which is likely to mean saying some things that may seem trivial.

Changing the numbers in a batch--or in some more complicated data structure--by multiplying or dividing all of them by the same constant usually does not affect the ease with which they can be looked at--either directly or after most standard arithmetic analyses. To make such a change is to do no more than to change feet to inches or meters, or to change kilowatts to megawatts. We usually regard this sort of change as quite trivial--both so far as easing our analysis goes (or making it more difficult) and in its effects on what the answers will be. (We do need to keep careful track of what we have done, so that we can find our way back when the analysis is completed.) Indeed, we will call such changes

trivial re-expressions.

We noticed above that the sort of re-expression we are discussing is mainly used for numbers that are all of one sign. If all the numbers we have--or might reasonably have had--are of one sign, there is likely to be a special significance for zero.

There are many purposes for which adding or subtracting the same constant to each number in a batch--or in some other data structure--is just as trivial as multiplying or dividing by a common constant. The most important exception is this:

⋄ if there is a natural zero for the initial numbers, then, once we have added or subtracted a constant, shifting the zero, we are not likely to take a power, a root, or a log of the modified numbers--unless we are dealing with small counts, where shifting "none counted" to a small positive value is often followed by taking a root or a log.

Surely the simplest important change in how something is expressed in numbers is to replace the raw numbers by the same simple power of each of them. It is not usual for this power to be a square or a cube. Because of the ways people choose to make instruments and write numbers down--as we have noticed--we much more frequently change them in the opposite direction, as by taking:

⋄ square roots (1/2 powers).

⋄ reciprocals (−1 powers).

⋄ reciprocals of square roots (−1/2 powers).

(For all these but one we have already seen critical tables.)

A look at these powers

$$-1, \quad -1/2, \quad +1/2, \quad +1$$

(where the last entry refers to the raw data itself), reveals an obvious gap. What about the zero power?

We were almost all once taught that anything raised to the zero power was 1. Our teachers were not wrong. But that does not mean that there can be nothing to fill the gap.

> **It turns out that the role of a zero power is, for these purposes of re-expression, neatly filled by the logarithm**.

Exhibit **18** shows how $y = $ (const) $\cdot$ (log $x$) fits neatly in among the simple powers. Note that negative powers need negative signs to make $y$ increase as $x$ increases.

**A central point is that moving up the scale**--whether from $x$ to $x^2$ to $x^3$, or from $-1/x^2$ to $-1/x$ to log $x$--**corresponds to emphasizing differences among larger $x$'s in comparison with differences among smaller $x$'s.** To do the opposite, emphasizing more strongly differences among smaller $x$, we need to move down the scale--as from $x$ to log $x$ to $-1/x$.

In re-expressing data, we ordinarily work with a narrower range of powers than those just illustrated. Not $+3$ to $-3$ but $+1$ to $-1$ is what we usually cover. Exhibit **19** shows similar information for the narrower range. All the same remarks apply. (Note that the horizontal scale for this plot was re-expressed to make the overall appearance clear.)

### quick reciprocal roots

To complete a set of break tables for the simple ladder requires filling in the (negative) reciprocal square root. We could give a break table for this form of re-expression. Since it seems to be very rarely useful--and since we can do well enough by using two break tables in succession, one for roots and one for (negative) reciprocals--we shall omit it.

### quick squares

The frequent pencil- and-paper user of more classical statistical procedures soon learns the table of squares up quite a way. We can also do fairly well by using the break table for roots backwards. (Details left to the reader.)

### review questions

What re-expressions are considered trivial? Always? Sometimes? What simple powers are rarely used for re-expression? What did (or should) our teachers tell us about zero powers? Were they right or wrong? Is there any paradox on logs playing the role that zero powers fail to play? What happens as we move up the scale of powers? Down the scale? How can we find semiquick reciprocal roots when needed?

## exhibit **18** of chapter 3: powers and logs

### The "shape" of simple functions of x

(Expressions of the form $A + B \cdot f(x)$, with $A$ and $B$ chosen to make the result tangent to $x - 1$ at $x = 1$.) Labels give $f(x)$. For $A$ and $B$ see below.

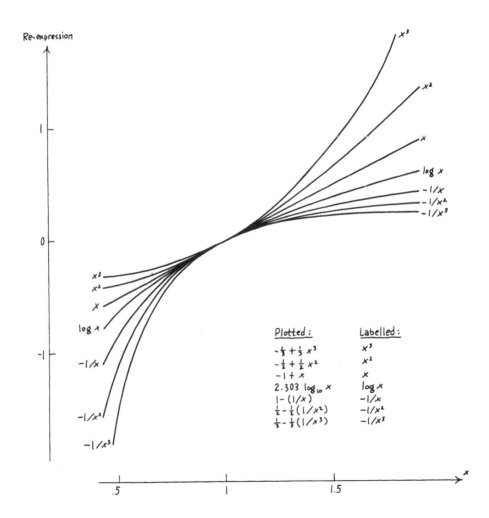

Plotted:
$-\frac{1}{3} + \frac{1}{3} x^3$
$-\frac{1}{2} + \frac{1}{2} x^2$
$-1 + x$
$2.303 \log_{10} x$
$1 - (1/x)$
$\frac{1}{2} - \frac{1}{2}(1/x^2)$
$\frac{1}{3} - \frac{1}{3}(1/x^3)$

Labelled:
$x^3$
$x^2$
$x$
$\log x$
$-1/x$
$-1/x^2$
$-1/x^3$

## exhibit 19 of chapter 3: powers and logs

**Logs and the usual powers**

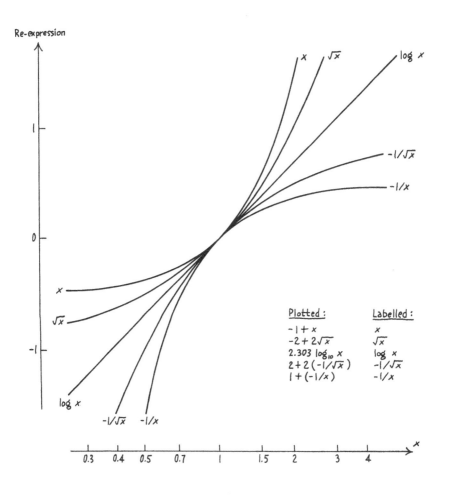

## 3H. How far have we come?

Data can be broadly divided into:

◊ amounts and counts. These are likely to need re-expression unless the ratio of largest to smallest value present is close enough to 1 for re-expression not to matter. Re-expression will usually be by logs, though other roots or powers can sometimes help.

◊ balances. Here re-expression is less frequent, and must usually be done before the balances are formed, not afterward. (Logs of anything are always balances.)

◊ counted fractions, percentages, and grades. Here re-expression is often essential, but detailed methods for doing it are going to have to wait for several chapters.

Break tables make re-expression in the commonest ways:

◊ first, logarithms,

◊ second, square roots,

◊ third, (negative) reciprocals,

◊ and even, (negative) reciprocal square roots,

both easy and quick, so long as few-digits answers will do, as is usually the case.

Counted data, unless we are dealing with quite small counts, are handled in the same way as amounts. We rarely lose by going as far as roots, and often find it pays to go to logs.

Whenever it matters, we do well to treat logs as playing the role of the zeroth power, so far as re-expression goes.

We are ready:

◊ to use break tables for few-digit re-expression, planning to occasionally come back and use more precision (efficient laziness).

◊ to "start" counts--and amounts that can be zero--by adding the same constant to each count that we are dealing with BEFORE we re-express them. (When counts are all large, starting is not likely to be worth the trouble.) If the counts include zeros, and we are to take logs, starting is essential.

◊ to calculate midsummaries, and to look toward their trends for more delicate indications of asymmetry than are given by just looking at a stem-and-leaf display of the values.

We now regard re-expression as a tool, something to let us do a better job of grasping data. In this chapter we have only begun: The grasping is with the eye and the better job is through a more symmetric appearance. As we go on, we shall learn about a variety of other ways in which re-expression can help us.

We have begun to realize that taking a firm hand with the data, before we display it or make detailed calculations, is often either the best or the only thing to do.

### 3K. How to think about logs (background)

We want to use logarithms--"logs"--freely as a tool. The points that follow make such use easier and are recalled for reference.

1. Logs can be looked up in tables.
2. Logs come "to" different bases; however, logarithms to different bases differ only by multiplicative constants. As a result, in data analysis, any base is almost (usually exactly) as good as any other.
3. We shall use "$\log_{10}$" for logs to the base 10 when we need to be sure (and will later use "$\log_e$" or "ln" for logs to the base $e$).
4. The log of 1 is 0.
5. The log of zero is not defined.
6. The log of a negative number can be defined, but is not a real number. (We shall not try to use logs of negative numbers or zeros, though sometimes we may write "L" in their place, meaning "treat as lower than any number.")
7. The log of a product is the sum of the logs of the factors, thus

$$\log uvw = \log u + \log v + \log w$$

8. The log of a ratio is the difference of the logs of numerator and denominator, thus:

$$\log(s/t) = \log s - \log t$$

9. The log to base 10 of 10 is 1 (that of 100 is 2, of 1000 is 3, of 0.1 is $-1$, of 0.01 is $-2$, of 0.001 is $-3$, etc.).

### 3P. Additional problems

See exhibits **20** to **22**.

exhibit **20** of chapter 3: data and problems

**A) Tonnage shipped from ports from Maine to Key West in 1964, (B) Largest lakes in each U.S. State**

```
       A) TONNAGE--nearest 1000 tons        B) AREAS in SQUARE MILES

      8* | 4,9    Newport, Rockland       0*  | 0‡,1‡,4,4,4,5,6,7,7,7,8,9
      9* | 6      Cambridge, Md.          1*  | 1,5
                                          2   | 0,5,9
     1** | 63,65,68,71,78                 3   | 1,1,8
      2  | 86                             4   | 7
      3  | 18,63,64,71,87                 5*  | 1,5,8,9
      4  | 33,58                          6   | 0,3
     5** | 66                             7   | 2,6,8
      6  | 24                             8   | 0
      7  | .                              9*  | 4
      8  | 14,70
     9** | 09,                           1**  | 05,08,11,11,17
                                         1..  | 23,33,36,37,40
    1*** | 127,207,249,358               1..  | 49,57,60,88,92
    1... | 460,771,826                   2**  | 15,47,52
    2*** | 127,223,349,397,699            3   | 60,82
      3  | 162,717                        4   | 30,51
      4  | 158,643,992                   5**  | .
    5*** | 107                            6   | 09,30
      6  | .                              7   | 00            Okeechobee
      7  | 161
      8  | 192,937                      1*** | 033‡,500‡,
    9*** | 232
                                        0****| 9910          Erie
   1****| 2206,3206,5206                   1 | .
   1... | 7622,7841,8830                   2 | 2400          Michigan
   2****| 0011              Boston       3****| 1800         Superior
      3 | .
   4****| 3949†,7042†,8220†

   0*****| .
      1  |     49151      New York
         †Norfolk, Philadelphia, Baltimore      ‡0.31 Candlewood in Connecticut
                                                0.66 Koloa in Hawaii
                                                1033 Iliamna in Alaska
                                                1500 Great Salt in Utah
```

**P) PROBLEMS**

20a) Panel A gives the amounts of tonnage shipped in 1964 from ports from Maine to Key West. What expression is indicated? Make the corresponding stem-and-leaf. Comment.

20b) Panel B gives the area of the largest lake in each state. What expression is indicated? Make the corresponding stem-and-leaf. Comment.

**S) SOURCE**

The World Almanac, 1966, (A) page 766, (B) page 285.

exhibit **21** of chapter 3: data and problems

**(A) Areas of 4 oceans and 21 seas (including "Malayan seas"), (B) Lengths of "important" rivers of the U.S. (some multiply named)**

```
       A) AREAS--in 1000 sq. miles        B) LENGTHS--in miles

   1*  |                              1** |10,28,10,00,31,00,12
   2   |9         Bass Strait         1•• |37,44,40,40,48,50,64,60,64
   3   |                              1•• |71,78,83,75,69,85,90
   4   |                              2** |02,30,10,17,33,10,15,37,46,50
   5*  |                              2•• |65,60,59,60,55,50,50,73,68
   6   |3†,9†                         2•• |82,90,86,80,92,91,81,80,87,76
   7   |                              3** |15,25,27,06,00,32,10,00,14,01
   8   |                              3•• |30,29,36,40,38,50,52,50,50,50
   9*  |2†,2†                         3•• |60,60,60,92,75,80,80,82
                                      4** |07,09,20,24,25,31,35,44,48
  1**  |63,69,78,80                   4•• |50,60,75
   2   |22                            5** |20,00,05,38,40,45,50,50,60
   3   |08,89                          6  |00,05,18,25,30,52,87,90
   4   |76,82                          7  |10,30,35,60
  5**  |90,96                          8  |40,60,62,70,90
   6   |•                             9** |06,81
   7   |•
   8   |76                           1*** |018,038,171,214,306,360
  9**  |67                           1••• |450,450,885‡
                                     2*** |315‡,348‡,553‡
 1***  |063,146,998                    3  |710
 2***  |1

0**** |5440        Arctic
   1  |•
   2  |8356        Indian
   3  |1839        Atlantic
   4  |•
5**** |•
   6  |3802        Pacific
```

```
   † 63 Gulf of California         ‡ 1450 Arkansas
     69 Engl. Channel & Irish Sea    1450 Colorado (of Texas)
     92 Gulf of St. Lawrence         1885 Rio Grande
     92 Persian Gulf                 2315 Missouri
                                     2348 Mississippi
                                     2533 Missouri-Red Rock
                                     3710 Mississippi-Missouri-Red Rock
```

**P) PROBLEMS**

21a) Panel A gives the areas of 4 oceans and 21 seas. What expression is indicated? Make the corresponding stem-and-leaf. Comment.

21b) Panel B gives the lengths of "important" rivers of the U.S. What expression is indicated? Make the corresponding stem-and-leaf.

**S) SOURCES**

The World Almanac, 1966: (A) page 275, (B) pages 266–267.

exhibit **22** of chapter 3: data and problems

**(A) heights of "famous" waterfalls (both highest fall and total fall may appear for one waterfall) and (B) reservoir capacity of major world dams**

```
    A) Heights--in feet              B) CAPACITY--in 1000's of acre feet
0** |40,54,65,66,68,                 0*  |8                    Speicheri
0·· |70,70,75,90,96,98               1*  |·
1** |01,09,15,20,25,25               2   |·
1·· |30,30,30,32,40,44,              3   |2                    Curnera
1·· |50,51,65,68,86                  4   |1                    Zeuzier
1·· |93,95,98                        5*  |4                    Alpe Gera
2** |00,07,07,13,14,18
2·· |20,30,40,45,51,51               0** |61,70,70,81,81
2·· |56,66,70,75,68                  1** |14,37,39,46,48
3** |00,00,08,11,15,17               1·· |52,61,86,95
3·· |20,30,30,35,44,45               2   |19,65
3·· |55,60,70,70,94,94               3   |24
4** |00,00,06,27                     4   |87
4·· |50,59,59,70                     5** |49
5** |00,05,08,18,25
5·· |40,42,90,94,97                  0***|600,602,746,756,930
6   |00,20,26,30,40,50,56            1***|261,325,375,405,586,709
7   |00,26,41                        2***|000,030,030,092,095
8   |20,30,48,80,89,90               2···|106,367,446,717
9** |74,84,84                        3***|024,453,468,484,648,789
                                     4   |413,493,500,500
1***|000,100,170,218,250,312         5***|·
1···|312,312,325,350,385             6   |100,550,600
1···|430,535,600,612                 7   |055,060
1···|640,650,696,904                 8   |000,000,512
2***|000,425,600,648                 9***|171,402,730,890
3   |110†,212†
                                     1****|0945,2940,4755
    †3110 Tugela (5 falls),          1····|9715,9000,9400
     3212 Angel                      2    |3600,4500,4800,7000,7160
                                     3    |1618,2471
                                     4    |7020                Kuibyshev
                                     5****|·

                                     0*****|62000              Portage Mt
                                     1*****|15000‡,27281‡,45115‡

                                     ‡115000 Manicouagan #5.
                                      127281 Sadu-El-Aali (High Aswan)
                                      145115 Bratsk
```

**P) PROBLEMS**

**22a)** Panel A gives the heights of "famous" waterfalls. What expression is indicated? Make the corresponding stem-and-leaf. Comment.

**22b)** Panel B gives the reservoir capacity of major dams. What expression is indicated? Make the corresponding stem-and-leaf. Comment.

**S) SOURCE**
The World Almanac, 1966: (A) page 286; (B) page 260.

# Effective comparison, including well-chosen expression   4

chapter index on next page

In dealing with batches, we have already given some attention to two reasons for choosing one form of expression rather than another:

◊ symmetry of spreading within each batch separately.

◊ agreement from batch to batch in amount of spreading.

Neither is a "big deal."

Symmetry of spread is, by itself, probably a "little deal."

Fortunately for all of us, however, conflict between these reasons is infrequent. A choice good for one is, much more often than not, a choice good for the other.

Agreement of spread is rather more important; it deserves to be a "middle deal".

Indeed, in situations where more important reasons apply, what is good for the present minor reasons is, again much more often than not, good for the major reasons. There are exceptions, and when we recognize one, the major reason will have to take control, but such instances are not common. Choice of expression is rarely a balancing of conflicting reasons. Much more often it is a matter of stretching the information the data offers about alternative choices far enough to make a choice.

**Most batches of data fail to tell us exactly how they should be analyzed. In making our most careful choices, we typically have to depend upon other bodies of data on similar subjects and upon experience.**

> **Choice of expression is only one of such choices.**

**This does not mean that we cannot do relatively well on the basis of the data before us; often, indeed we can do tremendously better than if we used the data in a raw form, or made other choices without thought.**

We will not meet "big deals" until we come to deal with more structured data--data with more "handles" for us to work with.

# index for chapter 4

review questions 99

**4A. Alternative forms of display of summaries** 99
review questions 101

**4B. Comparing several batches (continued)** 102
review questions 105

**4C. A more extensive example** 105
a condensed approach 109
review questions 110

**4D. The meaning of comparison** 110
review questions 110

**4E. Adjustments, rough and exact** 110
rough adjustments 110
exact adjustment 112
review questions 113

**4F. Residuals** 113
review questions 114

**4H. How far have we come?** 115

**4P. Additional problems** 116

| EXHIBIT | PAGE |
|---|---|
| 4A | |
| 1★ | 99 |
| 2★ | 100 |
| 4B | |
| 3★ | 102 |
| 4 | 104 |
| 4C | |
| 5★ | 106 |
| 6 | 108 |
| 4D | |
| 4E | |
| 7★ | 111 |
| 8★ | 112 |
| 4F | |
| 9★ | 113 |
| 4H | |
| 10★ 4P | 116 |
| 11★ | 117 |
| 12★ | 118 |
| 13★ | 119 |
| 14★ | 120 |
| 15★ | 121 |
| 16★ | 122 |
| 17★ | 123 |

## review questions

What are two reasons for choice of expression? How important are they? Do they usually agree with each other? With major reasons about which we have not yet learned? Does a body of data usually indicate clearly how it should be analyzed?

## 4A. Alternative forms of display of summaries

Suppose we know how we want to analyze our data and what sorts of summaries we want to suggest or present. There is always still a question of presenting them. There are choices among alternatives, sometimes among more alternatives than we think of at first.

Rainfall (including snow converted to rain) at New York City offers a simple and convenient example. Exhibit 1 shows the data for the first six years of each of seven decades. Exhibit 2 shows four ways to display the data in more or less summarized form:

⋄ as stem-and-leaf displays.

⋄ as medians (unadorned).

⋄ as slightly graphic medians.

⋄ as schematic plots, modified to emphasize medians.

exhibit 1 of chapter 4: New York City rainfall

**Precipitation\* in New York City (to nearest inch) in years ending with 0, 1, 2, 3, 4, and 5**

A) DATA

|        | 189- | 190- | 191- | 192- | 193- | 194- | 195- |
|--------|------|------|------|------|------|------|------|
| −0     | 52   | 42   | 36   | 49   | 35   | 45   | 45   |
| −1     | 41   | 47   | 40   | 34   | 36   | 36   | 47   |
| −2     | 39   | 47   | 38   | 43   | 39   | 50   | 46   |
| −3     | 53   | 49   | 44   | 37   | 50   | 40   | 38   |
| −4     | 44   | 42   | 34   | 38   | 45   | 52   | 43   |
| −5     | 36   | 44   | 41   | 37   | 33   | 46   | 41   |
| median | 42h  | 45h  | 39   | 37h  | 37h  | 45h  | 44   |

\* Rain plus rain equivalent of snow.

P) PROBLEM

1a) Find the data for 1960 to 1965, and use it to extend exhibit 2, below, to this additional decade.

S) SOURCE

Available from a variety of sources, including The World Almanac.

**exhibit 2** of chapter 4: New York City rainfall

## Four kinds of summaries of New York City precipitation compared

### A) STEM-and LEAF COMPARISON--unit 1 inch

|      | 1890–95 | 1900–05 | 1910–15 | 1920–25 | 1930–35 | 1940–45 | 1950–55 |
|------|---------|---------|---------|---------|---------|---------|---------|
| 5·   |         |         |         |         |         |         |         |
| 5*   | 23      |         |         |         | 0       | 02      |         |
| 4·   |         | 779     |         | 9       | 5       | 56      | 567     |
| 4*   | 14      | 224     | 014     | 3       |         | 0       | 13      |
| 3·   | 69      |         | 68      | 778     | 569     | 6       | 8       |
| 3*   |         |         | 4       | 4       | 3       |         |         |

### B) MEDIANS--unadorned

|  42.5 | 45.5 | 39.0 | 37.5 | 37.5 | 45.5 | 44.0 |

### C) PARTLY GRAPHIC MEDIANS

| 44:45 |      | 45h |     |     |     | 45h | 44 |
| 42:43 | 42h  |     |     |     |     |     |    |
| 40:41 |      |     |     |     |     |     |    |
| 38:39 |      |     | 39  |     |     |     |    |
| 36:37 |      |     |     | 37h | 37h |     |    |

### D) SCHEMATIC PLOTS--modified to stress medians

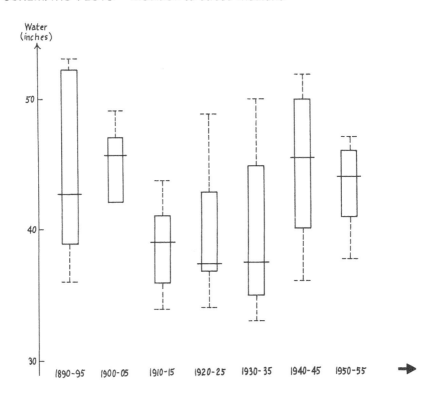

All four displays tell the same story:

⋄ in the 1890's, 1900's, 1940's, and 1950's, high typical annual rainfalls, near 44 inches;

⋄ in the three intervening decades, lower typical rainfalls, near 38 inches.

There is now an understandable message--1910 to 1935 seems to have been a period of lowered rainfall. Whether or not this ought to be regarded as accidental is a confirmatory question, a question we have not tried to answer-- and will not try to answer here. However, it should be regarded--since anything that happens once may happen twice--this earlier low-rainfall period might well have been a warning to prepare for the "drought" of the 1960's. Our concern is that use of any one of these displays uncovers a message in this instance--and that their use in other instances may also uncover messages.

How well does each display do? There will undoubtedly be some differences of opinion, as well there may be. If we compare the stem-and-leaf (panel A) with the schematic plot (panel D), I find little to choose. (If there were 600 values in each batch, rather than 6, we would almost surely prefer the schematic plot.) It is easier to compare the other pair of panels, in which the information about each batch is pared down to one number--the median-- with each other. The expenditure of a few extra lines to make the medians partly graphic has shown us directly--with one *coup d'oeil*--most of what is going on, yet there has been no loss of detail in panel C as compared with panel B. We should be on a constant lookout for opportunities to use partly graphic displays.

Comparison of one pair of displays with the other is--and should be-- harder. Sometimes we need the additional detail about the spread of the batches. Sometimes it merely confuses us slightly. We have to be guided by what we know or feel about the users, be they us or be they someone else.

**review questions**

What were the four kinds of display? What story did each tell? How did they pair off? How do the two displays in each pair compare? How did one pair compare with the other?

exhibit **2** of chapter 4 (continued)

**P) PROBLEM**

2a) Make a display like panel C, showing values for hinges (light) as well as for medians (bold), omitting hinges when they would overlap. How well do you like the result? Compare its effectiveness with panel C.

## 4B. Comparing several batches (continued)

When medians alone do not tell us enough, we want at least to look at schematic summaries, probably in the form of schematic plots. We want these summaries to tell us the story as clearly and as simply as they can. Both symmetry of spread within batches and, especially, balance of spread between batches, will help.

Exhibit 3 gives--in numbers--summaries of three batches from Winsor and Clarke's (1940) analysis of the catch of plankton using different kinds of nets. Three choices of expression are compared in panels B and C--raw count, root count, and log count. Whether we concentrate on symmetry--looking for midsummaries that do not drift within a batch--or on balanced spread--looking for spreads that agree from batch to batch--we come to the same position:

◇ raw counts are unacceptable.

◇ we are torn between root counts and log counts, and might even like to find a compromise.

exhibit 3 of chapter 4: Plankton hauls

**Estimated number of three kinds of plankton caught (in six hauls of each of two nets)**

### A) ESTIMATED COUNTS

Kind I: 387, 428, 470, 497, 537, 540, 620, 760, 845, 895, 1020, 1050
Kind IV: 6060, 7600, 7900, 8260, 8600, 8900, 9250, 9830, 10200, 11000, 15500
Kind III: 189, 223, 278, 281, 288, 290, 314, 328, 328, 346, 395, 433 (hundreds)

### B) RAW COUNTS--summaries (rounded) in units

|     | Kind I |     |     |     | Kind IV |     |     |     | Kind III |     |     |     |
|-----|-----|-----|-----|-----|-----|-----|-----|-----|-----|-----|-----|-----|
| M6h | 580 | 580 |     |     | 9075 | 9075 |     |     | 302** | 302** |     |     |
| H3h | 677 | 484 | 870 | 386 | 9048 | 8080 | 10,015 | 1935 | 308** | 280** | 337** | 57** |
| 1   | 718 | 387 | 1050 |    | 10,780 | 6060 | 15,500 |   | 311** | 189** | 433** |    |

H-spreads INcrease rapidly from left to right; mids INcrease from M toward 1.

### C) ROOT COUNTS--summaries (rounded) in units

|     | Kind I |     |     |     | Kind IV |     |     |     | Kind III |     |     |     |
|-----|-----|-----|-----|-----|-----|-----|-----|-----|-----|-----|-----|-----|
| M6h | 24  | 24  |     |     | 96  | 96  |     |     | 172 | 172 |     |     |
| H3h | 26  | 22  | 29h | 7h  | 95  | 90  | 100 | 10  | 176 | 168 | 184 | 16  |
| 1   | 26  | 20  | 32  |     | 101 | 78  | 124 |     | 172 | 136 | 208 |     |

H-spreads INcrease slowly from left to right; mids INcrease very slightly from M towards 1. ➜

exhibit 3 of chapter 4 (continued)

**D) LOG COUNTS**--summaries in 0.01 (rounded)

```
         Kind I              Kind IV              Kind III
M6h | 276|  276           |396|  396          |448|  448
H3h | 281| 268  294 |26   |396|391  400 |9    |449|445  453|8
  1 | 280| 259  302       |398|378  419       |446|428  464
```

H-spreads DEcrease moderately slowly from left to right; mids nearly neutral.

**E) HOW MUCH RE-EXPRESSION for SIMILAR SPREAD?**

| Kind | log M | log Hspr |
|------|-------|----------|
| I    | 2.76  | 2.59     |
| IV   | 3.96  | 3.29     |
| III  | 4.48  | 3.76     |

| Diff    |       |      | Ratio |
|---------|-------|------|-------|
| III − I | 1.72  | 1.17 | .7    |
| IV − I  | 1.20  | .70  | .6    |
| III − IV| .52   | .47  | .9    |

Ratios between .5 and 1.0; hence look at $\sqrt{\ }$ and log.

**P) PROBLEMS**

3a) Extend panel A to include eighths and mideighths. Any change in conclusions?

3b) Extend panel B to include EH-spreads. Any change in conclusions?

3c) Extend panel C to include log EHspr's and corresponding differences and ratios. Any change in conclusions?

3d) Invent a set of batches where logs will make spreads reasonably constant. Do the calculations of panel C. Are the ratios what you would expect?

3e) Invent a set of batches where roots will make spreads reasonably constant. Do the same.

**S) SOURCE**

C. P. Winsor and G. L. Clarke (1940). *Journal of Marine Research* (Sears Foundation) **3**: 1.
Also used by: G. W. Snedecor (1946). *Statistical Methods*, 4th ed., page 451.

exhibit 4/4: Effective comparison, including well-chosen expression

Exhibit 4 shows the schematic plots graphically. The leftmost, using raw counts, is of little use as a set of schematic plots--we can see hardly more than if the medians were shown. The schematic plot for kind I is too small to see any detail. That for kind III is so large as to claim much undeserved attention. Either of the other two expressions gives about the same information, makes about the same impression. We cannot say that one is clearly better. We can say that either will clearly do.

We ought not to have to calculate through all three cases in order to find which one or two are likely to be satisfactory. Panel E of exhibit 3 introduces a way of learning a fair amount from much less work. If we look at the relation between median and H-spread--also, in suitably large batches, between median and E-spread, etc.--we can learn roughly what change of expression is likely to help. This is easier to do in terms of logs.

exhibit 4 of chapter 4: plankton hauls

**Schematic plots for the data of exhibit 3, according to three choices of expression**

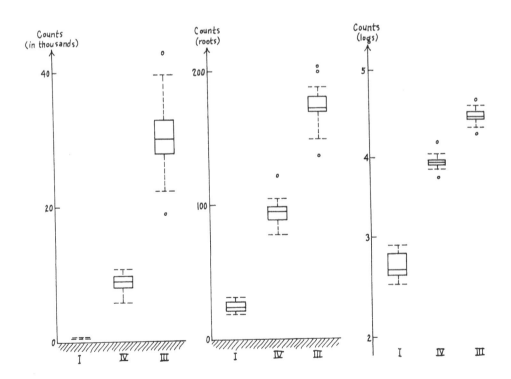

Accordingly, panel E gives first log median and log H-spread of each batch of raw values, and then the differences in these logs as we step from one batch to another. As the ratios show, the change in log H-spread is rather more than 1/2 the change in log median and rather less than 1 times that change.

Had these ratios been close to 1/2, we would expect roots to help us. Had they been close to 1, we would expect logs to help us. Since they are in the wishy-washy in-between, we may expect either to help us, usefully but not perfectly.

**review questions**

What do we ask of schematic plots? Why? Where in exhibit 3 do we look to see how well what we ask is provided? Does exhibit 4 convey the same message about choice of expression for these particular data? How can we use the results of analyzing one choice of expression to guide us in what expression is likely to do well for us?

### 4C. A more extensive example

In 1950, Bruner, Postman, and Mosteller reported the detailed results of a simple experiment in which subjects looked at a flat picture--the Schroeder staircase--which can be easily seen in perspective in two ways. They were given different instructions as to how they were to try to manage their perspective, and the number of changes of perspective were counted in each of 10 successive minutes under each instruction. Setting aside the first two minutes, to protect somewhat against "starting-up" effects, there are 19 batches (one per subject) of 8 counts under the "alternate" instruction.

The early columns of exhibit 5 show the median M and hinges H for each of the 19 batches. To make both looking at and plotting easier, the subjects (rows in the exhibit) have been arranged in increasing order of median counts. These early columns also give log M, H-spread, and log Hspr. The logs are plotted in exhibit 6 and clearly trend upward.

If we ask how fast they trend upward, the easiest way to get a rough answer is to pick out some "typical" points near each end, and then to hold your clear plastic triangle or ruler across the picture, turning and slipping the edge to what seems to you to be a good fit. In the upper panel of exhibit 6, it seems natural to take the two points at the far lower left and the one point at the far upper right. The lower portion of panel A of exhibit 5 calculates the tilts of the two lines thus fixed as

$$\frac{134 - 48}{216 - 108} = \frac{86}{108} = .8 \quad \text{and} \quad \frac{134 - 30}{216 - 122} = \frac{104}{94} = 1.1.$$

## 4C: A more extensive example/exhibit 5(A–S)

exhibit **5** of chapter 4 (continued)

**Panel A continued**

| root counts (*p) | | | | log counts | | | |
|---|---|---|---|---|---|---|---|
| | | H spread | | | | H spread | |
| ⌊H⌋ | ⌈H⌉ | ⌊raw⌋ | ⌊log⌋ | ⌊H⌋ | ⌈H⌉ | ⌊raw⌋ | ⌊log⌋ |
| 33 | 37 | 4 | 60 | 104 | 115 | 11 | 104 |
| 40 | 42 | 2 | 30 | 120 | 126 | 6 | 78 |
| 46 | 48 | 2 | 30 | 132 | 136 | 4 | 60 |
| 45 | 47 | 2 | 30 | 130 | 134 | 4 | 60 |
| 47 | 69 | 22 | 134 | 134 | 168 | 34 | 153 |
| 49 | 57 | 8 | 90 | 138 | 151 | 13 | 111 |
| 57 | 59 | 2 | 30 | 151 | 154 | 3 | 48 |
| 57 | 60 | 3 | 48 | 152 | 156 | 4 | 60 |
| 55 | 62 | 7 | 85 | 148 | 158 | 10 | 100 |
| 55 | 64 | 9 | 95 | 148 | 161 | 13 | 111 |
| 58 | 62 | 4 | 60 | 153 | 158 | 5 | 70 |
| 58 | 66 | 8 | 90 | 153 | 184 | 11 | 104 |
| 61 | 64 | 3 | 48 | 157 | 161 | 4 | 60 |
| 66 | 69 | 3 | 48 | 163 | 168 | 5 | 70 |
| 65 | 68 | 3 | 48 | 162 | 166 | 4 | 60 |
| 74 | 82 | 8 | 90 | 173 | 183 | 10 | 100 |
| 80 | 99 | 19 | 128 | 181 | 200 | 19 | 128 |
| 93 | 98 | 5 | 70 | 193 | 198 | 5 | 70 |
| 115 | 124 | 9 | 95 | 212 | 219 | 7 | 85 |
| | | 5 | 35 | | | −4 | −19 |
| | | 7 | 65 | | | +1 | 7 |
| | | | (.3) | | | | (−.2) |
| | | | (.7) | | | | (.1) |

PROBLEMS (continued)

5h) In view of (5f) and (5g), do you think it would have mattered if the center and lower panels of exhibit 6 had used logs of median root counts and log counts, respectively, in place of logs of median row counts? Check your answer for the lower panel.

5i) Do you think such a choice will ever matter? If not, why not? If yes, give an example.

5j) Look up the Bruner, Postman, and Mosteller paper, and apply the condensed approach (see text) to the data for the "natural" instruction.

5k) Find a set of data of about the same size that interests you and carry through a similar analysis.

**S) SOURCE**

J. S. Bruner, L. Postman, and F. Mosteller (1950). "A note on the measurement of reversals of perspective." Psychometrika 15: 63–72. Table 1 on page 65.

exhibit **6** of chapter 4: Schroeder staircase

**Plots of log H-spread against log median for the perspective reversal examples**

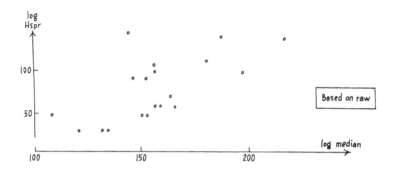

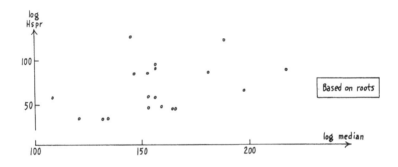

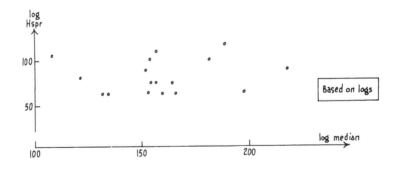

Clearly, the slope is about 1. If the rule of thumb mentioned above works, this would lead us to expect that logs will do well as an expression that keeps the H-spread from trending.

Later columns of exhibit 5 show calculations of hinges, H-spread, and log H-spread for both expression in roots and expression in logs. The center and lower panels of exhibit 6 show the corresponding plots. The trend is clearly smaller for roots and negligible for logs.

For reasons pretty well explained by the answers to problems (5f) to (5i), we have not bothered to change the horizontal scales in the center and lower panels of exhibit 6. Instead we have used log median (raw count) throughout. (This makes the three pictures easier to compare; but this is a little thing, since our next step is to learn how to make one picture do.)

### a condensed approach

If we had wanted to hold our calculation to a reasonable minimum, we could have avoided most of exhibits 5 and 6. Since we are dealing with counts, we expect to use either roots or logs. So we may go ahead as follows:

◇ find raw median and hinges; find log raw median.

◇ find root hinges; find H-spread for roots; find logs of same.

◇ make picture of log H-spread (for roots) vs. log M (for raws).

◇ select typical points; find tilts for same.

◇ working with the typical points only, find hinges for other expressions; find H-spreads; find logs of same.

◇ compare tilts of log H-spread for typical points.

◇ choose expression. (If we're lucky, it will be the one we started with.)

In this example, the tilt (against log median of raw counts) drops by almost 1/2 at each step of the re-expression scale (1.1 for raws, 0.7 for roots, 0.1 for logs). Dropping by about 1/2 happens very often.

Note also the unusually high position (on any of the three plots) of one subject who turns out to be #12. Turning back to the original data, we find that his ten counts, in order of occurrence were:

30, 22, 14, 22, 24, 32, 18, 48, 52, 53.

It would appear that after seven minutes this subject suddenly learned how to make his perspective reverse rapidly

**review questions**

To speed our calculations what kind of points should we select from our first picture? How can we use them to find a tilt? What did the rule of thumb say about tilts? How well did it apply in this example? What expressions should we try first with counted data? Was any one of the 19 subjects used by Bruner, Postman, and Mosteller unusual? How? Are you surprised? Why/ why not?

### 4D. The meaning of comparison

We have said much about comparison. We have used many pictures and many numbers to make comparisons. We have not yet said enough about comparison itself. It is time we did.

**Two kinds of comparisons come up in the simplest of common language:** "Bill is a head taller than Jim." "George weighs twice as much as his brother Jack." Each of these statements says what you must do to one person to make him match the other. The first statement bases itself on a plus sign, and says how much we would have to add to Jim's height to match Bill's. The second bases itself on a times sign, and says what we would have to multiply Jack's weight by to match George's.

There are many ways in which addition is simpler than multiplication; doing hand arithmetic and sliding things bodily along graphs are only two. Logarithms were invented centuries ago to reduce multiplication to addition, thus making hand arithmetic much easier. We can, do, and should use them to avoid comparisons by multiplication. When multiplication would otherwise seem needed, taking logs will let us deal with comparison by addition.

When we think of comparison, then, we want to think of what needs to be added or subtracted to the data--as carefully expressed for analysis--in order to make one thing match another. If the data is originally, or commonly, expressed in another way, we will try to translate back to that expression, too.

Thus, if boys' weights are really better thought of as "twice" or "three-quarters" of one another, we will do our analysis in log weight, finding, perhaps, that we have to add 0.30 to Jack's to match George's. Having found this, we would also convert back to raw weight, and say that Jack's had to be doubled in order to match George's. Both forms of answer help.

**review questions**

What kinds of comparison come up in common language? How ought we think of such comparisons? How many forms of answer help?

### 4E. Adjustments, rough and exact

**rough adjustments**

If we want to put several batches under the same microscope at the same time, we need to have them centered at about the same level. If they are not naturally somewhat similarly centered, we will have to adjust them so that they will be.

Thanks to the flexibility of the human eye, it is not important--if we are to look at a picture--that the centering be exactly the same. Rough adjustments will serve us essentially as well as precise ones.

Exhibit 7 returns to the catch-of-plankton example. For variety, we use an expression that compromises between roots and logs. (Recall that logs play the

exhibit 7 of chapter 4: plankton hauls

**Rough residuals for the plankton counts plotted in terms of $\sqrt[4]{count} = \sqrt{\sqrt{count}}$ (see exhibit 1 for data)**

A) PLOT of ROUGH RESIDUALS--of plankton counts

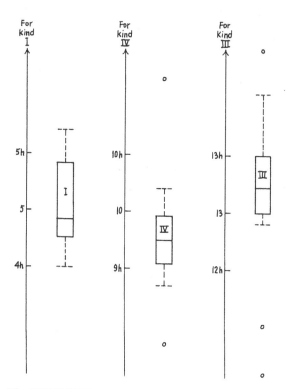

B) PROBLEMS

The fourth roots used to make the plots above were NOT obtained by two applications of exhibit 7 of chapter 3.

7a) Obtain them by two such applications, when the values will differ somewhat, and make the plots analogous to the above.

7b) Do you think the difference is important?

role of a zero power; clearly 1/4 is a natural compromise between 1/2 and 0.) Thus expression is

$$\sqrt[4]{\text{count}} = \sqrt{\sqrt{\text{count}}}$$

As exhibit 7 shows, when expressed in this way, the catches are in the vicinity of 5, 10, and 13 for Kinds I, IV and III, respectively.

In exhibit 7, then, we have shifted each schematic plot by a different amount. We have chosen round amounts, so that the three schematic plots do not exactly line up. We do have them under a common microscope, however, and can see clearly their general similarity in spread, both from hinge to hinge and from extreme to extreme.

### exact adjustment

Round values and approximate adjustment save arithmetic if we are going to make a plot. If we want to present the same information numerically, we need to do subtractions to get small numbers. Since it is about as easy to subtract one number as another, it usually pays to line something up exactly. Exhibit 8 shows what happens in the catch-of-plankton example if we line up the medians exactly. Clearly the results are rather easy to compare, or even to combine.

**exhibit 8 of chapter 4: plankton hauls**

**Results of lining medians up by adjustment (subtracting a constant) in the catch-of-plankton example expressed in log counts (all logs in .01's)**

**A) LETTER-VALUE DISPLAYS--for logs in 0.01**

| Kind I | Kind IV | Kind III |
|---|---|---|
| 276 | 396 | 448 |
| 268  294 | 391  400 | 445  453 |
| 259  302 | 378  419 | 428  464 |

**B) SAME for SHIFTED LOGS**

| Kind I down 276 | Kind IV down 396 | Kind III down 448 |
|---|---|---|
| 0 | 0 | 0 |
| −8   18 | −5   4 | −3   5 |
| −17  26 | −18  23 | −20  16 |

**P) PROBLEMS**

8a) Imitate panel B for the catches of plankton expressed in roots (see exhibit 3 for original form). What seems to be going on?

8b) Do the same for the raw counts.

8c) Can you find the median of the three displays in panel B? Why/why not?

**review questions**

Can rough adjustments serve as well as exact ones? What light does exhibit 7 shed on this question? If we are not making pictures, should we use rough adjustments? Why?

## 4F. Residuals

If we wanted to know about the variations in plankton catches in rather general terms, we might like to combine all the knowledge that we can gain from the 3 batches of 12 at hand. If we define

$$\text{residual} = \text{given value} - \text{summary value}$$

we can turn each given value into a residual, for example by using the median of the corresponding batch as the summary value.

For the catch-of-plankton data, we dare not combine residuals based on raw counts, but we might try doing this for either root counts or log counts.

Exhibit **9** gives the log counts, their residuals from the median, and--in stem-and-leaf form--the result of pooling the three sets together. (A tendency to straying may be suggested by the E-spread being considerably more than twice the H-spread).

While residuals are useful as an aid to pulling together information from several batches (a rudimentary form of what might be called mustering and borrowing strength), **we will soon learn much more important uses for them:**

◇ as keys to the successive step-by-step improvement of our analyses.

◇ as keys to turning an investigator's eye toward the adequacy of our current analysis.

exhibit **9** of chapter 4: plankton hauls

**Combining residuals for the catches of plankton expressed in logs (data and source, exhibit 3)**

**A) LOGS of COUNTS in 0.01's--two in median bold**

Kind I:   259, 263, 267, 270, 273, **273, 279,** 288, 293, 295, 301, 302

Kind IV: 378, 388, 390, 392, 393, **395, 397,** 398, 399, 401, 404, 419

Kind III: 428, 435, 444, 445, 446, **446, 450,** 452, 453, 454, 460, 464

**B) RESIDUALS from MEDIAN--for logs of counts in 0.01's.**

Kind I:   −17, −13, −9, −6, −3, −3, 3, 12, 17, 19, 25, 26

Kind IV: −18, −8, −6, −4, −3, −1, 1, 2, 3, 5, 8, 23

Kind III: −20, −13, −4, −3, −2, −2, 2, 4, 4, 6, 12, 16

## review questions

What is a residual? Can there be more than one set of residuals for a given set of data? What are three ways in which residuals can serve us? Which are the more important?

exhibit **9** of chapter 4 (continued)

**C) RESIDUALS POOLED**

```
  2    2·  | 56
  3    2*  | 3
  6    1·  | 796
  8    1*  | 22
 11    0·  | 586
 18    0*  | 3123244
 ──
 18   -0*  | 334314322
  9   -0·  | 9686
  5   -1*  | 33
  3   -1·  | 78
  1   -2*  | 0
```

36 pooled residuals

adj:  −20(III), 23(IV)

out:  26(I)

|       | (mid) |     | (spr) |    |
|-------|-------|-----|-------|----|
| M 18h | 0     |     | 0     |    |
| H  9h | 1     | −5  | 7     | 12 |
| E  5  | 2     | −13 | 17    | 30 |
|    1  | 3     | −20 | 26    | 46 |

|   |      |     |
|---|------|-----|
|   |      | 18  |
| f | −23  | 25  |
|   | xxx  | one |
| F |      | 43  |
|   |      | xxx |

**P) PROBLEMS**

9a) Make a comparative plot of 4 batches, the 3 separate batches of residuals above and the pooled batch.

9b) Pool the residuals for the data and mode of expression of (5j) above.

9c) Same for (5k) above.

9d) Same for a new set of self-selected data.

exhibit **5** of chapter 4: Schroeder staircase

**Calculations for perspective reversals (all logs in 0.01)**

**A) BASIC CALCULATIONS for THREE EXPRESSIONS**

|  | | | | | raw counts | |
|---|---|---|---|---|---|---|
|  | | | | | H spread | |
| # | M | log M | H | H | raw | log |
| 2 | 12 | 108 | 11 | 14 | 3 | 48 |
| (*) 9 | 16h | 122 | 16 | 18 | 2 | 30 |
| 13 | 22 | 134 | 21 | 23 | 2 | 30 |
| 19 | 22 | 134 | 20 | 22 | 2 | 30 |
| 12 | 28 | 145 | 22 | 48 | 26 | 142 |
| 4 | 29h | 147 | 24 | 32 | 8 | 90 |
| 16 | 33h | 152 | 32 | 35 | 3 | 48 |
| 15 | 34 | 153 | 33 | 36 | 3 | 48 |
| 11 | 34 | 153 | 30 | 38 | 8 | 90 |
| 6 | 36 | 156 | 30 | 41 | 11 | 104 |
| 3 | 36h | 156 | 34 | 38 | 4 | 60 |
| 18 | 36h | 156 | 34 | 44 | 10 | 100 |
| 5 | 38h | 159 | 37 | 41 | 4 | 60 |
| 14 | 44 | 164 | 43 | 48 | 5 | 70 |
| 7 | 45 | 165 | 42 | 46 | 4 | 60 |
| 8 | 64 | 181 | 54 | 67 | 13 | 111 |
| 1 | 74h | 187 | 64 | 98 | 34 | 153 |
| 10 | 92 | 196 | 86 | 95 | 9 | 95 |
| (*) 17 | 144h | 216 | 132 | 154 | 22 | 134 |
| 17 − 2 |  | 108 |  |  | 19 | 86 |
| diffs 17 − 9 |  | 94 |  |  | 20 | 104 |
| (ratios) (17 − 2) |  |  |  |  |  | (0.8) |
| (tilts) (17 − 9) |  |  |  |  |  | (1.1) |

➜

**P) PROBLEMS**

5a) Plot the raw H-spread for raw counts against raw medians (also for raw counts). Is there a trend? Which points near each end of the point cloud would you choose as reasonably typical?

5b) Do the same for the raw H-spread for root counts.

5c) Do the same for the raw H-spread for log counts.

5d) Look up the source and apply a similar analysis to the last 8 minutes under the "natural" instruction.

5e) Do the same for the "hold" instruction.

5f) Find logs of median root counts, and plot against logs of median raw counts. What do you see? Should you have expected it?

5g) Find logs of median log counts and plot against logs and median row counts. What do you see? Should you have expected it? ➜

### 4H. How far have we come?

In this chapter we have begun to take seriously one of the most important things we can do with data--comparison. Most of our attention has been devoted to learning how to strengthen comparisons--so that more nearly the full value of what the data offers us is put to use. We had already learned many of the techniques we need to do this--re-expression, useful display, subtraction. We now put them more carefully to work.

We are now ready:

◇ to make some kinds of slightly graphic summaries--displays that show forth the main issue without loss of numerical detail.

◇ to use plots of log spread against log level to help us judge what re-expression to try next.

◇ to calculate residuals by subtracting summary values from individual values. (We will learn about more general kinds of residuals in later chapters.)

Both **symmetry** of spreading-out and **similar extent** of spread make for useful comparison. While these are little and middle deals, respectively, the re-expression they point to is likely to be one that a BIG DEAL (when we have one) will ALSO prefer. We can seek such an expression by trial and error. With more data, though, it helps to use a systematic approach, plotting the log of a simple measure of spread against the log of a simple measure of level, and using the apparent tilt as a guide to both direction and amount of change.

If we want to compare in detail--rather than looking at the broadest questions--we want to adjust what is being compared so that the gross differences are taken out of the way.

Comparison, in common language, is either a matter of **difference** or a matter of **ratio**. Ratios almost always are of amounts or counts, which are never negative. Logs were invented to bring such ratios to differences. (Having values that are not fenced in, in particular are not kept from being negative, generally is likely to be helpful.)

Most importantly, we have just begun to learn that NO BODY OF DATA TELLS US ALL we need to know ABOUT ITS OWN ANALYSIS. It always takes information and insight gained from other, parallel bodies to let us analyze our body of data as well as we can. (If we don't have it, we do as well as we can.)

We are now well started on the analysis of the simplest kinds of data. We have a good chance of making an effective analysis: expressing the given values usefully, summarizing batches well enough for exploration, calculating residuals so that we can try to look deeper.

## 4P. Additional problems

See exhibits **10** through **17**.

exhibit **10** of chapter 4: data and problems

**Percent of major party vote for Republican presidential candidates in eight southwestern states**

A) DATA

|            | % for Nixon in 1960 | % for Goldwater in 1964 |
|------------|---------------------|-------------------------|
| Arizona    | 55.6%               | 50.5%                   |
| California | 50.3%               | 40.8%                   |
| Colorado   | 54.9%               | 38.4%                   |
| Nevada     | 48.8%               | 36.1%                   |
| New Mexico | 49.6%               | 40.6%                   |
| Oklahoma   | 59.0%               | 44.3%                   |
| Texas      | 50.5%               | 36.6%                   |
| Utah       | 54.8%               | 45.3%                   |
| (All U.S.) | (49.9%)             | (38.7%)                 |

P) PROBLEMS

10a) The snowfalls in New York City for 20 consecutive winters from 1918–19 to 1937–38 were (in inches) 3.5, 55.4, 18.2, 29.7, 55.2, 26.3, 27.9, 35.8, 21.9, 14.3, 13.3, 13.5, 9.7, 5.1, 24.5, 53.1, 29.0, 32.8, 11.9, 13.9. Those from 1938–39 to 1957–58 were: 31.9, 22.2, 35.0, 10.2, 27.6, 26.0, 26.7, 26.6, 33.2, 61.5, 43.0, 10.4, 10.9, 14.4, 9.1, 17.1, 10.9, 29.8, 19.1, 37.9. Compare these two batches in all the ways used in exhibit 2.

10b) Panel A gives, for two elections, percent of the major party vote for Republican presidential candates in eight southwestern states. Make schematic plots for:

⋄ the 1960 data.
⋄ the 1964 data.
⋄ the change (swing) from 1960 to 1964.

Discuss your results.

10c) Choose another group of eight states, find the data, and make the same plots as for (10b).

S) SOURCE

Richard M. Scammon. America at the Polls: A Handbook of Presidential Election Statistics, 1920–1964. Pittsburgh: University of Pittsburgh Press (1965) 521 pp.

exhibit 11 of chapter 4: data and problems

**Seasonal snowfall in Buffalo, New York and Cairo, Illinois, from 1918–19 to 1937–38 (inches)**

A) DATA

|         | Buffalo | Cairo |
|---------|---------|-------|
| 1918–19 | 25.0    | 1.8   |
| 1919–20 | 69.4    | 4.5   |
| 1920–21 | 53.5    | 13.9  |
| 1921–22 | 39.8    | 4.0   |
| 1922–23 | 63.6    | 1.2   |
| 1923–24 | 46.7    | 6.8   |
| 1924–25 | 72.9    | 7.2   |
| 1925–26 | 79.6    | 11.5  |
| 1926–27 | 83.6    | 6.2   |
| 1927–28 | 80.7    | 0.4   |
| 1928–29 | 60.3    | 11.5  |
| 1929–30 | 79.0    | 12.4  |
| 1930–31 | 64.8    | 11.3  |
| 1931–32 | 49.6    | 2.9   |
| 1932–33 | 54.7    | 7.4   |
| 1933–34 | 71.8    | 2.7   |
| 1934–35 | 49.1    | 1.6   |
| 1935–36 | 103.9   | 14.1  |
| 1936–37 | 51.6    | 5.4   |
| 1937–38 | 81.6    | 3.0   |

P) PROBLEMS

11a) The snowfalls in Buffalo, New York and Cairo, Illinois for the 20 consecutive winters from 1918–19 to 1937–38 are given in Panel A. Make schematic plots for both places, for this time period.

11b) What light do these two batches throw on how they should be expressed?

S) SOURCE

Report of the Chief of the Weather Bureau, 1918–1919 to 1934–1935, and U.S. Meteorological Yearbook 1935 to 1938.

exhibit **12** of chapter 4: data and problems

**Consistency of drape measurements**

A) DRAPE, measured by AREA in SQUARE INCHES

| Sample A | Sample B |
|---|---|
| 28.92 | 50.04 |
| 28.82 | 49.94 |
| 28.96 | 40.08 |
| 28.89 | 50.27 |
| 28.96 | 50.03 |
| 28.85 | 50.06 |
| 28.97 | 50.00 |
| 28.89 | 49.99 |
| 29.00 | 49.75 |
| 28.99 | 49.94 |

P) PROBLEMS

12a) In 1950, Chu, Cummings, and Teixeira gave the results of drape tests in panel A. Adjust schematic summaries to equal medians. Does expression in square inches seem to give equal spread?

12b) Compare log H-spreads with log medians. What expression is suggested? Repeat (12a) for this expression.

S) SOURCE

C. C. Chu, C. L. Cummings, and N. A. Teixeira, 1950. "Mechanics of elastic performance of textile materials. Part V: A study of factors affecting the drape of fabrics--The development of a drape meter," Textile Research Journal 20: 539–548.

exhibit **13** of chapter 4: data and problem

## Microdetermination of carbon monoxide in air

**A) PARTS PER MILLION of CARBON MONOXIDE**

| Sample | Standard* | Trial results † |
|--------|-----------|-----------------|
| A | 90.5 | 95,96,92,102,103,93,101,92,95,90 |
| B | 184.6 | 184,202,215,204,195,201,201,169,182,192, |
| C | 44.8 | 40,54,42,49,64,62,50,67,64,43, |
| D | 320 | 261,279,281,278,269,264,266,261,266,276, |
| E | 244.7 | 215,214,197,216,215,208,226,208,216,214, |
| F | 25.8 | 26,23,25,25,21,22,27,27,21,25, |
| G | 66.2 | 56,55,61,57,60,57,65,55,60,61, |
| H | 137.8 | 128,119,119,123,117,122,127,121,122,119, |
| I | 137.8 | 155,142,146,149,149,146,152,159, |

\* $I_2O_5$ method
† Beckman–McCullough method, in order of analysis

**P) PROBLEM**

13a) In 1948, Beckman, McCullough, and Crane gave the results in panel A of microdeterminations of carbon monoxide in air. Adjust schematic summaries to equal medians.

**S) SOURCE**

A. O. Beckman, J. D. McCullough, and R. A. Crane, 1948. "Microdetermination of carbon monoxide in air," **Analytical Chemistry** 20: 674–677.

exhibit 14 of chapter 4: data and problem

**Polarographic determination of aluminum**

A) % $Al_2O_3$ -- oxygen-containing samples -- in 0.01%

| Certified value | Sample | Determinations | |
|---|---|---|---|
| 416 | Limestone | 3**\| | 92,94 |
| | | 4 \| | 16,36 |
| 196 | Silica brick | 19*\| | 02336 |
| | | 20*\| | 04 |
| 191 | Magnetite ore | 171,175,177,190,210 | |
| 189 | Soda-lime glass | 159,161,214,214 | |
| 103 | Iron ore | 10*\| | 336799 |
| | | 11*\| | 236678 |
| 6.7 | Dolomite | 56,56 (in 0.001%) | |

B) % $Al_2O_3$ -- alloys -- in 0.01%

| Certified value | Sample | Determinations | | |
|---|---|---|---|---|
| 113 | Manganese bronze 62 | 11*\| | 444468 | |
| 097 | Manganese bronze 62b | 8*\| | 9 | |
| | | 9*\| | 4799 | |
| 106 | Nitralloy steel 106 | 9*\| | 5 | |
| | | 10*\| | 88 | |
| | | 11*\| | 01 | |
| 107 | Nitralloy steel 106a | 10*\| | 08 | |
| | | 11*\| | 022 | |
| 026 | High silicon steel | 25*\| | 18 | (in 0.001%) |
| | | 26*\| | 009 | |
| | | 27*\| | 0 | |

P) PROBLEM

14a) In 1950, Willard and Dean gave the results in panel A on the measurements of aluminum content. Find residuals. How should they be pooled? Combine them and discuss the results.

S) SOURCE

H. H. Willard and J. A. Dean (1950). "Polarographic determination of aluminum: Use of an organic reagent." **Analytical Chemistry** 22: 1264–1267. Table II on p. 1266.

exhibit **15** of chapter 4: data and problem

**Kjeldahl ultramicrodetermination of nitrogen**

A) The BATCHES--hundredths of microliters of 0.01N Hcl for a 10.42 microliter sample

| Sample | Material | Determinations | |
|---|---|---|---|
| A | Acetanilide | 6**\| | 60,78,81, |
|   |             | 7**\| | 03,57,60,72, |
| B | Acetanilide | 6**\| | 13,37,39,46, |
|   |             | 7**\| | 47,59,75, |
| C | $C_{14}H_{22}N_2O_2S$ | 6**\| | 25,63,67 |
|   |             | 7**\| | 32,43,63, |
| D | $C_{13}H_{20}N_2O_2S$ | 6*\| | 56,59,65, |

P) PROBLEM

15a) In 1950, Kuck, Kingsley, Kinsey, Sheehan, and Swigert gave the results in exhibit 15 on the ultramicromeasurement of nitrogen. Calculate and pool residuals. Can these data show how they should best be expressed? If not, why? If yes, show how.

S) SOURCE

J. A. Kuck, A. Kingsley, D. Kinsey, F. Sheehan, and G. F. Swigert 1950. Kjeldahl ultramicrodetermination of nitrogen: Applications in the industrial laboratory. **Analytical Chemistry** 22: 604–611.

exhibit **16** of chapter 4: data and problems

**Precipitation of platinum sulfide by various methods**

A) AMOUNTS of PRECIPITATE--calculated as milligrams of platinum

| Method | Platinum recovered from 10.12 mg Pt |
|---|---|
| Treadwell and Hall | 10.2* \| 89 |
|  | 10.3* \| 13334 |
| Gilchrist and Wichers | 10.2* \| 2334789 |
| Hillebrand and Lundell | 10.2* \| 25 |
|  | 10.3* \| 03688 |

P) PROBLEMS

16a) In 1942, Geoffrey Beall ("The transformation of data from entomological field experiments so that analysis of variance becomes applicable," *Biometrika* **32**: 243–262) counted the number of *Phlegethontius quinquemaculata* (an insect) per plot after various insect control treatments. Slight simplifications of his results include (letters are treatments): A; 10, 7, 20, 14, 14, 12, 10, 23, 17, 20, 14, 13. B; 11, 12, 21, 11, 16, 14, 17, 17, 19, 21, 7, 13. C; 0, 1, 7, 2, 3, 1, 2, 1, 3, 0, 1, 4. D; 3, 5, 12, 6, 4, 3, 5, 5, 5, 5, 2, 4. E; 3, 5, 3, 5, 3, 6, 1, 1, 3, 2, 6, 4. F; 11, 9, 15, 22, 15, 16, 13, 10, 26, 26, 24, 13. What choice of expression is indicated if we take the counts for each treatment as a batch?

16b) In 1950, Jackson and Beamish gave the results in panel A on the precipitation of platinum sulfide. Compare the quality of the three methods. Find the residuals. Should they be pooled?

16c) In 1952, S. P. Hersh and D. J. Montgomery ("Electrical resistance measurements on fibers and fiber assemblies," *Textile Research Journal* **22**; 805–818) gave these resistivities (in $10^9$ ohm-cm$^2$/cm) for the fibers indicated (at the relative humidities indicated in parentheses): Nylon monofilament--340 denier (60%): 12, 13, 12, 12, 16, 13, 12, 13. Nylon monofilament--30 denier 64%): 32, 26, 26, 29, 29, 41, 32. Nylon two--3 denier (85%): 1.00 (six times), 0.95 (three times), 1.10, 0.90. Human hair ~0.001 diameter (85%): 5.7, 6.8, 6.2, 6.2. Wool--Columbia 58's (85%): 0.095, 0.092, 0.089, 0.075, 0.050, 0.087. Make the best graphical presentation and comparison you can. (What do you think "resistivity," "denier," and "relative humidity" mean?)

S) SOURCE

D. S. Jackson and F. E. Beamish, 1950. "Critical examination of platinum sulfide precipitation," **Analytical Chemistry 22**: 813–817.

exhibit **17** of chapter 4: data and problems

**Head breadths of termites (in hundredths of mm)**

A) The DATA

|  | Nest 668 | Nest 670 | Nest 672 | Nest 674 | Nest 675 |
|---|---|---|---|---|---|
| Small soldiers | 227.3 | 247.9 | 249.4 | 244.7 | 245.6 |
|  | 233.2 | 260.3 | 245.7 | 238.8 | 262.6 |
|  | 237.5 | 261.3 | 245.2 | 251.5 | 263.3 |
|  | 237.3 | 255.7 | 239.6 | 244.5 | 248.7 |
|  | 231.8 | 237.7 | 277.9 | 231.2 | 241.0 |
| Large workers | 214.2 | 231.5 | 235.9 | 234.9 | 236.0 |
|  | 213.9 | 246.3 | 234.4 | 230.0 | 247.8 |
|  | 222.5 | 248.5 | 239.4 | 243.6 | 253.9 |
|  | 235.2 | 247.7 | 236.4 | 242.3 | 246.1 |
|  | 226.9 | 233.4 | 226.9 | 230.2 | 231.1 |

P) PROBLEMS

17a) In 1948, an anonymous correspondent Query 60, *Biometrics* **4**: 213–214) gave weights (in grams) of ducks (at age six weeks) after various protein supplements as follows (protein source in parentheses): (Horsebean) 179, 160, 136, 227, 217, 168, 108, 124, 143, 140. (Linseed oil meal) 309, 229, 181, 141, 260, 203, 148, 169, 213, 257, 244, 271. (Soybean oil meal) 243, 230, 248, 327, 329, 250, 193, 271, 316, 267, 199, 171, 158, 248. (Sunflower seed oil meal) 423, 340, 392, 339, 231, 226, 320, 295, 334, 322, 297, 318. (Meat meal) 325, 257, 303, 315, 380, 153, 262, 242, 206, 344, 258. (Casein) 368, 390, 379, 260, 404, 318, 352, 359, 216, 222, 283, 332. What expression is indicated?

17b) Make an effective graphical presentation and comparison of the data given in (17a). (What do you think "sunflower seed oil meal" is?)

17c) In 1909, Warren gave the head measurements on termites in panel A. Find residuals. Should they be combined? If not, why? If yes, show how.

S) SOURCE

Ernest Warren, 1909. "Some statistical observations on termites, mainly based on the work of the late Mr. G. D. Haviland." **Biometrika: 6**: 329–347.

# index for chapter 5

*a residual* 125
*a response* 125
*a factor* 126
*a circumstance* 126
*(factor, response)* 126
review questions 126

## 5A. How to plot y against x 126
choice of ruling 127
choice of scale units 127
consequences of our purpose 127
kinds of grids 128
shape of plot 129
ticks and numbers along axes 129
review questions 131

## 5B. Looking at subtraction 131
untilting 134
review questions 135

## 5C. Subtracting straight lines 135
finding the line 136
an example 136
subtracting different straight lines 137
*sum of the two lines* 139
back to the example 139
*supplementary line* 139
review questions 141

## 5D. Plotting the population of the U.S.A. 141
the later decades 142
coming to details 143
the earlier details 145
review questions 148

## 5E. Plotting the ratio of births to deaths 148
another try 149
going to the map 151
review questions 154

## 5F. Untilting defines "tilt" 154
*untilted* 154
*tilt* 154
review questions 156

## 5H. How far have we come? 156

## 5P. Additional problems 157

| EXHIBITS | PAGE |
|---|---|
| 5A | |
| 1 | 130 |
| 5B | |
| 2 | 132 |
| 3 | 133 |
| 4 | 133 |
| 5 | 134 |
| 5C | |
| 6 | 137 |
| 7 | 138 |
| 8 | 140 |
| 5D | |
| 9 | 141 |
| 10 | 143 |
| 11 | 144 |
| 12 | 144 |
| 13 | 145 |
| 14 | 146 |
| 15 | 147 |
| 16 | 147 |
| 5E | |
| 17 | 149 |
| 18 | 150 |
| 19 | 150 |
| 20 | 152 |
| 21 | 152 |
| 22 | 153 |
| 5F | |
| 23 | 155 |
| 24 | 155 |
| 5H | |
| 5P | |
| 25★ | 158 |
| 26★ | 160 |
| 27★ | 161 |
| 28★ | 162 |
| 29★ | 163 |
| 30★ | 164 |
| 31★ | 165 |
| 32★ | 166 |
| 33★ | 167 |

# Plots of relationship    5

We have learned something about scratching down batches--sets of similar numbers. We found more things to do--saw that more things could be learned about data by doing them--than we might have expected. We turn now to the use of simple plots, to the plotting of y against x, and find, again, more things to do and more gains from doing them than we might expect.

The most quoted passage of all the Sherlock Holmes stories concerns the unusual behavior of the dog in the night time. Holmes called attention to its unusualness--Watson couldn't see what was unusual--Holmes pointed out that the dog did nothing.

The moral is clear--we ought to judge each occurrence against the background of--or a background derived from--other "nearby" occurrences. We do not ordinarily think of "zero" as unusual--yet a 24-hour period when no one died anywhere in the world, or a winter with no snow at all in the White Mountains, would be recognized by each of us as strikingly unusual.

As in the large, so in the small. Once we have plotted some data, and found out how it behaves generally, our next step is to look at each of its elements--at each "point"--and ask how it seems to deviate from the general behavior of all the "points". To do this it very often helps to make a new plot, one that focuses on such deviations--in brief, "it very often pays to plot residuals".

As we have begun, so will we continue. Here, as well as later, we will be repeatedly engaged in splitting up data according to one version or another of the key relation

> given = fit PLUS residual.

Here the fit is our current description--always incomplete, always approximate--of the overall behavior of the data. Each individual observation is split up into a sum of this fit and what is left over, called

> a residual.

Residuals are our main tool in going further. They are to the data analyst what powerful magnifying glasses, sensitive chemical tests for bloodstains, and delicate listening devices are to a story-book detective. They permeate all sorts of data analysis and appear in many guises.

The most conventional plots follow a simple pattern. A quantity, habitually shown on a vertical scale, which we think of as

> a response.

Another quantity, habitually shown on a horizontal scale, which we call

a factor

which is usually

a circumstance

and which we think of as possibly explanatory or descriptive. The data--in whole or part--will consist of pairs of numbers of the form

(factor, response)

which will be plotted as points using these scales.

Most plots--at least in the popular press--concentrate on the fit. Too often, however, such plots are only to remind us that the anticipated relationship is there--that, for example, the population of the U.S.A. still increases. As such they are usually matters of elementary exposition rather than analyses. They show us "the big picture" that we already knew about. (In the sense of Linus in the familiar comic strip, they are just "security blankets".) At other times, such plots "in the large" are to show us the unexpected--either an unsuspected relationship or an unsuspected strength or weakness of an anticipated relationship. For these purposes, plots "in the large" are part of data analysis. They tell us of investigation's successes, perhaps very effectively, but they still need to be supplemented by pictures of residuals--pictures that tell us whether there is yet more to investigate.

For us the most useful plot will be one that might reveal the unexpected or the unobvious. Sometimes a plot "in the large" will do this. Usually, however, it is the plot of residuals that has the greatest use and the greatest impact.

review questions

How ought we judge occurrences or numbers? What is a "residual"? What is the key relation involving residuals? What is a fit? Is it final, complete, or exact? What is a "response"? A "factor"? A "circumstance"? How are they usually plotted? How is data involving one factor and one response usually written down? Plotted? When are plots "in the large" useful? When are they part of data analysis? How do we ask whether there is more to investigate? What kind of plot is most likely to help us?

5A. How to plot y against x

Plots are important. Great differences in ease of construction and great differences in effectiveness of use depend on apparently minor, purely technical procedures. The remarks in this section are intended to finish getting you over most of the elementary hurdles. NOW GO BACK and read the parts of section 2C, dealing with "tracing paper", "scale values," and "plotting without graph paper". All this still applies here.

### choice of ruling

If plotting is to be easy, you want graph paper with at least three different thicknesses of ruling:

- light lines for "units".
- medium lines for "fives".
- heavy lines for "tens".

A wide variety of graph papers can be obtained with these characteristics. (Some papers have extra heavy lines for "twenties". You will have to learn for yourself whether these help you more than they get in your way.)

DO NOT use papers ruled in "fours" and "eights", or in "sixes" and "twelves" for plotting data given in ordinary (decimal) numbers. (For monthly data, of course, "twelves" in one direction do help.)

If you want to plot fast, easily, and accurately, avoid "dime-store" sheets with only two thicknesses of ruling and, above all, avoid quadrille sheets with only one. (Quadrille sheets are very useful, for almost everything except plotting graphs.) (If you need to save money, see below.)

### choice of scale units

When you come to plot, you must choose units. Don't try to make one step of ruling (light, medium, or heavy) equal to 3 units, 7 units, 0.03 units, 0.007 units, or any such uncomfortable number. Stick to one step being 1, 2, or 5 times a power of ten. (One square = 20,000 or one step = 0.05 are examples that can work out quite well.)

You will find it hard enough to learn to be a fast and accurate plotter with two three-speed gear shifts (1, 2, or 5 across the graph and 1, 2, or 5 up the graph).

**If you need to use an abnormal scale--three units to the square, for example--convert your numbers into plotting units--as by dividing by 3--by slide rule, hand arithmetic, or what have you, BEFORE you start to plot. It will take less time, overall, to make a good picture.**

### consequences of our purpose

Throughout this account we shall be interested in graphs to be looked at, not to be used to find numbers. Our graphs are means for looking at the data, not stores of quantitative information. This means:

- we will want to keep our eyes on the points.
- we will usually **not** connect one point to the "next" one. (We are likely to draw in fitted lines or curves.)
- we will want to suppress the rulings of graph paper, at least from our mind's eye.

⋄ we will need only a few numbered ticks along the axes, horizontal or vertical, of the final picture.

⋄ we will want to use symbols large enough to stand out (and if we need more than one kind, different kinds should be both clearly different and--usually--almost equally noticeable).

> **There cannot be too much emphasis on our need to see behavior.**

We must play down or eliminate anything that might get in the way of our seeing what appears to go on.

### kinds of grids

The use of tracing paper makes it possible for any of us to really arm himself with an easy-graphing capability of broad scope. A tracing pad and one sheet of each of 20 kinds of graph paper puts us in business to do many things. (Actually, it usually seems worthwhile to have a reasonable number of sheets, perhaps in the form of a pad, of at least one or two of the most used kinds of graph paper. Many will wish to first plot on the graph paper and then copy on the tracing paper.)

"Ordinary" graph paper frequently comes 8 small squares to the inch (beware; it may be lined in "4's" or "8's" which you should never, never use--unless what you are plotting comes in eighths of an inch, or, like stock market prices, in eighths of a dollar), 10 to the inch, 12 to the inch (beware, it may be ruled in "6's" or "12's"), 20 to the inch, and 10 to the centimeter. The writer likes the coarser rulings at least as well as the fine; it is up to you to learn your own preference.

Beware of 12-to-the-inch rulings, since too many are *ruled* in "sixes" and "twelves"; but bear in mind that 12-to-the-inch ruled in fives and tens--like Codex Nos. 31,253 and 32,253, for example--is a very good base for graphs where the final product is a typed page. (Typewriters like 12 to the inch.) It has little special advantage for most other purposes, although it is quite satisfactory.

Semilogarithmic (uniform scale one way, log scale the other) and full logarithmic (log scales both ways) graph papers come in a variety of scales and patterns. To have a reasonable selection is to save time and encourage inquiry. One CAN always look up the logs and then plot on uniform-scale graph paper, but WILL one? If chapter 3 is at hand, probably yes. (We rarely get much more than two-decimal accuracy out of a plot on log paper, so exhibit 2 of chapter 3 will usually do as well as special graph paper.) Otherwise? As we saw above, logs are often the way to make data reveal itself.

A variety of other graph papers are useful, as we shall see later. We note here that paper with two square-root scales is available (Codex No. 31,298 or No. 32,298) and that trilinear or isometric paper--with three sets of rulings at 120° to one another--is often quite useful. (It is available from various manufacturers.) Reciprocal (one-way) paper is available both from Codex and from Keuffel and Esser.

### shape of plot

Naturally enough, graph paper usually comes in the same general shape and size (letter size) as ruled writing pads and typing paper. Clearly it can be used in two ways:

⋄ narrow edge at the top of the plot, or

⋄ broad edge at the top.

By analogy with the use of ruled writing pads, it is natural to make plots with the narrow edge up. Full use of the area then usually makes a plot that is taller than it is wide.

There are some purposes for which this is a good shape. The ever-faster-rising curve of early growth is an example. We can quite well use taller-than-wide plots for such simple pictures, most of which tell us that we haven't yet made an analytically useful plot. Sometimes, indeed, such simple plots can be clearer when they are taller than wide, rather than wider than tall.

Most diagnostic plots involve either a more or less definite dependence that bobbles around a lot, or a point spatter. Such plots are rather more often better made *wider* than tall. Wider-than-tall shapes usually make it easier for the eye to follow from left to right.

Perhaps the most general guidance we can offer is that smoothly-changing curves can stand being taller than wide, but a wiggly curve needs to be wider than tall (sometimes after a smooth part has been taken out).

When a plot is made wider than tall, convention says it should be turned in the direction illustrated by many plots in this chapter, even if this makes a down-to-up legend upside down when we first see the plot as we turn the pages.

### ticks and numbers along axes

We use marks along axes for two quite different purposes: (1) to plot the points; (2) to look at them. Different purposes call for different techniques, and the graph-paper–tracing-paper combination makes separate techniques easy. If one begins by plotting on the graph paper, it helps to have many numbers and ticks on the scale. (Remember to have the horizontal scale **above** the plot, and the vertical scale to the left*--both out from under the plotting hand.) To look effectively at the traced result, it helps to have ONLY a few numbers and ticks. Exhibit 1 shows five pairs of vertical axes, one in each pair for plotting and one for looking. Notice the technique, in each PLOTTING scale, of:

⋄ putting 4 ticks--or dots--between consecutive numbers (often useful: NOT to be religiously followed; sometimes no tick between, or one tick between, is better).

---

* To the right, for southpaws.

◇ putting a tick--or dot--for every one or two steps of some digit in what is plotted (sometimes it pays to do this at every five, instead).

(Some even standardize on dots for each one and ticks for each two.) All this is focused on making it as easy as possible to find where to put the point. Doing less than this slows down our plotting and wastes both effort and temper.

Notice also the technique, in each LOOKING scale, of using:

◇ only four or five ticks.
◇ only two or three numbers.

Doing more than this distracts our attention from what we ought to see. (If the scale is irregular, we may need more ticks and values. Scales for dates, where individual values like 1066, 1776, or 1929 are well remembered, often deserve more ticks and values.)

exhibit **1** of chapter 5; illustrative

**Pairs of scales, showing the difference between scales for plotting (marked "draw") and for looking (marked "show")**

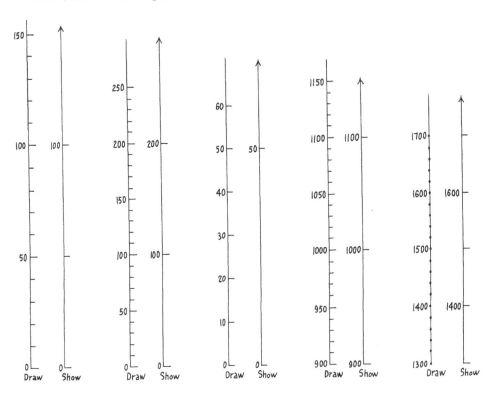

**Clearly, separating plotting from looking can be a great help. Three cheers for tracing paper or transparent plastic.**

### review questions

What is the minimum for well-ruled graph paper? What is optional? What choices of scale units are reasonable? What should we do if we need an unreasonable choice? Should we connect our points with lines or curves? Do graph-paper grids help us to plot points? To see what our points look like? How many ticks and numbers for looking? For plotting? Where should we put the scales for plotting? How do we use tracing paper? What are some conveniently available kinds of graph paper? Who should use graph paper with heavy 4's and 8's? With heavy 6's and 12's? What shapes of plot are desirable? Does this book follow the rule? How can we plot without graph paper?

## 5B. Looking at subtraction

Undoubtedly, the form of graphical representation we have been most exposed to in school is one involving two variables called $x$ and $y$, in which $y$ is said to **depend on** $x$.

**In data analysis, a plot of $y$ against $x$ may help us when we know nothing about the logical connection from $x$ to $y$--even when we do not know whether or not there is one--even when we know that such a connection is impossible.**

Before we can make full use of such plots, we need to understand--in terms both of doing and of feeling--certain things about such a plot, including:

◇ how to subtract one "curve" from another.

◇ how to find a numerical formula for a straight line drawn on graph paper.

◇ what effect subtracting different--two or more--straight lines from the same data points has.

◇ how to try to re-express either $y$ or $x$, or both, so as to make the data appear more nearly straight.

◇ why graph paper--on which we plot easily--expresses the essence of how points represent numbers much better than the kind of picture we will usually want to look at--which has axes with ticks and numbers.

When we have $x$ alone, subtraction is simply and easily represented by sliding arrows along. Subtracting 3 from 5 is a matter of drawing an arrow starting at +3 and extending to +5, and then sliding this arrow until it starts at 0. Its new endpoint, +2, is the result of subtraction. Exhibit **2** shows this example and three others that involve various combinations of minus signs.

In dealing with data we usually deal with subtraction of $y$'s rather than of $x$'s. For a hypothetical ABC Corporation, in 1960, sales were 44 million and expenses were 32 million. We can easily subtract expenses from sales to find profit before taxes of 12 million, as shown in exhibit **3**. Again, we are sliding the arrow to put its base at zero. This time we have to be careful to keep it at 1960.

In this exhibit and the next--for clarity of what we are doing, rather than for clarity of result--we have used two time scales placed side by side. Ordinarily we would use one time scale and slide each arrow along a single vertical line--as for the broken arrows on the left sides of these exhibits.

Exhibit **4** retains the pattern and shows sales, expenses, and profits for 12 consecutive years. To avoid confusing detail, the arrows and their sliding are shown for only three years, 1951, 1957, and 1960.

exhibit **2** of chapter 5: illustrative

**Four examples of subtraction**

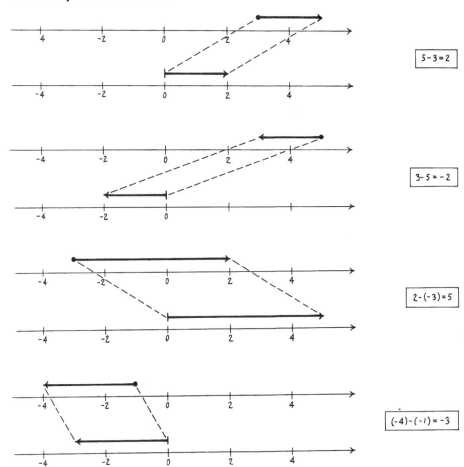

exhibit **3** of chapter 5: ABC Corp.

**The ABC Corporation in 1960**

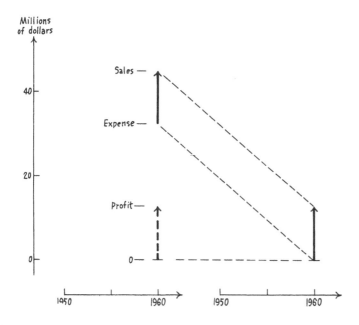

exhibit **4** of chapter 5: ABC Corp.

**Twelve years of the ABC Corporation**

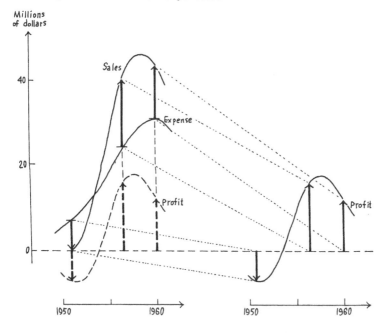

The case where we will be most concerned with graphical subtraction is the case where we solve the basic relation

$$\text{data} = \text{incomplete description PLUS residual}$$
$$= \text{fit PLUS residual}$$

for the residual, finding

$$\text{residual} = \text{data MINUS incomplete description}.$$

**untilting**

Exhibit 5 shows an example of this where the incomplete description is a straight line. Here we have slid the vertical arrows down (or up) the vertical on which they lie. Again we have shown only three arrows of the many possible.

**exhibit 5 of chapter 5: illustrative**

**Subtraction of an incomplete description from data to form residuals**

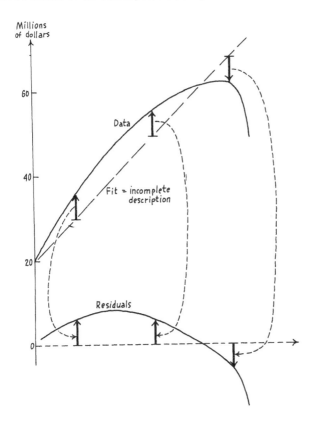

It is natural to think--and speak--of such a subtraction of a straight-line partial description as an *untilting*. Natural and useful. Yet such use of words seems likely to distort our thinking a little if we are not wary. Untilting suggests, at least to some of us, a rigid motion in which the final curve comes from the initial curve by a rigid motion--a rotation about some point. This is quite wrong.

There are many ways to see this. An easy one is to look at the length of the curve between the two crossings of the straight line. This distance is obviously greater for "data" than for "residuals". Thus we do not have rigid motion.

A convenient picture involves a deck of cards, on whose edges--the edges on one side--we have drawn both "data" and "partial description". Now let us:

⋄ clamp the deck together and saw off its bottom edge on a slant--a slant that is parallel to, because it is a constant vertical distance from, the partial description.

⋄ unclamp the deck and strike it on a table to line up its new bottom edges horizontally.

⋄ clamp the deck again.

The marks for "partial description" will now lie along a horizontal line, since they are a constant distance from the new bottom edges. If we call this upper line zero, the marks for "data" will now show us the "residuals".

The sliding of the cards with respect to one another exactly corresponds to sliding the various vertical arrows with respect to one another. This is a proper mechanical picture of graphical subtraction of one y from another. It works--rigid motion does not.

Of course, all this works for curved fits as well as for straight ones. (At least if we can make a curved cut in our deck of cards.)

### review questions

Why would we want to subtract one curve from another? A curve from some point? How does subtracting a straight line behave? Does it have anything to do with rotation? What is a mechanical model for subtracting a straight line?

### 5C. Subtracting straight lines

If we have taken our data, plotted it, and drawn a straight line through it, and now we wish to use the straight line as an incomplete description, our next task is to subtract the line from the data. Sometimes, as in exhibit 4, we can do the subtraction graphically. Often, however, this is more work than we like.

### finding the line

To do the subtraction by arithmetic, we have to turn the straight line into numbers. The easy way to do this is to choose two points on the line--let us call them $(x_1, y_1)$ and $(x_2, y_2)$--read off their coordinates, and say that an equation for the line is

$$y = y_1 + b(x - x_1)$$

where the slope, $b$, of the line is given by

$$b = \frac{y_2 - y_1}{x_2 - x_1}$$

Clearly, where $x = x_1$, we have $x - x_1 = 0$ and the equation gives $y = y_1$. When $x = x_2$, the second term on the righthand side of the equation is

$$\frac{y_2 - y_1}{x_2 - x_1}(x_2 - x_1) = y_2 - y_1$$

and we have

$$y = y_1 + (y_2 - y_1) = y_2$$

as we should.

To make the arithmetic of using such an equation easy, we should choose $x_1$ so that the values of $x - x_1$ are as simple as possible. To make finding the equation easy, we should choose $x_2$ so that $x_2 - x_1$ is a simple number. (Of course, we have to balance the advantages of simplicity against the advantages of getting well out toward the end.) There need never be great difficulty in turning a straight line into an equation.

### an example

Exhibit **6** shows a plot of the population of England and Wales at every decade from 1801 to 1931. An eye-fitted line has been drawn in and the necessary simple arithmetic--shown at the lower right--performed.

In carrying out the eye-fitting of a line and its conversion into numbers it is important:

⋄ to eye-fit the line on a picture without excessive background--use the tracing paper (or transparent plastic) version WITHOUT graph paper underneath.

⋄ to find two points through which to pass the line, put the transparent version back on the graph paper, and locate the points with its aid.

It takes two different looks at a point-swarm to do all we want do.

Exhibit **7** shows the calculation of the residuals, not only from the line of exhibit 6 (line 1) but also from another line (line 2). (The rest of this exhibit will be discussed in a moment.)

exhibit **6** of chapter 5: England and Wales

**Population in millions at successive censuses (1801–1931)**

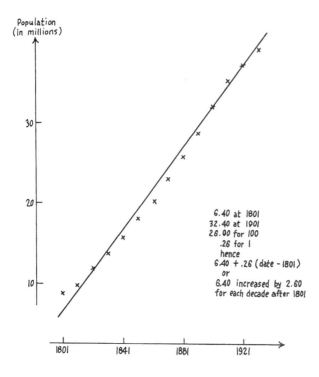

```
6.40 at 1801
32.40 at 1901
26.00 for 100
 .26 for 1
hence
6.40 + .26 (date - 1801)
or
6.40 increased by 2.60
for each decade after 1801
```

When, as in this example, the x-values step along in steps of constant size, one easy way to find the y-values is to begin at the low end and add the constant difference. For line 1, this means beginning with 6.40 and adding 2.60 repeatedly.

When the steps in x are not all the same (with the possible exception of short gaps), we have to do a little more arithmetic.

### subtracting different straight lines

If we look at data, and draw what looks like a reasonable straight line through it, we are not likely to draw exactly the line that will serve us best. What are the consequences of a somewhat unsatisfactory line?

◇ what will it do to the residuals?
◇ how hard will it be to move to a better choice?

An easy way to approach both these questions is to ask what happens if we subtract first one straight line from y and then a second straight line from the

exhibit 7/5: Plots of relationship

residuals thus formed. Algebra is easier than geometry here. If we begin by subtracting $a + bx$ we will form

$$y - (a + bx)$$

as our residual, where $(x, y)$ is the data point and $a + bx$ is some fit or other. We have just seen examples of this (in exhibit 7). If, from the residuals thus formed, we subtract $A + Bx$, we will form

$$[y - (a + bx)] - (A + Bx) = y - [(a + A) + (b + B)x].$$

exhibit 7 of chapter 5: England and Wales

**The population of England and Wales, with residuals from various lines (populations in millions)**

A) DATA and CALCULATIONS

| year | pop'n | line 1* fit | line 1* resid | line 2* fit | line 2* resid | | supplementary line* fit† | supplementary line* resid‡ |
|---|---|---|---|---|---|---|---|---|
| 1801 | 8.89 | 6.40 | 2.49 | 6. | 2.89 | s | 1.73 | 1.16 |
| 11 | 10.16 | 9.00 | 1.16 | 8.5 | 1.66 | t o p | 1.64 | .02 |
| 21 | 12.00 | 11.60 | .40 | 11. | 1.00 | h | 1.56 | −.56 |
| 31 | 13.90 | 14.20 | −.30 | 13.5 | .40 | e r | 1.48 | −1.08 |
| 41 | 15.91 | 16.80 | −.89 | 16. | −.09 | e | 1.39 | −1.48 |
| 1851 | 17.93 | 19.40 | −1.47 | 18.5 | −.57 | u n | 1.30 | −1.87 |
| 61 | 20.07 | 22.00 | −1.93 | 21. | −.93 | t i | 1.22 | −2.15 |
| 71 | 22.71 | 24.60 | −1.89 | 23.5 | −.79 | l | 1.14 | −1.93 |
| 81 | 25.97 | 27.20 | −1.23 | 26. | −.03 | a f | 1.05 | −1.08 |
| 91 | 29.00 | 29.80 | −.80 | 28.5 | .50 | t e | .96 | −.46 |
| 1901 | 32.53 | 32.40 | .13 | 31. | 1.53 | r | .88 | .65 |
| 11 | 36.07 | 35.00 | 1.07 | 33.5 | 2.57 | e x | .80 | 1.77 |
| 21 | 37.89 | 37.60 | .29 | 36. | 1.89 | h | .71 | 1.18 |
| 31 | 39.95 | 40.20 | −.25 | 38.5 | 1.45 | 8 | .62 | .83 |

* Line 1: 6.40 + .26 (date − 1801); Line 2: 6 + .25 (date − 1801).
   Supplementary line: 1.73 − .0085 (date − 1801).
† One more decimal kept in additions.
‡ These residuals are from the fit of the supplementary line to the residuals from line 2.

The result of two subtractions--of $a + bx$ and $A + Bx$--is always the same as the result of a single subtraction, of

$$(a + A) + (b + B)x$$

--of a subtraction of the

sum of the two lines.

Those who wish can show the same fact geometrically.

This fact about subtracting lines goes far toward answering our two questions:

◇ if we subtract one line and, on looking at the residuals, find them still tilted, we are free to draw a line among these residuals and subtract it further, thus finding new residuals. The new residuals correspond to subtracting a single line--the sum of the two actually subtracted. We need not go back and start again. This is particularly handy when the residuals are much smaller numbers than are the data.

◇ if we subtract an unsatisfactory line, and discover that we have done this by looking at the residuals, we could always correct this by a further subtraction. Accordingly, our first residuals will differ from the better residuals by being somewhat tilted. Since slight tilts do little to hide what we are looking at the residuals for--evidence of further structure or of unusual values--it will rarely be necessary to do the second subtraction (unless we want to publish the residuals). We can see what we need to see in the slightly tilted residuals. And, if we want to find the equation of the better line, we can draw a correction line and sum the expressions of the original and correction lines.

**Subtracting lines is simple and convenient in many ways.**

### back to the example

Exhibit **8** shows the residuals from line 2 of exhibit 7, plotting them against date. One eye-fitted line is shown. (Clearly, there could be considerable debate about which line to fit to this sequence of points.) This is naturally called a

supplementary line

since it is fitted to the residuals from a first fit. The result is

$$1.73 - .0085(\text{year} - 1801).$$

Since the residuals fitted came from the line

$$6 + .25(\text{year} - 1801)$$

the 2nd residuals--visible in exhibit 8 as deviations from the line, given numerically in the righthand column of exhibit 7--are residuals from the sum of these two lines, namely

$$(1.73 + 6) + (-.0085 + .25)(\text{year} - 1801)$$

or, simplified

$$7.73 + .2415(\text{year} - 1801).$$

Taking out first a preliminary line and then a supplementary one often, as in this example, helps us keep our (hand-done) arithmetic simpler. If we use "easy numbers" in the preliminary fit, we can often do the two fittings for less than the price of one. In addition--in fact a more important consideration--we get a look at a picture of some residuals, something that can have a variety of advantages, and often does.

**exhibit 8 of chapter 5: England and Wales**

**The residuals--from line 2 of exhibit 7--plotted against date and eye-fitted with a line.**

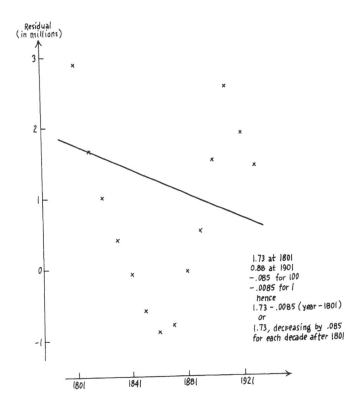

### review questions

How does one fit a straight line to two points? How does this help us in eye-fitting a line? What did we choose as an example? What happens if we subtract two lines, one after the other, from either some points or a curve? Can this make our arithmetic simpler? Why?

### 5D. Plotting the population of the U.S.A.

Many people would think that plotting $y$ against $x$ for simple data is something requiring little thought. If one only wishes to learn but little, it is true that a very little thought is enough, but if we want to learn more, we must think more.

A convenient example of how thinking more lets us learn more is given by the population of the U.S.A. as counted by the census every ten years from 1800 to 1950. Exhibit **9** shows the "little thought" version, in which millions of people are plotted against date. What can we see in this graph?

exhibit **9** of chapter 5: U.S.A. population

**Population of the U.S.A. (in millions; linear scale)**

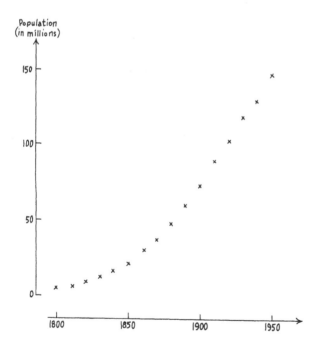

For something like the first half of the period the curve is hollow upward, so that over this period the population was growing at a steadily increasing rate. In fact, it might have been growing at something like a constant percentage each decade. For something like the second half of the period, the growth seems to approximate a straight line. Beyond this, the value for 1940 seems to be somewhat low.

All this is helpful. If we had never before looked at the population of the United States as a function of time, we would have rightly felt that we had learned quite a lot from exhibit 9. But once we have come this far, must we stop? Let us use what we have so far learned to help us look more deeply into the growth of the U.S.A.'s population.

In the present instance this is easy to do. What do we have as bases for further steps? Two things come in sight in exhibit 9:

⋄ the early years were years of accelerated growth, possibly at a constant percent per year.

⋄ in the later years, the population grew by about the same number of people each decade.

We can check up on these appearances and, more importantly, try to use them to go further.

To check up on constant-percent growth, the easy thing is to look up the logs of the population sizes, and then plot them against date. (If we didn't expect to go further, we could just plot the raw values on semilogarithmic graph paper instead.) Exhibit **10** shows the result of making such a plot. The earlier part of the plot now looks quite straight, even if we put our eye close to the paper so that we can look right along this hypothetical line. This looks very much as if constant percentage growth per decade is a good description of U.S.A. population growth in the early 1800's. Let us keep this appearance in mind, and plan to come back to it.

### the later decades

Before going further with the early decades, however, we shall turn to the apparent linearity of population growth seen in the original graph, exhibit 9, for the later decades. Exhibit **11** shows the result of drawing in a comparison line. To find an equation for this line, note that in 1870 the line has a height of about 35--35 million people--and in 1950 it has a height of somewhat less than 150, say 147. the line through (1870, 35) and (1950, 147) has slope

$$\frac{147 - 35}{1950 - 1870} = \frac{112}{80} = 1.4$$

and thus the equation is

$$y = 35 + 1.4(x - 1870)$$

This calculation is made by simple steps in the form of writing down words and numbers in the lower right of the exhibit. We recommend such a form (in general with "at" a value of x rather than "in" a date) whenever we eye-fit a line.

This line follows the data quite well, leaving us with a confirmed feeling about both approximate straightness and the dip in 1940. Need we stop here? Surely not. The straight line is an incomplete description of how the data behaves in later years. One of the great arts of data analysis consists of subtracting out incomplete descriptions and examining the residuals that are left. Let us do just this.

**coming to details**

Exhibit 12 shows the residuals from the line of exhibit 11, for the period from 1800 to 1950. (For example, at 1880 the census population is 50.2 millions and the fit is 35 + 1.4(10) = 49, so that the residual is +1.2 millions.) The detailed behavior of the population in the later half of this period is now fairly well revealed. The earlier half of the period is, however, telling us little. (Especially since the comparison line

$$35 + 1.4(\text{date} - 1870)$$

gives rather large residuals before 1840, making the comparison in earlier years unlikely to be helpful.) If we gave up looking at the early half, and

exhibit **10** of chapter 5: U.S.A. population

**Population of the U.S.A. (in millions; logarithmic scale)**

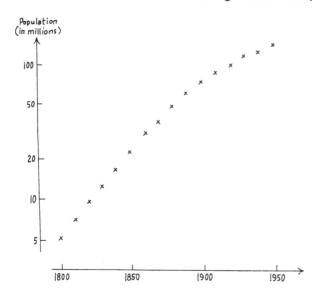

exhibit **11** of chapter 5: U.S.A. population

**Population of the U.S.A. (linear scale with comparison line)**

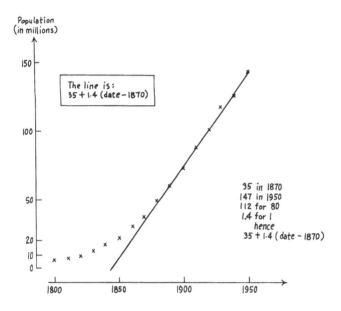

exhibit **12** of chapter 5: U.S.A. population

**Population of the U.S.A. (residuals = deviations from the specified straight line)**

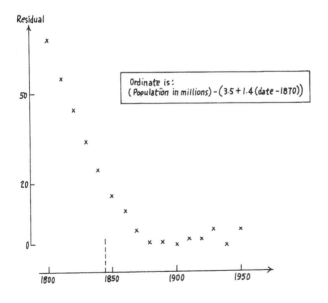

focused on the later half, we could put the data under a very much more powerful microscope. Why not do just this?

Exhibit **13** shows the same values as the righthand side of exhibit 12 but at 15 times the vertical scale. We can now see that the 1940 population was about five million less than would fit smoothly into the adjacent values. (Why do you think this happened?) And more, for we can now see that 1920 was a couple of millions low also (unless 1930 is thought to be unduly high).

Our magnifying glass is now working at full capacity, at least until we identify some further partial description and arrange to subtract it out also. To learn more about the later years of U.S.A. population growth, we would need either to get year-by-year estimates or to study the mechanisms that are involved.

### the earlier details

So much for the later years--what of the earlier ones? We left unfinished business when we said that the lefthand side of exhibit 10 seemed quite straight. We can attack this logarithmic straightness in the early years just as we attacked the linear straightness in the later ones.

exhibit **13** of chapter 5: U.S.A. population

**Later population of the U.S.A. (expanded deviations from the specified straight line)**

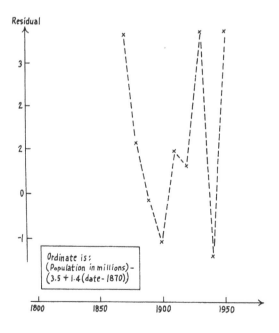

Ordinate is:
(Population in millions) −
$(3.5 + 1.4(\text{date} - 1870))$

Exhibit **14** shows the result of drawing a straight line on exhibit 10. It is encouraging enough for us to hasten to the residuals, shown in exhibit **15**. (In 1880, a population of 50.2 millions gives a log of 7.70, while the fit is 6.75 + .012(80) = 7.71, so that the residual is −0.01. More decimals were used in calculating the points for exhibit 15.) As we ought to have expected, the residuals prove useful in the early years but of very dubious value in the later ones. We can again afford to use the magnifying glass on the relatively flat section, as is done in exhibit **16**.

Exhibit 16, once we realize that ±0.01 in log is about ±2.3% in size, gives us a quite delicate view of U.S.A. population growth during the nineteenth century. After 1860, population growth was not as fast as before. Moreover, 1800 appears to have been additionally depressed by 3 or 4 per cent. Why? Again, we have gone as far with our microscope as seems reasonable without further inputs.

Experts believe many of the detailed fluctuations now so clear to us are due to variations in completeness of the census, rather than to changes in population growth. Clearly, the data cannot contradict the experts. The fluctuations are in the numbers, whatever their source. It is worthwhile to find them, whether they tell us about the growth of U.S. population or the deficiencies of U.S. censuses.

exhibit **14** of chapter 5: U.S.A. population

**Population of the U.S.A. (logarithmic scale with comparison line)**

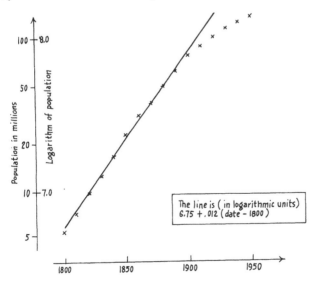

The line is (in logarithmic units)
6.75 + .012 (date − 1800)

exhibit **15** of chapter 5: U.S.A. population

**Population of the U.S.A. (deviations from logarithmic straight line)**

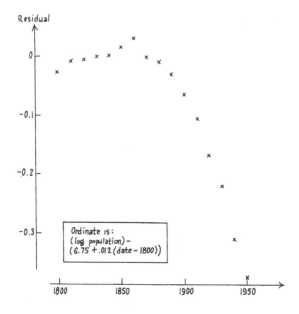

Ordinate is:
(log population) −
(6.75 + .012 (date − 1800))

exhibit **16** of chapter 5: U.S.A. population

**Earlier population of the U.S.A. (expanded deviations from given logarithmic line)**

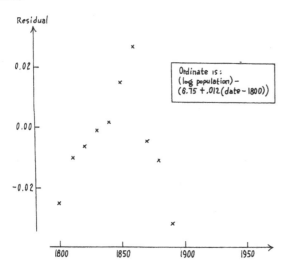

Ordinate is:
(log population) −
(6.75 + .012 (date − 1800))

If we had to choose a set of pictures to summarize U.S.A. population growth as completely as we now can, we would probably choose to show all four of the following:

⋄ exhibits 14 and 11, to show the general patterns of growth.

⋄ exhibits 16 and 13, to show local behavior.

Together, these four would be responsive to the request: Make useful plots of U.S. population against date. (If we were population specialists, we would know about logistic functions and be able to fit a single incomplete description all the way from 1800 to 1950. This would simplify the plotting of a single graph of residuals, and would probably allow us to summarize the situation in two plots, one showing the fit and the other the residuals.)

What are the lessons to be learned from this example? Not merely that thought can help us see deeper. We have seen specific examples of very general principles, including these:

⋄ **choosing scales to make behavior roughly linear always allows us to see local or idiosyncratic behavior much more clearly.**

⋄ **subtracting incomplete descriptions to make behavior roughly flat always allows us to expand the vertical scale and look harder at almost any kind of remaining behavior.**

---

**Whatever the data, we can try to gain by straightening or by flattening. When we succeed in doing one or both, we almost always see more clearly what is going on.**

---

### review questions

Can we plot without thinking? How much will we learn? What can we see from exhibit 9? How can we use what we see? What does exhibit 11 tell us to do next (go forward as far as you can)? Did we do it? Where? What does exhibit 14 tell us to do next (go forward as far as you can)? Did we do it? Where? Which pictures would we choose to "tell all" about the U.S. population? Why? What are two important lessons to be learned from this example?

### 5E. Plotting the ratio of births to deaths

The *County and City Data Book* of the U.S. Bureau of the Census contains much varied information. In particular, the 1962 edition gives for each state the number of live births for 1959, the number of deaths for 1959, and the density of population for 1960. One who believed in the "wide open spaces" might feel that the ratio of births to deaths would appear to be influenced by the density of population, at least if the South and the Atlantic coast states were set aside. A plot of the ratio of births to deaths against

exhibit **17** of chapter 5: births and deaths

**Births/deaths and population density**

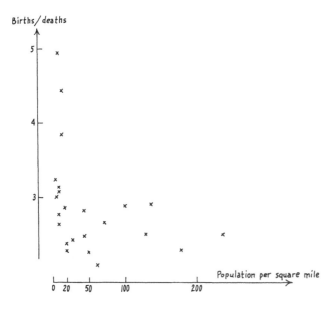

population density for the remaining states appears as in exhibit **17**. About all we can say from this plot is that the point spatter seems crudely L-shaped. The use of a linear scale for population per square mile has squeezed so many of the states up against the vertical axis that we can't be sure what is going on. If we are to see what, if anything, is going on, we must adjust the left-to-right scale so that the states are less jammed together.

Exhibit **18** shows the result of using a log scale for population density. We now see that the three states with unusually high ratios of births to deaths have low, but not very low densities, and do not appear to be typical of very low-density states. Looking only at the other states, there may be a faint tendency for a higher ratio of births to deaths to go with lower population density. (By setting aside the three states, we could double the vertical scale for the rest, putting this tendency under a slightly stronger microscope. Doing this teaches us almost nothing new, as the reader may verify.)

### another try

If we are to give up on population density, what next? The three unusual states (and their birth/death ratios) were: New Mexico (4.95), Utah (4.46), and Arizona (3.83). Looking again in the *County and City Data Book*, we discover

exhibit **18** of chapter 5: births and deaths

**Births/deaths and population density by states (density on logarithmic scale)**

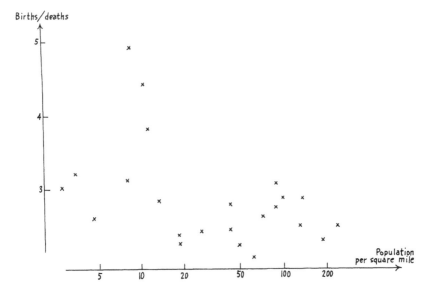

exhibit **19** of chapter 5: births and death

**Births/deaths and median age by state**

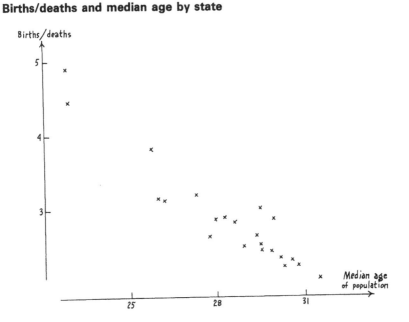

exhibit **20** of chapter 5: births and deaths

**Births/deaths (logarithmic scale) and median age by state**

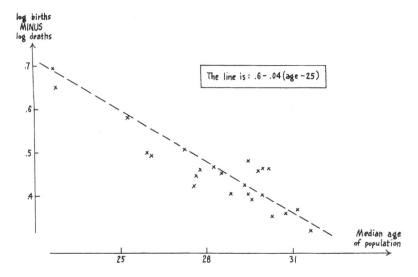

exhibit **21** of chapter 5: births and deaths

**Residuals of log $\frac{\text{births}}{\text{deaths}}$ against median age by state**

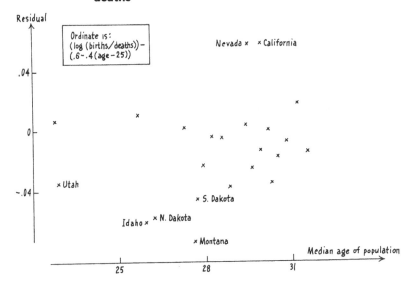

In this example, we have again seen the same main points as in the previous one:

◇ **changing scales to make dependences roughly linear usually helps.**
◇ **flattening by subtraction makes it much easier to see what is going on at more subtle levels.**

The fact that our dependences were approximate rather than exact did not alter these main points at all.

Approximate dependence did, however, bring in one new aspect, as we saw in exhibits 17 and 18:

◇ **the usefulness of changing scales to reduce confusion caused by crowding.**

In both examples, it is clear that we never expect the data to be ON the line--only that it might, if we are lucky, be NEAR the line. Once we put the first example under the microscope, what we saw is not intrinsically different from the other example. If we had contracted out the 1850 census to two

exhibit **22** of chapter 5: births and deaths

**The residuals of exhibit 17 mapped**

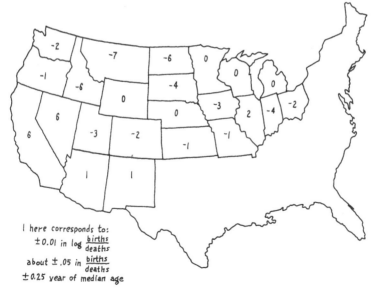

I here corresponds to:
±0.01 in log $\frac{births}{deaths}$
about ±.05 in $\frac{births}{deaths}$
±0.25 year of median age

different contractors, we would have gotten two different numbers for the population of the U.S.A. The potential values of the U.S.A. population--those we might reasonably have found--do not lie ON a curve; they merely lie QUITE NEAR one. We have happened not to buy more than one census value at a time. This makes the first example LOOK a little different, but will not keep us, in the next chapter, from interchanging the axes used for date and population. Once we face the uncertainties of "what the numbers might have been," almost all data is at best "just NEAR a line or curve".

### review questions

Where did we get the data for the example of this section? What did our first try teach us? Why did it pay to use a log scale for population density? What did we try next? Why? How well did it work? Why did we try a log scale for births/deaths? What two exhibits combine to tell the story (no map yet)? Why did we go to a map? How well did it work?

### 5F. Untilting defines "tilt"

We are now well aware of how much we can gain by flattening our picture, because this lets us expand its vertical scale. It is well to have things straight before flattening them; we can then "blow up" the picture even more, but flattening of even unstraightened pictures can help too.

We want a procedure that flattens the data out, whether or not it is straight. Here "flattening" must refer to the general run of the data, and not to its detailed behavior. Something that looks like a plausible set of residuals from a straight line is "flattened", though it may not be flat.

Exhibit **23** shows a rather extreme example of curvature, and exhibit **24** the results of "flattening" exhibit 23. Clearly, exhibit 24 is far from being flat, so "flattening" is a poor term. Equally, it is poor use of words to talk about "the slope" of exhibit 23, since there is a very small slope toward the left of the plot and a very large slope to the right.

Tilt is a short word, and conveys an appropriate feeling. We shall say that exhibit 24 is

untilted,

and that the slope of the line whose subtraction converts exhibit 23 into exhibit 24--thus clearing away an important part of what is going on so that we can see better what remains--is the

tilt

of exhibit 23.

exhibit **23** of chapter 5: illustrative

**A tilted set of data**

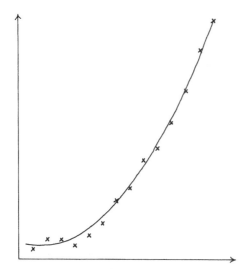

exhibit **24** of chapter 5: illustrative

**The data of exhibit 23 untilted**

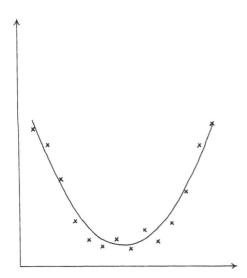

Thus, we take the question "How much is y tilted against $x$?" to mean: "What value of $b$ leaves $y - bx$ apparently untilted?" Equally, we shall take the $b$ thus found as our assessment of tilt, and the values of $y - bx$ as what we have been able to do about freeing y of tilt against x.

There can be a variety of different--often only slightly different, usually not very much different--definitions of "apparently untilted". For each such definition there is a--comparably slightly different--value of $b$ and thus both an assessed value for the tilt and a set of apparently untilted values. These differ--from definition to definition--by only comparably slight amounts.

The existence of such alternatives--and the need to choose between them--usually bothers us not at all.

In the great majority of cases, our concern with tilt is like the woodsman's concern with bushes and fallen trees on a trail he plans to use:

◇ we may be concerned to know that there appears to be some tilt (though we may have known that there would be a tilt in this particular direction long before we collected the data).

◇ we may even want to know how much tilt there seems to be.

◇ **we are almost certain,** whatever else, **to want to clear the tilt out of our way.**

We have already, in specific instances, cleared our way by using an eye-judged tilt to flatten our graphs. We frequently need to do something similar in fixing numbers for future analysis. Especially because we are going to use any tilt we assess in such a way, we will meet our needs quite well enough if we do a reasonably good job of assessing our tilt. (We need not worry as to in what way, if any, our assessment might be **best possible**--or even whether anyone can define "best possible" sensibly.)

### review questions

What is a tilt? Why use a special word? What is it to be untilted? How do we define tilt? Are there many definitions, or few, or only one definition of "untilted"? Do we need a "best possible" tilt? Can you define a "best possible" tilt?

### 5H. How far have we come?

In this chapter we have met plots of y against x. Perhaps we have even come to know them a little.

The state of our progress is not measured by the specific techniques we have seen or understood--though that kind of progress is essential. Rather, our progress is measured by our acceptance of such propositions as these:

1. Graphs are friendly.

2. Arithmetic often exists to make graphs possible.

3. Graphs force us to note the unexpected; nothing could be more important.

4. Different graphs show us quite different aspects of the same data.

5. There is no more reason to expect one graph to "tell all" than to expect one number to do the same.

6. "Plotting $y$ against $x$" involves significant choices--how we express one or both variables can be crucial.

7. The first step in penetrating plotting is to straighten out the dependence or point scatter as much as reasonable.

8. Plotting $y^2$, $\sqrt{y}$, log $y$, $-1/y$ or the like instead of $y$ is one plausible step to take in search of straightness.

9. Plotting $x^2$, $\sqrt{x}$, log $x$, $-1/x$ or the like instead of $x$ is another.

10. Once the plot is straightened, we can usually gain much by flattening it, usually by plotting residuals (with regard to the partial description implied by the straight line we may not have quite drawn in yet).

11. When plotting point scatters, we may need to be careful about how we express $x$ and $y$ in order to avoid concealment by crowding.

In particular, we have learned--or been reminded--how to:

⋄ understand about subtracting one curve from another.
⋄ use two points to find an equation for a straight line.

Our two examples differed in one way. The years of census are fixed by law, and at each a population is measured. It is rather easy to try to think of the points as having been selected from a curve with one population at each possible date. Births/deaths, on the other hand, whether compared with population density or with median age of population, provides a much more symmetric situation. The boundaries of states are fixed by law, and for each state two things happen, an $x$ and a $y$. There is no possibility of assuming that all the data are ON a curve; the most we can hope is that the data are NEAR a line or curve.

## 5P. Additional problems

See exhibits **25, 26, 27, 28, 29, 30, 31, 32, 33**.

exhibit 25(A–B)/5: Plots of relationship

exhibit 25 of chapter 5: data and problems

**Data for the 15 smallest counties of 3 states (from 1962 *County and City Data Book*)**

A) SOUTH CAROLINA

| (*) | Name | Total | Population (1960) % less than 5 yrs school† | % at least 65 | per sq. mile | Families (1959) % less than $3000 income | Local Gov't (1957) % budget for education | Total budget ($1000's) |
|---|---|---|---|---|---|---|---|---|
| 2424 | McCormick | 8,629 | 28.2 | 8.8 | 23 | 59.8 | 62.0 | 566 |
| 2130 | Allendale | 11,362 | 34.1 | 7.9 | 27 | 60.1 | 63.1 | 1082 |
| 2051 | Jasper | 12,237 | 37.1 | 7.5 | 19 | 60.2 | 79.7 | 1082 |
| 2050 | Calhoun | 12,256 | 26.1 | 7.9 | 33 | 68.2 | 76.6 | 659 |
| 1852 | Saluda | 14,554 | 17.0 | 9.7 | 33 | 50.6 | 68.9 | 656 |
| 1753 | Edgefield | 15,735 | 23.1 | 8.0 | 33 | 55.3 | 70.1 | 785 |
| 1720 | Bamberg | 16,274 | 26.6 | 8.3 | 41 | 58.5 | 63.1 | 1340 |
| 1622 | Hampton | 17,425 | 31.8 | 7.2 | 31 | 58.0 | 57.8 | 1312 |
| 1608 | Barnwell | 17,659 | 23.2 | 7.4 | 32 | 47.5 | 82.9 | 1565 |
| 1405 | Fairfield | 20,713 | 30.8 | 7.8 | 30 | 54.2 | 56.8 | 1596 |
| 1356 | Abbeville | 21,417 | 23.0 | 8.7 | 42 | 42.1 | 56.4 | 1717 |
| 1339 | Lee | 21,832 | 31.7 | 6.4 | 53 | 68.6 | 75.8 | 1671 |
| 1211 | Dorchester | 24,383 | 23.1 | 6.6 | 43 | 49.6 | 67.9 | 1684 |
| 1092 | Colleton | 27,816 | 29.3 | 7.5 | 27 | 57.9 | 46.7 | 3090 |
| 1066 | Marlboro | 28,529 | 29.6 | 6.8 | 59 | 58.3 | 63.4 | 1948 |
| (Whole state) | | ((34,262)) | (20.3) | (6.3) | (79) | (39.5) | (63.4) | -- |

State has 46 counties.

B) GEORGIA

| | | | | | | | | |
|---|---|---|---|---|---|---|---|---|
| 3077 | Echols | 1,876 | 28.0 | 8.5 | 4 | 55.6 | 59.5 | 122 |
| 3051 | Quitman | 2,432 | 34.8 | 9.3 | 14 | 70.0 | 79.5 | 205 |
| 3028 | Glascoch | 2,672 | 35.6 | 56.4 | 19 | 61.1 | 67.9 | 252 |
| 2979 | Webster | 3,247 | 34.5 | 8.7 | 17 | 71.2 | 73.3 | 225 |
| 2978 | Schley | 3,256 | 23.0 | 10.3 | 20 | 67.4 | 69.7 | 221 |
| 2970 | Taliaferro | 3,370 | 27.9 | 13.4 | 17 | 68.5 | 48.8 | 391 |
| 2950 | Dawson | 3,590 | 22.4 | 8.6 | 17 | 64.7 | 71.3 | 293 |
| 2934 | Long | 3,874 | 22.7 | 7.5 | 10 | 60.1 | 52.0 | 408 |
| 2875 | Towns | 4,538 | 11.3 | 10.2 | 27 | 63.7 | 33.4 | 724 |
| 2674 | Baker | 4,543 | 33.2 | 8.9 | 15 | 74.1 | 56.4 | 622 |
| 2873 | Clay | 4,551 | 26.8 | 10.5 | 20 | 66.6 | 59.2 | 417 |
| 2832 | Lanier | 5,097 | 30.1 | 7.5 | 31 | 57.4 | 64.5 | 686 |
| 2810 | Charlton | 5,313 | 26.5 | 6.2 | 7 | 44.0 | 49.1 | 703 |
| 2808 | Heard | 5,333 | 22.7 | 10.7 | 18 | 56.4 | 57.1 | 652 |
| 2805 | Wheeler | 5,342 | 29.4 | 9.5 | 18 | 63.7 | 54.4 | 418 |
| (Whole state) | | ((12,038)) | (17.6) | (7.4) | (68) | (35.6) | (45.0) | -- |

State has 159 counties.

exhibit **25** of chapter 5 (continued)

### C) ALABAMA

| | | | | | | | | |
|---|---|---|---|---|---|---|---|---|
| 2201 | Coosa | 10,726 | 18.4 | 10.7 | 18 | 51.5 | 39.5 | 1174 |
| 2179 | Cleburne | 10,911 | 21.5 | 9.4 | 19 | 52.4 | 48.2 | 1088 |
| 2034 | Clay | 12,400 | 14.4 | 13.0 | 21 | 54.0 | 48.3 | 1347 |
| 1945 | Bullock | 13,462 | 32.7 | 11.7 | 22 | 69.4 | 42.8 | 1727 |
| 1934 | Greene | 13,600 | 38.0 | 9.8 | 21 | 74.0 | 57.9 | 1514 |
| 1876 | Lamar | 14,271 | 14.7 | 11.0 | 24 | 51.4 | 37.5 | 2507 |
| 1870 | Bibb | 14,357 | 22.7 | 9.6 | 23 | 54.4 | 54.7 | 1405 |
| 1828 | Winston | 14,858 | 17.3 | 9.8 | 24 | 53.8 | 55.7 | 1072 |
| 1824 | Crenshaw | 14,909 | 23.9 | 11.2 | 24 | 69.5 | 51.4 | 1332 |
| 1796 | Henry | 15,286 | 27.8 | 9.0 | 27 | 63.8 | 56.0 | 1313 |
| 1794 | Washington | 15,372 | 23.3 | 7.8 | 14 | 51.7 | 54.5 | 1353 |
| 1790 | Lowndes | 15,417 | 37.2 | 9.2 | 22 | 72.1 | 55.4 | 1243 |
| 1729 | Fayette | 16,148 | 16.9 | 10.8 | 26 | 54.7 | 43.0 | 1425 |
| 1715 | Cherokee | 16,303 | 16.6 | 8.9 | 27 | 49.1 | 53.7 | 1498 |
| 1628 | Perry | 17,358 | 28.7 | 10.5 | 24 | 69.2 | 49.4 | 1689 |
| (Whole state) | | ((25,738)) | (16.3) | (8.0) | (64) | (39.1) | (45.8) | -- |

State has 67 counties.

   \* **National rank by population.**
   † **Of those 25 or over.**
   ( ) **Median.**

### P) PROBLEMS

**25a)** Panels A to C contain selected information about the 15 smallest counties in South Carolina, Alabama, and Georgia. Plot

$$y = \% \text{ with less than five years schooling}$$

against

$$x = \% \text{ with less than \$3000 income}$$

for at least two states. Continue the analysis. Comment.

### S) SOURCE
**1962** *County and City Data Book.*

that all three have young populations, for instance, as measured by median age. Thus it is natural to plot births/deaths against median age.

Exhibit **19** shows the result. Clearly, median age does a much better job of appearing to explain the ratio of births to deaths than did population density, though the apparent dependence is, of course, not perfect. If we are to go further, we will need to find and subtract out some partial description of this apparent dependence.

When we "put eye to paper" and look along the point spatter in exhibit 19, we seem to see a definite tendency to curvature (hollow above). If we could eliminate this, we could reasonably compare the individual points with a straight line. How might we approach such a simpler partial description?

One thing to try is changing the down-to-up scale. It is easy to see that using squares to fix this scale would make curvature worse, so we may as well try going in the opposite direction by using logs, a choice which is attractive because of the more symmetric way it treats births and deaths. Notice how the identities

$$\log\left(\frac{\text{births}}{\text{deaths}}\right) \equiv \log \text{births} - \log \text{deaths}$$

$$\equiv -(\log \text{deaths} - \log \text{births}) \equiv -\log\left(\frac{\text{deaths}}{\text{births}}\right)$$

exhibit this symmetry.

Exhibit **20** shows the result, complete with a convenient comparison line. The point spatter is now much more nearly straight. When we plot the corresponding residuals against median age and identify the more extreme states, we find the results shown in exhibit **21**.

### going to the map

The two states with notably high residuals are adjacent to one another on the map. The four states with middle median age and notably low residuals also touch one another. Clearly, we need to see the residuals on a map.

Exhibit **22** shows the residuals spread across a map of the United States. The roughly regular structure of these residuals is rather easily seen--adjacent states are clearly more often similar than are distant ones. (Might the somewhat surprisingly positive residual for Illinois be due to the unusual size of Chicago?) To go further here requires either:

⋄ more careful allowance for the age of the population, or
⋄ more knowledge about the mechanisms affecting birth and death rates in general. (Would state-by-state information on economic conditions help?)

Indeed, both of these are likely to be needed!

exhibit 26 of chapter 5: data and problems

**Some problems**

26a) The following data was obtained in preparing a standard curve for the determination of formaldehyde by the addition of chromatropic acid and concentrated sulphuric acid and the reading of the consequent purple color on a Beckman Model DU Spectrophotometer at 570 m$\mu$.

| Amount of $CH_2O$ Used | Optical Density |
|---|---|
| 0.1 | 0.086 |
| 0.3 | 0.269 |
| 0.5 | 0.446 |
| 0.6 | 0.538 |
| 0.7 | 0.626 |
| 0.9 | 0.782 |

Analyze graphically, using at least two graphs. Comment. (Bennett & Franklin, p. 216; from Roberts.)

26b) The relation between the amount of $\beta$-erythrodine dissolved in water and the turbidity of the solution--as read on a colorimeter--is not quite as simple. Some data gives:

| Concentration (in mg/ml) | Colorimeter reading |
|---|---|
| 40 | 69 |
| 50 | 175 |
| 60 | 272 |
| 70 | 335 |
| 80 | 390 |
| 90 | 415 |

Analyze graphically. Comment. (Bennett & Franklin, p. 217, from Woislawski.)

26c) Find two different collections of (x, y) points that interest you, and make useful plots.

**S) SOURCE**

See exhibit 27.

exhibit **27** of chapter 5: data and problem
**Carbon content of 36 clays measured directly and estimated indirectly**

**A) DATA**

| Clay # | Direct measurement | Indirect estimate |
|---|---|---|
| 1 | 1.53 | 2.46 |
| 2 | 0.87 | 1.54 |
| 3 | 0.28 | 0.70 |
| 4 | 0.27 | −0.40 |
| 5 | 3.07 | 4.82 |
| 6 | 0.25 | 0.30 |
| 7 | 0.25 | 0.64 |
| 8 | 0.29 | 0.78 |
| 9 | 0.12 | 0.12 |
| 10 | 1.50 | 2.36 |
| 11 | 1.31 | 2.14 |
| 12 | 0.31 | 0.08 |
| 13 | 0.14 | −0.01 |
| 14 | 2.98 | 4.53 |
| 15 | 6.84 | 9.94 |
| 16 | 2.15 | 3.68 |
| 17 | 1.35 | 1.84 |
| 18 | 0.40 | 0.97 |
| 19 | 4.18 | 6.14 |
| 20 | 0.22 | 0.52 |
| 21 | 0.38 | 0.40 |
| 22 | 0.24 | 0.46 |
| 23 | 1.79 | 2.80 |
| 24 | 0.58 | 2.09 |
| 25 | 6.55 | 9.68 |
| 26 | 2.54 | 4.08 |
| 27 | 1.43 | 2.80 |
| 28 | 2.74 | 3.93 |
| 29 | 6.08 | 8.22 |
| 30 | 0.75 | 0.28 |
| 31 | 0.16 | 0.35 |
| 32 | 5.06 | 7.49 |
| 33 | 0.86 | 1.41 |
| 34 | 0.16 | −0.50 |
| 35 | 11.43 | 15.80 |
| 36 | 0.19 | 0.18 |

exhibit **27** of chapter 5 (continued)

**P) PROBLEM**

27a) The amount of carbon in a clay can be measured directly by heating the clay until all the carbon compounds are burned, collecting the carbon dioxide thus formed, and measuring its amount. The amount of carbon can be estimated by combining the amounts of its constituents in a suitable standard way. The results of such measurements on 36 clays from South Devonshire, England, are given in exhibit 27. Analyze graphically. Comment.

**S) SOURCE**
C. A. Bennett and N. L. Franklin 1954, Statistical Analysis in Chemistry and the Chemical Industry. John Wiley, New York. Table 6.3 on page 218.

exhibit **28** of chapter 5: data and problems

**Percentage Democratic in 12 presidential elections for 24 Northeastern and Central States (percentage Democratic of major party vote)**

**A) DATA**

|  | 1920 | 1924 | 1928 | 1932 | 1936 | 1940 | 1944 | 1948 | 1952 | 1956 | 1960 | 1964 |
|---|---|---|---|---|---|---|---|---|---|---|---|---|
| Colorado | 37.7 | 27.8 | 34.4 | 57.0 | 61.9 | 48.7 | 46.6 | 52.7 | 39.3 | 39.5 | 45.1 | 61.6 |
| Connecticut | 34.5 | 30.9 | 45.9 | 49.4 | 57.8 | 53.6 | 52.7 | 49.2 | 44.1 | 36.3 | 53.7 | 67.9 |
| Delaware | 43.0 | 38.9 | 33.9 | 48.8 | 54.9 | 54.8 | 54.6 | 49.4 | 48.1 | 44.7 | 50.8 | 61.1 |
| Illinois | 27.3 | 28.4 | 42.6 | 56.8 | 59.2 | 51.2 | 51.7 | 50.4 | 45.0 | 40.4 | 50.1 | 59.5 |
| Indiana | 42.3 | 41.2 | 39.9 | 56.0 | 57.5 | 49.3 | 47.1 | 49.6 | 41.4 | 39.9 | 44.8 | 56.2 |
| Iowa | 26.4 | 23.0 | 37.8 | 59.1 | 56.0 | 47.8 | 47.7 | 51.4 | 35.8 | 40.8 | 43.3 | 62.0 |
| Kansas | 33.4 | 27.7 | 27.3 | 54.8 | 53.9 | 42.7 | 39.4 | 45.4 | 30.7 | 34.3 | 39.3 | 54.6 |
| Maine | 30.2 | 23.3 | 31.1 | 43.6 | 42.8 | 48.8 | 47.5 | 42.7 | 33.8 | 29.1 | 43.0 | 68.8 |
| Maryland | 43.3 | 47.7 | 42.6 | 63.1 | 62.7 | 58.8 | 51.9 | 49.3 | 44.2 | 40.0 | 53.6 | 65.5 |
| Massachusetts | 28.9 | 28.5 | 50.5 | 52.1 | 55.1 | 53.4 | 52.9 | 55.9 | 45.6 | 40.5 | 60.4 | 76.5 |
| Michigan | 23.4 | 14.8 | 29.1 | 54.1 | 59.2 | 49.8 | 50.5 | 49.1 | 44.2 | 44.2 | 51.0 | 66.8 |
| Minnesota | 21.6 | 11.7 | 41.4 | 62.3 | 66.6 | 51.9 | 52.8 | 58.9 | 44.4 | 46.2 | 50.7 | 63.9 |
| Nebraska | 32.6 | 38.5 | 36.4 | 64.1 | 58.4 | 42.8 | 41.4 | 45.8 | 30.8 | 34.5 | 37.9 | 52.6 |
| New Hampshire | 39.7 | 36.7 | 41.2 | 49.3 | 50.9 | 53.2 | 52.1 | 47.1 | 39.1 | 33.9 | 46.6 | 63.9 |
| New Jersey | 29.6 | 30.6 | 40.0 | 51.0 | 60.1 | 51.8 | 50.7 | 47.7 | 42.5 | 34.6 | 50.4 | 66.0 |

➡

exhibit **28** of chapter 5 (continued)

| | | | | | | | | | | | | |
|---|---|---|---|---|---|---|---|---|---|---|---|---|
| New York | 29.5 | 34.3 | 48.8 | 56.7 | 60.2 | 51.8 | 52.5 | 49.5 | 44.0 | 38.7 | 52.6 | 68.7 |
| North Dakota | 18.9 | 12.7 | 44.8 | 71.3 | 69.2 | 44.7 | 45.8 | 45.4 | 28.6 | 38.2 | 44.5 | 58.1 |
| Ohio | 39.8 | 28.9 | 34.7 | 51.5 | 60.8 | 52.2 | 49.8 | 50.1 | 43.2 | 38.9 | 46.7 | 62.9 |
| Pennsylvania | 29.2 | 22.6 | 34.2 | 47.1 | 58.2 | 53.5 | 51.4 | 48.0 | 47.0 | 43.4 | 51.2 | 65.2 |
| Rhode Island | 33.9 | 37.9 | 50.3 | 56.0 | 56.8 | 56.8 | 58.7 | 58.2 | 49.1 | 41.7 | 63.6 | 80.9 |
| South Dakota | 24.5 | 21.2 | 39.4 | 64.9 | 56.0 | 42.6 | 41.7 | 47.6 | 30.7 | 41.6 | 41.8 | 55.6 |
| Vermont | 23.5 | 16.7 | 33.0 | 41.6 | 43.4 | 45.1 | 42.9 | 37.5 | 28.3 | 27.8 | 41.4 | 66.3 |
| West Virginia | 43.9 | 47.1 | 41.3 | 55.1 | 60.7 | 57.1 | 54.9 | 57.6 | 51.9 | 45.9 | 52.7 | 67.9 |
| Wisconsin | 18.5 | 17.9 | 45.3 | 67.0 | 67.8 | 50.9 | 49.1 | 52.3 | 38.8 | 38.1 | 48.1 | 62.2 |

### P) PROBLEMS

Panel A gives the % Democratic vote in each of 24 northeastern and central states for 12 presidential elections, 1920 to 1964. Plot the following:

28a) 1964 against 1956

28b) 1960 against 1920

28c) 1952 against 1932

28d) Any one against any other that you think will have a close relationship.

exhibit **29** of chapter 5: data and problems

**More problems**

29a) Determination of ethylene chlorohydrin. (264, 270) translates as: "With 26.4 milligrams of ethylene chlorohydrin present, 27.0 milligrams were found". Data-sets (6): (264, 270), (595, 594), (1173, 1183), (1777, 1780), (2355, 2370), (3578, 3576). SOURCE: K. Uhrig 1946. Determination of ethylene chlorohydrin. *Industrial and Engineering Chemistry, Analytical Edition* **18**: 369 only. Table 1 on page 369. PROBLEM: Choose a plot that is likely to be revealing by thinking hard. Explain the reasons for your choice. Make the plot.

29b) Polarographic behavior of ions containing vanadium. (94, 35) translates as: "For a concentration of vanadite ion of 0.094 millimoles per liter, the anodic diffusion constant in microamperes was 0.35 microamperes". Data-sets (8): (94, 35), (278, 98), (508, 178), (880, 309), (1548, 563), (1840, 696), (352, 1285), (505, 1813). (Last point: 5.05 millimoles/liter, 18.13 microamperes.) SOURCE: J. J. Lingane 1945, "Polarographic characteristics of vanadium in its various oxidation states," *J. Amer. Chem. Soc.* **67**: 182–188. Table I on page 186. PROBLEM: Make useful plots of diffusion current against vanadite concentration. ➡

exhibit **29** of chapter 5 (continued)

29c) Amount of desired product in a chemical reaction after different reaction times and under different circumstances. (1; 32, 54; 87, 159, 226) translates as: "In run 1, the amount of desired product in moles per liter was 0.032 after 80 minutes, 0.054 after 160 minutes, 0.087 after 320, 0.159 after 640, and 0.226 after 1280 minutes." Data-sets (16 runs under 16 different sets of conditions): (1; 32, 54; 87, 159, 226), (2; 147, 234; 343, 342, 203), (3; 48, 108; 225, 346, 420), (4; 232, 390; 556, 634, 416), (5; 37, 38; 172, 200, 239), (6; 179, 283; 405, 342, 216), (7; 86, 133; 259, 398, 508), (8; 309, 514; 722, 764, 389), (9; 74, 99; 200, 309, 249), (10; 253, 343; 391, 284, 75), (11; 133, 271; 430, 580, 494), (12; 508, 756; 842, 570, 115), (13; 96, 158; 276, 339, 230), (14; 308, 444; 467, 249, 29), (15; 228, 372; 579, 691, 539), (16; 626, 880; 895, 434, 58). SOURCE: G. E. P. Box and W. G. Hunter 1962, "A useful method for model-building," *Technometrics* 4: 301–318. Table 1 on page 304. PROBLEM: Make useful plots of concentration at 640 minutes against concentration at 160 minutes.

29d) Make useful plots for one or more other pairs of reaction times. (Data-sets in problem (29c) above)

exhibit **30** of chapter 5: data and problems

**Yet more problems**

30a) Analysis of samples for chrysanthenic acid. (0, 23) translates as: "When 0 micrograms of synthetic racemic chrysanthenic acid were added, the colorimeter scale reading was 23". Data-sets (13): (0, 23), (5, 32), (10, 40), (20, 54), (40, 86), (60, 118), (80, 146), (100, 179), (120, 212), (140, 240), (160, 272), (180, 300), (200, 330). SOURCE: A. A. Schreiber and D. B. McClellan 1954. Estimation of microquantities of pyrethroids. *Analytical Chemistry* **26**: 604–607. Table I on page 605. PROBLEM: Make useful plots of colorimeter reading against amount of chrysanthenic acid.

30b) Residual strength of 8-oz. cotton duck attacked by 4 different kinds of fungus. (3; 97, 105; 103, 101) translates as: "After 3 hours of incubation, the strengths--referred to initial strength = 100--of the sample exposed to Thielaria was 97, that exposed to Humicola was 105, for Chaetomium was 103, for Myrothecium was 101." Data-sets (24): (3; 97, 105; 103, 101), (6; 98, 106; 101, 105), (9; 95, 107; 99, 95), (12; 96, 105; 95, 95), (15; 97, 106; 90, 100), (18; 98, 102; 91, 97), (21; 97, 101; 78, 98), (24; 97, 90; 74, 93), (27; 90, 81; 71, 82), (30; 96, 78; 71, 76), (33; 89, 73; 65, 67), (36; 88, 69; 58, 64), (39; 89, 63; 53, 59), (42; 86, 59; 47, 54), (45; 82, 55; 44, 50), (48; 79, 53; 44, 42), (51; 73, 52; 42, 41), (54; 73, 41; 40, 40), (57; 73, 42; 40, 39), (60; 68, 41; 39, 35), (63; 59, 36; 38, 37), (66; 57, 37; 37, 33), (69; 57, 31; 35, 34), (72; 55, 34; 36, 31). SOURCE: E. Abrams 1950, "Microbiological deterioration of cellulose during the first 72 hours of attack," *Textile Research J.* **20**: 71–86. Table 2 on page 75. PROBLEM: Plot helpful curves for loss of strength from at least two kinds of fungus. ➔

exhibit **30** of chapter 5 (continued)

30c) Rapid analysis for caffeine. (257, 131) translates as: "For a caffeine concentration of 0.257 milligrams in 100 milliliters, the average optical density was 0.131." Data-sets (20): (257, 131), (498, 262), (506, 265), (514, 263), (747, 384), (760, 393), (770, 396), (996, 512), (1013, 518), (1027, 523), (1245, 633), (1266, 643), (1284, 650), (1494, 760), (1519, 768), (1541, 775), (1798, 903), (2054, 1040), (2311, 1160), (2568, 1290). SOURCE: N. H. Ishler, T. P. Finucaine, and E. Borker 1948, "Rapid spectrophotographic determination of caffeine," *Analytical Chemistry* **20**:1162–1166. Table 1 on page 1162. PROBLEM: Make useful plots of optical density against caffeine concentration.

exhibit **31** of chapter 5: data and problems

**Still more problems**

31a) Survival of automobiles and trucks in use by a public utility. (0h, 990) translates as: "After 1/2 year, 0.990 of all vehicles were still in service." Data-sets (8): (0h, 990), (1h, 972), (2h, 944), (3h, 895), (4h, 784), (5h, 679), (6h, 593), (7h, 497). SOURCE: S. A. Krane 1963, "Analysis of survival data by regression techniques," *Technometrics* **5**: 161–174. Table on page 168. His source: H. A. Cowles, Jr., 1957. Prediction of mortality characteristics of industrial property groups. Ph.D. Thesis, Iowa State University. PROBLEM: Make helpful plots of fraction surviving against age.

31b) Heat and entropy contents of a sodium silicate. (400, 3080, 885) translates as: "The increases from 'room temperature' (298.16°K) to an absolute temperature of 400°K were 3,080 calories per mole for the heat content of $Na_2SiO_3$ and 8.85 calories/degree/mole for its entropy content." Data-sets (17): (400, 3080, 885), (500, 6300, 1604), (600, 9650, 2214), (700, 13190, 2760), (800, 16910, 3256), (900, 20730, 3708), (1000, 24700, 4124), (1100, 28770, 4511), (1200, 32940, 4874), (1300, 37210, 5216), (1361, 39870, 5416), (1361, 52340, 6332), (1400, 54010, 6453), (1500, 58390, 6748), (1600, 62570, 7024), (1700, 66850, 7284), (1800, 71130, 7528). (The last point, at 1800°K, gives a heat content change of 71,130 calories/mole and an entropy content change of 75.28 calories/degree/mole.) SOURCE: B. F. Naylor 1945, "High-temperature heat contents of sodium metasilicate and sodium disilicate," *J. Amer. Chem. Soc.* **67**: 466–467. Table II on page 467. PROBLEM: Make helpful plots of the increase in heat content against temperature.

31c)' For Naylor's data (immediately above) make helpful plots of increase in entropy content against temperature.

exhibit **32** of chapter 5: data and problems

**And yet more**

32a) Equilibrium splitting of plutonium tribromide by water (gases at high temperature). (911, 153) translates as' "For an absolute temperature of 911°K, the observed equilibrium constant was 0.0153/atmosphere." Data-sets (11): (911, 153), (914, 156), (919, 149), (920, 163), (882, 246), (876, 282), (875, 247), (883, 243), (815, 704), (817, 502), (816, 692). SOURCE: I. Shift and N. R. Davidson 1949. Equilibrium in the vapor-phase hydrolysis of plutonium tribromide. Paper 6.24, at pages 831–840 of *The Transuranium Elements*, edited by Seaborg, Katz, and Manning. National Nuclear Energy Series IV-14B. McGraw Hill. Table 2 on page 835. PROBLEM: Make useful plots of equilibrium constant against temperature. Which three of the 11 data-sets do you think the authors rejected?

32b) Sales of Swiss bond issues since World War II. (46, 527) translates as: "In 1946, total sales of Swiss bonds--governmental and private--were 527 million francs". Data-sets (23): (46, 527), (47, 276), (48, 472), (49, 342), (50, 174), (51, 434), (52, 333), (53, 249), (54, 242), (55, 492), (56, 613), (57, 1148), (58, 827), (59, 686), (60, 890), (61, 1023), (62, 1124), (63, 2091), (64, 2503), (65, 2523), (66, 2292), (67, 2446), (68, 2648). SOURCE: *Swiss Statistical Abstract*, issued by the Swiss Credit Bank, November 1969. (Title also in French and German.) Table on page 46. PROBLEM: Make useful plots based on the data from 1950 to 1968.

32c) Comparison of two ways of measuring the water content of samples of the sea bed. (0 to 3; 76, 76) translates as: "For a sample from 0 to 3 inches below the surface of the sea level, measurement of % water by drying in an oven gave 76%, measurement by analyzing for chloride--and using the known concentration of chloride in deep sea water--gave 76%". Data-sets (14): (0 to 3; 76, 76), (3 to 6; 68, 72), (6 to 9; 69, 69), (9 to 12; 67, 67), (12 to 15; 60, 64), (15 to 18; 62, 62), (18 to 21; 60, 60), (21 to 24; 58, 59), (24 to 27; 57, 57), (27 to 30; 55, 56), (30 to 33; 55, 55), (33 to 36; 55, 55), (36 to 39; 53, 54), (39 to 42; 54, 54). SOURCE: L. J. Anderson 1948, "Conductometric titration of chloride in sea water and marine sediments," *Analytical Chemistry* **20**: 618–619. Table II on page 619. PROBLEM: Find differences between water by chloride and water by oven. Make a stem-and-leaf display of these differences. Comment on the appearance of this display. Find and plot residuals from a straight-line fit of water content against depth, separately for each method of finding water content. What do you conclude about the two methods of measuring water?

exhibit 33 of chapter 5: data and problems

**Median ages of urban and rural populations and estimated colonial population**

A) MEDIAN AGE, URBAN and RURAL POPULATIONS--at U.S. Censuses

| Year | Median age | Urban population | Rural population |
|---|---|---|---|
| 1950 | 30.4 | 88,927,464 | 61,769,897 |
| 40 | 29.5 | 74,923,702 | 57,245,573 |
| 30 | 27.1 | 68,954,823 | 53,830,223 |
| 20 | 26.1 | 54,157,973 | 51,552,647 |
| 10 | 24.9 | 41,998,932 | 49,973,334 |
| 1900 | 23.8 | 30,159,921 | 45,834,654 |
| 1890 | 22.9 | 22,106,265 | 40,841,449 |
| 80 | 21.6 | 14,129,735 | 36,026,048 |
| 70 | 20.6 | 9,902,361 | 28,656,010 |
| 60 | 20.2 | 6,216,518 | 25,226,803 |
| 50 | 19.5 | 3,543,716 | 19,648,160 |
| 40 | 17.9 | 1,845,055 | 15,224,398 |
| 30 | 17.2 | 1,127,247 | 11,738,773 |
| 20 | 16.5 | 693,255 | 8,945,198 |
| 10 | 15.9 | 525,459 | 6,714,422 |
| 1800 | 15.7 | 322,371 | 4,986,112 |
| 1790 | 15.9 | 201,655 | 3,727,559 |

Median age = A90 Median age of white males
Urban population = A195 Population of "urban territory"
Rural population = A206 Population of "rural territory"

B) ESTIMATED POPULATION OF AMERICAN COLONIES

| Year | Estimated population |
|---|---|
| 1780 | 2,780,369 |
| 70 | 2,148,076 |
| 60 | 1,593,625 |
| 50 | 1,170,760 |
| 40 | 905,563 |
| 30 | 629,445 |
| 20 | 466,185 |
| 10 | 331,711 |
| 1700 | 250,888 |
| 1690 | 216,372 |
| 80 | 151,507 |
| 70 | 111,935 |
| 60 | 75,058 |
| 50 | 50,368 |
| 40 | 26,634 |
| 1630 | 4,646 |

exhibit **33** of chapter 5 (continued)

**P) PROBLEMS**

33a) Analyze the median ages of panel A carefully.

33b/c) Analyze the urban/rural populations of panel A carefully.

33d) Take logs of the populations in panel B, and compare them with the extension of the fit given in text for 1790 to 1860.

33e) Fit the data of panel B, re-expressing it if necessary.

**S) SOURCES**
Historical Statistics of the U.S. Colonial times to 1957. Washington 1960.
(Panel A entries are from series A90, A195, A206 as indicated.

# Straightening out plots (using three points) 6

chapter index on next page

We are now sure that we want to first straighten out and then flatten out plots. Straightening out is important--in the language of the opening of Chapter 4, a "big deal"--we want to learn to do it as easily as we reasonably can. This chapter is devoted to techniques and examples.

The sort of re-expression that concerns us is--as we have said--almost entirely re-expression of amounts, including large counts. (Here, counts that are never smaller than 3 are surely "large", and others may be.) These are the kinds of numbers where it is natural to take **powers, roots,** and **logs.**

In dealing with them, we will want to be sure that our origins are reasonably chosen. It is only for amounts measured from reasonable origins that we are likely to get full value from changing to a power, a root, or a logarithm.

Powers and roots of amounts are again amounts. Logs of amounts are **balances**. For these purposes--as for so many others--large counts are merely a special kind of amount. Accordingly, (nonzero) powers and roots of large counts are amounts, while logs of large counts are balances.

In thinking about our problems of re-expression--which need not be the same sort of thing as analyzing the data involved--we need to think about whether $x$--or $y$--varies much or little. So long as we deal with amounts, the natural way to make comparisons is by ratios--including percents. Thus we are interested in such facts as:

$$\frac{\text{largest } x}{\text{smallest } x} = 3,$$

All $x$ are within ±50% of a middle value,

or

$$\frac{\text{largest } x}{\text{smallest } x} = 1.1,$$

largest $x$ = smallest $x$ plus 10%.

When we work in logs, we are concerned with balances, not amounts, and the natural way of expressing spread is either in log units or--occasionally--by the ratios into which **differences** in log units can be re-expressed.

# index for chapter 6

review questions 171

**6A. Looking at three points** 171
review questions 172

**6B. Re-expressing y alone** 173
U.S. population again 173
fitting lines to three points 174
review questions 175

**6C. Re-expressing x alone** 175
back to the population of the U.S. 176
a caveat 180
review questions 181

**6D. A braking example** 181
using our knowledge 182
review questions 186

**6E. The vapor pressure of H$_2$O** 187
review questions 190

**6F. Re-expressing the second variable** 191
another try 192
review questions 193

**6F. Wise change of origin as a preliminary** 193
a radioactive decay example 194
review questions 197

**6H. How far have we come?** 198

**6P. Additional problems** 199

| EXHIBITS | PAGE |
|---|---|
| 6A | |
| 6B | |
| 6C | |
| 1 | 176 |
| 2 | 179 |
| 3 | 179 |
| 4 | 180 |
| 6D | |
| 5 | 182 |
| 6 | 183 |
| 7 | 183 |
| 8 | 184 |
| 9 | 185 |
| 10 | 185 |
| 11 | 186 |
| 6E | |
| 12 | 187 |
| 13 | 188 |
| 14 | 189 |
| 15 | 189 |
| 16 | 190 |
| 6F | |
| 6G | |
| 17 | 195 |
| 18 | 196 |
| 19 | 196 |
| 20 | 197 |
| 6H | |
| 21 | 198 |
| 6P | |
| 22★ | 199 |
| 23★ | 200 |
| 24★ | 201 |
| 25★ | 202 |
| 26★ | 203 |

#### review questions

What will we try to do in this chapter? What sort of re-expression concerns us? Need we bother about choice of origin? What is important about how much $x$--or $y$--varies? When it is an amount or count? When it is a balance?

### 6A. Looking at three points

We have seen various illustrations of straightening out data. So far, either these have come from simple rational considerations, or else they have come-- "out of the air"--with little apparent reason for the choice. How are we to behave when we have some other kind of data? Do we have to try all possible combinations of an expression of $y$ with an expression of $x$ on the whole data? Or can we save most of the effort this would involve?

The purpose of these changes of expression is to straighten out the data. If the data looks curved, in some overall way, we can make this obvious by picking out three representative points. For the early growth of the U.S.A. population, for example, we could choose the points corresponding to 1800, 1850, and 1890. These three points are:

(1800, 5.3)

(1850, 23.2)

(1890, 62.9)

A simple way to see that three points do not lie on a single straight line is to find the slopes of the straight lines through the first pair and the second pair, respectively; this gives:

$$\frac{23.2 - 5.3}{1850 - 1800} = \frac{17.9}{50} = 0.36$$

and

$$\frac{62.9 - 23.2}{1890 - 1850} = \frac{39.7}{40} = 0.99$$

which are quite different. The second slope is greater, so the curve is hollow upward. (Draw yourself a sketch.)

If any pair of choices of expression is going to straighten out the early portion of the U.S.A. population curve, these same choices will have to do a reasonably good job in straightening out these three points. We can save a lot of effort by screening our pairs of expressions on these three points. We will then need to try only the one best--or perhaps the few best--on the whole data.

We can often ease the task a little by choosing the spacing of the three points to simplify matters further. Had we chosen

(1810, 7.2)
(1850, 23.2)
(1890, 62.9)

we could have compared

$16.0 = 23.2 - 7.2$

with

$39.7 = 62.9 - 23.2$

since both changes in $x$ are the same. $(1850 - 1810 = 40' = 1890 - 1850)$.

Sometimes we can go almost this far by picking points so that changes in $x$ have a simple ratio--1 to 2, 1 to 3, 3 to 2, etc.--instead of being equal.

Even trying all reasonably possible pairs of expressions on three points is an effort. Can we save some of this by looking into the direction in which changing an expression shifts curvature?

### review questions

What relation ought three selected points have to all the available points? How do we ask three points about curvature? How can we make asking easier?

## 6B. Re-expressing y alone

Exhibit 19 of chapter 3 shows various expressions of $y$ plotted against $y$. We see at once that the higher curves are hollow upward, while the lower curves are hollow downward.

To say that a curve is **hollow upward** means that if we take three points on the curve, the middle point is **below** the line joining the other two. Similarly, **hollow downward** means that the middle point is **above** the line joining the other two. Since we are trying to get a middle point **onto** the line joining the outer two, these are just the sort of facts that matter to us.

We can say more about the simple ladder of ways of expression, which includes

$$y^3$$
$$y^2$$
$$y$$
$$\sqrt{y}$$
$$\log y$$
$$-\frac{1}{\sqrt{y}}$$

$$-\frac{1}{y}$$

$$-\frac{1}{y^2}$$

$$-\frac{1}{y^3}$$

than we just have. We stated one special case of the following:

⋄ **if one expression is straight, those above it are hollow upward, those below it are hollow downward.**

Exhibit 19 of chapter 3 has already shown us that this is true when y is straight. Exhibit 20 of chapter 3 shows us that this is true when log y is straight. The appearance of both those exhibits makes it plausible that the statement is true if any expression of the simple ladder is straight. (The doubting reader should try to use the following two statements to construct a general proof.)

⋄ **if our three points are hollow UPward, we look further DOWN the ladder for straightness.**

⋄ **if our three points are hollow DOWNward, we look further UP the ladder for straightness.**

These must be true. Consider the first: If a new expression is to be straight, and our present expression is hollow upward, the present expression has to be higher up the ladder than the new expression. To find the new expression, we must move down the ladder from the present expression.

So far as re-expressing y goes, the rule is simple:

⋄ **move on the ladder as the bulging side of the curve points.**

### U.S. population again

Applying this to the early U.S. population--since we have seen that the curve is hollow upward--moves us down the ladder. Let us try this, trying $-1/y$ first.

Turning to exhibit 6 of chapter 3 we can find

$$-1/5.3 = -0.188$$
$$-1/23.2 = -0.043$$
$$-1/62.9 = -0.016$$

**174** /6: Straightening out plots (using three points)

The three points and the two slopes are now

$$(1800, -0.188)$$
$$(1850, -0.043)$$
$$(1890, -0.016)$$

$$\frac{-0.043 - (-0.188)}{1850 - 1800} = \frac{0.145}{50} = 0.0029$$

$$\frac{-0.016 - (-0.043)}{1890 - 1850} = \frac{0.027}{40} = 0.0007$$

The slope for the first pair of points is now four times that for the second pair. When we used y, that for the second was three times that for the first. We would like equal slopes, so it is natural to try next about halfway from y to $-1/y$. Thus we are led to try log y. Turning to exhibit 3 of chapter 3, we see that

$$\log 5.3 = 0.72$$
$$\log 23.2 = 1.37$$
$$\log 62.9 = 1.80$$

so that the three points and two slopes are

$$(1800, 0.72)$$
$$(1850, 1.37)$$
$$(1890, 1.80)$$

$$\frac{1.37 - 0.72}{1850 - 1800} = \frac{0.65}{50} = 0.013$$

$$\frac{1.80 - 1.37}{1890 - 1850} = \frac{0.43}{40} = 0.011$$

Now the slopes agree with each other rather well. We should now go ahead--calculating log y and plotting x and log y for either many more or all of the points. Exhibits 7 and 10 of chapter 5 have already shown us how well this choice works.

### fitting lines to three points

Once we have our three points fairly well on a line, we may as well fit a line to them. The three points just fixed offer a reasonable example.

To fit a line to three reasonably-spaced points, we usually do well to fit a slope to the two endpoints, and then take the mean of the three adjusted values to find the constant. For (1800, 0.72), (1850, 1.37), (1890, 1.80), this leads to

$$\frac{1.80 - 0.72}{1890 - 1800} = \frac{1.08}{90} = 0.012$$

and since it seems easy to work with

$$x = \text{date} - 1800$$

we form the three values of

$$y - .012(\text{date} - 1800)$$

namely

$$0.72 - .012(0) = 0.72$$
$$1.37 - .012(50) = 0.77$$
$$1.80 - .012(90) = 0.72$$

for which the mean is 0.74 to two decimals. (The agreement of the two 0.72's is an important check on our arithmetic.) Thus our fit is

$$\text{Population}(\log \text{ of millions}) = 0.74 + 0.012(\text{date} - 1800)$$

> **Three points can take us a long way. If they are well chosen, they can do very well for us.**

### review questions

How do three points indicate hollow upward? Hollow downward? In which direction are expressions hollow that fall above (on the ladder of ways of expression) a straight expression? Those below? How do these rules help us in straightening the growth of the U.S. population? How do we fit a line to three (reasonably spaced) points? Why did we not use the median of the three adjusted values? Should we be surprised that 0.012 (in the 3-point fit) falls between 0.011 and 0.013 (for the two pairs of points)?

## 6C. Re-expressing x alone

We now know how to be guided in choosing a better expression for $y$. What if we wish to leave $y$ alone and re-express $x$?

If we flip our picture over, interchanging the $y$- and $x$-axes, we convert one problem into the other. (We have to look through the back side of the paper after the flip, but a straight line stays a straight line.) Since it is now the same problem, all the same arguments apply.

**exhibit 1/6: Straightening out plots**

This means that if the curve bulges toward large *x* and we are to re-express *x*, we ought to move *x* up the ladder, while if the curve bulges toward small *x*, we should move *x* down the ladder.

**back to the population of the U.S.**

Exhibit 1 shows three points from the population curve after the interchange of *x* and *y*. We see that the bulge is toward larger *x* (upward on this interchanged plot), so if we are to re-express *x* we would have to move upward on the *x* ladder.

Should we try re-expressing *x* in this situation? No, because we can see that re-expression is unlikely to help us. As amounts, 1810, 1850, and 1890 are very similar, the outer values differing only a few percent from the middle one.

Re-expressing a variable that changes by only a few percent rarely gets rid of more than a barely detectable curvature.

One way to make the *x*'s less alike as amounts would be to figure dates from some origin later than the birth of Christ. To figure from 1776 would be quite unrealistic, since there had been much immigration before that date. To figure from 1600 would be more realistic, but even this would be neglecting a substantial Amerind population.

exhibit **1** of chapter 6: U.S. population

**The early population of the U.S., x and y interchanged**

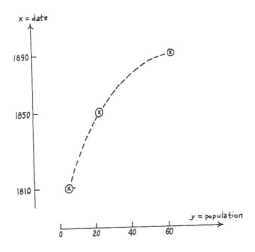

Let us try "date − 1600" anyway. An example should be useful, so long as it is not quite foolish. To keep our numbers of convenient size, we may as well measure "date − 1600" in centuries.

Our new $x$ values for the last 3 points are 2.1, 2.5, and 2.9. They now vary by about ±15%, which is much more than before. We have some hope, but we cannot expect to find it easy to take care of substantial curvature, such as we still face. Since the change from date in years to $[(1/100)(\text{date}) - 16]$ is a trivial re-expression--one that involves only multiplication by, and addition of, constants, our need to move up the $x$ ladder is unchanged.

Let us begin by going to the cube of the new $x$. We find

$$(2.1)^3 = 9.261$$
$$(2.5)^3 = 15.625$$
$$(2.9)^3 = 24.389$$

The three points are:

$$(\ 9.3,\ \ 7.2)$$
$$(15.6,\ 23.2)$$
$$(24.4,\ 62.9)$$

and the two slopes are:

$$3.2 = \frac{23.2 - 7.2}{15.6 - 9.3}$$

$$4.5 = \frac{62.9 - 23.2}{24.4 - 15.6}$$

The slope is still larger for the second interval, though only by about 2 to 1 rather than by 3 to 1. We have made progress, but clearly need to go further still.

Trying $x^6$, which is easy to find by squaring $x^3$, gives

$$(\ 85,\ \ 7.2)$$
$$(244,\ 23.2)$$
$$(501,\ 62.9)$$

and

$$.102 = \frac{23.2 - 7.2}{244 - 85}$$

$$.114 = \frac{62.9 - 23.2}{591 - 244}$$

This is a lot closer, but we are not there yet.

## /6: Straightening out plots (using three points)

Trying $x^8$ gives

$$( 378, 7.2)$$
$$(1526, 23.2)$$
$$(5002, 62.9)$$

and

$$.0139 = \frac{23.2 - 7.2}{1526 - 378}$$

$$.0114 = \frac{62.9 - 23.2}{5002 - 1526}$$

We seem at last to have gone too far.

Trying $x^7$ should come quite close. We find

$$( 180, 7.2)$$
$$( 610, 23.2)$$
$$(1725, 62.9)$$

$$.0372 = \frac{23.2 - 7.2}{610 - 180}$$

$$.0357 = \frac{62.9 - 23.2}{1725 - 610}$$

Agreement is now fairly good.

If we are to re-express

$$(\text{year} - 1600),$$

we find that our three points suggest the use of either

$$(\text{year} - 1600)^7$$

or, equivalently,

$$\left(\frac{\text{year} - 1600}{100}\right)^7.$$

We ought at least look at the results of doing this. Exhibit **2** shows the numbers, exhibit **3** the gross picture, exhibit **4** some differences. As either exhibit 2 or exhibit 4 shows, the deviations of the U.S. population, expressed in millions, from

$$0.72 + 0.036 \left(\frac{\text{date} - 1600}{100}\right)^7$$

are less than 150,000 from 1800 to 1830 and from 1870 to 1890, with an "extra" million or two counted in 1840, 1850, and 1860.

exhibit **2** of chapter 6: U.S. population

**The results of re-expressing $x$ in dealing with U.S. population in the nineteenth century**

| Date | $z$ $\dfrac{(\text{date} - 1600)^7}{100}$ | $y$ Population (in millions) | $.036z$ | Diff. |
|---|---|---|---|---|
| 1800 | 128 | 5.31 | 4.61 | 0.70 |
| 1810 | 180 | 7.24 | 6.48 | 0.76 |
| 1820 | 249 | 9.64 | 8.96 | 0.68 |
| 1830 | 340 | 12.87 | 12.24 | 0.63 |
| 1840 | 459 | 17.07 | 16.52 | 1.55 |
| 1850 | 610 | 23.19 | 21.96 | 1.23 |
| 1860 | 803 | 31.44 | 28.91 | 2.53 |
| 1870 | 1046 | 38.58 | 37.66 | 0.92 |
| 1880 | 1349 | 50.16 | 48.56 | 1.60 |
| 1890 | 1725 | 62.95 | 62.10 | 0.85 |
| 1900 | 2187 | 76.0 | 78.7 | −2.7 |
| 1910 | 2751 | 92.0 | 99.0 | −7.0 |
| 1920 | 3436 | 105.7 | 123.7 | −18.0 |

exhibit **3** of chapter 6: U.S. population

**The early U.S. population plotted against** $z = \left(\dfrac{\text{date} - 1600}{100}\right)^7$

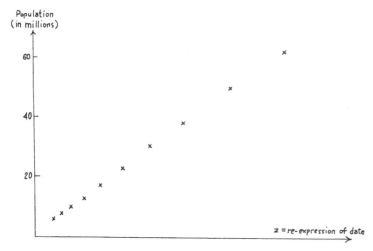

exhibit **4** of chapter 6: U.S. population

**The result of flattening exhibit 3**

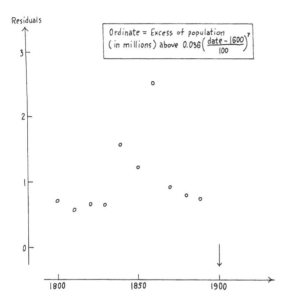

### a caveat

So far, either as quality of fit or ability to set residuals in clear view goes, plots such as exhibits 3 and 4 are as good as the plots of the last chapter, which related log population to date in years. If our purpose is to examine residuals--as it so often is--either plot is effective and useful.

When then should we make a distinction between the two plots? Surely we do need to make a distinction when we want to find an easily communicated description. That the U.S. population grew about 2.8% per year from 1800 to 1890 is relatively easy both to communicate and to understand. This is a major advantage, since we cannot say the same of the $y$ vs. $x^7$ relationship. (Both, of course, apply over a limited span of years. Both fail--faster and faster--as we move on beyond 1900.) For communication, there is no doubt that log $y$ vs. $x$ is a more useful description.

Like any good fit, either the log $y$ vs. $x$ or the $y$ vs. $x^7$ plot is subject to dangers of overvaluation. We see that each fits closely, though they cannot be exactly alike. Particularly if we have found only one of the two, there is a very natural tendency to convert "a good fit" into "this is how it had to be" or "a basic law of population growth". **One example of a close fit, by itself, is far** from representing evidence for such strong statements. The fact that we have found two close fits of quite different form emphasizes our need to learn to avoid this sort of jumping at conclusions. **Conversely, we can make many good uses of a close fit, whether or not it is "a basic law".**

### review questions

If we are to re-express $x$, which way ought we move on the ladder? Why must this be so? What happens if we exchange $x$ and $y$ axes in the U.S. population example? If $x$ varies by only a few percent, what then? What trials did we make in re-expressing dates? Do we expect only one choice of re-expression to straighten a given set of points out thoroughly? What if we find several? Can we infer a "basic law" from one close fit? Why is one straightening of U.S. population growth easier to communicate than the other?

### 6D. A braking example

Let us next look at an example where re-expressing $x$ seems to be the natural way to make the data more orderly and more describable.

Exhibit 5 shows the speed and distance to stop for 50 cars. Exhibit 6 plots the data. We could fit most of the data points with a straight line. However, a fitted line would give zero stopping distance at a speed between 5 and 10 miles per hour. The one thing we are sure of in this example is that zero stopping distance goes with zero speed, and vice versa. We didn't have to test cars to know that--or to put a point at (0, 0). We must face curvature, and try to eliminate it.

Three reasonable points to take are, then,

$$(0, 0)$$

and the two marked by large $x$'s in exhibit 6,

$$(15, 35) \quad \text{and} \quad (25, 90).$$

The two slopes are

$$\frac{35 - 0}{15 - 0} = 2.3 \quad \text{and} \quad \frac{90 - 35}{25 - 15} = 5.5.$$

To re-express $x$, since the curve bulges toward large $x$, we ought to move toward $x^2$, $x^3$, etc. Trying $x^2$ gives

$$(0, 0) \quad (225, 35) \quad (625, 90)$$

with slopes of

$$.15 = \frac{35 - 0}{225 - 0} \quad \text{and} \quad .14 = \frac{90 - 35}{625 - 225}$$

Clearly, using $x^2$ is a reasonable try.

**exhibit 5/6: Straightening out plots**

Exhibit **7** shows the plot of y against $x^2$, which now looks quite straight. If we take

$$(0, 0) \quad \text{and} \quad (600, 80)$$

as representative points, we are led to try

$$y - .133x^2$$

as a reasonable flattened quantity.

Exhibit **8** shows the result. It is far from wonderful, but seems to be reasonably flat, although the behavior for $x^2$ near 0 does not fit too well with the known point at $(0, 0)$.

**using our knowledge**

We need to try something further. We believe in $y = 0$ when $x = 0$. Perhaps we should use this belief in choosing what to plot. How can we do this?

exhibit **5** of chapter 6: braking distances

**Speed and distance to stop**

| Speed, x (mph) | Distance to stop, y (in feet) |
|---|---|
| 4 | 2, 10, |
| 7 | 4, 22, |
| 8 | 16, |
| 9 | 10, |
| 10 | 18, 26, 34, |
| 11 | 17, 28, |
| 12 | 14, 20, 24, 28, |
| 13 | 26, 34, 34, 46, |
| 14 | 26, 36, 60, 80, |
| 15 | 20, 26, 54, |
| 16 | 32, 40, |
| 17 | 32, 40, 50 |
| 18 | 42, 56, 76, 84, |
| 19 | 36, 46, 68, |
| 20 | 32, 48, 52, 56, 64, |
| (21) | |
| 22 | 66, |
| 23 | 54, |
| 24 | 70, 92, 93, 120, |
| 25 | 85, |

**S) SOURCE**

Mordecai Ezekiel, 1930. Methods of Correlation Analysis. New York, John Wiley. Table 11, **page 41**.

Also Table 10, page 43, of 2nd edition 1943. Table 4.1, page 45 of 3rd edition 1959, by Mordecai Ezekiel and Karl A. Fox, has similar but different data.

exhibit **6** of chapter 6: braking distances

**Plot of exhibit 5s data**

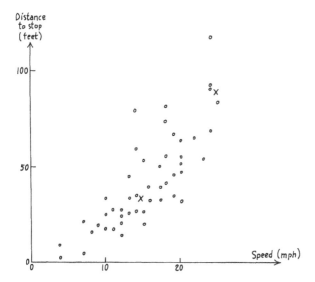

exhibit **7** of chapter 6: braking distances

**Speed² and distance to stop for 50 motor cars**

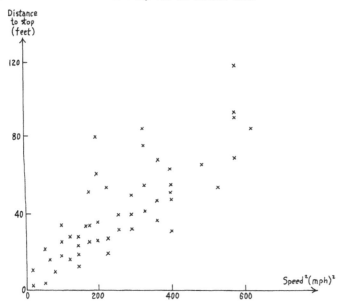

**184** /exhibit 8/6: Straightening out plots

One way is to plot $y/x$ -- rather than $y$ -- against $x$. Any finite value for $y/x$ will make $y$ go to zero as $x$ goes to zero. Thus any reasonable behavior of our $y/x$ vs. $x$ plot will lead to a fit that makes $x = 0$ go with $y = 0$.

Exhibit **9** gives the values of $y/x$. Exhibit **10** plots $y/x$ against $x$. The behavior of the point cloud is quite reasonable, so we select two representative points as shown. They are

$$(5, 1.4) \quad \text{and} \quad (25, 3.7).$$

The corresponding line is

$$\frac{y}{x} = .115x + .8.$$

The residuals are written down in exhibit **11** and plotted in exhibit **12**.
We now see:

⋄ 10 wandering points on the high side -- 9 between 1.1 and 1.9, one very high -- these presumably represent bad brakes or slow-responding drivers.

⋄ a mass of other points which, if taken by themselves, seem quite level but placed about 0.5 too low.

exhibit **8** of chapter 6: braking distances

**Flattening by use of $y - 1.33x^2$**

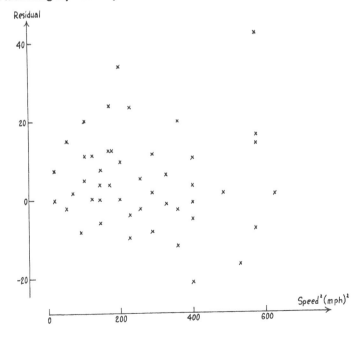

exhibit **9** of chapter 6: braking distances

**The values of y/x**

| \|x\| | y/x |
|---|---|
| 4 | 0.5, 2.5 |
| 7 | 0.6, 3.1 |
| 8 | 2.0 |
| 9 | 1.1 |
| 10 | 1.8, 2.6, 3.4 |
| 11 | 1.6, 2.6 |
| 12 | 1.2, 1.7, 2.0, 2.3 |
| 13 | 2.0, 2.6, 2.6, 3.5 |
| 14 | 1.9, 2.6, 4.3, 5.7 |
| 15 | 1.3, 1.7, 3.6 |
| 16 | 2.0, 2.5 |
| 17 | 1.9, 2.4, 3.0 |
| 18 | 2.3, 3.1, 4.2, 4.7 |
| 19 | 1.9, 2.4, 3.6 |
| 20 | 1.6, 2.4, 2.6, 2.8, 3.2 |
| (21) | |
| 22 | 3.0 |
| 23 | 2.3 |
| 24 | 2.9, 3.8, 3.9, 5.0 |
| 25 | 3.4 |

exhibit **10** of chapter 6: braking distances

**y/x against x**

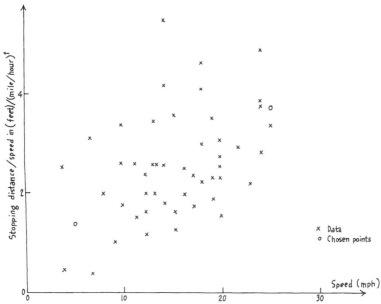

† 1 foot/(mile/hour) = $\frac{30}{44}$ seconds = $\frac{15}{22}$ second

Thus, our final description can run as follows:

40 points fairly well described by:

$$\frac{y}{x} = .115x + .8 - .5$$

or

$$y = .115x^2 + .3x,$$

accompanied by 9 points with y greater than this expression by about $2x$ and one point greater by about $4x$.

Careful and repeated analysis can lead to effective description.

**review questions**

What example did we use in this section? What fact were we entitled to be sure of? How does exhibit 6 behave? What three points is it natural to choose? What re-expression do they lead to? How effective does this re-expression seem to be? What did we then think of trying? Why was it natural to think of it? How did the resulting plots look? Do there seem to be stray values? Are you surprised?

**exhibit 11 of chapter 6: braking distances**

**The values of $(y/x) - 0.115x - 0.8$**

| $\lvert x \rvert$ | $\lvert 0.115x + 0.8 \rvert$ | $y/x - 0.115x - 0.8$ |
|---|---|---|
| 4 | 1.3 | −0.8, 1.2, |
| 7 | 1.6 | −1.2, 1.5, |
| 8 | 1.7 | 0.3, |
| 9 | 1.8 | −0.7, |
| 10 | 2.0 | −0.2, 0.6, 1.4, |
| 11 | 2.1 | −0.5, 0.5, |
| 12 | 2.2 | −1.0, −0.5, −0.2, 0.1, |
| 13 | 2.3 | −0.3, 0.3, 0.3, 1.2, |
| 14 | 2.4 | −0.5, 0.2, 1.9, 3.3, |
| 15 | 2.5 | −1.2, −0.8, 1.1, |
| 16 | 2.6 | −0.6, −0.1, |
| 17 | 2.8 | −0.9, −0.4, 0.2, |
| 18 | 2.9 | −0.6, 0.2, 1.3, 1.8, |
| 19 | 3.0 | −1.1, −0.6, 0.6, |
| 20 | 3.1 | −1.5, −0.7, −0.5, −0.3, 0.1, |
| (21) | | |
| 22 | 3.3 | −0.3 |
| 23 | 3.4 | −1.1 |
| 24 | 3.6 | −0.7, 0.2, 0.3, 1.4 |
| 25 | 3.7 | −0.3 |

## 6E. The vapor pressure of $H_2O$

The vapor pressure of water or ice is the highest pressure of water vapor (steam, if you like) that can exist in equilibrium with the water or ice at a given temperature. Its values are well known, and of considerable practical importance. In terms of temperatures Celsius (once called centigrade, giving 0°C = 32°F and 100°C = 212°F) and pressures in mm of mercury (760 mm is one standard atmosphere), these vapor pressures at different temperatures are given in exhibit **13**.

Just a look at the first two columns of this table shows that a direct plot of $p$ against $t$ will show us only a very rapidly rising pressure. (If 139893.20 can be plotted, even 760.00 can hardly be seen.) Accordingly, we plan to do something to at least partly straighten out this plot. Taking the logarithm of $p$ looks as if it might help.

exhibit **12** of chapter 6: braking distances

**Plot of $(y/x) - 0.115x - 0.8$ against $x$**

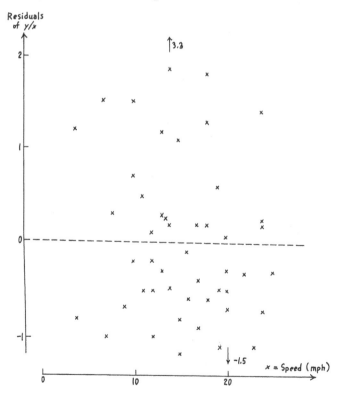

## exhibit 13/6: Straightening out plots

Column (3) of exhibit 13 contains the values of log p. This looks conceivably plottable, so we proceed to exhibit **14**, which can be looked over, but is far from straight.

Another step needs to be taken. We could try to find some empirical approach, but we can go more directly to our goal if we recall that plotting log pressure against the reciprocal of the absolute temperature is rather common in physical chemistry. (Or if we use the closing example of the next section.) Here, one kind of absolute temperature is the Celsius temperature increased by about 273.1°C.

Column (4) of exhibit 13 contains values of $-1000(1/T)$, where $T = t + 273.1°C$. (The factor 1000 has been used to avoid unsightly and confusing zeros.) The result of plotting log p against column (4) of exhibit 13 is shown in exhibit **15**. At last we have a reasonably good straight line, one worth calculating residuals about.

Column (5) of exhibit 13 contains values of $\log p - 2.25(-1000/T)$ which, as we see, is reasonably constant. Subtracting 8.8 clearly gives residuals well

exhibit **13** of chapter 6: vapor pressure

**The vapor pressure of H$_2$O and some associated quantities**

| t(°C) | p(mm Hg) | log p | (4) | (5) |
|---|---|---|---|---|
| −40 | 0.105 | −.9788 | −4.2900 | 8.6737 |
| −20 | 0.787 | −.1040 | −3.9510 | 8.7858 |
| 0 | 4.5687 | .6598 | −3.6617 | 8.8986 |
| 20 | 17.363 | 1.2396 | −3.4118 | 8.9162 |
| 40 | 54.865 | 1.7393 | −3.1939 | 8.9256 |
| 60 | 148.88 | 2.1728 | −3.0021 | 8.9275 |
| 80 | 354.87 | 2.5501 | −2.8321 | 8.9223 |
| 100 | 760.00 | 2.8808 | −2.6802 | 8.9113 |
| 120 | 1489.14 | 3.1729 | −2.5439 | 8.8966 |
| 140 | 2710.92 | 3.4331 | −2.4207 | 8.8797 |
| 160 | 4636.00 | 3.6661 | −2.3089 | 8.8611 |
| 180 | 7520.20 | 3.8762 | −2.2070 | 8.8420 |
| 200 | 11659.16 | 4.0667 | −2.1137 | 8.8225 |
| 220 | 17395.64 | 4.2404 | −2.0280 | 8.8034 |
| 240 | 25100.52 | 4.3397 | −1.9489 | 8.7847 |
| 260 | 35188.00 | 4.5464 | −1.8758 | 8.7670 |
| 280 | 48104.20 | 4.6822 | −1.8080 | 8.7502 |
| 300 | 64432.80 | 4.8091 | −1.7449 | 8.7351 |
| 320 | 84686.80 | 4.9278 | −1.6861 | 8.7215 |
| 340 | 109592.00 | 5.0398 | −1.6311 | 8.7097 |
| 360 | 139893.20 | 5.1458 | −1.5795 | 8.6997 |

(4) = $-1000(1/T)$, where $T = t + 273.1°C$
(5) = $\log p - 2.25(-1000/T)$

exhibit **14** of chapter 6: vapor pressure

**Vapor pressure of H₂O and temperature (logarithmic scale for pressure, which is in mm Hg)**

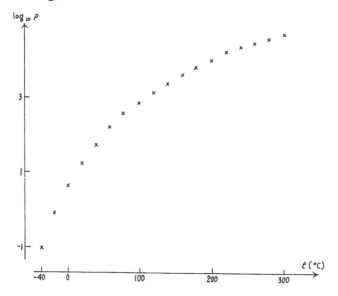

exhibit **15** of chapter 6: vapor pressure

**Vapor pressure of H₂O and temperature (logarithm of pressure, reciprocal of absolute temperature)**

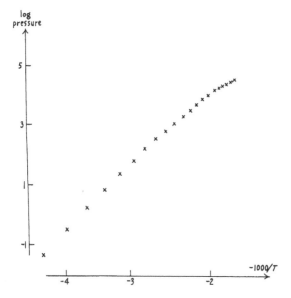

worth plotting. Exhibit **16** shows the results. The most striking result is the apparent split of our curve into two parts with a corner at the third point from the left. What might this mean?

When we recall that this point comes at 0°C (= 32°F), which is the freezing point of water, we see that a break at this point is quite reasonable. Below 0°C we are dealing with the vapor pressure of solid water (ice), while above this point we are dealing with the vapor pressure of liquid water.

**Note that this fact was not forced upon us by our data until we both straightened and flattened the plot.**

### review questions

What data did we look at in this section? Could we plot it at all in its original form? Did we re-express y? Why? Was this enough? Could we go further? How? How well did we then do? Could we see anything new when we looked at the residuals? Did such a look start off any ideas?

exhibit **16** of chapter 6: vapor pressure

**Residuals of log vapor pressure from a straight line in − 1/T (water)**

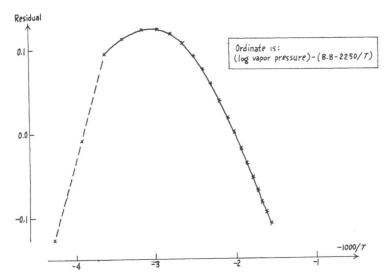

Ordinate is:
(log vapor pressure) − (8.8 − 2250/T)

Abscissa is : 1000 times the negative reciprocal of the absolute temperature (°K)

## 6F. Re-expressing the second variable

As we have just seen, it will not always be wise--or satisfactory--to change only one of the two expressions. Sometimes it pays to re-express **both** $x$ and $y$. In such a situation we may need to do "cut and try". However, once we have agreed to try a specific expression for one of the variables, be it $x$ or $y$, we can use the same rules to narrow down our choice of the second expression.

As an example, let us take three points from the example just treated in the last section, and see how we might have been guided. From exhibit 13 we have

$$(0, \quad 4.5687)$$
$$(100, \quad 760.00)$$
$$(200, \quad 11659.16).$$

Clearly, the middle point is below the line that joins the other two--and also to the right of that line. If we are to re-express $y$ we should look toward log $y$ and $-1/y$. Trying logs gives

$$(0, 0.6598) \quad (100, 2.8808) \quad (200, 4.0667)$$

with slopes of

$$\frac{2.8808 - 0.6598}{100 - 0} = \frac{2.2210}{100} = .02221$$

and

$$\frac{4.0667 - 2.8808}{200 - 100} = \frac{1.1859}{100} = .01186.$$

Thus the middle point is now above and to the left of the line. We could try going only as far as $\sqrt{y}$, which is about halfway to log $y$. If we do this, say by using exhibit 5 of chapter 3, we find

$$(0, \quad 2.12)$$
$$(100, \quad 27.6)$$
$$(200, \quad 108)$$

and we see that $\sqrt{y}$ goes nowhere nearly far enough. To go on to

$$\sqrt{\sqrt{y}} = y^{1/4} = \sqrt[4]{y}$$

is easy. If we do this, we find

$$(0, \quad 1.44)$$
$$(100, \quad 5.3)$$
$$(200, \quad 10.4)$$

for which the slopes are

$$\frac{5.3 - 1.44}{100 - 0} = \frac{3.86}{100} = .0386$$

$$\frac{10.4 - 5.3}{200 - 100} = \frac{5.1}{100} = .0510$$

so that we see that we are not yet far enough. Those who love to repeat square roots will now try

$$\sqrt{\sqrt[4]{y}} = y^{1/8} = \sqrt[8]{y},$$

finding

(0, 1.20)
(100, 2.32)
(200, 3.20)

for which the slopes are

$$\frac{2.32 - 1.20}{100 - 0} = \frac{1.12}{100} = .0112$$

$$\frac{3.20 - 2.32}{200 - 100} = \frac{.88}{100} = .0088.$$

This time we have gone too far. As a result, if we are to change only y, we should try something between $y^{1/4}$ and $y^{1/8}$, perhaps $y^{1/6}$ or $y^{1/7}$.

### another try

This is, however, not our only reasonable choice. Going to log y did much to straighten out our three points. True, it went a little too far, but perhaps we could do something by keeping log y and re-expressing x. Before we do this, however, we should stop and think for a moment. As we have written it, x is temperature in °C (freezing water = 0°C, boiling water = 100°C). If we are to re-express simply, we ought not to tie our zero to a property of so special a substance as water. (We are looking at water's vapor pressure, but water has a vapor pressure below 0°C, too.) Rather we should figure our temperatures so that we run from the so-called absolute zero, which is just below −273°C.

Our intermediate starting point then should be

(273.1, 0.6598)
(373.1, 2.8808)
(473.1, 4.0667)

with slopes of .02221 and .01186, and a middle point above and to the left of the line. If we are to re-express $x$, we should move toward $\log x$ or $-1/x$. Let us try $\log x$. We find:

$$(2.44, 0.6598)$$
$$(2.57, 2.8808)$$
$$(2.67, 4.0667)$$

$$\frac{2.8808 - 0.6598}{2.57 - 2.44} = \frac{2.2210}{.13} = 17.1$$

$$\frac{4.0667 - 2.8808}{2.67 - 2.57} = \frac{1.1859}{.10} = 11.8$$

Using logs has helped, but not enough. Let us try $-1/x$, for which we have

$$(-.00366, 0.6598)$$
$$(-.00268, 2.8808)$$
$$(-.00211, 4.0667)$$

$$\frac{2.8808 - 0.6598}{-.00268 - (-.00366)} = \frac{2.221}{.00098} = 2266$$

$$\frac{4.0667 - 2.8808}{-.00211 - (-.00268)} = \frac{1.186}{.00057} = 2080$$

for which the slopes agree better--have a ratio nearer 1--than with any other combination so far tried. Accordingly

$$\log y \quad \text{and} \quad -1/x$$

where $x$ is absolute temperature, seems to be a good choice. As we know from the last section, it proves to be one.

### review questions

Does it ever pay to re-express both $x$ and $y$? Do we have to do this blind? How is guidance for re-expressing the second coordinate to be found? How much is it like guidance when only one coordinate is re-expressed? If we tackle the vapor pressure of water, re-expressing only $y$, what are we led to? If we pick $\log y$, what are we led to for $x$?

### 6G. Wise change of origin as a preliminary

We have already seen two examples where a change of origin was part of a sensible approach to straightening: (1) the early population of the U.S. as a function of time (where a time origin at the birth of Christ--1600 years before the beginning of European immigration to North America--was far from sensible--once we thought about it! Why 1600 and not 600 or 2600?) and (2) the vapor pressure of water as a function of centigrade temperature. (Why should we take the origin at freezing?)

The same sort of need to think about a sensible origin arises in other problems, sometimes with regard to $x$ and sometimes with regard to $y$. In one broad class of cases, the question arises because the amount we measure is the sum of contributions, some of which change slowly if at all, while the remaining contribution changes in such a way that re-expression could flatten the plot if this contribution could be measured separately.

The natural approach to this sort of data is to form

observation MINUS background

where "background" is a constant chosen to allow for the slowly changing components. We are not likely to be able to pick the value for "background" either on the basis of general insight or on the basis of doubtfully-related historical data. We expect to learn about it from the same set of data that we are trying to flatten. We are likely to approach choosing a plausible value for "background" by trying various values for it and seeing which one leads to data that can be more thoroughly flattened by further re-expression.

Observation of radioactive decay yields many problems of this sort. Any single kind of radioactive atom decays on a steady percentage basis--so many percent is gone every so many days, years, or millennia. Many processes of isolating--or making--some one kind of radioactive atom also isolate--or make--one or more other kinds. If these other kinds decay--but decay more slowly--their presence can often be adequately allowed for by a constant background.

### a radioactive decay example

In 1905, the study of radioactive substances was in its infancy. Meyer and von Schweidler reported the relative activities for an experimental object set out in exhibit **17**. If we plot activity against time, the result is exhibit **18**, which is far from being straight. It is again natural to take logarithms, especially since, in simple radioactive decay, the logarithm of activity should decrease linearly with time.

Exhibit **19** shows the behavior of log activity against time. A noticeable curvature remains. In this situation the most plausible source of curvature is contamination by some other radioactive substance that decays much more slowly than the one of central interest. If such there be, it will not have contributed more than two units of activity (since at 45 days the observed activity is down to 2.1 units).

It is reasonable to explore the consequences of assuming the presence of 1.0 to 1.5 units of activity from such a contaminant by, successively:

⋄ plotting against time either

$$\log(\text{activity} - 1.0)$$

or

$$\log(\text{activity} - 1.5)$$

which could represent the log of the rapidly-decaying activity (and which are given in columns (4) and (5) of exhibit 17).

⋄ fitting a straight line, perhaps roughly.
⋄ plotting the residuals.

exhibit **17** of chapter 6: radioactivity

**The decay of radioactivity and associated quantities**

| time (days) | activity (relative) | (3) | (4) | (5) | (6) | (7) |
|---|---|---|---|---|---|---|
| 0.2 | 36.0 | 1.556 | 1.544 | 1.538 | 1.551 | 1.545 |
| 2.2 | 26.0 | 1.415 | 1.398 | 1.389 | 1.471 | 1.466 |
| 4.0 | 23.1 | 1.364 | 1.344 | 1.334 | 1.476 | 1.474 |
| 5 | 18.9 | 1.276 | 1.253 | 1.241 | 1.418 | 1.416 |
| 6 | 17.8 | 1.250 | 1.225 | 1.212 | 1.423 | 1.422 |
| 8 | 14.7 | 1.167 | 1.137 | 1.121 | 1.401 | 1.401 |
| 11 | 13.4 | 1.127 | 1.093 | 1.076 | 1.456 | 1.461 |
| 12 | 11.3 | 1.053 | 1.013 | .991 | 1.409 | 1.411 |
| 15 | 8.5 | .929 | .875 | .845 | 1.370 | 1.370 |
| 18 | 5.9 | .771 | .690 | .643 | 1.284 | 1.273 |
| 26 | 5.0 | .699 | .602 | .544 | 1.460 | 1.454 |
| 33 | 3.4 | .531 | .380 | .279 | 1.469 | 1.434 |
| 39 | 2.4 | .380 | .146 | −.046 | 1.433 | 1.319 |
| 45 | 2.1 | .322 | .041 | −.222 | 1.526 | 1.353 |

(3) = log(activity)
(4) = log(activity − 1.0)
(5) = log(activity − 1.5)
(6) = .033t + (column 4)
(7) = .035t + (column 5)

**S) SOURCE**

S. Meyer and E. von Schweidler, 1905. Sitzungsberichte der Akademie der Wissenschaften zu Wien, Mathematisch–Naturwissenschaftliche Classe, p. 1202 (Table 5).

exhibit **18** of chapter 6: radioactivity

**The data of exhibit 17**

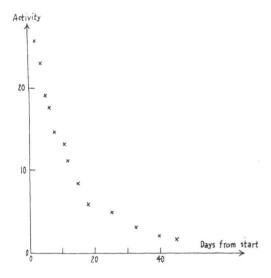

exhibit **19** of chapter 6: radioactivity

**The same data on a log scale**

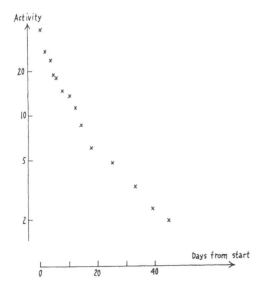

The results are shown in exhibit **20.** (Columns (6) and (7) of exhibit 17 give these residuals increased by 1.45, calculated as "$.033t$ + (column 4)" and "$.035t$ + (column 5)", respectively.) Of the two choices, allowance for 1.0 unit of contamination seems to give a more nearly horizontal set of residuals--to lead to a closer fit.

(Perhaps the residuals for 1.0 do trend upward a little. The reader may wish to try 1.1 or 1.2.)

### review questions

Ought we expect to make changes of origin? What are three examples? What is a background? When we measure radioactive substances, whether made or isolated, are backgrounds common? What did a plot of raw activity against raw time show? What two things did we do then? Did it all work out well?

exhibit **20** of chapter 6: radioactivity

**Residuals from decay line allowing for long-lived contamination (two versions)**

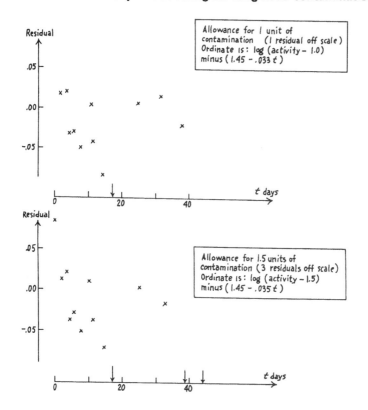

**6H. How far have we come?**

This chapter has been devoted to guidance about how to re-express $x$ and $y$ with a view to straightening a plot. This problem arises most simply when one or both of $x$ and $y$ is an amount--or, really just a special case of an amount, a large count. The natural re-expressions are by powers, roots, and logs.

To keep computation down, we routinely begin with three well-chosen points. The basic rule of thumb is then simple:

◇ **move on the ladder of expressions in the direction in which the curve bulges.**

Exhibit 21 shows the four possible cases, and the natural steps along both $x$- and $y$-ladders. (We can move along either or both.) We try new expressions on more of the data only when the three points have already responded well to them.

In certain cases, a fresh choice of origin--before going to powers, roots, or logs--is valuable. Sometimes the new choice is a matter of common sense, sometimes of how flat the final result proves to be.

**exhibit 21 of chapter 6: indicated behavior**

**How to move for each variable separately; the four cases of curve shape**

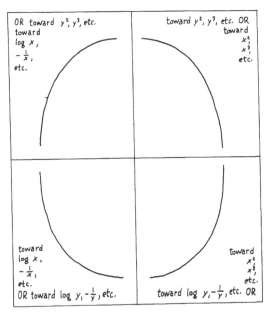

As we use this technique, we will need to remember that:

◇ straightening by re-expressing x is not the same as re-expressing y.

◇ just because re-expression makes things quite straight, there is no guarantee that we have found a new natural law.

We are now ready:

◇ to approach the analysis of (x, y) data in terms of re-expression followed by straight-line fitting (perhaps in two or more steps) and the examination of residuals.

◇ to turn to another class of important problems.

### 6P. Additional problems

See exhibits **22 through 26**.

### exhibit 22 of chapter 6: data and problems

**Three examples of radioactive decay**

#### A) DATA

| Days* | Activity† | Days* | Activity† | Days* | Activity† |
|---|---|---|---|---|---|
| 0.8 | 6.70 | 0.8 | 2.82 | 0.8 | 2.05 |
| 2.8 | 6.40 | 2.8 | 2.34 | 1.0 | 2.03 |
| 6.9 | 5.70 | 4.8 | 1.90 | 1.8 | 1.79 |
| 8.9 | 5.10 | 6.8 | 1.80 | 2.1 | 1.77 |
| 13.1 | 4.30 | 11.8 | 1.34 | 3.9 | 1.54 |
| 15.2 | 4.00 | 13.9 | 1.24 | 5.9 | 1.35 |
| 16.8 | 3.95 | 16.8 | 1.03 | 7.1 | 1.29 |
| 20.1 | 3.40 | 19.8 | 1.00 | 9.1 | 1.23 |
| 20.8 | 3.40 | 23.8 | 0.80 | 12.1 | 1.01 |
| 20.9 | 3.20 | 31.8 | 0.55 | 12.9 | 0.96 |
| 21.9 | 3.20 | 32.5 | 0.11 | 16.9 | 0.82 |
| 31.1 | 2.42 | | | 19.9 | 0.68 |
| 36.8 | 2.30 | | | 22.8 | 0.59 |
| 43.8 | 2.11 | | | 25.9 | 0.51 |
| 49.8 | 2.00 | | | 33.8 | 0.40 |
| 53.8 | 1.99 | | | 44.1 | 0.32 |
| 58.8 | 1.98 | | | | |
| 65.9 | 1.90 | | | | |
| 73.8 | 1.80 | | | | |
| 87.1 | 1.65 | | | | |

\* Since start.
† In volts/minute.

exhibit **22** of chapter 6 (continued)

**P) PROBLEM**

22a) In a later paper, Meyer and Von Schweidler reported the decay of radioactivity for three samples as in panel A. Analyze at least two of these. Comment.

**S) SOURCE**

S. Meyer and E. von Schweidler 1907. "Untersuchungen über radioaktive Substanzen. VIII Mitteilung: Über ein radioaktives Produkt aus dem Aktinium." **Sber. Ak. Wiss. Wien. Math–Nat. Classe 116 IIA1**, pp. 315–322 (especially 316–317).

exhibit **23** of chapter 6: data and problems

**Vapor pressure of mercury**

**A) DATA**

| Temperature (°C) | Pressure (mm Hg) |
|---|---|
| 0 | 0.0004 |
| 20 | 0.0013 |
| 40 | 0.006 |
| 60 | 0.03 |
| 80 | 0.09 |
| 100 | 0.28 |
| 120 | 0.8 |
| 140 | 1.85 |
| 160 | 4.4 |
| 180 | 9.2 |
| 200 | 18.3 |
| 220 | 33.7 |
| 240 | 59. |
| 260 | 98. |
| 280 | 156. |
| 300 | 246. |
| 320 | 371. |
| 340 | 548. |
| 360 | 790. |

**P) PROBLEMS**

23a) The vapor pressure of mercury is stated to be as in panel A. Analyze graphically. Comment. ➡

exhibit 23 of chapter 6 (continued)

23b) Given the three points found partway through section 6F,

(0, 0.6598)
(100, 2.8808)
(200, 4.0667),

what change in expression of this (once new) y will come close to straightening out these points?

23c) Apply the result of the last problem to the data of exhibit 13, fit a straight line, and plot the residuals. How do the results compare with exhibit 16?

23d) Apply the re-expression

$y$ becomes $y^{1/6}$

(suggested by the analysis of section 6E) to the data of exhibit 13, fit a straight line, and find residuals. How do the results compare with exhibit 16? Complete the graphical analysis. Comment.

exhibit 24 of chapter 6: data and problems

**More problems**

24a) Rates of mortality from breast cancer in different latitudes. (50; 1025, 513) translates as: "In latitude 50°N the mortality index for breast cancer is 102.5 and the mean annual temperature is 51.3." Data-sets (16): (50; 1025, 513), (51; 1045, 499; 52; 1004, 500), (53; 959, 492), (54; 870, 485), (55; 950, 478), (56; 886, 473), (57; 892, 451), (58; 789, 463), (59; 846, 421), (60; 817, 442), (61; 722, 435), (62; 651, 423), (63; 681, 402), (69; 673, 318), (70; 525, 340). SOURCE: A. J. Lea, 1965, "New observations on distribution of neoplasms of female breast in certain European countries," *British Medical J.* **1** (for 1955): 486–490. Table II on page 489. PROBLEM: Can we straighten this plot? (Mortality index against mean annual temperature.)

24b) Plasticity of wool: slow stretching of a single fiber. (1, 321) translates as: "After 1 minute under load, a fiber 53.3 microns in average diameter (coefficient of variation of diameter = 5.4%) had stretched 32.1% beyond its unloaded length." Data-sets (34): (1, 321), (3, 330), (5, 334), (8, 337), (16, 342), (32, 348), (50, 352), (110, 361), (240, 377), (440, 394), (740, 413), (1310, 442), (1460, 449), (1630, 458), (1900, 469), (2090, 478), (2760, 499), (2950, 505), (3080, 509), (3460, 519), (4280, 540), (4970, 556), (5720, 572), (6000, 579), (6320, 586), (7120, 600), (7360, 604), (7540, 607), (8520, 623), (9020, 629), (9230, 633), (9950, 643), (10260, 647), (10680, 654). (The last data set is for 10,680 minutes under load and 65.4% stretch.) SOURCE: O. Ripa and J. B. Speakman, 1951. "The plasticity of wool." *Textile Research J.* **21**: 215–221. Table I on page 217. PROBLEM: How can we straighten the plot?

exhibit **24** of chapter 6 (continued)

24c) Amount of interstitial space in young chickens. ("Interstitial space" is a volume defined by the amount of thiocyanate ion taken up outside the blood within 10 minutes after injection and measured by its ratio to blood volume.) (1, 52) translates as: "For chickens 1 week old, the interstitial space accounted for 52% of body weight." Data-sets (8): (1, 52), (2, 42), (3, 39), (4, 38), (6, 37), (8, 36), (16, 25), (32, 22). SOURCE: W. Medway and M. R. Kare, 1959, "Thiocyanate space in growing fowl," *Amer. J. of Physiology* **196**: 873–875. Table 1 on page 874. PROBLEM: How can we straighten the plot?

24d) A more detailed look at interstitial volume (1, 55, 52) translates into: "At age 1 week, the average body weight of 6 chickens was 55 grams, of which the interstitial space was 52%." Data-sets (8): (1, 55, 52), (2, 108, 42), (3, 175, 39), (4, 242, 38), (6, 372, 37), (8, 527, 36), (16, 1137, 25), (32, 1760, 22). SOURCE: As for problem (24c). PROBLEM: How can we straighten the plot? How does the straightness here compare with that for problem (24c)? What expression of amount (not percent) of interstitial space in terms of weight corresponds to our fit?

exhibit **25** of chapter 6: data and problems

**Some more problems**

25a) Vapor pressure of a boron analog of mesitylene. (130, 29) translates as: "At a temperature of 13.0°C, the vapor pressure of B-trimethylborazole was 2.9 millimeters of mercury." Data/(sets) (13): (130, 29), (195, 51), (225, 85), (272, 103), (318, 146), (384, 213), (457, 305), (561, 514), (644, 745), (714, 1002), (805, 1437), (857, 1769), (915, 2169). SOURCE: E. Wiberg, K. Hertwig, and A. Bolz, 1948, "Zur kenntnis der beiden symmetrischen Trimethyl-borazole ("anorganisches Mesitylen")," *Zeitschrift für Anorganische Chemie* **256**: 177–216. Table on page 191. PROBLEM: How can we straighten this plot? Check your answer!

25b) Effects of small amounts of biotin on the mobility of a microorganism. (5E–7, 1354) translates as: "At a biotin concentration of $5 \times 10^{-7}$, the mobility of *Lactobacillus casei* was 1.354 units". Data-sets (7): (0, 1415), (5E–7, 1354), (1E–6, 1311), (5E–6, 1230), (1E–5, 1234), (5E–5, 1181), (1E–4, 1188). SOURCE: V. R. Williams and H. B. Williams, 1949, "Surface activity of biotin," *Journal of Biological Chemistry* **177**: 745–750. PROBLEM: How can we straighten the plot?

25c) Electric current produced by heating aluminum phosphate. (880, 1) translates as: "At a temperature of 880°C--an absolute temperature of 880 + 273 = 1153°K--positive electrification produced a current of 1 unit, where 1 unit = $2 \times 10^{-9}$ amperes". Data-sets for run A (8): (880, 1), (950, 4), (970, 7), (995, 15), (1030, 35), (1030, 35), (1055, 49), (1110, 126). Data-sets for run B (9): (1036, 1), (1088, 8), (1135, 5), (1160, 8), (1195, 15), (1230, 34), (1245, 35), (1295, 74), (1330, 168). SOURCE: A. E. Garrett, 1910, "Positive electrification due to heating aluminium phosphate," (London, Edinburgh, and Dublin) *Philosophical Magazine* **20**: 571–591. Table on page 581. PROBLEM: How can we straighten the plots?

exhibit **25** of chapter 6 (continued)

25d) Growth-promoting effect of a purine for a deficient strain of red bread-mold. (0, 112) translates as: "When 0 moles of guanine per mole of adenine were included with 0.1 milligrams of adenine in 25 milliliters of basal medium, the dray weight of mycelium produced by a purine-deficinet strain of *Neurospora* was 11.2 milligrams". Data-sets (9): (0, 112), (0.25, 135), (0.59, 152), (0.75, 185), (1, 196), (1.5, 203), (2, 203), (2.5, 243), (3, 224). SOURCE: J. L. Fairley, Jr., and H. S. Loring, 1949, "Growth-promoting activities of guanine, guanosine, guanylic acid, and xanthine for a purine-deficient strain of *Neurospora*," *Journal of Biological Chemistry* **177**: 451–453. Table I on page 453. PROBLEM: How can we straighten the plot?

exhibit **26** of chapter 6: data and problems
**Still more problems**

26a) Measurement of a certain impurity in DDT; change of scale factor with temperature. (21, 248) translates as "At 21°C, the rate of crystallization (in microns per 5 minutes) is 24.8 times the log of the percent of this particular impurity." Data-sets (14): (21, 248), (22, 308), (23, 388), (24, 465), (25, 569), (26, 678), (27, 806), (28, 959), (29, 114), (30, 139), (31, 168), (32, 202), (33, 236), (34, 270). SOURCE: W. McCrone, A. Smedal, V. Gilpin, 1946, "Determination of 2,2, bis-p-chlorophenyl-1,1,1-trichloroethane in technical DDT: A microscopical method," *Industrial and Engineering Chemistry* **18**: 578–582. Table IV on page 582. PROBLEM: How can this plot be straightened?

26b) Demand deposits in post-office savings accounts in Switzerland. (37, 458) translates as: "In 1937, there were 458 million francs in post-office savings accounts". Data-sets (29): (37, 458), (38, 498), (39, 523), (40, 643), (41, 701), (42, 787), (43, 839), (44, 927), (45, 1001), (46, 1079), (47, 1007), (48, 1033), (49, 1090), (50, 1125), (51, 1212), (52, 1248), (53, 1334), (54, 1393), (55, 1443), (56, 1720), (57, 1720), (58, 1896), (59, 2050), (60, 2268), (61, 2643), (62, 3140), (63, 3353), (64, 3513), (65, 3810). SOURCE: *Swiss Statistical Abstract*, issued November 1969 by the Swiss Credit Bank. (Title also in French and German.) Tables at pages 24 and 25. PROBLEM: How can we best straighten a plot for this data? Make the fit and plot the residuals. Summarize all results.

26c) Revenue passenger miles on U.S. passenger airlines. (37, 412) translates as: "In 1937, there were 412,000 revenue passenger miles on U.S. domestic scheduled airlines". Data-sets (24): (37, 412), (38, 480), (39, 683), (40, 1052), (41, 1385), (42, 1418), (43, 1634), (44, 2178), (45, 3362), (46, 5948), (47, 6109), (48, 5981), (49, 6753), (50, 8003), (51, 10566), (52, 12528), (53, 14760), (54, 16769), (55, 19819), (56, 22362), (57, 25340), (58, 25343), (59, 28269), (60, 30514). (The last translates as: "In 1960, 30,514,000 revenue passenger miles".) SOURCE: Robert G. Brown, 1963. *Smoothing, Forecasting and Prediction of Discrete Time Series*, Prentice-Hall. Table C.7 on page 427. His source: *F.A.A. Statistical Handbook of Aviation.* PROBLEM: How can this plot be straightened? Plot residuals from a well-chosen straight-line fit.

26d) Find, from other sources, 2 batches of (x, y) data-sets that interest you and deserve to have their plots straightened by the methods of this chapter. Straighten the plots. Plot the residuals.

# index for chapter 7

rainfall, wheat, and gold  205
smoothing  205
smoothing gold, wheat, and rainfall  205
what approach?  207
*sequence*  208
review questions  210

## 7A. Medians of 3  210
*running medians*  210
repeated medians  213
notation  214
a comment  214
bituminous coal; suspended payments  214
review questions  214

## 7B. Eye resmoothing  214
judgment = choices  215
review questions  216

## 7C. Looking ahead  216
a comparison of four pictures  216
blurring, advance notice  218
agenda  218
review questions  220

## 7D. Copying-on--and more, usually  221
end-value smoothing  221
for those who prefer verbal arithmetic  221
3R or 3R′  222
a comment  223
review questions  223

## 7E. Blurring the smooth--and setting the fences  223
modified hinges, medians, sizes, etc.  223
exact zero  223
*-letter values  223
proper laziness  225
outside values  226
*smooth supported by the rough*  227
review questions  227

## 7F. Splitting peaks and valleys  227
smoothing short subsequences  227
going farther (optional)  230
a comment  231
review questions  231

## 7G. Hanning  231
*skip means*  232
comments  234
going a little farther  234
going a lot farther  234
what of the rough?  234
review questions  235

## 7H. How far have we come?  235

| EXHIBITS | | PAGE |
|---|---|---|
| 1 | | 206 |
| 2 | | 206 |
| 3 | | 207 |
| 4 | | 207 |
| 5 | | 209 |
| 6 | | 209 |
| | 7A | |
| 7★ | | 211 |
| 8★ | | 212 |
| | 7B | |
| 9 | | 215 |
| | 7C | |
| 10 | | 216 |
| 11 | | 217 |
| 12 | | 219 |
| 13 | | 220 |
| | 7D | |
| 14 | | 222 |
| | 7E | |
| 15★ | | 224 |
| 16 | | 226 |
| | 7F | |
| 17★ | | 228 |
| 18 | | 230 |
| | 7G | |
| 19★ | | 232 |
| | 7H | |

# Smoothing sequences   7

"Why do I want to smooth?" This is not an unlikely question in a reader's mind when he or she first comes to this chapter. A few examples may show why it often helps to smooth.

### rainfall, wheat, and gold

Exhibit **1** shows the annual precipitation in New York City, year by year, from 1872 to 1958. What we see is mainly a blur--quite spread out but apparently horizontal. Can we look under this blur and see anything more?

Exhibit **2** shows the production of wheat in the U.S., year by year, from 1897 to 1958. What we see here is a rising blur, perhaps with some peaks on it. Can we look under this blur and see anything more?

Exhibit **3** shows the production of gold in the U.S. (Alaska included), year by year, from 1872 to 1956. This is less blurred, with various ups and downs moderately clear. Can we see all that we would like to see here? Can we unhook our eyes from the irregularities?

### smoothing

Most of us could do a fair job of looking at exhibit 3 and drawing a "smooth" curve "through" the points. Doing a similar thing with exhibit 2 would be more difficult, and doing anything useful about exhibit 1 by eye alone would be very hard indeed.

### smoothing gold, wheat, and rainfall

The techniques we will learn in this chapter, when applied to exhibit 3, give the result shown in exhibit **4**. While this "smooth" curve has some wiggles left that we might--or might not--like to smooth off, in general it does a pretty good job of showing us how gold production has behaved, details set aside. We may feel that the 1873-4 peak should be shown as going higher, for example, but for many (if not most) purposes, this "smooth" tells us well about the general run of gold production in the U.S. The value of smoothing --here, as elsewhere--is the clearer view of the general, once it is unencumbered by detail.

The same techniques applied to the wheat-production data in exhibit 2 gives the "smooth" of exhibit **5.** Now we see not only the general rise but some swings up and down. The low patch in the middle thirties and the high patch(es) in the teens and the late 40's are now quite striking. We see quite a lot more from the smooth than we could from the data.

Finally, exhibit **6** shows a "smooth" of the data of exhibit 1. Where once we could see almost nothing, we can now see a lot. Periods of wetness in 1884 to 1904 and 1934 to 1953 seem to be surrounded by dry spells, some of which last 25 to 30 years. If the water supply of New York City were your concern, you would learn much from exhibit 6 that would be hard to gain from exhibit 1.

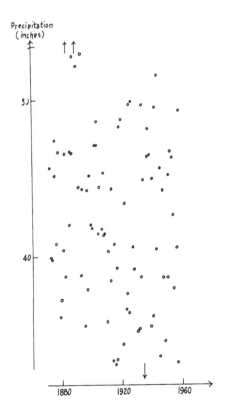

exhibit **1** of chapter 7:
N.Y.C. precipitation

**Annual precipitation in New York City, 1872–1958--actual**

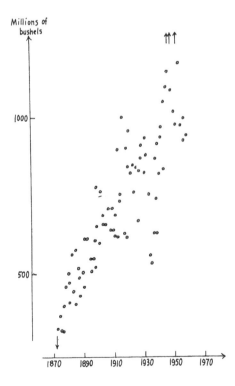

exhibit **2** of chapter 7:
wheat production

**U.S. annual wheat production, 1872–1957--actual (in millions of bushels)**

### what approach?

Now that we understand something of what we want to do, how do we proceed? In earlier chapters, we gave straight lines a special place, first by working hard to straighten plots, and then by emphasizing residuals from lines and working with schematic plots based on them. Straight lines are as important as their treatment in those chapters would suggest.

But not everything *is* a straight line. We need techniques that will help us see what the data are saying, even when they don't follow a straight line.

One approach is to try to climb the polynomial ladder. (Bah, it isn't straight; try a quadratic! Double bah, it isn't quadratic; try a cubic--and so on.) Occasionally this works well--especially when (1) the curve is nearly

exhibit **3** of chapter 7:
gold production

**U.S. gold production, 1872–1956
-- actual (in millions of fine
troy ounces)**

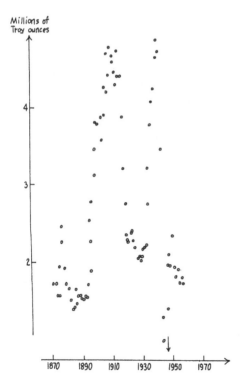

exhibit **4** of chapter 7:
gold production

**U.S. gold production, 1872–1956
-- smoothed (in millions of fine
troy ounces)**

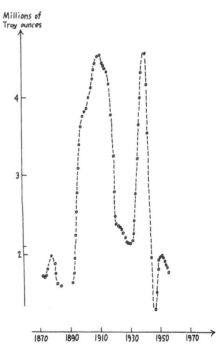

straight or (2) there is some principle that points to a particular kind of polynomial. These cases are nice--when they happen. Counting on their happening to you, however, is unsafe.

While it was profitable to start with a family of straight lines and then help the data choose one as "the fit", this is not a good pattern for smoothing. Often the data can be usefully thought about in terms of some relatively smooth curve--even though it would be difficult, unpleasant, or just impossible to specify a family of relatively smooth curves from which the one we use came. If we have no selected family of curves, how can we go ahead?

The two keys are reliance on nearby values for guidance and a number of steps of smoothing. At each step, we compare each value with a few adjacent values--often only 2 or 3--and alter it as they suggest. Continuing this process with steps of various kinds, we may hope that the result gets smoother and smoother. (It may tend to drift away a little, too. We will learn to deal with this when we come to chapter 16.)

Techniques that lead us to "smoothings" of the data which we can think of as reflecting smooth curves can be useful, particularly since their use preserves much of our flexibility. Because we have smooth curves in mind, the process is called smoothing. As a fitting process, it leads to the general relation

$$\text{data} = \text{fit PLUS residuals}$$

which here can be written

$$\text{given data} = \text{smooth PLUS rough}$$

Since the **"smooth"** is intended to be smooth, we will usually show its points **connected.** Sometimes we join the points of the smooth sequence with line segments, obtaining a chain of line segments. Sometimes we just draw a "smooth curve" through these points. Since the **"rough"** is rough, we show its residuals as **separated points.**

The simplest kind of $(x, y)$ data has $x$ proceeding in order by equal steps, as when we have a value every year--or every mile. We can smooth such data without any specific reference to the values of $x$--our attention can be focussed on the

$$\text{sequence}$$

made up of the successive values of $y$. For equal $x$-spacing we start with a given sequence--and from it we find two sequences, a smooth--a sequence of smooth values--and a rough--a sequence of residuals. This is the simplest and neatest case, and we can learn to deal with it effectively in a simple way.

In this chapter we learn about simple and easy smoothing, leaving elegance of appearance, which is often valuable, to a later chapter and to other accounts.

In this chapter, where we deal only with sequences, we can forget about the $x$'s; similarly, we will often use "date" to describe "when" an input value belongs. We do this when the input values come once a year. We would do it equally freely if they came once every million years, once every 7.543 days, or once every tenth of a second.

We might well do it, also, when the input values come once a mile down a road from a given starting point. So long as we have equal spacing, we usually may as well use the words "time" and "date", for what we do--and how well

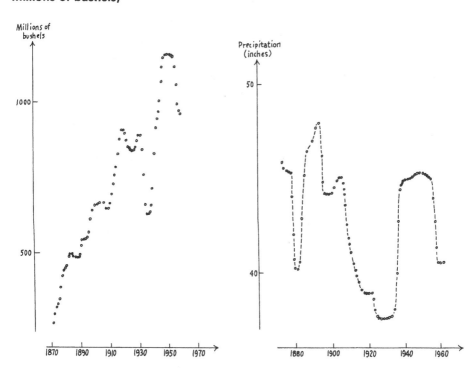

exhibit **5** of chapter 7:
wheat production

**U.S. annual wheat production, 1872–1957--smoothed (in millions of bushels)**

exhibit **6** of chapter 7:
N.Y.C. precipitation

**Annual precipitation in New York City, 1872–1958--smoothed**

it serves us—will both be very close to what would happen if we did have values equally spaced in time. The words "time series" are often used in similar situations. Since their use often goes with assumptions of more detailed structure for the data and/or use of more complex techniques, we will stick to "sequence", which is equally precise.

**review questions**

What examples did we start with? How did each set of data look? What did their smoothing do for us? What do we want to look under? What do we want to unhook our eyes from? Are straight lines important? Are they always good enough? What about polynomials? Can we look for a fit without explicitly specifying what we are fitting? What is the role of repeated steps? Of adjacent values? What is smoothing? What is a sequence? What is a smooth? What is a rough?

## 7A. Medians of 3

Given a sequence of numbers like this:

4, 7, 9, 3, 4, 11, 12, 1304, 10, 15, 12, 13, 17

we easily see that a reasonable smoothing of this sequence will increase slowly from near 5 to near 15, paying as little attention as reasonable to 1304.

1304 may be real, indeed very important, but it has nothing to do with a smooth curve. It ought to generate a very large residual, and leave the smooth fit as nearly alone as we can conveniently arrange to have it do. Whether OR not it is an error, we do NOT want it to disturb the smooth fit.

Let us take the values 3 at a time, rearrange each set of three in order of value, and take their median, as in exhibit **7**. The first three have the same order in value as in sequence, so little happens. The 2nd, 3rd, and 4th in sequence order have a different order when put in order of value. The 7, which is 2nd (middle) in value—and thus the median—is not the middle one in sequence.

It is easy to learn to look at three successive values in a sequence, find the middle one in order of value, and write it down—across from the middle one in sequence order. (A moving finger, covering up the values on one side or another of the three being medianed, often helps.) Exhibit 7 shows how this works.

Because the blocks whose medians we take can be thought of as sliding along, such sequences of medians are often called moving medians or

**running medians**

We have made a good start on smoothing this particular sequence, but we have yet to learn what to do about the missing values at the ends.

exhibit **7** of chapter 7: illustrative

**Overlapping blocks of three, their medians, and the result**

**A)** The INPUT

4, 7, 9, 3, 4, 11, 12, 1304, 10, 15, 12, 13, 17

**B)** EXAMPLES of RUNNING MEDIANS--we will never bother with this much detail again

| Three successive values || |Median| |
|---|---|---|
| As given (sequence order) | In order (value order) | |
| 4  7  9  | 4  7  9  | 7 |
| 7  9  3  | 3  7  9  | 7 |
| 9  3  4  | 3  4  9  | 4 |
| 3  4  11 | 3  4  11 | 4 |
| 4  11 12 | 4  11 12 | 11 |

**C)** The "3" SMOOTH and its ROUGH

| Given | Smooth | Rough |
|---|---|---|
| 4 | ? | ? |
| 7 | 7 | 0 |
| 9 | 7 | 2 |
| 3 | 4 | −1 |
| 4 | 4 | 0 |
| 11 | 11 | 0 |
| 12 | 12 | 0 |
| 1304 | 12 | 1292 |
| 10 | 15 | −5 |
| 15 | 12 | 3 |
| 12 | 13 | −1 |
| 13 | 13 | 0 |
| 17 | ? | ? |

Examples: 7, 7, 4, 4, 4, etc., as in **(B)**. $0 = 7 - 7$, $2 = 9 - 7$, $-1 = 3 - 4$, etc.

**P)** PROBLEMS

7a) Smooth the smooth of (C) above by running medians of 3. Is the result smoother? Why/why not?

7b) Smooth the rough of (C) above by running medians of 3. Comment on the result.

7b2) Smooth the result of (7b) again by running medians of 3. Comment on the result.

**repeated medians**

We are now ready to see what happens when we take repeated medians of 3. Starting with an extension of the same example by two more items (20 and 24), the first two smoothings by 3's are:

```
4 7 9 3 4 11 12 1304 10 15 12 13 17 20 24
? 7 7 4 4 11 12    12 15 12 13 13 17 20  ?
? ? 7 4 4 11 12    12 12 13 13 13 17     ?
```

In this instance, after the final smoothing by 3's--here the 2nd, since nothing more will happen if we smooth by 3's again--the behavior of the smoothed sequence is put together from these kinds of (overlapping) parts: steady upward, flat tops (of length at least 2), steady downward and flat bottoms (of length at least 2). All such parts will remain untouched by further median smoothing by 3's, so we know we were right to say that we were done.

exhibit **8** of chapter 7: coal production; suspended deposits

**Bituminous coal production (in millions of net tons per year) in the United States, 1920–1968, and deposits in suspended banks (in millions of dollars) in the United States, 1921–1967**

| Year | Coal production | 3* | 3R | Suspended deposits | 3R |
|---|---|---|---|---|---|
| 1920 | 569 | ? | | | |
| 1 | 416 | 422 | | 196 | ? |
| 2 | 422 | 422 | | 111 | 189 |
| 3 | 565 | 424 | | 189 | 189 |
| 4 | 484 | 520 | | 213 | 189 |
| 25 | 520 | 520 | | 173 | 213,194 |
| 6 | 573 | 520 | | 272 | 194 |
| 7 | 518 | 518 | 518 | 194 | 194 |
| 8 | 501 | | | 139 | 194 |
| 9 | 505 | | | 230 | 230 |
| 30 | 468 | | | 853 | 230 |
| 1 | 382 | 382 | | 1691 | 853,116 |
| 2 | 310 | 334 | | 716 | 1691,853 |
| 3 | 334 | 334 | | 3599 | 716 |
| 4 | 359 | 359 | | 37 | 37 |
| 35 | 372 | | | 10 | 11 |
| 6 | 439 | | | 11 | 11 |
| 7 | 446 | | | 11 | 11 |
| 8 | 349 | | | 19 | |
| 9 | 395 | | | 13 | |
| 40 | 461 | | | 34 | |
| 1 | 511 | | | 5.9 | |
| 2 | 583 | | | 3.7 | |
| 3 | 590 | | | 1.7 | |
| 4 | 620 | | | 6.2 | |

exhibit **8** of chapter 7 (continued)

| Year | Coal production | 3* | 3R | Suspended deposits | 3R |
|---|---|---|---|---|---|
| 45 | 578 | | | .40 | |
| 6  | 534 | | | 0   | |
| 7  | 631 | | | 0   | |
| 8  | 600 | | | .17 | |
| 9  | 438 | | | 0   | |
| 50 | 516 | | | .04 | |
| 1  | 534 | | | 3.1 | |
| 2  | 467 | | | 1.4 | |
| 3  | 457 | 457 | | 44.4 | |
| 4  | 392 | 457 | 457 | 2.9 | |
| 55 | 467 | 467 | 467 | 6.5 | |
| 6  | 500 | 493 | 493 | 11.9 | |
| 7  | 493 | 493 | | 12.9 | |
| 8  | 410 | | | 6.3 | |
| 9  | 412 | | | 2.0 | |
| 60 | 416 | | | 8.0 | |
| 1  | 403 | | | 7.5 | |
| 2  | 422 | | | 1.2 | |
| 3  | 459 | | | 23.3 | |
| 4  | 467 | | | 22.0 | |
| 65 | 512 | | | 45.9 | |
| 6  | 534 | 534 | | 0.7 | |
| 7  | 552 | 545 | | 11.8 | |
| 8  | 545 | ? | | | |

* "3" means here "smoothed by medians of 3".

P) PROBLEMS

8a) Complete the "3" column for coal production.

8a2) Would anything happen if this "3" column were smoothed by medians of 3 again? Why or why not?

8a3) What would a "3R" column look like for coal production?

8b) Complete the "3R" column for suspended deposits.

8c) Find a different sequence of data (at least 20 entries long) that interests you and do smoothing by medians of 3.

8c2) Complete the result to "3R" smoothing.

8d/8d2) As problem 8c/8c2, but for at least 50 entries.

S) SOURCES

**Coal Production;** *World Almanac,* 1946 page 635, 1957 page 715, 1965 page 721, 1970 page 149, 1969 page 128.
Their source: Bureau of Mines.

**Suspended deposits:** *World Almanac;* 1931 page 306, 1965 page 749, 1970 page 99.
Their source: Federal Reserve System.

**notation**

We write "3" for a single smoothing by medians of 3 and "3R" for repeated smoothing by medians of 3, a smoothing continued until there is no further change.

**a comment**

Notice that we are now well into the pattern suggested in the opener for this chapter:

⋄ we are using repeated steps.

⋄ each step is applied separately to each value.

⋄ what happens to a value in a step depends only on that value and a few neighboring values (here one on each side).

**bituminous coal; suspended payments**

Exhibit **8** gives the data, and part of the smoothing, for forty-odd years of: (a) U.S. bituminous coal production and (b) total deposits for those U.S. banks suspending payments in any year. For coal production we show the given values, part of a "3" smooth by medians of 3, and a smaller part of a "3R" smooth by repeated medians of three. For the suspended deposits, we show the given values, and the condensed calculation of its "3R" smooth. Notice, in the suspended-deposits example, at 1925, how we write "213,194" to mean that smoothing by 3's gave 213, while resmoothing by 3's gave 194. Notice also that, in these examples, once we have done a first smooth by medians of 3, going on to 3R involves little effort (and causes little change).

**review questions**

What is a running median? How do we calculate one? What partial behaviors cause values to be unchanged by running medians of 3? What behaviors ensure change? What happens when running medians of a given length are repeated? Are repeated many times? How can our fingers help? How does what we do fit the pattern announced in the opener of this chapter?

## 7B. Eye resmoothing

How can we make as effective use as possible of the 3R smooths we have just learned to make?

Such smooths are much better behaved than the raw data. If we had only the raw data, and wanted to tell ourselves about what it seemed to say, we could try to draw a smooth curve "through" it. It is much easier to draw a smooth curve through the points of our 3R smooth than through the raw data. Exhibit **9** shows what can happen when we do draw a smooth curve through the 3R? smooth of the bituminous coal data.

**judgment = choices**

Once we bring in judgment, we have choices. Exhibits **10** and **11** show two more eye smooths of the same 3R smooth we saw in exhibit 9. In exhibit 10 we have chosen the long view--one so long that 10-year-long peaks are not of interest. In exhibit 11 we have gone in the opposite direction, following the 3R-smooth points **more** closely than in exhibit 9 rather than less.

Different users are always likely to prefer different choices. A great advantage of judgment is its ability to meet different needs. (The corresponding disadvantage is shown by the arguments among those who want to make different judgments.) Those who eye-smooth their 3R smooths are using a minimum of arithmetic and a maximum of judgment.

Those who want to use more arithmetic and somewhat less judgment can use their judgment to pick out one of the smooths that we shall learn to do in later sections of this chapter (or even what we will learn when we come to Chapter 16).

exhibit **9** of chapter 7: coal production

**A moderate eye smooth of the numerical NE smooth of exhibit 3 (the first of three eye smooths of the 3R? smooth)**

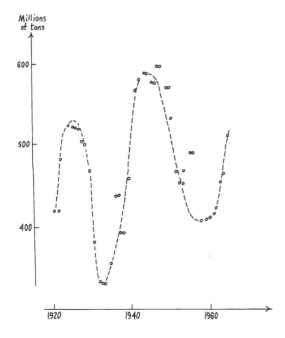

**exhibit 10/7: Smoothing sequences**

**review questions**

Is there a place for eye-fitting combined with numerical fitting? Why/why not? Can we use snap judgment for arithmetic? Why/why not?

## 7C. Looking ahead

We now know how to do the very simplest and easiest smooths. Let us look to see how much better we could do if we wanted to, and how we can build easily on what we now know.

**a comparison of four pictures**

Exhibit **12** pictures, for the bituminous-coal example,

⋄ the raw data (in NW corner).
⋄ the data smoothed as in section 7A (3R? smooth, in NE corner).

exhibit **10** of chapter 7: coal production

**A very general eye smooth of the NE smooth of exhibit 3 (the second of three eye smooths of the 3R? smooth)**

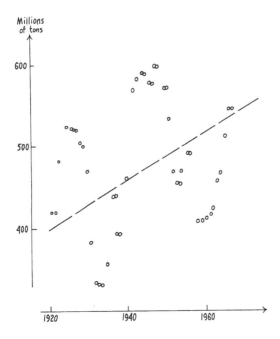

⋄ the data smoothed as of the end of section 7E (3RSS smooth, in SW corner).

⋄ the data smoothed quite heavily, as in section 7F (3RSSH3RSSH3 smooth, in SE corner).

Here the "?" in "3R?" refers to our postponed treatment of end values, to be settled in section 7D, the "S" in "3RS" to splitting, which will be explained at the end of section 7F, and "H" to "hanning", which will be explained in section 7G.

These pictures deserve careful looking at and careful thinking over. Notice:

⋄ that, as we go from NW to NE to SW to SE, the general behavior becomes clearer.

⋄ that, as we go back any step, we can see, by looking carefully, that whatever we saw in the later picture can in fact be seen in the earlier one--but with greater effort.

exhibit **11** of chapter 7: coal production

**A detailed eye smooth to the NE smooth of exhibit 3 (the third of three eye smooths of the 3R? smooth)**

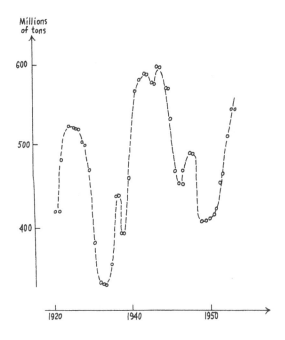

○ how hard it is to see the general behavior in the NW version, which is so irregular as to make attempts to join the points unproductive.

○ how MANY actual details are gone in even the NE version, while essentially ALL details are gone in the SE version.

○ how close the eye-smoothing of exhibit 9 came to the final (SE) version.

This example shows clearly many of the main points about smoothing. The analyst who needs the actual details cannot depend on the smooth (though he can often do well with smooth and rough combined). The analyst who needs general behavior naturally turns to a smooth (as often he MUST). The purpose of the smooth is to show off general behavior. It is NOT to show most of the actual details. And, at least in its present form, it is NOT to give an impression of variability, NOT to show how precise we ought to think the smooth is. (We make it smooth to ease the eye's task and keep the eye from getting stuck on trivia. We often want it to be smooth enough so that if we compared several different smooths, all reasonable enough, they would look quite different.)

### blurring, advance notice

Fortunately for those who do not want to go to more complicated arithmetic, and who find-eye smoothing "too indefinite," there is another way to unhook the viewer's eye from the minor wiggles of a less-than-polished smooth. We can do this by blurring the smooth, something easily done by replacing points by short vertical lines of appropriate length. Exhibit **13** shows a blurring of the SW version of exhibit 3, in comparison with the more highly polished SE version (of the same exhibit). Clearly we have gained by blurring a lot--probably most--of what additional arithmetic could bring us.

We will want to do such blurring with a view to the size of the rough. Details will wait until we reach section 7E below.

### agenda

What do we now know we want to learn? Surely these things:

○ how to deal with end values   (7D).

○ how to blur   (7E).

○ how to deal with the flat peaks and valleys only two entries wide (noticeable in exhibit 7, for instance)   (7F).

○ how to do a still smoother smooth   (7G).

In addition, we will eventually come, in chapter 16, to ways for following ups and downs more closely, thus leaving little systematic behavior in our roughs and thus getting still better smooths. In this chapter, however, we will learn no further new principles, only techniques for doing a smoother job than 3R? does. (In the optional sections, chapter $7^+$, these same principles will be used more thoroughly.)

exhibit **12** of chapter 7: coal production

**Raw data and three smooths--U.S. bituminous coal production in millions of tons (data in exhibit 8)**

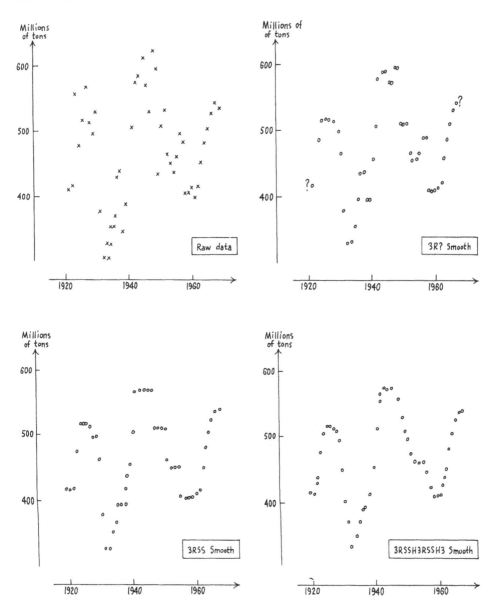

**review questions**

What happened when we compared various smooths of the same data? What example did we choose? What did we show as ways to unhook eyes from wiggles and corners? What do we need to learn now?

exhibit **13** of chapter 7: coal production

**Two different ways to unhook the viewer's eyes from the wiggles and corners of a moderately easy smooth**

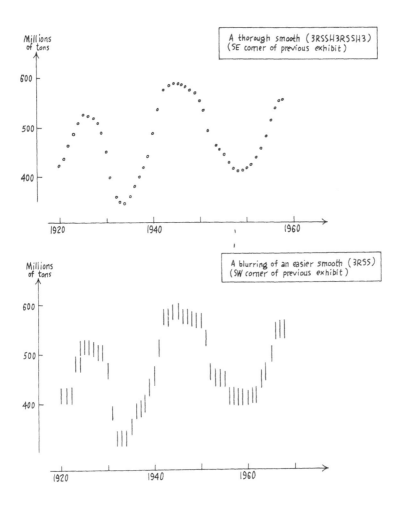

## 7D. Copying-on--and more, usually.

When we introduced repeated medians of 3 (in section 7A) we were purists, and let the question-marks creep in from each end. We did this so we could talk about one idea, repeated medians of 3, by itself. In the real world we will not let the ?'s creep in; instead we will "copy on" the end values. Thus the example on page 212 really looks like this:

```
4 7 9 3 4 11 12 1304 10 15 12 13 17 20 24
4 7 7 4 4 11 12    12 15 12 13 13 17 20 24
4 7 7 4 4 11 12    12 12 13 13 13 17 20 24
```

This will at least keep our sequences from shrinking, but is it enough? Occasionally, yes, when we think an end value may have a special message; but usually no. So we need some end-value smoothing. We turn now to the simplest end-value smoothing that seems to do what we need.

### end-value smoothing

We have now to be slightly complicated. Exhibit **14** shows some pictures. All show the same two end input values and the same five smoothed values. What we are to do uses only the extreme end input value and the next two smoothed values. The NW panel emphasizes the actual end value with a square box. The NE panel emphasizes the next-to-end smooth, also with a square box. The SW panel emphasizes the result of straight-line extrapolation of the next-to-end and next-but-one-to-end smoothed values--extrapolation NOT just to the end, but one step BEYOND.

The SE panel emphasizes not only all three of the points emphasized in the other panels but especially the median of these three, which is what we take as our smoothed value at the end.

### for those who prefer verbal arithmetic

A more arithmetic way to put it is this:

⋄ the change from the end smoothed-value to the next-to-end smoothed-value is between 0 and +2 times the change from the next-to-end smoothed-value to the next-to-end-but-one smoothed value.

⋄ subject to this being true, the end smoothed-value is as close to the end input-value as possible.

This means that we can look at two differences:

end input-value MINUS next-to-end smoothed-value

and

next-to-end smoothed value MINUS next-but-one-to-end smoothed value

and if the first is between 0 and +2 times the second, we can copy on.

Otherwise we make

> end smoothed-value MINUS next-to-end smoothed value

either zero or two times

> next-to-end smoothed-value MINUS next-but-one-to-end smoothed value.

3R    or    3R'

Since copying-on during the 3's, followed by end-value smoothing after completing the 3's, is the most usual case, we will use 3R to label it, saving 3R'

exhibit **14** of chapter 7: purely illustrative

**End-value smoothing: parts and result**

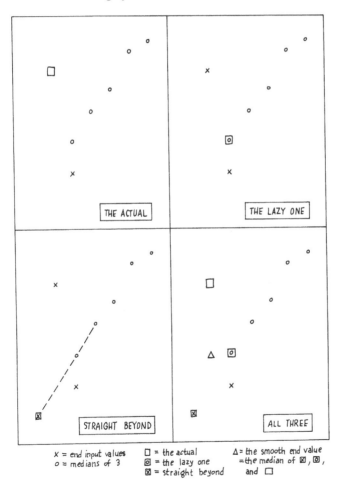

x = end input values    □ = the actual    △ = the smooth end value
o = medians of 3    ⊡ = the lazy one    = the median of ⊠, ⊡,
                              ⊠ = straight beyond    and □

to label the use of copying-on only (during repeated smoothing by medians of 3).

#### a comment

Notice that both copying-on and end-value smoothing again represent steps where a value is, at most, compared with a few nearby values.

#### review questions

How were we purists in 7A? What do we do in the real world? What is copying on? Do we allow smoothing by medians of 3 to shorten our sequences? Is copying-on usually enough? What else might we do? How often? When not?

### 7E. Blurring the smooth--and setting the fences

We want next to fix the lengths of the vertical lines we use to blur our smooth. Clearly these lines should go equally far above and below our 3R smoothed values. It seems natural to use the sizes of the residuals as a basis. But how?

In our schematic plots (chapter 2), we used a box to cover half the points. So let us start by going up-and-down far enough to cover half the rough. This means going as far as the median of the sizes of the residuals. Panels A and B of exhibit **15** show stem-and-leaf displays of the sizes of residuals for the two smooths of the bituminous-coal production that have concerned us most. Panel C calculates the ordinary medians, hinges, and eighths. Are we happy?

#### modified hinges, medians, sizes, etc.

For the rough that goes with the more highly polished smooth, whose rough had only 8 zeros out of 49, we can be fairly happy. To know that half the roughs have sizes of 11 or less is helpful. For the 3RSS rough--the one that goes with the 3RSS smooth--we are far from happy. Being told that the median size is 7 is some help--we get the idea, correctly, that many, many residuals are zero; but we really don't get an idea how large the residuals are. Perhaps we should modify our definitions.

One thing we could do easily would be to treat each

#### exact zero

as some fraction of a value, not a whole one. The easiest fraction is 1/2. Since it seems to work, we will use it. The median sizes of residual for the two examples are now 13 (instead of 11) and 12h (instead of 7). Both are informative.

We indicate such modification by putting a star, *, in front of each and any of M, H, E, ..., f, F, #, etc. that we calculate. We then call these

#### *-letter values.

exhibit **15** of chapter 7: coal production

**Looking at sizes of roughs, here also sizes of residuals**

**A)** The SE (polished) ROUGH--"Z" means zero

```
49  Z|ZZZZZZZZ
41  0|1112223333445778
25  1|13337
20  2|34
18  3|0145
14  4|1469
10  5|36
 8  6|289
 5  7|88
 3  8|1
 2  9|0
 1  H|(147)
```

**B)** The SW (3RSS) ROUGH--"Z" means zero

```
49  Z|ZZZ...(18 in all)...ZZZ
31  0|2444677
24  1|002233
18  2|445
15  3|06
13  4|44
11  5|1336
 7  6|05
    7|
 5  8|118
    9|
 2  H|(147, -138)
```

**C)** SIMPLE MEDIANS, HINGES, etc., of SIZES

(for the SE)      (for the SW)

#49 (all)         #49 (all)
M25|11            M25| 7
H13|44            H13|44
E 7|68            E 7|60

**D)** MODIFIED MEDIANS, HINGES, etc., of SIZES

(for the SE)           (for the SW)

*#45 = 41 + 4          *#40 = 31 + 9
*M23 | 13              *M20h | 12h
*H12 | 46              *H10h | 52
*E 6h | 68h            *E 5h | 73

→

From now on the use of *-medians (and other *-letter values) will be standard with sizes of residuals. That is why the vertical lines of exhibit 10 went up-and-down ±12h--the *M of the sizes of residuals is 12h.

**proper laziness**

Only one who loves either arithmetic or self-torture will make this kind of picture by adding 12h to and subtracting 12h from each smoothed value, and then plotting these two ends of each vertical line. It is much easier to do this:

⋄ plot all the smoothed values.

⋄ mark the edge of a piece of paper or card with three marks 12h apart.

⋄ put the center mark on each plotted point in turn, and draw the vertical between the endpoints. (If you use first graph paper and then tracing paper, as you should, it is usually easier to put only center and endpoints on the graph paper, and draw the lines only on the tracing paper.)

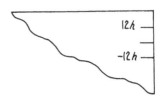

This keeps the work to a reasonable level.

exhibit **15** of chapter 7 (continued)

E) PARTS of STEM-and-LEAF DISPLAYS for the RESIDUALS--not their sizes

```
(for the SE)           (for the SW)
-0|874433221           -0|7642
-0|331                 -1|320
-2|4                   -2|4
-3|40                  -3|6
-4|4                   -4|4
-5|                    -5|6
-6|2                   -6|50
 L|(-78)                L|(-138)
```

Note: These are needed for some of the problems.

P) PROBLEMS

15a/b  Make up full stem-and-leafs for the residuals (SE for (a), SW for (b) and find (i) their simple medians, hinges, etc., including fences, and (ii) their modified medians, hinges, etc., including fences.

15a2/b2) Find what values of residuals are (i) outside and which are (ii) far out.

15c) Find raw and modified medians of the residuals of the 3R? smooth (found when exhibit 8 was completed; note that residual at each end is zero).

#### outside values

If we are to make good use of our roughs to support our smooths, just using the modified median of their sizes to fix how far our blurring lines extend is not enough. We also need to identify outside--and far out--points.

The problems of exhibit 15 included doing this for our two usual examples, using both simple hinges and modified hinges. With fences hung on simple hinges, the situation for the 3RSS example (SW version) is bad. With 18 zeroes, the hinges were at 10 and 0. Of 31 nonzero values, **only 8** were **not** outside, and 16 were far out. To call for identification and special treatment for 19 out of 49 residuals--or, if you prefer, 19 out of the 27 nonzero ones--is to dilute unusualness till it is not useful. (If we try to do the same with the 3R? rough, we will not find it much better.)

As we had to do with medians of sizes, so too will we have to do with fences. We shall use *-fences--fences hung on *-hinges--routinely in dealing with 3R roughs.

exhibit **16** of chapter 7: coal production

**Outside and far out points added to the blurred smooth of exhibit 4 (the 3RSS smooth)**

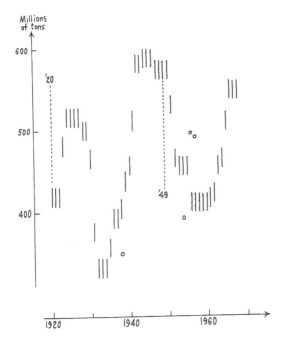

For the 3RSS case, then, we have *-hinges at −8h and 24, so that the *-fences are: first at −59 and 74, second at −107h and 123h, leaving 5 values outside and 2 values far out. Exhibit **16** shows these 7 added to the blurred smooth of exhibit 10.

When both blurring and outside points appear, we refer to the

smooth supported by the rough.

Notice that we have used a moderately weak circle for outside values, and an only moderately strong string of dots for far out values, and that we have tagged (here dated) only the far out values. This has been done so as to leave the individual points somewhat noticeable, without letting them interfere too much with the general impressions to be gained from the blurred smooth itself.

**review questions**

What earlier case guided us in deciding how broadly we should blur? What aspect of rough and smooth do we use to fix the breadth of the blur? What are modified hinges? Modified median sizes? Other modified letter values? (Does it make sense to modify a median? Why/why not? How do we show that letter values have been modified? How can we save effort in showing blurs? How do we hang fences when there are many zeroes? Where can we best show outside and far out points? What does "the smooth supported by the rough" mean? How can we keep outside (and far out) points from distracting our attention from the rest of the rough?

## 7F. Splitting peaks and valleys

The most annoying aspect of the kinds of smooths we are now getting is probably the little flat-topped hills and valleys, each 2 points wide, that are scattered through our smooth. How can we conveniently deal with these "bumps"? A simple approach is to treat each such pair as if they were end values--first copying on, and then using end-value smoothing--and then smooth once more by 3R. After the first 3R, then, we first break the result up into subsequences and then bring these subsequences back together and use 3R again to smooth over roughness at the joints.

Panels A and B of exhibit **17** illustrate, on some pieces of the 3R smooth of the bituminous-coal production data, just how such a calculation goes. Panel C carries out the calculation for a 25-year stretch of that data.

**smoothing short subsequences**

The last subsequence, formed by splitting the 457 2-wide valley, contains only two entries: 457, 467. It is too short to think of smoothing, though there is no reason not to use the followup 3R in the hope of pulling the subsequence smooths together. The first subsequence has three values: 468, 382, 334. These are unaffected by the first 3R. We are willing to continue with end-value smoothing, for which we have the fewest number of entries that will work. In the case of these three, nothing happens.

If we had had 468, 454, 334, end-value smoothing would have given 468, 454, 426. If we had had 468, 354, 334, it would have given 394, 354, 334. For a subsequence of three entries, end-value smoothing will change at most one.

exhibit **17** of chapter 7: bituminous coal

**Splitting the peaks and valleys illustrated on the 3R smooth**
**A) A VALLEY**

| ⌊3R⌋ | ⌊Split⌋ | ⌊Evs*⌋ | ⌊Reun*⌋ | ⌊3R⌋ | |
|---|---|---|---|---|---|
|  | 468 |  |  |  | Examples: |
| 468 |  |  | 468 | N |  |
|  | 382 |  |  | o | $210 = 382 - 2(468-382)$ |
| 382 |  |  | 382 |  | $334 = \text{median}(382, 210, 334)$ |
|  | 334 | 334 |  | c | $333 = 359 - 2(372-359)$ |
| 334 |  |  | 334 | h | $334 = \text{median}(359, 333, 334)$ |
|  |  |  |  | a |  |
| 334 |  |  | 334 | n |  |
|  | 334 | 334 |  | g |  |
| 359 |  |  | 359 | e |  |
|  | 359 |  |  |  |  |
| 372 |  |  | 372 |  |  |
|  | 372 |  |  |  |  |

\* Evs = end-value smoothing
Reun = reuniting

**B) A HILL**

| ⌊3R⌋ | ⌊Split⌋ | ⌊Evs*⌋ | ⌊Reun*⌋ | ⌊3R⌋ |
|---|---|---|---|---|
|  | 359 |  |  |  |
| 359 |  |  | 359 |  |
|  | 372 |  |  |  |
| 372 |  |  | 372 |  |
|  | 439 | 398 |  |  |
| 439 |  |  | 398 | 395 |
|  |  |  |  |  |
| 439 |  |  | 395 | 398, 395 |
|  | 439 | 395 |  |  |
| 395 |  |  | 439** | 395 |
|  | 395 |  |  |  |
| 395 |  |  | 395** |  |
|  | 395 |  |  |  |

\*\* These values come from 395, 395 being a valley, and corresponding calculations not shown in this panel. ➡

exhibit **17** of chapter 7 (continued)

**C** Above included in a 25 YEAR STRETCH--working style

|  |  | | 1st split | | | 2nd split | |
|---|---|---|---|---|---|---|---|
| Year | 3R | Raw | 3R | Raw | 3R |
| 1930 | 468 |  | 468 |  |  |
| 1 | 382 |  | 382 |  |  |
| 2 | 334⎤ | 334 | 334⎤ | 334 | 334 |
| 3 | 334⎦ | 334 | 334⎦ | 334 | 334 |
| 4 | 359 |  | 359 |  |  |
| 35 | 372 |  | 372 |  |  |
| 6 | 439⎤ | 398 | 395 |  |  |
| 7 | 439⎦ | 395 | 395 |  |  |
| 8 | 395⎤ | 439 | 398 |  |  |
| 9 | 395⎦ | 395 | 439 |  |  |
| 1940 | 461 |  | 461 |  |  |
| 1 | 511 |  | 511 |  |  |
| 2 | 583 |  | 583 |  |  |
| 3 | 590⎤ | 590 | 583 |  |  |
| 4 | 590⎦ | 578 | 590 |  |  |
| 45 | 578⎤ | 590 | 590 |  |  |
| 6 | 578⎦ | 600 | 590 |  |  |
| 7 | 600⎤ | 578 | 578 |  |  |
| 8 | 600⎦ | 516 | 516 |  |  |
| 9 | 516 |  | 516 |  |  |
| 1950 | 516 |  | 516 |  |  |
| 1 | 516 |  | 516 |  |  |
| 2 | 467 |  | 467 |  |  |
| 3 | 457⎤ | 457 | 457⎤ | 457 | 457 |
| 4 | 457⎦ | 457 | 457⎦ | 457 | 457 |
| 55 | 467 |  | 467 |  |  |

Notice:
1. Use of ] to mark 2-wide peaks and valleys.
2. Use of end-value smoothing to find all values in the "Raw" columns.
3. Repetition of the S process exactly twice.
4. That, in this example, splitting the second time had no effect.

**P) PROBLEMS**

**17a)** Extend this calculation to the entire 3R smooth for this data, finding a 3RSS smooth.

**17b)** The same for the suspended bank deposits example.

Notice that splitting is planned to be done exactly twice. Accordingly, we symbolize its use, after 3R, by writing

3RSS

as the label of our new smooth.

**going farther (optional)**

Two uses of "splitting" is one good standard--in many examples it seems to be a good place to stop. Sometimes, however, we can gain by going farther, or not so far.

When we do go farther, it sometimes seems useful to "split it to death" just as we are used to "3-ing it to death" when we do 3R. When we do, we

exhibit **18** of chapter 7: N.Y.C. precipitation

**The 1872–1945 portion of the precipitation record and three of its smooths compared**

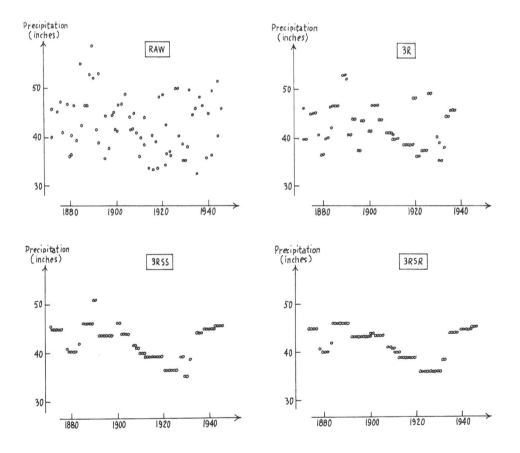

write "SR". Exhibit **18** shows, for the New York City precipitation example of the opener (exhibits 1 and 6) of this chapter, a comparison between

- the given data up to 1945 (NW corner).
- a 3R smooth (NE corner).
- a 3RSS smooth (SW corner).
- a 3RSR smooth (SE corner).

In this example, it clearly is worthwhile to go all the way to SR.

Notice also that, just as 3R smoothings are often the same as 3 smoothings would be, so SR smoothings are often the same as SS smoothings would be.

**a comment**

We are still proceeding step-by-step, comparing each value with nearby values. (We have to look at 4 adjacent values to identify a peak or valley that is only 2 points wide. We have to look at only 3 values to make the end-value adjustment.)

**review questions**

What problem concerns us in this section? How do we approach it? What happens when a 2-wide peak is adjacent to a 2-wide valley? Which end-value rule do we use? What do we do after "splitting"? How many times do we do all this? What do we use to tag the new smooth?

## 7G. Hanning

A steadily increasing sequence, such as

$$4, \ 5, \ 6, \ 8, \ 11, \ 74, \ 78, \ 79, \ 80, \ 81$$

will not be affected by the most powerful smoothers we have so far found it necessary to use--3RSS or 3RSR. (So long as the values increase [or decrease] steadily, the median by value of three adjacent numbers will be the same as the median in sequence. Thus "3R" will have no effect and "S" will not be called on to do anything.)

We may not find such a sequence smooth enough for our task. How then are we to make it smoother? A little thought leaves us with the belief that nothing simpler than old-fashioned arithmetic will do. Surely, however, we want to keep the effort involved as small as we can. The one piece of arithmetic that is simplest for us is that involved in taking the median of a batch of 2, 4, or some other even count of numbers, namely

$$\frac{\text{(one number)} + \text{(another number)}}{2}$$

How can we use this in our smoothing?

Exhibit **19** shows us two ways to reach the same result. In panels A and B we take the means of adjacent values, not once but twice. (Notice that it is natural to write the results of the first step half-way between the lines we used for the start.)

In panels C and D, designed to avoid half-lines, we do

$$\frac{(\text{one number}) + (\text{another})}{2}$$

twice, also, but in a different pattern. We start by forming

skip means

in which the two values combined lie one line above and one line below the line on which the skip mean is entered. Thus 11 and 78, no matter what

exhibit **19** of chapter 7: illustrative

**Introduction to hanning (or how to use $(a + b)/2$ to make things smoother)**

A) MEANS OF PAIRS OF ADJACENT NUMBERS

| Start | Mean of adjacent |
|---|---|
| 4 | |
| | 4h |
| 5 | |
| | 5h |
| 6 | |
| | 7 |
| 8 | |
| | 9h |
| 11 | |
| | 42h |
| 74 | |
| | 76 |
| 78 | |
| | 78h |
| 79 | |
| | 79h |
| 80 | |
| | 80h |
| 81 | |

B) The SAME REPEATED

| Start | Mean of adjacent | Again |
|---|---|---|
| (4) | | |
| | 4h | |
| 5 | | 5 |
| | 5h | |
| 6 | | 6 |
| | 7 | |
| 8 | | 8 |
| | 9h | |
| 11 | | 26 |
| | 42h | |
| 74 | | 59 |
| | 76 | |
| 78 | | 77 |
| | 78h | |
| 79 | | 79 |
| | 79h | |
| 80 | | 80 |
| | 80h | |
| (81) | | |

Examples:
4h = (4 + 5)/2,  5h = (5 + 6)/2
7 = (6 + 8)/2,   6 = (5h + 7)/2
8 = (7 + 9h)/2, 26 = (9h + 42h)/2

→

separates them in the start, come out with a skip mean thus:

$$\begin{array}{cc} & 11 \\ xy & 44h \\ & 78 \end{array}$$

We head a column of skip means with ">".

Once we have a full set of skip means, we set about forming means of the last two values in each line, so that 74 and 44h always give 59, as in the pattern

$$\begin{array}{ccc} tu & vw & \\ 74 & 44h & 59 \\ za & bc & \end{array}$$

no matter what comes above or below them in either column. Again the only difficulty--here not at all serious--is end values. With only one value in the

exhibit **19** of chapter 7 (continued)

| C) ISOLATED SKIP MEANS | | D) A FULL SET of SKIP MEANS followed by LINE MEANS | | | | |
|---|---|---|---|---|---|---|
| ⌊In⌋ | ⌊>⌋ | ⌊In⌋ | ⌊>⌋ | ⌊H⌋ | ⌊(copy on)⌋ | |
| 4 | | 4 | | | (4) | |
| 5 | 5 | 5 | 5 | 5 | | |
| 6 | | 6 | 6h | 6 | | |
| 8 | 8h | 8 | 8h | 8 | | |
| 11 | | 11 | 41 | 26 | | |
| 74 | 44h | 74 | 44h | 59 | | |
| 78 | | 78 | 76h | 77 | | |
| 79 | | 79 | 79 | 79 | | |
| 80 | 80 | 80 | 80 | 80 | | |
| 81 | | 81 | | | (81) | |

Examples:

5 = (4 + 6)/2, 8h = (6 + 11)/2

44h = (11 + 78)/2, 80 = (79 + 81)/2, ...

5 = (5 + 5)/2, 6 = (6 + 6)/2, 8 = (8 + 8h)/2

26 = (11 + 81)/2, 59 = (74 + 44h)/2, ...

P) PROBLEMS

19a ) Hann the result of (17a), which was a 3RSS smooth of the bituminous coal production example.

19a2) "3" the result of (19a).

19a3) Smooth the result of (19a2), using 3RSSH3 again.

19a4) Find the rough, corresponding to the result of (19a2), and smooth it with 3RSSH3; comment on the result.

line, about all we can do sensibly is to copy on (as shown in parentheses). We head a column of such means with "H", for hanning.

Notice that, except for end values, the two procedures--that of panel B, and that of panel D--give the identical result. That means that we can leave the choice among them to the reader. Printers and those who like to save on paper--or the cost of books--will be likely to prefer the skip and line means of panel D, with its avoidance of half lines. The writer has decided that he finds this approach easier to use by hand, also. So all examples from here on will use the ⟩ (for skip mean), H (for line mean) approach. (Readers need not follow this choice.)

The name hanning is after Julius von Hann, an Austrian meteorologist of the last century, who liked to smooth weather information--temperatures, pressures, etc.--in this way.

### comments

Again we add a new kind of step that involves only a few adjacent values --three for hanning. Again we apply this step separately to each and every value (end values treated specially). Again we assemble a variety of such steps to make a smoother that is better than any one individual step could provide.

If we fear the influence of stray values, as we almost always should, we would not dare hann the raw data. An isolated stray point, if hanned directly, will spread and become three stray points. While they will not stray as far, their three-ness is likely to trouble us. We trust to the medians of 3R--and the short peak and short valley removal of SS--to bring most strays to values that will not trouble us too much, **before** we hann.

### going a little farther

Especially in some of the uses in the next chapter, we find that 3RSSH has some residual local roughnesses that catch our eye when we wish this didn't happen. An easy thing to do is one more smoothing by running medians of 3. We naturally call the result 3RSSH3. (See problem (19a2).)

### going a lot farther

Anything we can do once, we can do twice. So having applied 3RSSH to the given data, we can apply 3RSSH to the 3RSSH-smooth. The result is a 3RSSH3RSSH smooth, and we can sprinkle another "3" on the tail end, thus making 3RSSH3RSSH3, if we wish. (See problem (19a3).)

### what of the rough?

Problem (19a4) asked us to smooth the rough for the 3RSSH3 smooth of the coal-production example. Those who did this found that this smooth did not come out to be zero. Clearly, this means that there is more that we should subtract from the rough before we look at the rough in too much detail. When we come to chapter 16 we will do just that.

Sometimes the change in the smooth from being careful about what was left in the rough is important. Often it is not. We have had enough details in this chapter to make it wise for us to postpone such matters.

**review questions**

What kind of sequences that escape 3RSS smoothing might we want to smooth further? How did we decide to tackle them? Are there alternative ways to hann? What are their advantages/disadvantages? Can you describe one in detail? how do we head which columns? Why is it called "hanning"? How does hanning fit into the general pattern of this chapter? Would we dare to hann raw data? How can we go a little farther? A lot farther? What are we postponing to Chapter 16?

### 7H. How far have we come?

This chapter has introduced us to the smoothing of sequences. Sometimes such smoothing will let us see regularities where without it there was only a fog. More often, perhaps, smoothing lets us see easily what we might have seen by patient and detailed examination and re-examination.

A sequence from which to start, and a smooth found from it, implies the existence of a rough, satisfying

$$\text{input} = \text{smooth} + \text{rough}$$

the current version of

$$\text{data} = \text{fit} + \text{residual}.$$

We have postponed to chapter 16 any detailed consideration of roughs. (This has, in particular, let us keep our smoothing techniques simpler.)

Running medians of 3 have been our prime tool. (Surely nothing could be simpler for hand computation.) Almost always we used them repeatedly, until no further change occurs.

We have a wide choice as to how thoroughly we smooth a sequence. If we smooth it too hard, we may miss detail we wanted to see. If we don't smooth it enough we may find our eyes hung up on irrelevant detail. What is "too much" or "not enough" will depend upon the specific example.

In particular, we are ready:

⋄ to use, but not always, end-value smoothing, thus finding a smooth for the extreme ends of the sequence, where running medians of 3 do not apply.

⋄ to blur the smooth carefully, thus detaching our eyes from the smooth and saving us from wanting or needing to smooth further.

◊ to use end-value smoothing to help us deal with the narrowest peaks and valleys (those made up of only two points) that running medians of three leave behind (splitting).

◊ to use hanning, conveniently represented in terms of split means and line means, to smooth still further.

# Optional sections for chapter 7    7+

chapter index on next page
___

The three sections that follow take the considerations of chapter 7 somewhat further. Exhibit numbers and section letters are continued.

### 7I. Breaking a smooth

Sometimes the given values of a sequence show a clear break. When this happens we have a choice:

◇ we may smooth across the break.

◇ we may treat the data as two sequences, applying the end-value smoothing on either side of the break.

Which we do depends on our purposes and our understanding of the data.

If we believe that there was a very unusual change between two adjacent values--whether from an extra-heavy dose of whatever drives the values of our particular sequence up and down OR from an isolated event that happens quite rarely--we are not likely to want to smooth across the gap with any smoother more polishing than 3RSS. If we go on to 3RSSH or 3RSSHS, we will surely find the values next to the supposed break pulled toward each other. Indeed, if we use only 3R, we may at times see this happen. To have these adjacent values pulled toward each other is a way to hide the "break". If we think the break is "real", we usually do not want to hide it.

If, on the other hand, we think the possible "break" is only a somewhat unusually large change, we will not want to preserve a break--rather we will want to smooth over it.

Whether we want to show a "break" or a smoother appearance will also be affected by whether a noticeable break in the smooth will seriously distract attention from other aspects that we particularly need to see--or to show to others--in this particular picture.

#### presidential New Hampshire

Exhibit **20** starts from the Presidential vote in the two northernmost of the counties of New Hampshire that adjoin the state of Maine, noting the

## index for chapter 7+

**7I. Breaking a smooth** 237
    presidential New Hampshire 237
    back to suspended deposits 241
    review questions 243

**7J. Choice of expression** 247
    first 3R 248
    second, end values 248
    third, residuals as a basis of choice 248
    comparing expressions 250
    fourth, picturing the smooth 250
    counted data 250
    suspended bank counts 256
    review questions 256

**7K. A 2-section example** 259
    review questions 263

**7M. How much more may we have learned?** 264

| EXHIBITS | PAGE |
|---|---|
| 7I | |
| 20★ | 239 |
| 21 | 241 |
| 22 | 242 |
| 23 | 243 |
| 24 | 244 |
| 25★ | 245 |
| 26 | 246 |
| 27 | 247 |
| 7J | |
| 28★ | 249 |
| 29★ | 251 |
| 30 | 257 |
| 31 | 258 |
| 32★ | 259 |
| 33 | 263 |
| 7M | |

percent of the vote of each county for the Democratic candidate in each election from 1896 to 1972. The difference in percent Democratic, Coos MINUS Carroll, starts small, rises somewhat in 1916 to 1924, and breaks upward in 1928, and then remains close to 25%. The 3R column of exhibit 20 shows the result of the broken smooth. (The parenthetic entries in the "(unbroken)" column show what would have happened had we not broken the data before smoothing.)

Exhibit **21** shows the broken smooth and hints at the unbroken version. Clearly we have a wide choice between appearances. We can have a wide gap-- about 15% change--or we can have a smooth which shows almost no unusual change at all. Which should we prefer (as amateur psephologists!)?

It seems to be a tradition among political scientists to stress critical changes in individual elections. Such a tradition would stress the change between '24 and '28 and thus lead to a preference for the broken smooth. Those who go this way would presumably describe behavior in two intervals: 1896–1924 and 1928–1972.

exhibit **20** of chapter 7[+]: Coos–Carroll

**Percent Democratic vote for President in two counties of New Hampshire and a split smooth of their difference**

A) DATA, BROKEN, SMOOTH, and ROUGH--"Z" is an exact zero

|      | % Demo. for Pres. ||| | | Unbroken | |
|      | Coos | Carroll | Diff | 3R | Rough | 3R | (Rough) |
|---|---|---|---|---|---|---|---|
| 1896 | 29.75 | 28.38 | 1.37 | .82 | .55 | | |
| 1900 | 41.33 | 40.54 | .79 | .82 | −.03 | | |
|      | 39.61 | 38.79 | .82 | .82 | Z | | |
| 1908 | 39.25 | 37.73 | 1.52 | .82 | .70 | | |
|      | 40.52 | 43.12 | −2.60 | 1.52 | −4.12 | | |
| 1916 | 51.99 | 46.65 | 5.34 | 5.34 | Z | ( 5.34) | (Z) |
|      | 44.40 | 35.01 | 9.39 | 6.55 | 2.84 | ( 6.55) | ( 2.84) |
| 1924 | 39.65 | 33.10 | 6.55 | 6.55 | Z | ( 9.39) | (−2.84) |
|      | 43.11 | 22.37 | 20.74 | 20.74 | Z | (16.94) | ( 3.80) |
| 1932 | 52.14 | 35.20 | 16.94 | 20.74 | −3.80 | (20.74) | (−3.80) |
|      | 55.67 | 33.26 | 22.41 | 22.41 | Z | (22.41) | (Z) |
| 1940 | 60.30 | 33.66 | 26.64 | 26.45 | .19 | | |
|      | 58.36 | 31.91 | 26.45 | 26.64 | −.19 | | |
| 1948 | 52.29 | 23.22 | 29.07 | 26.64 | 2.43 | | |
|      | 44.03 | 17.39 | 26.64 | 26.64 | Z | | |
| 1956 | 33.86 | 14.53 | 19.33 | 26.64 | −7.31 | | |
|      | 57.28 | 20.39 | 36.89 | 26.08 | 10.81 | | |
| 1964 | 71.09 | 45.01 | 26.08 | 30.15 | Z | | |
|      |       |       |       | 26.08 |    | | |
|      | 53.37 | 23.22 | 30.15 | 26.08 | 4.07 | | |
| 1972 | 37.45 | 21.54 | 15.91 | 26.08 | −10.17 | | |

**240** exhibit 20(B–P)/7⁺: Optional sections for chapter 7

A hard look at the broken smooth, however, makes us wonder. Its pattern can be generally described as follows:

◇ an initial flat at about 0.8%.
◇ a rise from 0.8% to 6.6%.
◇ a "jump" from 6.6% to 20.6%.
◇ a rise from 20.6% to 26.5%.

exhibit **20** of chapter 7⁺ (continued)

**B)** The ROUGH--from the BROKEN smooth

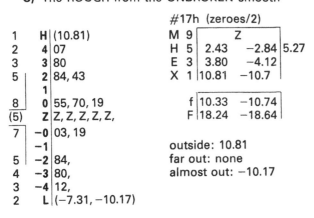

```
                              #16h (zeroes/2)         (sizes)
 1      H  (10.81)         M 8h│    Z              3 │ H│(10.81, 7.31, 10.17)
 2      4 07,              H 4H│ 1.56  -2.00       5 │ 4 07, 12,
        3                  E 2h│ 3.46  -5.72       6 │ 3 80
 4      2 84, 43           X  1│10.81 -10.17       8 │ 2 84, 43
        1                    f │ 6.90  -7.34       1 │
 7      0 55, 70, 19,        F │12.24 -12.68      (6)│ 0 55, 70, 19, 03, 19
(7)     Z Z,Z,Z,Z,Z,Z,Z                            7 │ Z Z,Z,Z,Z,Z,Z,Z,
       -0 03, 19
       -1                  outside: 10.81 and
       -2                            -10.17        median = 1.56
 4     -3 80,              far out: none          (at depth 8h)
 3     -4 12,              almost out: -7.31
 2     -L (-7.31, -10.17)
```

**C)** The ROUGH from the UNBROKEN smooth

```
                              #17h (zeroes/2)
 1      H (10.81)          M 9 │   Z
 2      4 07               H 5 │ 2.43  -2.84 │5.27
 3      3 80               E 3 │ 3.80  -4.12
 5      2 84, 43           X 1 │10.81 -10.7
        1                   f  │10.33 -10.74
 8      0 55, 70, 19        F  │18.24 -18.64
(5)     Z Z,Z,Z,Z,Z,
 7     -0 03, 19
       -1                  outside: 10.81
 5     -2 84,              far out: none
 4     -3 80,              almost out: -10.17
 3     -4 12,
 2      L (-7.31, -10.17)
```

**P)** PROBLEMS

**20a1)** Find a pair of counties in another state whose difference in % Democratic shows a comparable shift.

**20a2)** Carry through the analysis parallel to that above.

Of a total rise of about 26%, almost half occurred before '24 or after '28. Accordingly, it is not at all unreasonable to use the unbroken smooth and describe behavior in **three** intervals: 1896-1910, 1910-1938, 1938-1972. This choice is somewhat supported by the appearance of the raw data, shown in exhibit **22**. (Thinking about a change spread over 28 years is quite different from trying to think about a change over 4 years!)

Exhibits **23** and **24** show the result of bringing in the rough to assist both broken and unbroken smooths. The naive looker can surely be pardoned for seeing no strong case for a break.

### back to suspended deposits

For an example in which a broken smooth is suggested by the data AND turns out to be very much in order, we return to the total deposits in banks suspending payments in the U.S. each year, where, up to 1933, the amounts were large and generally increasing, but thereafter quite small. We use started logs, starting all values by 1 million dollars.

exhibit **21** of chapter 7[+]: Coos–Carroll

**Broken smooth of the difference in Presidential Democratic in the voting of two adjacent New Hampshire counties (unbroken smooth in faint dots)**

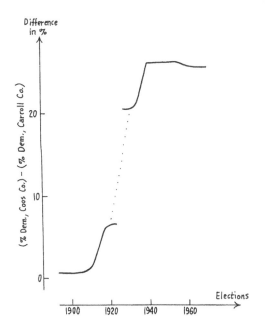

Breaking the sequence between 1933 and 1934 gives the values shown in the "3R/broken" column of exhibit **25**. It would seem that the break is quite worthwhile.

When we look up the facts, we find not only that "up to 1933" is different from "1934 on"--because of the establishment of the Federal Deposit Insurance Corporation--but that 1933 is different from both before and after-- because of banks not allowed to reopen after the bank holiday.

This leads us to the "double break", found on the extreme right of exhibit 25, where we break once between '32 and '33 and once between '33 and '34. The resulting analysis is shown--rough supporting the smooth--in exhibit **26**,

exhibit **22** of chapter 7[+]: Coos–Carroll

**The raw data: difference in % Democratic for President, Coos County MINUS Carroll County, 1896–1972**

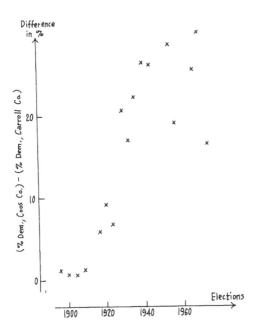

which uses one length of bar for all dates, and exhibit **27,** which also shows outside and far outside points.

### review questions

What might we do when the given values of a sequence seem to show a break? How do we decide? To what example did we turn? What choice were we faced with? What might subject-matter experts have preferred? To what second example did we then turn? Did we really have a choice? Why/why not? Do we ever need double breaks? How close to one another might double breaks reasonably be?

exhibit **23** of chapter 7⁺: Coos–Carroll

**The broken smooth supported by its rough**

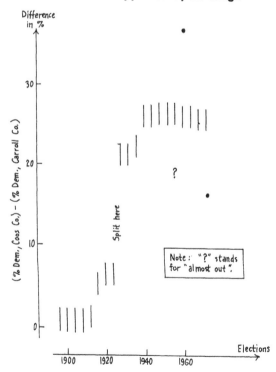

exhibit **24** of chapter 7⁺: Coos–Carroll

**The unbroken smooth supported by its rough**

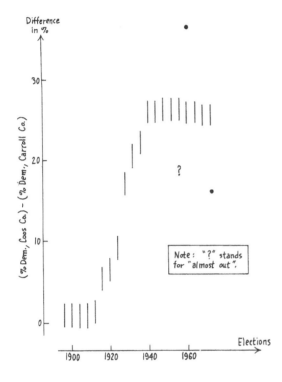

exhibit **25** of chapter 7[+]: suspended deposits

**Split smooth of 100 log (c + suspended deposits) for c = 1 million dollars**

A) The DATA SMOOTHED

| year | suspended bank deposits millions | same + 1 | log | broken 3R | rough (Z = zero) | 3R double break | rough | (outside) |
|---|---|---|---|---|---|---|---|---|
| 1921 | 196 | 197 | 229 | 229 | 1 | | | |
| 22 | 111 | 112 | 205 | 225 | −23 | s | | |
| 23 | 189 | 190 | 228 | 228 | Z | a | | |
| 24 | 213 | 214 | 233 | 228 | 5 | m | | |
| 25 | 173 | 174 | 224 | 233, 229 | −5 | e | | |
| 26 | 272 | 273 | 244 | 229 | 15 | | | |
| 27 | 194 | 195 | 229 | 229 | Z | | | |
| 28 | 139 | 140 | 215 | 229 | −14 | | | |
| 29 | 230 | 231 | 236 | 236 | Z | 236 | Z | |
| 30 | 853 | 854 | 293 | 293 | Z | 293 | Z | |
| 31 | 1691 | 1692 | 323 | 293 | 30 | 293 | 30 | (F) |
| 32 | 716 | 717 | 286 | 323 | −37 | 293 | −7 | |
| 33 | 3599 | 3600 | 356 | 356 | Z | | | |
| 34 | 37 | 38 | 158 | 108 | 50 | | | (F) |
| 35 | 10 | 11 | 104 | 108 | −4 | | | |
| 36 | 11 | 12 | 108 | 108 | Z | | | |
| 37 | 11 | 12 | 108 | 108 | Z | | | |
| 38 | 19 | 20 | 130 | 115 | 15 | | | |
| 39 | 13 | 14 | 115 | 130, 115 | Z | | | |
| 40 | 34 | 35 | 154 | 115 | 39 | t | | (F) |
| 41 | 5.9 | 6.9 | 84 | 84 | Z | h | | |
| 42 | 3.7 | 4.7 | 67 | 67 | Z | e | | |
| 43 | 1.7 | 2.7 | 43 | 67 | −24 | | | (f) |
| 44 | 6.2 | 7.2 | 86 | 43 | 43 | | | (F) |
| 45 | .4 | 1.4 | 15 | 15 | Z | s | | |
| 46 | 0 | 1.0 | 0 | 0 | Z | a | | |
| 47 | 0 | 1.0 | 0 | 0 | Z | m | | |
| 48 | .17 | 1.17 | 7 | 0 | 7 | e | | |
| 49 | 0 | 1.0 | 0 | 2 | −2 | | | |
| 50 | .04 | 1.04 | 2 | 2 | Z | | | |
| 51 | 3.1 | 4.1 | 61 | 38 | 23 | | | (f) |
| 52 | 1.4 | 2.4 | 38 | 61, 59 | −21 | | | (f) |
| 53 | 44.4 | 45.4 | 166 | 59, 61 | 105 | | | (F) |
| 54 | 2.9 | 3.9 | 59 | 88 | −29 | | | (f) |
| 55 | 6.5 | 7.5 | 88 | 88 | Z | | | |
| 56 | 11.9 | 12.9 | 111 | 111 | Z | | | |
| 57 | 12.9 | 13.9 | 114 | 111 | 3 | | | |
| 58 | 6.3 | 7.3 | 86 | 86 | Z | | | |
| 59 | 2.0 | 3.0 | 48 | 86 | −38 | | | (F) |
| 60 | 8.0 | 9.0 | 95 | 93 | 2 | | | |
| 61 | 7.5 | 8.5 | 93 | 93 | Z | | | |
| 62 | 1.2 | 2.2 | 34 | 93 | −59 | | | (F) |
| 63 | 23.3 | 24.3 | 139 | 136 | 3 | | | |
| 64 | 22.0 | 23.0 | 136 | 139, 136 | Z | | | |
| 65 | 45.9 | 46.9 | 167 | 136 | 31 | | | (f) |
| 66 | 0.7 | 1.7 | 23 | 111 | −88 | | | (F) |
| 67 | 11.8 | 12.8 | 111 | 111 | Z | | | |

→

exhibit **25** of chapter 7⁺ (continued)

**B) SOME INFORMATION about SIZE of ROUGH**

|  | *Hspr | *Espr | *M | *H | *f = 4(*M) | *F = 7(*M) |
|---|---|---|---|---|---|---|
| '21–'33 | 7 | 20 | 5 | 14 | | |
| '34–'67 | 19 | 68 | 11 | 38 | | |
| '21–'32 | 7 | 20 | 5 | 14 | | |
| all (split once) | 14 | 60h | 4h | 34 | | |
| all (split twice) | 13 | 56h | 4h | 30 | 18 | 31h |

**P) PROBLEMS (teasers)**

25a1) Try to find another sequence that needs splitting almost as much as this one.

25a2) Analyze it with splitting.

exhibit **26** of chapter 7⁺: suspended deposits

**Deposits in banks suspending payment, U.S., 1921–1967, rough supporting smooth; vertical scale = started logs, start = 1 million dollars**

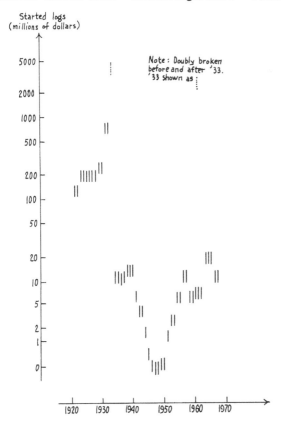

## 7J. Choice of expression

There are four places where the choice of expression might matter in the kind of smoothing we are so far using:

◇ first, in the main body of our 3R smoothing.
◇ second, in using the end-value rules.
◇ third, in calculating residuals, and in picturing the rough.
◇ fourth, in picturing the smooth.

Let us look at these four in turn.

exhibit **27** of chapter 7⁺: suspended deposits

**Deposits in banks suspending payments, U.S. 1921–1967, vertical scale in millions of dollars (started logs, start = 1 million dollars). Rough supporting smooth and outside points shown.**

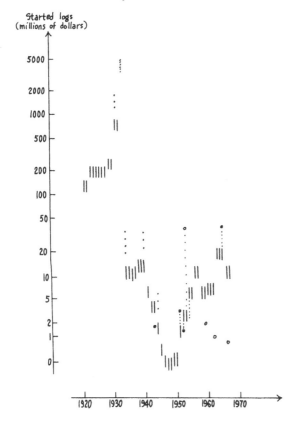

### first, 3R

In the main body of our 3R smoothing, choice of expression ordinarily makes no difference at all. So long as one choice rises whenever another does --and falls when it falls--the same "date" will be the one to provide the "median by value" for any set of three adjacent "dates". So our two smooths will be related by exactly the same change of expression as the values they come from. Exhibit **28** gives a simple example. Notice (in panel B) that, for example, wherever the raw smooth is 64, the root smooth is 8 and the log smooth is 1.8, just as these numbers would be related for input values. So we can forget about choice of expression--so far as 3R goes.

### second, end values

End values--and splitting--may be a different story. At the foot of exhibit 28, if we use raw or root values, end-value smoothing squeezes the end-value in (from 1024 to 160, or from 32 to 16, respectively). If we use log values, such smoothing does not squeeze in 3.0 at all.

### third, residuals as a basis of choice

When we look only at residuals--at the rough--our main concern is usually with answers to three questions:

◇ about how big are the residuals?

◇ roughly how does this size change--as "time" changes, or as the size of the smoothed values change, or both?

◇ which individual residuals look unusual, and roughly how unusual?

We want to answer these questions as well as we can. What effect will the choice of expression have on this?

Consider, as a simple example, the third line from the top and the fifth line from the bottom of exhibit 28. For raw expression, they give:

$$\text{residual} = \text{input} - \text{smooth}$$
$$768 = 1024 - 256 \quad \text{(3rd line)}$$
$$48 = 64 - 16 \quad \text{(-5th line)}$$

for root expression, they give

$$16 = 32 - 16 \quad \text{(3rd line)}$$
$$4 = 8 - 4 \quad \text{(-5th line)}$$

and for log expression, they give

$$0.6 = 3.0 - 2.4 \quad \text{(3rd line)}$$
$$0.6 = 1.8 - 1.2 \quad \text{(-5th line)}$$

In this instance, we see residuals of very different size for raw expression--ones much more nearly of a size for root expression--and equal-sized ones for log expression. Why?

**Because the input (and smoothed) values were of a very different size near one end than near the other.** So far as residuals go, changing from raw to root to log expression is like introducing a microscope of varying magnification, one where differences between little things are enlarged more than differences between large things. (To do this may be either good or bad.)

What *would* we like to do in our imaginary case? What can choice of expression do for us, either here or in general? Mainly it can make the answers to our three questions as simple as we can arrange to have them. How can such answers become simpler? Rather easily:

> ◇ if one expression makes the size of the residuals change less than another: (i) the first question--how big?--will be easier to answer AND its answer will tell us more; (ii) the answer to the second question--How changing?--can be much simpler.

exhibit **28** of chapter 7⁺: illustrative

**A simple example of 3R smoothing different choices of expression**

| raw | root | log | Raw | | Root | | Log | |
|---|---|---|---|---|---|---|---|---|
| | | | in | 3R | in | 3R | in | 3R |
| 4 | 2 | .6 | 4 | ? | 2 | ? | .6 | ? |
| 64 | 8 | 1.8 | 64 | 64 | 8 | 8 | 1.8 | 1.8 |
| 1024 | 32 | 3.0 | 1024 | 256 | 32 | 16 | 3.0 | 2.4 |
| 256 | 16 | 2.4 | 256 | 256 | 16 | 16 | 2.4 | 2.4 |
| 16 | 4 | 1.2 | 16 | 256 | 4 | 16 | 1.2 | 2.4 |
| 256 | 16 | 2.4 | 256 | 64 | 16 | 8 | 2.4 | 1.8 |
| 64 | 8 | 1.8 | 64 | 64 | 8 | 8 | 1.8 | 1.8 |
| 16 | 4 | 1.2 | 16 | 64 | 4 | 8 | 1.2 | 1.8 |
| 64 | 8 | 1.8 | 64 | 64 | 8 | 8 | 1.8 | 1.8 |
| 4096 | 64 | 3.6 | 4096 | 64 | 64 | 8 | 3.6 | 1.8 |
| 16 | 4 | 1.2 | 16 | 64 | 4 | 8 | 1.2 | 1.8 |
| 64 | 8 | 1.8 | 64 | 16 | 8 | 4 | 1.8 | 1.2 |
| 16 | 4 | 1.2 | 16 | 16 | 4 | 4 | 1.2 | 1.2 |
| 4 | 2 | .6 | 4 | 16 | 2 | 4 | .6 | 1.2 |
| 64 | 8 | 1.8 | 64 | 64 | 8 | 8 | 1.8 | 1.8 |
| 1024 | 32 | 3.0 | 1024 | ? | 32 | ? | 3.0 | ? |

**P) PROBLEMS**

28a) Invent an example of your own and work it through to show that the same sort of thing happens.

⋄ if one expression makes the residuals more symmetrically spread around zero, it is likely that fewer will seem unusual, and the answer to the third question will be simpler.

We seek the same kinds of simplicity for our residuals here that we sought in a seemingly quite different situation in chapter 4.

### comparing expressions

We can use the same techniques that we have already used to compare one expression with another:

⋄ plotting size of residual against fitted value to look for stability of size.

⋄ finding hinges, eighths, etc., to look for symmetry.

We leave this to problems (see exhibit **29**).

Sometimes we can profit by half-simplicity, by having the behavior of the rough simple, not in the data as a whole, but only in each of two or three sections.

### fourth, picturing the smooth

If we are to bring the rough to support the smooth--by blurring and showing outside points--we will ordinarily want to use the same expression for the smooth that made the behavior of the rough simple. (Otherwise we will get into blurring by bars of different length, for example, something that we can of course face when the different expression of the smooth is really worth the trouble.)

If we want to use a different expression for the smooth, we can do this easily enough so long as we confine ourselves to

⋄ the points of the smooth, and

⋄ the outside--and far out--points.

These can be re-expressed without difficulty.

### counted data

As usual, the analysis of counted data follows the same pattern as that of other data, subject always to:

⋄ we look first at either the logs or the (square) roots of counts.

⋄ if we start with roots of counts, we frequently drift in the direction of logs of counts.

⋄ if we go to logs, and have small counts, we tend to start the logs slightly --taking logs of "count + start" rather than of "count" alone--with +1/6, +1/4, and +1 likely initial choices (we should be prepared to revise our initial choice once this seems indicated).

exhibit **29** of chapter 7[+]: data and problems

**Some problems on choice of expression (from looking at the roughs [residuals] alone)** (H = hundreds, K = thousands, M = millions, T = tens, TM = tens of millions, c = counts, b = barrels, lb = pounds, t = tons)

**A)** FIRST SECTIONS (to 1825 or 1855) of FIRST GROUP OF SEQUENCES-- see panel 1 for identification

| Year | H541 (T#) | Q167 (KT) | U1/(U10) (MS) | Y369 (S) | Year | C89 (K#) | C90 (K#) | C91 (K#) | C92 (K#) | C94 (K#) | W76 (#) |
|---|---|---|---|---|---|---|---|---|---|---|---|
| 1790 | 58 | — | 20 | — | 1820 | 7.7 | 2.4 | 3.6 | .02 | 1.0 | — |
| 1 | 76 | — | 19 | 18.6 | 1 | 5.9 | 3.2 | 1.5 | .02 | .4 | — |
| 2 | 66 | — | 21 | 18.4 | 2 | 4.4 | 1.2 | 2.3 | .03 | .1 | — |
| 3 | 68 | — | 26 | 18.6 | 3 | 4.0 | 1.1 | 1.9 | .01 | .2 | — |
| 4 | 67 | 4 | 33 | 17.5 | 4 | 5.0 | 1.3 | 2.3 | .02 | .2 | — |
| 95 | 61 | 3 | 48 | 17.5 | 5 | 8.5 | 2.1 | 4.9 | .02 | .4 | — |
| 6 | 57 | 2 | 59 | 17.6 | 6 | 9.8 | 2.3 | 5.4 | .03 | .5 | — |
| 7 | 59 | 1 | 51 | 16.8 | 7 | 16.7 | 4.2 | 9.8 | .03 | .4 | — |
| 8 | 60 | 1 | 61 | 15.8 | 8 | 24.7 | 5.4 | 12.5 | .06 | 1.9 | — |
| 9 | 62 | 6 | 79 | 15.2 | 9 | 12.5 | 3.2 | 7.4 | .03 | .6 | — |
| 1800 | 65 | 3 | 71 | 15.7 | 1830 | 7.2 | 1.2 | 2.7 | .02 | 2.0 | — |
| 1 | 73 | 3 | 93 | 15.1 | 1 | 13.0 | 2.5 | 5.8 | .04 | 2.4 | — |
| 2 | 87 | 3 | 72 | 14.2 | 2 | 34.1 | 5.3 | 12.4 | .33 | 10.2 | — |
| 3 | 104 | 12 | 56 | 13.1 | 3 | 29.1 | 4.9 | 8.6 | .19 | 7.0 | — |
| 4 | 114 | 12 | 78 | 14.2 | 4 | 57.5 | 10.5 | 24.5 | .07 | 17.7 | — |
| 05 | 120 | 6 | 96 | 13.2 | 35 | 42.0 | 9.0 | 20.9 | .07 | 8.3 | — |
| 6 | 131 | 11 | 102 | 11.7 | 6 | 70.5 | 13.1 | 30.6 | .47 | 20.7 | 8 |
| 7 | 145 | 9 | 108 | 10.4 | 7 | 71.0 | 12.2 | 28.5 | 4.0 | 23.7 | 7 |
| 8 | 153 | 5 | 22 | 9.5 | 8 | 34.1 | 5.4 | 12.6 | .11 | 11.7 | 17 |
| 9 | 164 | 4 | 52 | 8.1 | 9 | 64.1 | 10.3 | 24.0 | .38 | 21.0 | 10 |
| 1810 | 175 | 4 | 67 | 7.4 | 1840 | 80.1 | 2.6 | 39.4 | .21 | 29.7 | 19 |
| 1 | 185 | 5 | 61 | 6.4 | 1 | 76.2 | 16.2 | 37.8 | .23 | 15.3 | 21 |
| 2 | 196 | 3 | 39 | 5.9 | 2 | 99.9 | 22.0 | 51.3 | .39 | 20.4 | 11 |
| 3 | 215 | 3 | 28 | 7.2 | 3 | 49.0 | 8.4 | 19.7 | 1.8 | 14.4 | 8 |
| 4 | 212 | 1 | 7 | 10.0 | 4 | 74.7 | 14.4 | 33.5 | 1.3 | 20.7 | 20 |
| 15 | 212 | 1 | 53 | 11.9 | 45 | 109.3 | 19.2 | 44.8 | 1.0 | 34.4 | 12 |
| 6 | 215 | 1 | 82 | 14.7 | 6 | 146.3 | 22.2 | 51.8 | 2.0 | 57.6 | 19 |
| 7 | 226 | 5 | 88 | 13.9 | 7 | 229.1 | 23.3 | 105.5 | 1.3 | 74.3 | 21 |
| 8 | 230 | 17 | 93 | 11.3 | 8 | 218.0 | 35.2 | 112.9 | 1.1 | 58.5 | 14 |
| 9 | 242 | 32 | 70 | 19.2 | 9 | 286.5 | 55.1 | 159.4 | 3.5 | 60.2 | 17 |
| 1820 | 258 | 36 | 70 | 9.5 | 1850 | 308.3 | 51.1 | 164.0 | 1.6 | 78.9 | 20 |
| 1 | 282 | 28 | 65 | 10.0 | 1 | 369.5 | 51.5 | 221.3 | 2.4 | 72.5 | 17 |
| 2 | 299 | 49 | 72 | 9.1 | 2 | 362.5 | 40.7 | 159.5 | 4.1 | 145.9 | 20 |
| 3 | 314 | 41 | 75 | 8.6 | 3 | 361.6 | 37.6 | 162.6 | 3.4 | 141.9 | 26 |
| 4 | 330 | 33 | 76 | 8.3 | 4 | 405.5 | 58.6 | 101.6 | 4.2 | 215.0 | 35 |
| 1825 | 342 | 35 | 100 | 7.5 | 1855 | 187.7 | 47.6 | 49.6 | 1.8 | 71.69 | 41 |

→

exhibit **29** of chapter 7⁺ (continued)

**B)** SECOND SECTIONS (1921 to 1956) of FIRST GROUP of SEQUENCES

| Year | H541 (T#) | Q167 (Kt) | U1 (MS) | Y369 (K#) | C89 (K#) | C90 (K#) | C91 (K#) | C92 (K#) | C94 (#) | W76 |
|---|---|---|---|---|---|---|---|---|---|---|
| 1921 | 6289 | *** | 4560 | 220.9 | 625.4 | 51.1 | 28.4 | 22.8 | 6.8 | 3963 |
| 2 | 6444 | * 1 | 3931 | 208.6 | 216.4 | 25.2 | 10.8 | 14.6 | 17.9 | 4455 |
| 3 | 6522 | * 8 | 4269 | 199.6 | 307.9 | 45.8 | 15.7 | 34.2 | 48.3 | 4133 |
| 4 | 6604 | * ! | 4753 | 186.2 | 364.3 | 59.5 | 17.1 | 35.6 | 75.1 | 4723 |
| 5 | 7066 | *** | 5272 | 177.1 | 148.4 | 27.2 | 26.7 | 16.8 | 46.1 | 5347 |
| 6 | 6830 | 42 | 5017 | 167.3 | 155.6 | 25.5 | 24.9 | 16.8 | 50.4 | 5103 |
| 7 | 7171 | 46 | 5142 | 155.5 | 168.4 | 23.7 | 28.5 | 16.9 | 48.5 | 4918 |
| 8 | 7248 | 55 | 5776 | 146.1 | 158.5 | 20.0 | 25.3 | 16.2 | 45.8 | 5218 |
| 9 | 7245 | 57 | 5491 | 149.0 | 158.6 | 21.3 | 19.9 | 17.4 | 46.8 | 5921 |
| 1930 | 7319 | 40 | 4013 | 131.5 | 147.4 | 31.0 | 23.4 | 6.9 | 26.6 | 6085 |
| 1 | 7247 | 83 | 2918 | 135.4 | 61.9 | 9.1 | 7.3 | 3.1 | 10.4 | 6897 |
| 2 | 7301 | 73 | 2434 | 156.1 | 20.6 | 2.1 | .5 | .9 | 27 | 7376 |
| 3 | 7153 | 102 | 2061 | 179.5 | 12.4 | 1.0 | .3 | .5 | 1.9 | 7170 |
| 4 | 7254 | 108 | 2202 | 214.1 | 17.2 | 1.3 | .4 | .6 | 4.4 | 6489 |
| 5 | 7320 | 98 | 2304 | 225.6 | 22.8 | 1.4 | .4 | .7 | 5.2 | 5980 |
| 6 | 7346 | 146 | 2495 | 263.8 | 23.5 | 1.3 | .4 | .6 | 6.3 | 5734 |
| 7 | 7387 | 129 | 3407 | 282.8 | 31.9 | 1.7 | .5 | 1.0 | 10.9 | 5638 |
| 8 | 7507 | 125 | 3107 | 286.3 | 41.5 | 2.3 | 1.1 | 1.4 | 17.2 | 5776 |
| 9 | 7590 | 132 | 3192 | 309.0 | 63.9 | 3.1 | 1.2 | 1.2 | 33.5 | 6338 |
| 1940 | 7360 | 137 | 4030 | 323.2 | 50.4 | 6.2 | .8 | 1.3 | 21.5 | 6148 |
| 1 | 7683 | 157 | 5153 | 367.1 | 26.5 | 7.7 | .3 | 1.1 | 4.0 | 5311 |
| 2 | 7838 | 152 | 8081 | 537.1 | 11.1 | .9 | .1 | .4 | 2.2 | 3943 |
| 3 | 7979 | 153 | 13028 | 999.8 | 4.9 | 1.0 | .2 | .2 | .2 | 2625 |
| 4 | 8046 | 169 | 15345 | 1452 | 4.5 | 1.3 | .1 | .3 | .2 | 2564 |
| 5 | 8084 | 191 | 10897 | 1849 | 5.9 | 3.0 | .4 | .2 | .2 | 2112 |
| 6 | 8430 | 187 | 9996 | 1905 | 52.9 | 33.6 | 1.8 | 1.3 | 2.6 | 1656 |
| 7 | 8568 | 194 | 14674 | 1792 | 83.5 | 23.8 | 2.6 | 4.9 | 13.9 | 1617 |
| 8 | 8651 | 193 | 12967 | 1721 | 103.5 | 26.4 | 7.5 | 6.1 | 19.4 | 1984 |
| 9 | 8293 | 180 | 12160 | 1695 | 129.6 | 21.1 | 8.7 | 6.7 | 55.3 | 3105 |
| 1950 | 8936 | 146 | 10816 | 1697 | 199.1 | 12.8 | 5.8 | 5.7 | 128.6 | 4408 |
| 1 | 9066 | 182 | 15672 | 1653 | 149.5 | 14.9 | 3.1 | 5.5 | 87.8 | 4888 |
| 2 | 9180 | 194 | 15262 | 1650 | 193.6 | 22.2 | 3.5 | 5.4 | 104.2 | 5635 |
| 3 | 9152 | 193 | 15827 | 1667 | 82.4 | 16.6 | 4.3 | 5.5 | 27.3 | 4331 |
| 4 | 9223 | 182 | 15136 | 1670 | 92.1 | 16.7 | 4.7 | 5.5 | 33.1 | 4433 |
| 5 | 9313 | 187 | 15563 | 1660 | 110.6 | 15.8 | 5.2 | 5.2 | 29.6 | 4065 |
| 6 | 9445 | 189 | 19124 | 1622 | 156.9 | 19.0 | 5.6 | 5.7 | 44.6 | 6646 |
| 7 | — | 191 | 20989 | 1580 | 169.6 | 24.0 | 8.2 | 6.2 | 60.4 | 6282 |

exhibit **29** of chapter 7⁺ (continued)

**C) FIRST SECTIONS (to 1919) of SECOND GROUP of SEQUENCES**—see panel J for identification.

| Year | D711 (#) | D772 (#) | J241 (in) | K77 $/mo | K78 $/mo | M178 (M1b) | M179 ($/b) | Q145 (#) | Q184 (Ht) | W74 (T#) | X113 (TM$) | X118 (M$) |
|---|---|---|---|---|---|---|---|---|---|---|---|---|
| 1882 | 353 | 38 | — | — | 19 | 3.2 | 2.25 | — | 940 | 86 | — | — |
| 3 | 372 | 55 | — | — | — | 4.2 | 2.15 | — | 1102 | 102 | — | — |
| 4 | 341 | 50 | — | — | 19 | 4.0 | 2.10 | — | 840 | 115 | — | — |
| 85 | 486 | 67 | — | — | 19 | 4.2 | 1.95 | — | 481 | 77 | — | — |
| 6 | 1073 | 210 | — | — | — | 4.5 | 1.95 | — | 306 | 59 | — | — |
| 7 | 836 | 299 | — | — | 19 | 6.9 | 1.95 | — | 240 | 95 | — | — |
| 8 | 540 | 163 | — | — | 19 | 6.5 | 1.95 | — | 338 | 83 | — | — |
| 9 | 662 | 173 | — | 13h | 19h | 6.8 | 1.67 | — | 400 | 72 | — | — |
| 1890 | 1039 | 318 | — | 13h | 19h | 7.8 | 2.09 | — | 786 | 89 | — | — |
| 1 | 867 | 334 | — | 13h | 20 | 8.2 | 2.13 | 293 | 1055 | 84 | — | — |
| 2 | 693 | 261 | — | 13h | 20 | 8.7 | 2.91 | 376 | 606 | 82 | — | — |
| 3 | 783 | 257 | 104 | 14 | 20 | 8.0 | 1.96 | 299 | 371 | 90 | — | — |
| 4 | 865 | 206 | 114 | 12h | 18h | 8.4 | 1.73 | 324 | 287 | 93 | — | — |
| 95 | 810 | 217 | 93 | 12h | 18h | 8.7 | 1.60 | 170 | 268 | 111 | — | — |
| 6 | 547 | 297 | 101 | — | — | 9.5 | 1.57 | 181 | 396 | 144 | 2 | 96 |
| 7 | 680 | 193 | 95 | — | — | 11.0 | 1.61 | 222 | 219 | 162 | 2 | 87 |
| 8 | 645 | 236 | 86 | 13h | 19 | 12.3 | 1.62 | 221 | 239 | 180 | 5 | 104 |
| 9 | 1014 | 471 | 114 | 14 | 20 | 15.8 | 1.43 | 239 | 688 | 214 | 8 | 98 |
| 1900 | 931 | 414 | 101 | — | — | 17.2 | 1.09 | 249 | 722 | 175 | 16 | 124 |
| 1 | 1413 | 1016 | 101 | — | — | 20.1 | .99 | 282 | 830 | 173 | 10 | 143 |
| 2 | 1604 | 1051 | 92 | 15h | 22 | 25.8 | 1.21 | 345 | 759 | 64 | 12 | 139 |
| 3 | 1778 | 1200 | 69 | — | — | 39.9 | 1.24 | 355 | 670 | 54 | 15 | 169 |
| 4 | 944 | 964 | 73 | — | — | 31.7 | .88 | 441 | 514 | 55 | 11 | 170 |
| 05 | 942 | 800 | 64 | — | — | 40.1 | .94 | 537 | 1193 | 49 | 8 | 225 |
| 6 | — | — | 69 | 18h | 26 | 51.0 | 1.13 | 359 | 323 | 62 | 9 | 239 |
| 7 | — | — | 61 | — | — | 52.2 | 1.11 | 610 | 444 | 59 | 18 | 314 |
| 8 | — | — | 72 | — | — | 52.9 | .85 | 381 | 709 | 76 | 13 | 202 |
| 9 | — | — | 75 | 22 | 28 | 66.7 | .81 | 253 | 272 | 68 | 7 | 219 |
| 1910 | — | — | 67 | 21 | 28 | 77.8 | .89 | 324 | 234 | 64 | 5 | 311 |
| 1 | — | — | 53 | 21h | 28 | 79.5 | .84 | 299 | 237 | 100 | 5 | 344 |
| 2 | — | — | 73 | 22 | 29h | 85.9 | .81 | 283 | 231 | 134 | 6 | 374 |
| 3 | — | — | 76 | 22h | 30 | 89.5 | 1.00 | 350 | 271 | 168 | 5 | 410 |
| 4 | 403 | 253 | 83 | 22h | 29h | 87.3 | .93 | 232 | 150 | 171 | 7 | 458 |
| 15 | 770 | 312 | 72 | 22h | 30 | 82.7 | 86 | 199 | 186 | 154 | 5 | 486 |
| 6 | 2036 | 721 | 78 | 25 | 33 | 95.4 | 1.10 | 239 | 376 | 174 | 4 | 561 |
| 7 | 2268 | 799 | 82 | 31 | 40h | 91.3 | 1.35 | 301 | 525 | 150 | 15 | 1029 |
| 8 | 1869 | 584 | 83 | 37h | 48h | 71.3 | 1.60 | 471 | 883 | 121 | 154 | 2221 |
| 1919 | 2036 | 609 | 74 | 143 | 56 | 86.1 | 1.71 | 273 | 1778 | 152 | 91 | 2517 |

➡

exhibit **29** of chapter 7⁺ (continued)

**D) REMAINDER (from 1920) of SECOND GROUP of SEQUENCES**

| Year | D771 (#) | D772 (#) | J241 (in) | K77 $/mo | K78 $/mo | M178 (M1b) | M179 ($/b) | Q145 (#) | Q184 (Ht) | W74 (T#) | X113 (TM$) | X118 (M$) |
|---|---|---|---|---|---|---|---|---|---|---|---|---|
| 1920 | 2038 | 622 | 90 | 51 | 65 | 97.1 | 2.02 | 229 | 2080 | 248 | 26 | 4540 |
| 1 | 1501 | 373 | 100 | 33h | 44h | 96.0 | 1.89 | 205 | 1507 | 327 | 40 | 3598 |
| 2 | 563 | 208 | 10 | 33 | 43h | 118.6 | 1.76 | 203 | 570 | 161 | 16 | 1805 |
| 3 | 721 | 308 | 71 | 37h | 47h | 137.2 | 1.90 | 143 | 131 | 193 | 30 | 2218 |
| 4 | 537 | 244 | 79 | 38 | 49 | 147.5 | 1.81 | 153 | 32 | 267 | 18 | 1644 |
| 25 | 537 | 219 | 71 | 38h | 49 | 159.0 | 1.77 | 176 | 56 | 282 | 18 | 1887 |
| 6 | 478 | 206 | 72 | 395 | 50 | 164.2 | 1.71 | 152 | 50 | 260 | 24 | 2157 |
| 7 | 273 | 240 | 83 | 39h | 50 | 174.0 | 1.62 | 88 | 66 | 239 | 23 | 2227 |
| 8 | 222 | 226 | 73 | 39h | 50 | 178.1 | 1.57 | 91 | 114 | 318 | 27 | 3364 |
| 9 | 373 | 382 | 50 | 40 | 51 | 172.0 | 1.48 | 114 | 128 | 290 | 38 | 3628 |
| 1930 | 284 | 207 | 69 | 37h | 48 | 160.8 | 1.44 | 61 | 186 | 271 | 30 | 2891 |
| 1 | 447 | 221 | 90 | 28h | 38 | 128.4 | 1.11 | 46 | 266 | 294 | 45 | 2358 |
| 2 | 560 | 162 | 93 | 20h | 29 | 81.4 | 1.01 | 27 | 521 | 294 | 43 | 2513 |
| 3 | 926 | 533 | 88 | 18 | 25h | 64.8 | 1.33 | 51 | 259 | 241 | 56 | 1516 |
| 4 | 717 | 835 | 82 | 20 | 28 | 76.6 | 1.54 | 38 | 9 | 292 | 174 | 848 |
| 35 | 760 | 945 | 80 | 22 | 30h | 76.2 | 1.51 | 30 | 19 | 386 | 82 | 620 |
| 6 | 256 | 1083 | 65 | 24 | 32h | 114.6 | 1.51 | 41 | 7 | 456 | 114 | 751 |
| 7 | 1410 | 2728 | 76 | 27h | 36h | 115.7 | 1.48 | 34 | — | 514 | 67 | 776 |
| 8 | 776 | 1385 | 61 | 27 | 36 | 108.2 | 1.48 | 81 | — | 503 | 60 | 601 |
| 9 | 699 | 1411 | 75 | 27 | 36 | 125.1 | 1.47 | 40 | — | 559 | 79 | 632 |
| 1940 | 753 | 1243 | 78 | 27h | 37h | 132.9 | 1.36 | 83 | — | 614 | 82 | 598 |
| 1 | 1535 | 2138 | 65 | 34h | 49h | 170.4 | 1.47 | 48 | — | 649 | 75 | 637 |
| 2 | 1423 | 943 | 59 | 45h | 59 | 187.8 | 1.53 | 122 | — | 373 | 184 | 128 |
| 3 | 1906 | 565 | 59 | 59 | 77 | 129.5 | 1.57 | 278 | — | 223 | 803 | 626 |
| 4 | 2146 | 808 | 73 | 71 | 91 | 95.6 | 1.59 | 267 | — | 291 | 1951 | 785 |
| 45 | 1956 | 946 | 84 | 79 | 101 | 107.8 | 1.63 | 156 | — | 351 | 2438 | 866 |
| 6 | 2238 | 1617 | 82 | 86 | 108 | 172.1 | 1.72 | 128 | — | 278 | 1341 | 1073 |
| 7 | 1707 | 1102 | 77 | 92 | 117 | 190.4 | 1.90 | 79 | — | 210 | 136 | 1164 |
| 8 | 1737 | 780 | 89 | 99 | 124 | 207.7 | 2.18 | 59 | — | 397 | 218 | 1353 |
| 9 | 1682 | 781 | 74 | 99 | 121 | 209.3 | 2.30 | 37 | — | 445 | 230 | 1386 |
| 1950 | 2559 | 919 | 102 | 99 | 121 | 231.0 | 2.35 | 180 | — | 472 | 380 | 1680 |
| 1 | 2102 | 888 | 89 | 113 | 137 | 244.6 | 2.54 | 150 | — | 416 | 633 | 2078 |
| 2 | 2447 | 839 | 69 | 119 | 146 | 254.8 | 2.54 | 24 | — | 296 | 612 | 2453 |
| 3 | 2825 | 745 | 92 | 122 | 151 | 264.3 | 2.67 | 49 | — | 271 | 394 | 2603 |
| 4 | 1726 | 588 | 86 | 120 | 151 | 278.4 | 2.76 | 30 | — | 254 | 589 | 2478 |
| 55 | 2154 | 844 | 81 | 123 | 154 | 310.2 | 2.86 | 24 | — | 271 | 541 | 2826 |
| 6 | 1821 | 744 | 79 | 128 | 161 | 325.6 | 3.05 | 57 | — | 298 | 553 | 3460 |
| 7 | 1730 | 751 | 72 | 133 | 168 | — | — | 15 | — | 236 | 362 | 5248 |

➡

exhibit **29** of chapter 7⁺ (continued)

**I) IDENTIFICATION of FIRST GROUP of SEQUENCES**

C--(see below)

H541 Membership of Methodist Church (in thousands) (pp. 228–229)

Q167 Gross tonnage of documented vessels engaged in whale fisheries (in thousands of tons) (pp. 444–445)

U1 Value of U.S. Exports (gold, silver, and merchandise) (in millions of dollars) (pp. 537–538). [series U10 used for 1790–1820.]

W--(see below)

Y369 Per capita public debt (in dollars) (pp. 720–721) [calculated from Y368 and A2 for 1791–1850.]

C89 Total Immigrants (in thousands per year) (pp. 56–57)

C90 Immigrants (in thousands per year) from Great Britain (pp. 56–57)

C91 Immigrants (in thousands per year) from Ireland (pp. 56–57)

C92 Immigrants (in thousands per year) from Scandinavia (pp. 56–57)

C94 Immigrants (in thousands per year) from Germany (pp. 56–57)

W76 Patents issued to residents of foreign countries (pp. 607–608)

**J) IDENTIFICATION of SECOND GROUP of SEQUENCES** (units are those used above)

D771 Number of work stoppages with major issue "wages and hours" (page 99)

D772 Number of work stoppages with major issue "union organization" (page 99)

J241 Annual total precipitation (in inches as rain) at the Weather Bureau Office, Tatoosh Island, Washington.

K77 Average wages (in dollars per month) for farm workers; when board included (pp. 280–281).

K78 Average wages (in dollars per month) for farm workers: when board NOT included (pp. 280–281).

M178 Shipments of cement (in millions of barrels) (page 364)

M179 Average values of same (in dollars per barrel) (page 364)

Q145 Number of passengers killed on railroad trains (page 437)

Q184 Merchant vessels built and documented on the New England Coast (in thousands of tons) (pp. 448–449)

W74 Number of design patents granted (in tens of patents) (pp. 607–608)

X113 U.S. Government deposits in all commercial banks (in tens of millions of dollars) (page 632)

X118 Other liabilities of all commercial banks (in millions of dollars) (page 632)

### suspended bank counts

While banks come in different sizes, the data we have been using on deposits in suspended banks is not very far from being counted data. And we can have real counted data by going to the same sources and finding the **number** of banks that suspended payment in each year. Exhibit **30** has the analysis when the started logs are started by one count. Exhibit **31** shows the resulting picture, with the rough supporting the smooth. Clearly we have shown the picture for bank suspensions at least as clearly by using COUNTS of suspending banks as we did earlier using AMOUNTS of suspended deposits (in millions of dollars).

### review questions

In what four places in the use of smoothing might choice of expression have made a difference? Which of these matter? When? How may residual behavior guide our choice of expression? How many general questions do we expect to ask about the rough? How can choice of expression simplify answers to these questions? How do we compare one choice of expression with another? What do we gain/lose by expressing the smooth with a choice different from the one that makes the rough simply behaved? What about smoothing sequences of counts? What did we take as an example? How is it related to an earlier example using amounts? What do we think is the moral?

exhibit **29** of chapter $7^+$ (continued)

**P) PROBLEMS**

29a/b/c/d/e/f/g/h/i/j) Smooth and examine rough for re-expression, starting from H541/A167/U1(10)/Y369/C89/C90/C91/C92/C94/W76.

29k/l/m/n/o/p/q/r/s/t/u/v) Do the same, starting with D771/D772/J241/K77/K78/M178/M179/Q145/Q184/W74/X113/X118.

29w) Do the same for some expression of D771 divided by D772.

29x) Take the difference of K78 and K77 as indicating the value of board (as provided to farm workers) and analyze it.

29pq2) Plot the rough of M178 against the rough of M179; discuss.

29y, on) In the earlier, 1945, version of the source, such sequences as A109, E113, and E114 are a few of those that could be considered.

**S) SOURCE**

Historical Statistics of the United States, Colonial times to 1957. Washington, 1960. Column headings are the labels used in the source.

exhibit **30** of chapter 7[+]: suspended banks

**Split smooths of log (1 + number of banks suspending payment)**

| | banks suspending | | | | | | same split | |
|---|---|---|---|---|---|---|---|---|
| year | # | #+1 | log (#+1) | 3R split | rough | out-side | double split | rough |
| 1921 | 505 | 506 | 270 | 270 | Z | | | |
| 22 | 367 | 368 | 257 | 270 | −13 | | | |
| 23 | 648 | 649 | 281 | 281 | Z | | | |
| 24 | 776 | 777 | 289 | 281 | 8 | | | |
| 25 | 618 | 619 | 279 | 289, 283 | −4 | | | |
| 26 | 976 | 977 | 299 | 283 | 16 | | | |
| 27 | 669 | 670 | 283 | 283 | Z | | | |
| 28 | 499 | 500 | 270 | 282 | −12 | | | |
| 29 | 659 | 660 | 282 | 282 | Z | | 282 | |
| 30 | 1352 | 1353 | 313 | 313 | Z | | 313 | Z |
| 31 | 2294 | 2295 | 336 | 316 | 20 | | 316 | 20 |
| 32 | 1456 | 1457 | 316 | 336 | −20 | | 316 | Z |
| 33 | 4004 | 4005 | 360 | 360 | Z | | 360 | |
| 34 | 57 | 58 | 176 | 165 | 11 | | | |
| 35 | 34 | 35 | 154 | 165 | −11 | | | |
| 36 | 44 | 45 | 165 | 165 | Z | | | |
| 37 | 59 | 60 | 178 | 175 | 3 | | | |
| 38 | 55 | 56 | 175 | 175 | Z | | | |
| 39 | 42 | 43 | 163 | 163 | Z | | | |
| 40 | 22 | 23 | 136 | 136 | Z | | | |
| 41 | 8 | 9 | 95 | 100 | −5 | | | |
| 42 | 9 | 10 | 100 | 95 | 5 | | | |
| 43 | 4 | 5 | 70 | 70 | Z | | | |
| 44 | 1 | 2 | 30 | 30 | Z | | | |
| 45 | 0 | 1 | 0 | 0 | Z | | | |
| 46 | 0 | 1 | 0 | 0 | Z | | | |
| 47 | 1 | 2 | 30 | 0 | 30 | | | |
| 48 | 0 | 1 | 0 | 30 | −30 | | | |
| 49 | 4 | 5 | 70 | 30 | 40 | f | | |
| 50 | 1 | 2 | 30 | 60 | −30 | | | |
| 51 | 3 | 4 | 60 | 60 | Z | | | |
| 52 | 3 | 4 | 60 | 60 | Z | | | |
| 53 | 4 | 5 | 70 | 60 | 10 | | | |
| 54 | 3 | 4 | 60 | 70, 60 | Z | | | |
| 55 | 4 | 5 | 70 | 60 | 10 | | | |
| 56 | 3 | 4 | 60 | 60 | Z | | | |
| 57 | 3 | 4 | 60 | 60 | Z | | | |
| 58 | 8 | 9 | 95 | 60 | 35 | f | | |
| 59 | 3 | 4 | 60 | 60 | Z | | | |
| 60 | 2 | 3 | 48 | 60 | −12 | | | |
| 61 | 9 | 10 | 100 | 48 | 52 | f | | |
| 62 | 2 | 3 | 48 | 48 | Z | | | |
| 63 | 2 | 3 | 48 | 48 | Z | | | |
| 64 | 8 | 9 | 95 | 90 | 5 | | | |
| 65 | 7 | 8 | 90 | 90 | Z | | | |
| 66 | 1 | 2 | 30 | 70 | −40 | f | | |
| 67 | 4 | 5 | 70 | 70 | Z | | | |

**S) SOURCE**

The World Almanac 1969/
144(1929–67), 1954/753(26–52),
1931/306(21–29), 1937/
288(1921–1935)

exhibit **31** of chapter 7⁺: suspended banks

**Number of banks suspending payments in the U.S., 1921–1967 (vertical scale = started log, start = 1 bank; see exhibit 30 for numbers).**

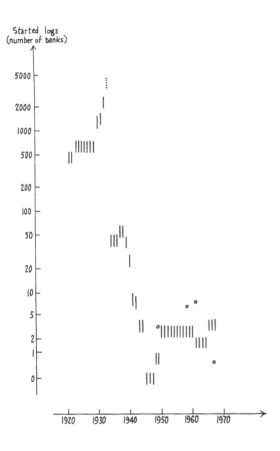

## 7K. A 2-section example

Let us turn back to suspended deposits as given in exhibit 8. We need something like logs to make the sizes of residuals about the same. Exhibit 32 lists the 3R smooth of the logs, and its rough, and gives the stem-and-leaf and modified letter values--both for residuals and for their **sizes**--doing this first for all years and then for two groups of years. Clearly, the residuals are bigger for the second group of years, the comparison being

|  | *-Spreads of residuals | | *-Letter values of sizes | |
|---|---|---|---|---|
|  | *Hspr | *Espr | *M | *H |
| 1921–44 | 20 | 57 | 8 | 26h |
| 1948–67 | 60 | 140 | 33 | 98 |

The apparent ratio of sizes varies from about 3:1 to about 4:1.

exhibit **32** of chapter 7⁺: suspended deposits

**The 3R smooth of the logs of suspended deposits and the corresponding residuals (deposits in 100 log of millions of dollars)**

**A)** The SMOOTH and the ROUGH

| Date | Susp. deposits | 3R | Rough | Size | f or F |
|---|---|---|---|---|---|
| 1921 | 229 | 228 | 1 |  |  |
| 2 | 205 | 228 | −21 | 23 |  |
| 3 | 228 | 228 | Z |  |  |
| 4 | 233 | 228 | 5 | 5 |  |
| 25 | 224 | 233, 229 | −5 | 5 |  |
| 6 | 243 | 229 | 14 | 14 |  |
| 7 | 229 | 229 | Z |  |  |
| 8 | 214 | 229 | −15 | 15 |  |
| 9 | 236 | 236 | Z |  |  |
| 30 | 293 | 293 | Z |  |  |
| 1 | 323 | 293 | 30 | 30 |  |
| 2 | 285 | 323, 293 | −8 | 8 |  |
| 3 | 356 | 285 | 71 | 71 | f |
| 4 | 157 | 157 | Z |  |  |
| 35 | 100 | 104 | −4 | 4 |  |
| 6 | 104 | 104 | Z |  |  |
| 7 | 104 | 104 | Z |  |  |
| 8 | 128 | 111 | 17 | 17 |  |
| 9 | 111 | 128, 111 | Z |  |  |
| 40 | 153 | 111 | 42 | 42 |  |
| 1 | 77 | 77 | Z |  |  |
| 2 | 57 | 57 | Z |  |  |
| 3 | 23 | 57 | −34 | 34 |  |
| 4 | 79 | 23 | 56 | 56 |  |

→

exhibit **32** of chapter 7⁺ (continued)

(Panel A continued)

| Date | Susp. deposits | 3R | Rough | Size | f or F |
|---|---|---|---|---|---|
| 45 | −40 | −40 | Z | | |
| 6 | L | L | ? | | ? |
| 7 | L | L | ? | | ? |
| 8 | −77 | L | H | | F |
| 9 | L | −140 | L | | F |
| 50 | −140 | −140 | Z | | |
| 1 | 49 | 15 | 34 | 34 | |
| 2 | 15 | 49,46 | −31 | 31 | |
| 3 | 165 | 46,49 | 116 | 116 | F |
| 4 | 46 | 81 | −35 | 35 | |
| 55 | 81 | 81 | Z | | |
| 6 | 108 | 108 | Z | | |
| 7 | 111 | 108 | 3 | 3 | |
| 8 | 80 | 80 | Z | | |
| 9 | 30 | 80 | −50 | 50 | |
| 60 | 90 | 88 | 2 | 2 | |
| 1 | 88 | 88 | Z | | |
| 2 | 8 | 88 | −80 | 80 | f |
| 3 | 137 | 134 | 3 | 3 | |
| 4 | 134 | 137,134 | Z | | |
| 65 | 166 | 134 | 32 | 31 | |
| 6 | −15 | 107 | −122 | 122 | F |
| 7 | 107 | 107 | Z | | |

L = "lower still" entered as the log of zero
F = "f" for analysis of panel E

B) STEM-and-LEAF for ALL residuals--?'s not counted

```
  3   H  | (71, H, 116)((?))
  4   5  | 6
  5   4  | 2
  8   3  | 041
      2  |
 10   1  | 47
 14   0  | 9322
 19   Z  | 000000000000000000
 12  −0  | 4584
  8  −1  | 5
     −2  |
  7  −3  | 415
     −4  |
  4  −5  | 0
  3   L  |(L, 3, −80, −122)((?))
```

→

exhibit **32** of chapter 7⁺ (continued)

**C)** *LETTER VALUES WITHOUT and WITH ?'s for ALL residuals

```
*#36    (zeros/2, no?)            *#38    (zeros/2, ?in)
*M18h    |    Z    |              *M19h†   |    Z    |
*H 9h   |15h  |-11h|              *H 10†  | 17  | -15 |32
*E  5   |42   |-35 |              *E  5h† | 49  |-42h |91h
              |40|                              |48|
        *f |55h  |-51h|                   *f | 65 |-63 |
        *F |95h  |-91h|                   *F |113 |-111|
outside: 71, -80                  outside: 71, -80
far out: h, 116, -122, L          far out: (?), H, 116, -122, L, (?)
```

† Since counts in (B) omit ?'s, reduce by one unit before using (B)'s counts.

**D)** STEM-and-LEAF and *-LETTER VALUE for SIZES of ALL RESIDUALS

```
 8   H|(71, 116, H, 80, 122, L, ?, ?)
10   5|06
11   4|2
16   3|01245
19   1|457
27   0|12334558
18   z|000000000000000000
```

*-median = 11 (at depth 19h, since * = 38h)

**E)** STEM-and-LEAF and LETTER-VALUE for RESIDUALS for 1921-45 and 1948-67

```
       1921-45                    1948-67
 1    H|(71)              2      H|(H, 116)
 2    5|6                         5|
 3    4|2                         4|
 4    3|0                 4       3|42
      2|                          2|
 6    1|47                        1|...
 8    0|15                7       0|323
11    z|00000000000       7       z|000000
 6   -0|458                      -0|
 3   -1|5                        -1|
 2   -2|3                        -2|
 1   -3|4                 7      -3|15
     -4|                         -4|
     -5|                  4      -5|0
      L|                  3       L|(L, -80, -122)
```

→

exhibit **32** of chapter 7⁺ (continued)

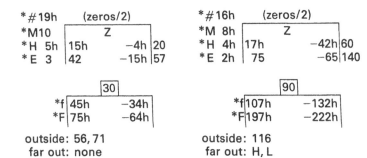

outside: 56, 71          outside: 116
far out: none            far out: H, L

**F)** STEM-and-LEAF and *-LETTER VALUES for SIZES of residuals in 1921-44 and 1948-67

```
       1921-45                     1948-67
 1  H |(71)              5   H |(H, 116, 80, 122, L)
 2    5|6                6    5|0
 3    4|2                     4|
 5    3|04              10    3|1245
 6    2|3                     2|
 9    1|457                   1|
14    0|14558          13    0|233
12    Z|00000000000     7    Z|0000000
*median = 8 (at depth 10)   *median = 33 (at depth 9)
*H = 26h (at depth 5h)      *H = 98 (at depth 5)
```

**P) PROBLEMS**

**32a)** Make stem-and-leaf displays, and find *M and *H for the **sizes** of residuals for periods 1921-33 and 1934-45. How do they compare with each other and with 1948-67? How many groups seem to be needed? What are they?

**32b)** (Triple value) Repeat Panels (A), (B), and (F) for started logs, using 100 log(1 + suspended deposits in millions).

**32c)** (After (b)) Do (32a) for this new analysis.

Exhibit **33** shows the corresponding plot, with the rough supporting the smooth. Here we have used two lengths of bars, one for each group of years. The result is quite effective.

**review questions**

What did we take as an example? How many sections seemed needed? How did the sizes behave?

exhibit **33** of chapter 7⁺: suspended deposits

**Deposits in banks suspending payment, U.S., 1921–1967 (rough supporting smooth: two median sizes, one up to 1945, one from 1968 on, 5 values off scale)**

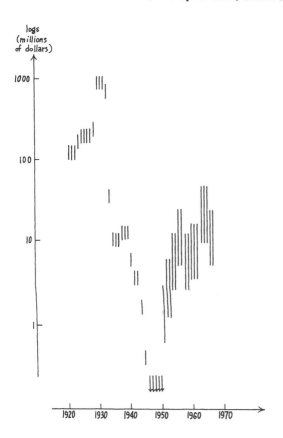

## 7M. How much more may we have learned?

We have extended what we knew at the end of chapter 7 in a variety of ways. We are now ready:

◇ to split a sequence whenever this is clearly indicated.

◇ to think hard about whether or not to split in a specific instance.

◇ to use our usual techniques for looking at residuals to judge which expression of the data yields the more reasonable smooth.

◇ to do all this equally freely for counted data.

◇ to assess size of rough and thence calculate a blurring, separately in two or more sections of the sequence (two or more subsequences).

◇ to show blurrings involving two or more sizes of bar.

We now understand more clearly:

◇ that re-expression does alter the smooth, but not by a tremendous amount, so that not bothering to re-express is often sensible.

◇ which steps in the smoothing are most subject to alteration by re-expression.

# Parallel and wandering schematic plots    8

265

chapter index on next page

Now that we are able to smooth sequences with moderate ease and effectiveness, we have the essential tool for extending to batches of points in the plane--to $(x, y)$ pairs--some of the techniques we used to summarize batches of numbers. To do this, we will need to replace a number for a median by something that tells us about the middle of one variable, say $y$, for different values of the other, here $x$. Similarly, we will have to replace numbers by curves--by "traces" as we will say--in the roles played by hinges and (inner) fences. The ideas of outside and far-out values will clearly go over with little change to outside and far out points, once we have fences. The idea of adjacent points will need a small amount of thought, but will in fact go over quite easily.

This chapter will close with the simplest analog for $(x, y)$ points of a schematic plot for values, which we will call a **wandering schematic plot.** In chapter 9, then, we can go on to other analogs, for $(x, y)$ pairs, of schematic plots for values.

### review questions

What will we be up to in this chapter? What tool are we now ready to use? What will we need to replace median with? Hinges? Fences? Outside and far-out values? Adjacent values? How far will we get in this chapter? How is this chapter related to the next?

### 8A. Parallel schematic plots

One of the simplest things we can do to look at a collection of $(x, y)$ pairs is:

◇ slice the data up according to its $x$-values.

◇ find a schematic plot for the $y$-values in each slice.

◇ set these plots up side by side.

Exhibit **1** sets out, in two different sorted ways, the paired usages of electricity and gas during a winter for 152 townhouses at Twin Rivers, New Jersey. The data was collected as part of a study of energy consumption, in which differences in energy use were to be compared with structural aspects of the individual buildings and behavioral aspects of their inhabitants. Panels A and B give the data arranged in order, once by increasing electrical use and once by increasing gas use. This prepares us to start from either coordinate.

# index for chapter 8

review questions  265

**8A. Parallel schematic plots  265**
letter cuts  270
review questions  273

**8B. Smoothing the cross-medians  274**
broken median  274
cross-medians  275
middle trace  276
median trace  276
review questions  276

**8C. Smoothing broken hinges  276**
cross-hinges  276
review questions  278

**8D. Dealing with two questions  279**
cross hinges  279
smoothing the separations  280
letter-value differences  280
hinge traces  281
review questions  281

**8E. Wandering schematic plots  283**
adjacent polygon  285
wandering schematic plot  285
interpretation  286
review questions  287

**8F. A more demanding example: Governor's salary and bank deposits  287**
first analysis  287
untilted analysis  291
combined analysis  294
review questions  298

**8G. Further questions/analysis in the example  298**
time changes  298
vertical width  298
slopes and levels  299
consolidated summary  299
moral  299
review questions  306

**8H. How far have we come?  306**

**8I. The need to smooth both coordinates (optional)  307**
review questions  308

| EXHIBITS | PAGE |
|---|---|
| **8A** | |
| 1★ | 267 |
| 2★ | 268 |
| 3 | 269 |
| 4 | 270 |
| 5★ | 271 |
| 6★ | 272 |
| 7 | 273 |
| **8B** | |
| 8 | 274 |
| 9★ | 275 |
| **8C** | |
| 10 | 277 |
| 11★ | 277 |
| **8D** | |
| 12 | 279 |
| 13 | 280 |
| 14 | 281 |
| 15★ | 282 |
| **8E** | |
| 16 | 283 |
| 17★ | 284 |
| 18 | 285 |
| 19 | 286 |
| **8F** | |
| 20 | 288 |
| 21 | 289 |
| 22 | 291 |
| 23 | 292 |
| 24 | 292 |
| 25 | 293 |
| 26 | 294 |
| 27★ | 295 |
| 28 | 296 |
| 29 | 296 |
| 30 | 297 |
| **8G** | |
| 31★ | 300 |
| 32★ | 302 |
| **8H** | |
| 33 | 307 |
| 34 | 308 |

exhibit of **1** of chapter 8: Twin Rivers

**The 152 (ElUse, GaUse) pairs (winter energy consumption in therms; ElUse = therms of electricity; GaUse = therms of gas; 1 therm approximately 30 kilowatt hours and exactly 100,000 btu)**

**A)** The DATA ordered by ElUse——(ElUse; GaUse) pairs

```
 7  (89; 424)   (108; 388)   (119; 778)   129; 780)   (134; 832, 1016)   143; 988)
14  (152; 1186) (153; 612)   (156; 668)   (157; 1108) (159; 752)   (163; 676)   (167; 968)
21  (170; 854) *  (171; 670)  (175; 530)  (177; 810, 766)  (178; 1008)  (180; 958)
28  (182; 686)  (186; 986)   (188; 824)   (189; 802)  (190; 1014, 1016)  (191; 1078)
34  (192; 978, 568) *  (193; 1144)  (195; 838)  (196; 1298)  (198; 612)
41  (204; 584, 908)  (206; 1102)  (207; 756, 810)  (209; 1028)  (211; 610)
49  (214; 1044) (217; 1128, 922, 980) *  (218; 830, 812)  (224; 1038, 1022)
56  (225; 832)  (227; 938)   (229; 742)   (231; 1152)  (233; 844, 1030)  (234; 790)
62  (235; 886, 1064)  (238; 860)  (239; 922) *  (240; 912)  (241; 892)
71  (242; 1068, 860) (243; 756) (244; 1068, 788) (245; 658, 876) (246; 992, 774)
 9  (247; 700)  (252; 1058, 1020, 1184, 1026) *  (254; 1050)  (255; 840, 912, 862)
72  (256; 1060) (257; 1018)  (258; 870, 824, 840, 1062)  (259; 1062)  (262; 812)
64  (263; 1000) (264; 922)   (265; 982) *  (266; 684, 816, 754)  (267; 748)  (268; 956)
56  (270; 928, 746)  (271; 1068, 834) (272; 1086)  (273; 988, 824)  (274; 940)
48  (275; 1074, 958) (276; 686) *  (277; 868, 852)  (279; 1158)  (280; 754, 782)
40  (281; 958)  (283; 796)   (284; 1030)  (285; 1012)  (287; 1070, 788, 1096)
33  (288; 654)  (290; 892)   (291; 1004) *  (292; 798, 766)  (293; 942)
27  (294; 926, 1120) (295; 882) (297; 1144) (298; 814) (299; 1214) (300; 766)
20  (301; 974)  (302; 924)   (305; 1064)  (306; 900)  (315; 956) *  (316; 828)
14  (319; 1024) (323; 892)   (324; 932)   (325; 662)  (333; 1150)  (334; 946)
 8  (347; 732)  (348; 956)   (352; 720)   (367; 762)  (371; 1088)  (422; 1222)
 2  (429; 1002) (435; 860)
```

Note: (134; 832, 1016) stands for (134; 832) and (134; 1016), etc.

**B)** The DATA ordered by GaUse——(GaUse; ElUse) pairs

```
 6  (388; 108)  (424; 89)   (530; 175)   (568; 192)   (584; 204)   610; 211)
13  (612; 198, 153)  (654; 288)  (658; 245)  (662; 325)  (668; 156)  (670; 171)
20  (676; 163)  (684; 266)  (686; 276, 182)  (700; 247)  (720; 352)  (732; 347)
28  (742; 229)  (746; 270)  (748; 267)  (752; 159)  (754; 280, 266)  (756; 243, 207)
36  (762; 367)  (766; 300, 292, 177)  (775; 246)  (778; 119)  (780; 129)  (782; 280)
44  (788; 287, 244)  (790; 224)  (796; 283)  (798; 292)  (802; 189)  (810; 177, 207)
52  (812; 218, 262)  (814; 298)  (816; 266)  (824; 188, 258, 273)  (828; 316)
60  (830; 218)  (832; 225, 134)  (834; 271)  (838; 195)  (840; 255, 258)  (844; 233)
68  (852; 277)  (854; 170)  (860; 238, 435, 242)  (862; 255)  (868; 277)  (870; 258)
76  (876; 245)  (882; 295)  (886; 235)  (892; 290, 323, 241)  (900; 306)  (908; 204)
76  (912; 240, 255)  (922; 217, 239, 264)  (924; 302)  (926; 294)  (928; 270)
68  (932; 324)  (938; 227)  (940; 274)  (942; 293)  (946; 334)  (956; 348, 315, 268)
60  (958; 281, 180, 275)  (968; 167)  (974; 301)  (978; 192)  (980; 217)  (982; 265)
52  (986; 186)  (988; 273, 143)  (992; 246)  (1000; 263)  (1002; 429)  (1004; 291)
45  (1008; 178)  (1012; 285)  (1014; 190)  (1016; 190, 134)  (1018; 257)  (1020; 252)
38  (1022; 224)  (1024; 319)  (1026; 252)  (1028; 209)  (1030; 284, 233)  (1038; 224)
31  (1044; 214)  (1050; 254)  (1058; 252)  (1060; 256)  (1062; 259, 258)
25  (1064; 305, 235)  (1068; 244, 271, 242)  (1070; 287)  (1074; 275)  (1078; 191)
17  (1086; 272)  (1088; 371)  (1096; 287)  (1102; 206)  (1108; 157)  (1120; 294)
11  (1128; 217)  (1144; 297, 193)  (1150; 333)  (1152; 231)  (1158; 219)  (1184; 252)
 4  (1186; 152)  (1214; 299)  (1222; 422)   1298; 196)
```

Note: (612; 153, 198) stands for (612; 153) and (612; 198), etc.

exhibit **1** of chapter 8 (continued)

**P) PROBLEMS**--suitable for sharing among a group.

1a) Insert *'s in panel B to make batches of 15 and 16.

1b/c/d/e/f/g/h/i/j/k) Find the y-letter values for the corresponding slice thus formed.

1l) Prepare a table like exhibit 2 for this case.

1m) Prepare a picture like exhibit 3 for this case.

1n) Prepare a picture like exhibit 4 for this case.

**S) SOURCE**

Personal communication from Professor R. H. Socolow. (Data from a Princeton University School of Engineering study of energy use and energy saving.)

exhibit **2** of chapter 8: Twin Rivers

**Letter values and (inner) fences for the GaUse batches defined by the *'s in exhibit 1**

| # | x = ElUse | | y = GaUse | | | | | | | |ElUse| |
|---|---|---|---|---|---|---|---|---|---|---|---|
| | min | max | lo out | lo adj | lo H | M | hi H | hi adj | hi out | med |
| 15 | 89 to 170 | ××× | | 388 | 672 | 780 | 978 | 1186 | ××× | 152 |
| 15 | 171 to 192 | ××× | | 530 | 726 | 824 | 997 | 1078 | ××× | 186 |
| 15 | 193 to 217 | ××× | | 584 | 783 | 922 | 1073 | 1298 | ××× | 207 |
| 15 | 218 to 239 | ××× | | 742 | 831 | 886 | 1026 | 1152 | ××× | 231 |
| 16 | 240 to 252 | ××× | | 658 | 781 | 902 | 1042 | 1184 | ××× | 245 |
| 15 | 254 to 265 | ××× | | 812 | 851 | 922 | 1034 | 1062 | ××× | 258 |
| 16 | 266 to 276 | ××× | | 684 | 751 | 881 | 973 | 1086 | ××× | 271 |
| 15 | 277 to 291 | ××× | | 654 | 792 | 892 | 1021 | 1158 | ××× | 284 |
| 15 | 292 to 315 | ×× | | 766 | 848 | 926 | 1019 | 1214 | ××× | 298 |
| 15 | 316 to 435 | ××× | | 662 | 795 | 932 | 1013 | 1222 | ××× | 347 |

**P) PROBLEMS**

2a/b/c/d/e) Write out the stem-and-leaf for the y-values (GaUse) in the 1st/2nd/3rd/4th/5th slice and verify the letter values above.

2f/g/h/i/j) Do the same for the x-values (ElUse) of the 6th/7th/8th/9th/10th slice.

In panel A of exhibit 1, the asterisks divide the 152 points into ten slices, each of 15 or 16 points. When we calculate letter values for y in each slice separately, we obtain the results set down in exhibit 2. Plotted as a set of parallel schematic plots, we get the picture of exhibit 3.

The fact that we have to write in numbers for the x-intervals defining the slices shows that we are not yet fully graphic. Exhibit 4 centers each schematic plot at the medians of the slice it represents, and shows, as light dotted lines, the cuts that define the slices. It also provides a crossing of line segments where the x-median and y-median of each slice cross each other. A hard look at this particular picture shows us one of its most important weaknesses: a lot of detail near the median of x, where nothing very interesting is likely to happen; little detail for extreme x, where something interesting is much more likely.

exhibit **3** of chapter 8: Twin Rivers

**Parallel schematic plots (regrettably equally spaced)**

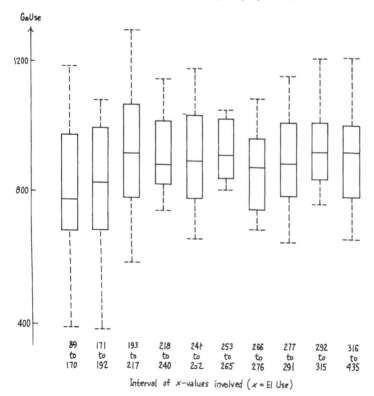

Interval of x-values involved ($x$ = El Use)

**exhibit 4/8: Parallel and wandering schematic plots**

While a first thought might be to go to slices of equal width in $x$, a little thought shows that this would tend to make the number of points in the end slices impractically small. A reasonable compromise is often to use

**letter cuts**

to:

◇ start with the letter values of $x$, moving them a shade inward where necessary to avoid any value exactly on a cut. (The median may well be moved outward a shade in both directions.)

Panel C of exhibit **5** shows the letter values and letter cuts for $x$, going out as far as B on each side. We are left with 4 slices of count 3 or less. Were we interested only in medians, slices of 3 would serve us, perhaps well. If we want hinges as well, we need at least 5 values. Accordingly, panel D shows the mark-up using C, D, ..., D, C.

exhibit **4** of chapter 8: Twin Rivers

**Parallel schematic plots (spaced to fit x-medians)**

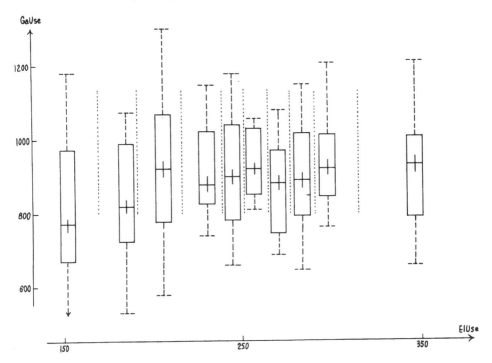

exhibit **5** of chapter 8: Twin Rivers

**Continuation of exhibit 1**

C) LETTER VALUES and LETTER CUTS

| #152 | letter values | | letter cuts | | slice counts | | | |
|---|---|---|---|---|---|---|---|---|
| M 76h | 253 | | M 252h | 253h | HM | 37 | 38 | MH |
| H 38h | 207 | 283h | H 207h | 283h | EH | 20 | 19 | HE |
| E 19h | 177h | 301h | E 177h | 301h | DE | 9 | 9 | ED |
| D 10 | 156 | 333 | D 156h | 332h | CD | 4 | 5 | DC |
| C 5h | 134 | 359h | C 134h | 359h | BC | 3 | 2 | CB |
| B 3 | 119 | 422 | B 119h | 421h | −B | 3 | 3 | B− |

Note that (1) each letter cut is at an h, the median having been split and moved outward if needed to reach an h, the other letter points having been moved inward when needed to reach an h (thus 177h and 301h were not changed). (2) Each piece has been tagged with the two letters, in order, of the cuts that define it—though we will often just use serial numbers. Note also that it would be a very helpful option to move the hiC cut inward one value to obtain at least 3 values in slice #11 (CB).

D) Panel A MARKED UP according to PANEL C--only to C's, with loD moved 1 in.

```
 7 (89;424) (108;388) (119;778) (129;780) (134;832,1016) C (143;988)
13 (152;1186) (153;612) (156;668) (157;1108) D (159;752) (163;676)
20 (167;968) (170;854) (171;670) (175;530) (177;810,766) E (178;1008)
27 (180;958) (182;686) (186;986) (188;824) (189;802) (190;1014,1016)
34 (191;1078) (192;978,568) (193;1144) (195;838) (196;1298) (198;612)
41 (204;584,908) (206;1102) (207;756,810) H (209;1028) (211;610)
50 (214;1044) (217;1128,922,980) (218;830,812) (224;1038,1022) (225;832)
56 (227;938) (229;742 (231;1152) (233;844,1030) (234;790)
64 (235;886,1064) (238;860) (239;922) (240;912) (241;892) (242;1068,860)
72 (243;756) (244;1068,788) (245;658,876) (246;992,774) (247;700)
 9 (252;1058,1020,1184,1026) M (254;1050) (255;840,912,862) (256;1060)
71 (257;1018) (258;970,824,840,1062) (259;1062) (262;812) (263;1000)
63 (264;922) (265;982) (266;684,816,754) (267;748) (268;956)
56 (270;928,746) (271;1068,834) (272;1086) (273;988,824) (274;940)
48 (275;1074,958) (276;686) (277;868,852) (279;1158) (280;754,782)
40 (281;958) (283;796) H (284;1030) (285;1012) (287;1070,788,1096)
33 (288;654) (290;892) (291;1004) (292;798,766) (293;942) (294;926,1120)
25 (295;882) (297;1144) (298;814) (299;1214) (300;766) (301;974) E
19 (302;924) (305;1064) (306;900) (315;956) (316;828) (319;1024)
13 (323;892) (324;932) (325;662) D (333;1150) (334;946) (347;732)
 7 (348;956) (352;720) C (367;762) (371;1088) (422;1222) (429;1002)
 1 (435;860)
```

P) PROBLEMS--suitable for sharing among a group

5a) Insert letter cuts in panel B, moving them as desirable.

5b/c/d/e/f/g/h/i/j/k) Find the y-letter values for the corresponding slice thus formed.

5l) Prepare a table like exhibit 6 for this case.

5m) Prepare a picture like exhibit 7 for this case.

The y-letter values for these slices are as in exhibit 6 giving the picture of exhibit 7. We have now paid the price of exploring the tails in more detail. Because we have 3 batches of the minimum size--5--for sensible schematic plots, we expected greater irregularity. And we received it. On the other hand, instead of having the extreme schematic plots coming in toward the middle of the $x$ values as far as 170 for the low extreme slice and as far as 316 for the high one, the extreme slices now before us extend to 134 below and 367 above. Clearly, we are looking near the extremes in more detail.

Whichever choice we make--exhibit 4, exhibit 7, or the result of some other cutting--we now need some smoothing, to which we will turn in the next section.

**exhibit 6 of chapter 8: Twin Rivers**

**Letter values (HMH) and adjacent values for the batches defined by the letter cuts of exhibit 5**

A) CALCULATIONS

| | ElUse | | | GaUse | | | | | | ElUse |
|---|---|---|---|---|---|---|---|---|---|---|
| | min | max | lo out | lo adj | lo H | M | hi H | hi adj | hi out | median |
| 6 | 89 to 134 | | ××× | 388 | 424 | 779 | 832 | 1016 | ××× | 124 |
| 5 | 143 to 157 | | ××× | 612 | 668 | 988 | 1108 | 1186 | ××× | 153 |
| 8 | 159 to 177 | | ××× | 530 | 673 | 759 | 832 | 968 | ××× | 170h |
| 20 | 178 to 207 | | ××× | 568 | 779 | 933 | 1015 | 1293 | ××× | 192 |
| 37 | 209 to 252 | | ××× | 610 | 830 | 922 | 1030 | 1184 | ××× | 235 |
| 38 | 254 to 283 | | ××× | 684 | 816 | 891 | 1000 | 1158 | ××× | 267h |
| 19 | 284 to 301 | | ××× | 654 | 806 | 942 | 1050 | 1214 | ××× | 292 |
| 9 | 302 to 325 | | one | 828 | 892 | 924 | 956 | 1024 | one | 316 |
| 5 | 333 to 352 | | ××× | 720 | 732 | 946 | 956 | 1150 | ××× | 347 |
| 5 | 367 to 435 | | ××× | 762 | 860 | 1002 | 1088 | 1222 | ××× | 422 |

The values outside are 662 and 1064. The H-spreads are, in order of appearance; 408, 440, 159, 236, 200, 184, 244, 64, 224, 228. The outside values are associated with the strikingly low H-spread of 64.

P) PROBLEMS

6a) Divide the pairs of panel A of exhibit 1 into slices of successive sizes 3, 5, 9, 17, 33, 18, 33, 17, 9, 5, 3.

6b/c/d/e/f/g/h/i/j/k (uses 6a)) Find as many analogs of the values in Panel A above for the 1st/2nd/3rd/.../10th slice as you can.

6l (uses all earlier)) Combine these results in an analog of exhibit 7.

## review questions

What is one of the simplest things we can do with (x, y) pairs? What are its three parts? What example did we try this on? What was the result? (Try to give yourself a description in words!) How did we make this more graphic? What seemed to be a major weakness? What did we decide to do about this? What are letter cuts? How did we use them? What was the result? What do we now need?

exhibit **7** of chapter 8: Twin Rivers

**Parallel schematic plots (using letter cuts)**

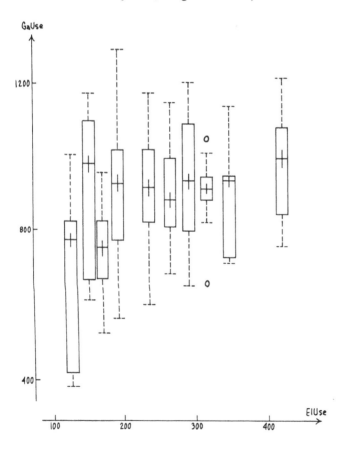

### 8B. Smoothing the broken median

If we want to summarize the "middle" of gas use as a function of electrical use, we can put together slice medians and letter cuts from exhibits 5 and 6, as shown in the top panel of exhibit **8**. While it is quite easy to say how we got this

broken median

it is also easy to see that most of what we have is "wiggle" and that our eye is distracted from any general trends or bends that may be present. Clearly we need to smooth something.

exhibit **8** of chapter 8: Twin Rivers

**Broken median--and the corresponding middle trace**

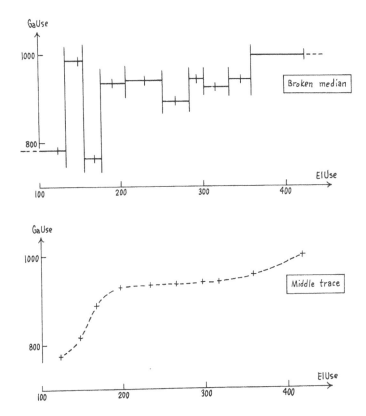

The easiest things to smooth are the

### cross-medians

the places where an x-median crosses a y-median for the same slice. (These have been shown as + in exhibits 4, 7, and 8.) We want at least initially to smooth lightly at the ends. After all, any extreme slice may be trying to tell a special story. So we plan to copy extremes on, and do only 3R'SS, where the ' in R' reminds us that we are copying on.

Exhibit **9** has the essentials of the arithmetic; the lower part of exhibit 8 has the picture. Notice that, for reasons to be discussed in an optional section (8I), the sequence of x-medians is also smoothed. (Since they are in increasing order anyway, hanning will be the first--and only--part of the smooth that affects them.)

exhibit **9** of chapter 8: Twin Rivers

**Smoothing the cross-medians of exhibit 6**

### A) SMOOTHING

| Slice # | ElUse median | H | GaUse median | 3R' | 3R'SS | 3R'SSH | 3R'SSH3 |
|---|---|---|---|---|---|---|---|
| 1  | 124  | 124  | 779  | 779  | T | 779  | T |
| 2  | 153  | 150  | 988  | 779  | h | 815  | h |
| 3  | 170h | 171  | 759  | 922  | e | 886  | e |
| 4  | 192  | 197h | 933  | 922  |   | 922  |   |
| 5  | 235  | 232h | 922  | 922  | s | 922  | s |
| 6  | 267h | 265h | 891  | 922  | a | 922  | a |
| 7  | 292  | 292  | 942  | 924  | m | 928  | m |
| 8  | 316  | 318  | 924  | 942  | e | 938  | e |
| 9  | 347  | 358  | 946  | 946  |   | 959  |   |
| 10 | 422  | 422  | 1002 | 1002 |   | 1002 |   |

### P) PROBLEMS

9a) Plot the broken median based on exhibits 1 and 2.

9b) Write out the smoothing for that case.

9c) Plot the middle trace based on this smoothing.

9d) Show the two broken medians (from exhibits 1 and 2, and from exhibits 5 and 6) in the same picture. Discuss.

9e) Show the two middle traces in the same picture. Discuss.

9f) Take the results of problems 6b to k and carry out the analog of panel A. Plot the resulting median trace.

/8: Parallel and wandering schematic plots

We call the result of smoothing a broken median a

### middle trace

whatever set of cuts was used to fix the slices and whatever smooth was used. It represents a good smooth idea of how the "middle" of $y$ seems to behave as $x$ changes. In it we take advantage of the ideas at hand: breaking into slices, medians, smoothing to get something that traces a possible middle--something reasonably smooth but not constrained to any such particular family of shapes as all quadratics, all exponentials, or all sine waves. We will feel free to also call it a

### median trace

whenever the context makes this term more natural.

#### review questions

What is an easy way to summarize the median of $y$ as a function of $x$? What is a broken median? How well does it serve us? What are its main deficiencies? What are we going to need to do? What are cross-medians? What are we going to do with them? Which section will discuss why we smooth $x$'s as well as $y$'s? What do we think about extreme slices? How does this affect our choice of smoother? What smoother do we choose? What happened when we tried it? What is a middle trace? What is a median trace?

## 8C. Smoothing broken hinges

We are ready to draw a broken median, and smooth the cross-medians to find a middle trace instead. Can we gain by doing something similar for hinges? Only if we are careful.

Exhibit **10** shows the broken hinges that result from taking letter cuts and $y$-hinges from exhibit 6--with $x$-medians crossed and $x$-hinges shown, for further reference, by little dots. Exhibit **11** shows smoothings over slices for upper and lower GaUse hinges ($y$-hinges)--and for ElUse hinges ($x$-hinges).

Exhibit **12** plots the smooths, which are candidates for being "hinge traces", together with circles showing where the median trace went. Two things are now relevant, one to be observed and the other to be remembered:

◇ particularly after the middle trace has been firmly drawn in exhibit 12, our attention focusses more on the spacing between the traces than on the exact location of the individual traces.

◇ we-are still unsettled as to whether to plot where (smoothed) $y$-hinges cross (smoothed) $x$-medians, or whether we should define

### cross-hinges

as the intersection of each $y$-hinge with an appropriate $x$-hinge. (Which one would be appropriate?)

exhibit **10** of chapter 8: Twin Rivers

**Broken hinges**

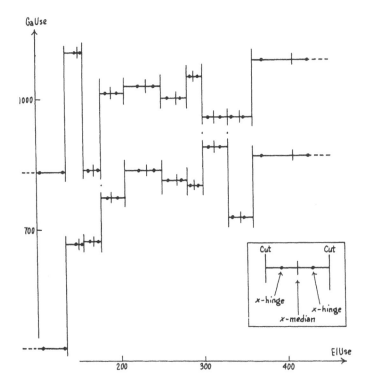

exhibit **11** of chapter 8: Twin Rivers

**Smoothing the y-hinges and x-hinges for (GaUse, ElUse) (mostly based on exhibit 6)**

A) UPPER HINGE--see exhibit 9 for smooth of ElUse median

| ElUse median | | GaUse upper hinge | | | | |
|---|---|---|---|---|---|---|
| H3 | raw | 3R' | 3R'SS | 3R'SSH | 3R'SSH3 | |
| 124 | 832 | 832 | | 832 | 832 | |
| 150 | 1108 | 832 | | 878 | 878 | |
| 171 | 832 | 1015 | | 969 | 969 | |
| 197h | 1015 | 1015 | | 1015 | 1015 | |
| 232h | 1030 | 1015 | | 1015 | 1015 | |
| 265h | 1000 | 1015 | | 1011 | 1011 | |
| 292 | 1050 | 1000 | | 996 | 996 | |
| 318 | 956 | 956 | 970 | 978 | 996 | |
| 358 | 956 | 956 | 970 | 1000 | 1000 | |
| 422 | 1088 | 1088 | | 1088 | 1088 | |

→

What are we to do about the two questions thus posed?

### review questions

What would we like to try to do? What picture did we start with? What list of things did we smooth? What did we plot? What observation then arose? About what were we still unsettled?

exhibit **11** of chapter 8 (continued)

**B) LOWER HINGE**

| ElUse median | | GaUse lower hinge | | | | | |
|---|---|---|---|---|---|---|---|
| H3 | raw | 3R' | 3R'SS | | 3R'SSH | 3R'SSH3 | |
| 124 | 424 | 424 | T | | 424 | | T |
| 150 | 668 | 668 | h | | 608 | | h |
| 171 | 673 | 673 | e | | 698 | | e |
| 197h | 779 | 779 |   | | 760 | |   |
| 232h | 830 | 816 | s | | 804 | | s |
| 265h | 816 | 816 | a | | 816 | | a |
| 292 | 806 | 816 | m | | 814 | | m |
| 318 | 892 | 816 | e | | 827 | | e |
| 358 | 732 | 860 |   | | 849 | |   |
| 422 | 860 | 860 |   | | 860 | |   |

**C) SMOOTHING the ElUse HINGE -- x-hinges**

| | lower ElUse hinges | | | upper ElUse hinges | | |
|---|---|---|---|---|---|---|
| Count | raw | H | H3 | raw | H | H3 |
| 6 | 108 | 108 | T | 134 | 134 | T |
| 5 | 152 | 144 | h | 156 | 155h | h |
| 8 | 165 | 167h | e | 176 | 177 | e |
| 20 | 188h | 191h |   | 201 | 205h |   |
| 37 | 224 | 223h | s | 244 | 241 | s |
| 38 | 258 | 257 | a | 275 | 272h | a |
| 19 | 287h | 285 | m | 296 | 297h | m |
| 9 | 306 | 308h | e | 323 | 322h | e |
| 5 | 334 | 336 |   | 348 | 362 |   |
| 5 | 371 | 371 |   | 429 | 429 |   |

**P) PROBLEMS**

11a/b/c) Do the calculation of panels A/B/C for the numbers of exhibit 2.

11d (uses a, b, and c)) Make a plot like exhibit 12 based on the results of 11a, 11b, and 11c.

11e/f/g) As 11a/11b/11c for the slicing of problem 9a.

11h (uses e, f, and g)) As (11d) for the results of (11e), (11f), and (11g).

## 8D. Dealing with the two questions

Our basic difficulty in the second question thus posed is "which $x$-hinge goes with a chosen $y$-hinge?"

Exhibit 13 shows what happens when the given points lie along a smooth curve. Since all the points are ON this curve, the curve itself ought--except for undesired byproducts of smoothing--to be not only the middle trace but also both hinge traces. The only way to get a good start on this is to define the

### cross-hinges

as the intersection of upper hinge with upper hinge--and lower hinge with lower hinge--WHEREVER the curve--or the sequence of cross-hinges--is RISING (as on the left of exhibit 13) BUT as the intersection of upper hinge with lower hinge--and of lower hinge with upper hinge--WHEREVER the curve--or the sequence of cross-hinges--is FALLING (as on the right of exhibit 13). Where the curve is sensibly flat, it does not matter which choice we make, so that, while either can be used, it is often reasonable to compromise and use the $x$-median rather than either $x$-hinge.

exhibit **12** of chapter 8: Twin Rivers

**Smooths of broken hinges (and the middle trace)**

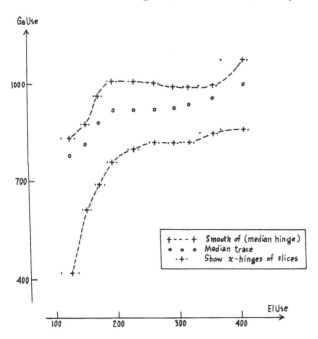

280 exhibit 13/8: Schematic plots

If we follow this lead, we will make similar matchings whenever a "hinge trace" appears to be rising--or be falling--or be nearly level. In the Twin Rivers example, exhibit 12 suggests that we may as well take both hinge traces as rising throughout. Exhibit **14** shows the result of doing this, with middle trace solid, and the "hinge traces" thus defined dashed. The dots show the ($x$-median, $y$-hinge) points, which would otherwise have defined the "hinge traces". In this example the "hinge traces" are brought appreciably together.

### smoothing the separations

The other thing we may want to do is to focus on the separations between "hinge trace" and middle trace, rather than on the position of the hinge traces themselves. The way to do this is clearly to begin by finding the corresponding

### letter-value differences

here "H-to-M", defined as

$$\text{median} \quad \text{MINUS} \quad \text{lower hinge}$$

and "M-to-H", defined as

$$\text{upper hinge} \quad \text{MINUS} \quad \text{median}$$

exhibit **13** of chapter 8: illustrative

**Assumed cuts, slice medians and slice hinges for a batch of points lying on a smooth curve**

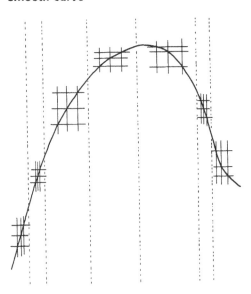

and then:

◇ smoothing the letter-value differences.

◇ combining them with the middle traces to produce smoothing of separated "hinge traces".

◇ smoothing once more, perhaps only with 3, to remove any newly-created irregularities.

Exhibit **15** gives the details for our example. Notice the use of 3R (end-value rule) instead of 3R′ (copying on).

Exhibit **16** plots the result, USING THE X-HINGES CHOSEN ABOVE, thus taking action about both the questions toward which this section is directed. The results are good enough, and the issues well-enough handled, for these traces to deserve being called

hinge traces.

**review questions**

What is the difficulty about y-hinges? What example did we look at? What did it tell us? What are cross-hinges? What do we do where the sequence of

exhibit **14** of chapter 8: Twin Rivers

**Using the x-hinges for the smoothed broken hinges, shown with middle trace**

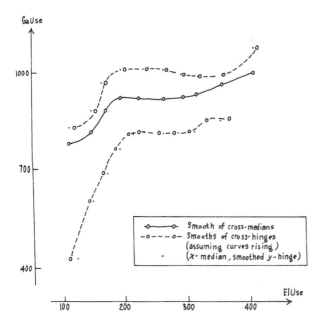

cross-hinges is rising? What where it is falling? What real example did we then consider? What did we do? What was the result?

To what problem did we then turn? What are letter-value differences? Which ones did we turn to? How did we use them? What did we add together? How did we show the result? What is a hinge trace?

exhibit **15** of chapter 8: Twin Rivers

**Smoothing and recombining the hinge–median differences (based mainly on exhibit 6)**

### A) SMOOTHING HINGE–MEDIAN DIFFERENCES

| | | GaUse HtoM | | | | | GaUse MtoH | | | |
|---|---|---|---|---|---|---|---|---|---|---|
| (count) | raw | 3R' | 3R'SS | 3R'SSH | 3R'SSH3 | raw | 3R' | 3R'SS | 3R'SSH | 3R'SSH3 |
| (6) | 355 | 355 | T | 355 | T | 53 | 53 | | 53 | T |
| (5) | 320 | 320 | h | 288 | h | 120 | 73 | | 76 | h |
| (8) | 86 | 156 | e | 181 | e | 73 | 108 | | 99 | e |
| (20) | 154 | 92 | | 108 | | 190 | 108 | | 108 | |
| (37) | 92 | 92 | s | 92 | s | 108 | 108 | | 108 | s |
| (38) | 75 | 92 | a | 92 | a | 109 | 108 | | 108 | a |
| (19) | 136 | 92 | m | 103 | m | 108 | 108 | | 108 | m |
| (9) | 32 | 136 | e | 126 | e | 32 | 32 | 108 | 102 | e |
| (5) | 214 | 142 | | 140 | | 10 | 32 | 86 | 92 | |
| (5) | 142 | 142 | | 142 | | 86 | 86 | | 86 | |

### B) RECOMBINING SMOOTHED SEQUENCES

| GaUse lower hinge | | | | GaUse upper hinge | | | |
|---|---|---|---|---|---|---|---|
| mid | HtoM | lo H | 3 | mid | MtoH | hi H | 3 |
| 779 | 355 | 424 | | 779 | 53 | 832 | T |
| 815 | 288 | 523 | | 815 | 76 | 891 | h |
| 886 | 114 | 705 | | 886 | 99 | 985 | e |
| 922 | 108 | 814 | | 922 | 108 | 1030 | |
| 922 | 92 | 830 | | 922 | 108 | 1030 | s |
| 922 | 92 | 830 | | 922 | 108 | 1030 | a |
| 928 | 103 | 825 | | 928 | 108 | 1036 | m |
| 938 | 126 | 812 | 819 | 938 | 102 | 1040 | e |
| 959 | 140 | 819 | | 959 | 92 | 1051 | |
| 1002 | 142 | 860 | | 1002 | 86 | 1088 | |

### P) PROBLEMS

15a/b/c) Do the calculations of panel A/B/C starting from the numbers of exhibit 2.

15d(abc)) Prepare a picture like exhibit 16 based on the results of (15a), (15b), and (15c).

15e/f/g(b)) Do the calculations of panel A/B/C starting from the numbers of problems (6b) to (6k).

15h(efg)) Prepare a picture like exhibit 16 based on the results of (15e), (15f), and (15g).

### 8E. Wandering schematic plots

We now have analogs of most of the parts of a schematic plot:

◇ the hinges become hinge traces.
◇ the median becomes a middle trace (median trace).
◇ and the outside points?
◇ and the adjacent points?

What we need next are fences, so as to help us distinguish outside and adjacent points.

In this case, the simplest calculation seems to be the best. We fix convenient values of $x$, read off (from the graph paper original) the corresponding $y$'s according to the hinge traces, and then proceed as usual. Exhibit **17** shows the arithmetic, making it clear that no one of the 152 points in the Twin Rivers example has even a faint chance of being outside. Thus all that remains is to locate the adjacent points and represent them appropriately.

exhibit **16** of chapter 8: Twin Rivers

**At last, hinge traces--and the middle trace**

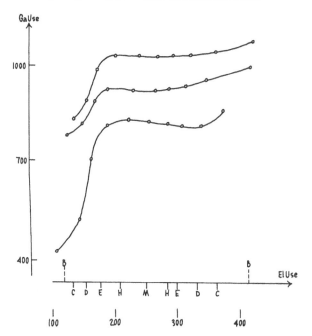

**exhibit 18** shows:

◇ hinge traces and median trace.

◇ each point adjacent in its slice.

◇ each (non-outside) point in the two extreme slices.

◇ a dashed-line convex polygon surrounding all these points just mentioned.

**exhibit 17** of chapter 8: Twin Rivers

**The fences**

**A) CALCULATION of FENCES**

| ElUse | GaUse Hinge · traces | | GaUse H-spread | | | | | Step | GaUse inner | fences |
|---|---|---|---|---|---|---|---|---|---|---|
| | | | raw | 3R | S | H | 3 | | | |
| 125 | 560 | 835 | 245 | 245 | T | 245 | T | 348 | 212 | 1183 |
| 150 | 640 | 875 | 235 | | h | 236 | h | 354 | 286 | 1229 |
| 175 | 750 | 980 | 230 | | e | 228 | e | 342 | 408 | 1322 |
| 200 | 825 | 1033 | 208 | 215 | | 218 | | 327 | 398 | 1360 |
| 225 | 830 | 1045 | 215 | 214 | s | 214 | s | 321 | 509 | 1366 |
| 250 | 828 | 1042 | 214 | | a | 214 | a | 321 | 507 | 1363 |
| 275 | 828 | 1042 | 214 | | m | 216 | m | 324 | 504 | 1366 |
| 300 | 823 | 1045 | 222 | | e | 220 | e | 330 | 493 | 1375 |
| 325 | 820 | 1048 | 228 | 224 | | 224 | | 336 | 484 | 1384 |
| 350 | 826 | 1050 | 224 | 224 | | 224 | | 336 | 490 | 1386 |
| 375 | (868) | 1056 | (188) | 224 | | 224 | | 336 | (532) | 1392 |

Note: Further smoothing of the fences thus found is optional

**B) POINTS MOST NEARLY OUTSIDE** -- see exhibit 2

| Slice | Low extreme | High extreme | Near fence |
|---|---|---|---|
| 1 | (108, 388) | | below 148 |
| 2 | | | |
| 3 | | (196, 1298) | above 1322 |
| 4 | | | |
| 5 | | | |
| 6 | | | |
| 7 | | | |
| 8 | | | |
| 9 | | | |
| 10 | | | |

Other points omitted as having no chance of approaching fences. [In first slice, (89, 424) was the only other possibility.]

**P) PROBLEMS**

**17a(15d*))** Use the graph-paper original of problem 15d to find the numbers needed for the calculation of panels A and B. Compare your result with that found above.

**17b(15h*))** Do the same from the graph-paper original of problem 15h.

If we omit the points, keeping the

### adjacent polygon

dashed and the hinge and median traces solid, we will obtain what we may as well call

### wandering schematic plot

as shown in exhibit **19**. This example is of course rather special, in that there are no outside points.

Here we have made the adjacent polygon convex. In other circumstances, for instance where all the traces and hence the fences were high at both ends and low in the middle, so that a crescent-shaped adjacent polygon would clearly make good sense, we do not require that the adjacent polygon be convex. Reasonable judgment should be used, always being sure that the adjacent polygon includes all points not outside.

exhibit **18** of chapter 8: Twin Rivers

**Hinge and median traces--candidate points--adjacent polygon**

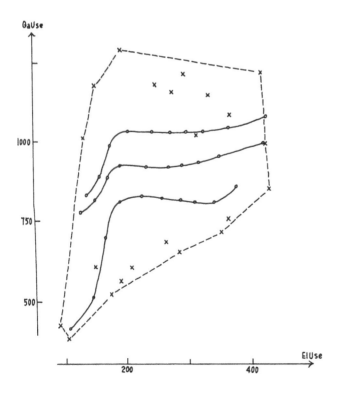

#### interpretation?

What interpretation can we place upon this particular wandering schematic plot? The largest part of each trace is roughly horizontal. At the left, however, there is tilt, strongest for the lower hinge trace. (At the right, there is a weak indication of tilt, also, but hardly strong enough to persuade us of its importance or reality.) This is just what we ought to expect, if

◇ for most townhouses there was no connection--not even a statistical one--between gas and electrical usages.

◇ a few townhouses had strikingly low values of both usages.

exhibit **19** of chapter 8: Twin Rivers

**The wandering schematic plot**

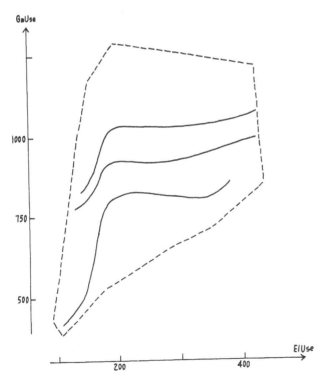

And how might this last have come about? Perhaps a few residents took winter vacations, turning the heat down and the lights off. Perhaps something else happened.

### review questions

What parts for a schematic plot had we when this section began? Which did we need? Did this lead us to search for fences? What seemed the simplest approach? What did we do? How did things come out in the Twin Rivers example? What list of plausible possibilities for adjacent points did we consider? With what did we surround them? Was this always to be convex? Why/why not? What is a wandering schematic plot? What are its pieces? What is an adjacent polygon?

## 8F. A more demanding example: Governor's salary and bank deposits

As so often happens, it is not large batches of points that call for attention in the construction of wandering schematic plots. Rather it is small batches that need special care. We devote this section to a batch of 50, one for each of the U.S.A.'s 50 states. So far as we can see, there is no way to gather any larger batch, for there are no political divisions elsewhere that have been operating in similar political and economic conditions. So all we can do is to analyze these 50 points with considerable care.

### first analysis

Exhibit **20** sets out, for about 1969, the total deposits in banks and the governor's salary for each of the 50 states, continuing with convenient multiples of the logarithms of each. As an amount with greatest/least close to 300, we should surely expect to start with logs of bank deposits. Since the corresponding ratio is only 5 for governor's salary, we could hardly be sure that logs were needed for this variable, but we will find them helpful.

Exhibit **21** continues with an arraying of the points, first (BaDep, GoSal) in BaDep order, then GoSal sorted within slice, with the medians and hinges displayed by circling. (The other two columns will be discussed below.) Here we have sliced to our own convenience, making slices of 5, 5, 9, 12, 9, 5, and 5. (The column of state names will help us to identify outside and adjacent states.)

When slicing for convenience, we often

◇ choose slice counts of 5, 9, 17, etc., when minimizing hand arithmetic, since these choices make letter depths without halves.

◇ choose slice counts of 4 (if only a median is wanted), 8, 16, 32, ..., when looking for a little more numerical stability in the letter values, since these choices gives h's to as many letter depths as possible, thus more often using means of two observations.

exhibit **20** of chapter 8: governor's salary

**Bank deposits and governor's salary for the 50 states ca. 1969**

A) DATA--and logs (a and b indicate how further digits would break ties. a < b)

|       | Bank deposits | | Governor's salary, 1969/70 | |
|-------|----------|-----------|-----------|------------------------|
|       | Billions | 100 log$_{10}$ | thousands | 1000 log$_{10}$ (raw/10) |
| Alab  | 4.287   | 63   | 25     | 398 |
| Alas  | .430    | −37  | 32     | 505 |
| Ariz  | 2.856   | 46   | 27.5   | 439 |
| Ark   | 2.625   | 42b  | 10     | 0   |
| Cal   | 45.422  | 166  | 44.1   | 644 |
| Colo  | 3.886   | 59   | 20     | 301 |
| Conn  | 9.328   | 97   | 35     | 544 |
| Del   | 1.456   | 16   | 35     | 544 |
| Fla   | 11.542  | 106  | 36     | 556 |
| Georg | 6.529   | 81   | 42.5   | 628 |
| Haw   | 1.380   | 14   | 33.5   | 525 |
| Idaho | 1.146   | 6    | 17.5   | 243 |
| Ill   | 31.587  | 150  | 45     | 653 |
| Ind   | 9.603   | 98   | 25     | 398 |
| Iowa  | 5.806   | 76b  | 30     | 477 |
| Kans  | 4.412   | 64   | 20     | 301 |
| Kent  | 4.652   | 67   | 30     | 477 |
| Louis | 5.708   | 76a  | 28.374 | 453 |
| Maine | 1.862   | 27b  | 20     | 301 |
| Maryl | 5.512   | 74   | 25     | 398 |
| Mass  | 20.006  | 130a | 35     | 544 |
| Mich  | 19.234  | 128  | 40     | 602 |
| Minn  | 9.088   | 96   | 27.5   | 439 |
| Miss  | 2.698   | 43   | 25     | 398 |
| Mo    | 10.933  | 104  | 37.5   | 574 |
| Mont  | 1.400   | 15   | 23.25  | 366 |
| Nebr  | 3.094   | 49   | 18     | 255 |
| Nev   | .956    | −2   | 25     | 398 |
| NH    | 1.855   | 27a  | 30     | 477 |
| NJ    | 15.776  | 120  | 35     | 544 |
| NM    | 1.217   | 9    | 20     | 301 |
| NY    | 121.905 | 209  | 50     | 699 |
| NC    | 6.326   | 80b  | 35     | 544 |
| ND    | 1.362   | 13   | 18     | 255 |
| Ohio  | 20.094  | 130b | 40     | 602 |
| Okla  | 4.854   | 69   | 25     | 398 |
| Ore   | 3.969   | 60   | 28.5   | 455 |
| Penn  | 29.648  | 147  | 45     | 653 |
| RI    | 2.609   | 42a  | 30     | 477 |
| SC    | 1.986   | 30   | 25     | 398 |
| SD    | 1.326   | 12   | 18     | 255 |
| Tenn  | 6.221   | 79   | 18.5   | 267 |
| Tex   | 23.475  | 137  | 40     | 602 |
| Utah  | 1.696   | 23   | 22     | 342 |

→

exhibit 20 of chapter 8 (continued)

| Verm | .994 | −0 | 25 | 398 |
|---|---|---|---|---|
| Virg | 6.876 | 84 | 30 | 477 |
| Wash | 6.283 | 80a | 32.5 | 512 |
| WVa | 2.298 | 36 | 25 | 398 |
| Wisc | 8.489 | 93 | 25 | 398 |
| Wyo | .703 | −15 | 20 | 301 |
| (Name) | | (BaDep) | | (GoSal) |

**S) SOURCE**

The 1970 World Almanac: Bank deposits (from FDIC) at page 91 and governor's salaries at page 7.

exhibit **21** of chapter 8: governor's salary

**Exhibit 20 continued**

**B) WORKING ARRAYS**--circled values are slice medians and slice hinges

| BaDep (sorted overall) | GoSal BaDep order | sorted in slices | GoSal MINUS 3(BaDep) BaDep order | sorted in slices | | Name BaDep order |
|---|---|---|---|---|---|---|
| −37 | 505 | 243 | 616 | 225 | | Alas |
| (−15) | 301 | (301) | 346 | (346) | | Wyo |
| (−2) | 398 | (398) | 404 | (398) | slice 1 | Nev |
| (−0) | 398 | (398) | 398 | (404) | | Verm |
| 6 | 243 | 505 | 225 | 616 | | Ida |
| 9 | 301 | 255 | 274 | 216 | | NM |
| (12) | 255 | (255) | 219 | (219) | | SD |
| (13) | 255 | (301) | 216 | (274) | slice 2 | ND |
| (14) | 525 | (366) | 483 | (321) | | Haw |
| 15 | 366 | 525 | 321 | 483 | | Mont |
| 30 | | | | | | |
| 16 | 544 | 0 | 496 | −126 | | Del |
| 23 | 342 | 301 | 273 | 220 | | Utah |
| (27) | 477 | (342) | 396 | (269) | | NH |
| 27b | 301 | 398 | 220 | 273 | Slice 3 | Maine |
| (30) | 398 | (398) | 308 | (290) | | SC |
| 36 | 398 | 398 | 290 | 308 | | WVa |
| (42) | 477 | (477) | 351 | (351) | | RI |
| 42b | 0 | 477 | −126 | 396 | | Ark |
| 43 | 398 | 544 | 269 | 496 | | Miss |

exhibit **21(B)/8**: Parallel and wandering schematic plots

Exhibit **22** completes the calculation of a middle trace and two hinge traces, as seen in exhibit **23**. We cannot be too happy with the crossings of these curves. (Nor would we be happy to plot the smoothed GoSal hinges at BaDep medians rather than at BaDep hinges.) The plausible cause of trouble is the steepness of the traces toward the righthand side.

exhibit **21** of chapter 8 (panel B continued)

| | | | | | | |
|---|---|---|---|---|---|---|
| 46 | 439 | 255 | 301 | 30 | | Ariz |
| 49 | 255 | 267 | 108 | 108 | | Nebr |
| (59) | 301 | (301) | 124 | (109) | | Colo |
| (60) | 455 | (301) | 275 | (124) | | Ore |
| 63 | 398 | 398 | 209 | 176 | Slice 4 | Alab |
| (64) | 301 | (398) | 109 | (191) | | Kans |
| (67) | 477 | (398) | 276 | (209) | | Kent |
| 69 | 398 | 439 | 191 | 225 | | Okla |
| (74) | 398 | (453) | 176 | (249) | | Maryl |
| (76a) | 453 | (455) | 225 | (275) | | Louis |
| 76b | 477 | 477 | 249 | 276 | | Iowa |
| 79 | 267 | 477 | 30 | 301 | | Tenn |
| | | | | | | |
| 80a | 512 | 398 | 272 | 104 | | Wash |
| 80b | 544 | 398 | 304 | 119 | | NC |
| (81) | 628 | (439) | 385 | (151) | | Ga |
| 84 | 477 | 477 | 225 | 225 | Slice 5 | Virg |
| (93) | 398 | (512) | 119 | (253) | | Wisc |
| 96 | 439 | 544 | 151 | 262 | | Minn |
| (97) | 544 | (544) | 253 | (272) | | Conn |
| 98 | 398 | 574 | 104 | 304 | | Ind |
| 104 | 574 | 628 | 262 | 385 | | Mo |
| | | | | | | |
| 106 | 556 | 544 | 238 | 154 | | Fla |
| (120) | 544 | (544) | 184 | (184) | | NJ |
| (128) | 602 | (556) | 218 | (212) | Slice 6 | Mich |
| (130) | 544 | (602) | 154 | (218) | | Mass |
| 130b | 602 | 602 | 212 | 238 | | Ohio |
| | | | | | | |
| 137 | 602 | 602 | 191 | 72 | | Tex |
| (147) | 653 | (644) | 212 | (146) | | Penn |
| (150) | 653 | (653) | 203 | (191) | Slice 7 | Ill |
| (166) | 644 | (653) | 146 | (203) | | Cal |
| 209 | 699 | 699 | 72 | 212 | | NY |

➔

### untilted analysis

Since the righthand side slope is close to 3, it is natural to consider working with

$$\text{GminB} = \text{GoSal} \text{ MINUS } 3 \text{ BaDep}$$

exhibit 22 of chapter 8: governor's salary

**Exhibit 21 continued**

#### C) SMOOTHING of GoSal and BaDep MEDIANS

| median GoSal | | | | | median BaDep | |
|---|---|---|---|---|---|---|
| raw | 3R' | S | H | 3 | raw | H |
| 398 |     | T | 398 | T | −2 | −2 |
| 301 | 398 | h | 398 | h | 13 | 13h |
| 398 |     | e | 398 | e | 30 | 35 |
| 398 |     | s | 424 | s | 65h | 63h |
| 512 |     | a | 494 | a | 93 | 95 |
| 556 |     | m | 569 | m | 128 | 125 |
| 653 |     | e | 653 | e | 150 | 150 |

#### D) SMOOTHING of LOW HINGES

| HtoM GoSal | | | | | loH | | loH BaDep | |
|---|---|---|---|---|---|---|---|---|
| raw | 3R | S | H | 3 | comb | 3 | raw | H |
| 97 | 56 |   |    | T | 342 | T | −15 | −15 |
| 46 | 56 |   |    | h | 342 | h | 12 | 9 |
| 56 |    |   |    | e | 342 | e | 27 | 31h |
| 97 | 73 | 56 | 46 | s | 380 | s | 59h | 57 |
| 73 |    | 18 | 26 | a | 468 | a | 81 | 85h |
| 12 |    |   | 13 | m | 556 | m | 120 | 117 |
| 9  |    |   | 9  | e | 644 | e | 147 | 147 |

Note: Values of "comb" are final smooths of "median GoSal" (see panel C) MINUS final smooths of "HtoM GoSal".

#### E) SMOOTHING of HIGH HINGES

| MtoH GoSal | | | | | hiH | | hiH BaDep | |
|---|---|---|---|---|---|---|---|---|
| raw | 3R | S | H | 3 | comb | 3 | raw | H |
| 0  | 65 | T | 65 | T | 463 |     | −0 | 0 |
| 65 | 65 | h | 65 | h | 463 |     | 14 | 17h |
| 79 | 65 | e | 63 | e | 461 | 463 | 42 | 43 |
| 56 | 56 | s | 56 | s | 480 |     | 75 | 72 |
| 32 | 46 | a | 46 | a | 539 |     | 97 | 100 |
| 46 | 32 | m | 26h | m· | 598 |    | 130 | 131 |
| 0  | 4  | e | 4  | e | 657 |     | 166 | 166 |

Note: Values of "comb" are final smooths of "median GoSal" (see panel C) PLUS final smooths of "MtoH GoSal".

The two further columns of exhibit 20, and exhibit 24, have all the necessary arithmetic, including, in panel G, the conversion of GminB back to GoSal. Exhibit 25 shows the median and hinge traces for GminB, according to panel F (of exhibit 23), against BaDep (using panels C to E of exhibit 22). In this exhibit, since the general trend is downward, we have paired low hinges for each of GaUse and ElUse with high hinges for the other. There is now no tendency for the traces to cross toward the right (although a crossing has appeared on the left).

exhibit **23** of chapter 8: governor's salary

**Median and hinge traces as computed directly (slopes of 2 and 3 shown for comparison)**

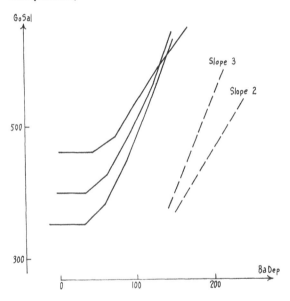

exhibit **24** of chapter 8: governor's salary

**Exhibit 22 continued**

F) SMOOTHING of GoSal MINUS 3BaDep--called below GminB

| median | | | | | HtoM | | | | | lo H | |
|---|---|---|---|---|---|---|---|---|---|---|---|
| \|raw\| | \|3R'\| | \|S\| | \|H\| | \|3\| | \|raw\| | \|3R\| | \|S\| | \|H\| | \|3\| | \|comb\| | \|3\| |
| 398 | 398 | T | 398 | T | 52 | 52 | 52 | 52 | T | 346 | T |
| 274 | 290 | h | 313 | h | 55 | 52 | 52 | 52 | h | 261 | h |
| 290 | 274 | e | 273 | e | 21 | 55 | 52 | 50 | e | 220 | e |
| 200 | 253 | s | 248 | s | 83h | 83h | 45 | 47 | s | 201 | s |
| 253 | 212 | a | 222 | a | 102 | 83h | 45 | 45 | a | 177 | a |
| 212 | 212 | m | 207 | m | 28 | 45 | 45 | 45 | m | 162 | m |
| 191 | 191 | e | 191 | e | 45 | 45 | 45 | 45 | e | 146 | e |

→

exhibit **24** of chapter 8 (continued)

| | MtoH | | | | hi H | |
|---|---|---|---|---|---|---|
| raw | 3R | S | H | 3 | comb | 3 |
| 6 | 19 | 19 | | | 417 | |
| 47 | 47 | 40 | | | 353 | |
| 61 | 47 | 44 | 40 | | 313 | |
| 62 | 33 | 33 | | | 281 | |
| 19 | 19 | 20 | | | 242 | |
| 26 | 12 | 14 | | | 221 | |
| 12 | 12 | 12 | | | 203 | |

**G) REASSEMBLING SMOOTH of GoSal -- as GminB PLUS 3 BaDep**

| median | | | | loH GoSal | | | | hiH GoSal | | | |
|---|---|---|---|---|---|---|---|---|---|---|---|
| med 3BaDep | med GminB | sum | 3 | hiH 3BaDep | loH GminB | sum | 3 | loH 3BaDep | hiH GminB | sum | 3 |
| −6 | 398 | 392 | | 0 | 346 | 346 | | −45 | 417 | 372 | T |
| 40h | 313 | 353h | 378 | 52h | 260 | 312h | 346 | 27 | 353 | 380 | h |
| 175 | 273 | 378 | | 129 | 220 | 349 | | 94h | 313 | 407h | e |
| 190h | 248 | 438h | | 216 | 196 | 412 | | 171 | 281 | 451 | s |
| 285 | 222 | 507 | | 300 | 174 | 474 | | 256h | 242 | 498h | a |
| 375 | 207 | 582 | | 393 | 162 | 555 | | 351 | 221 | 572 | m |
| 450 | 191 | 641 | | 498 | 146 | 644 | | 441 | 203 | 644 | e |

exhibit **25** of chapter 8: governor's salary

**Median and hinge traces for GminB against BaDep**

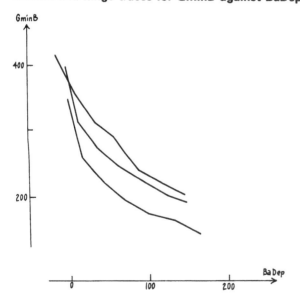

## exhibit 26/8: Parallel and wandering schematic plots

The result of plotting the complete conversion back to GoSal (see panel G) is shown in exhibit **26**.

**combined analysis**

We are now prepared to choose final hinge and median traces. For the three leftmost slices, the general run of GoSal seems about level. For these slices, then, the results of exhibits 20 and 21 seem appropriate. For the three rightmost slices, the general run of the GoSal values is slanted steeply upward. For these slices, the results of exhibits 22 and 24 seem appropriate. For the middle slice we could use either, and will take both. Combining then gives the values in panel A of exhibit **27** and the picture in exhibit **28**.

Proceeding to calculate fences in exhibit **29**, and to plot possibly outside and possibly adjacent points (based on panel B of exhibit 27), we finally reach the wandering schematic plot of exhibit **30**. Clearly, Arkansas is far out. We have shown New York as outside, although we did not have enough data to draw fences for BaDeps larger than 160. In fact, New York is outside for BaDep alone.

**exhibit 26** of chapter 8: governor's salary

**The hinge and median traces of exhibit 25 translated back in terms of GoSal**

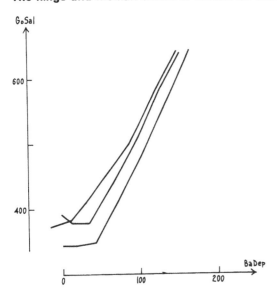

exhibit **27** of chapter 8

**Values needed in plotting**

A) CHOSEN TRACES

|   | median trace | | | lo hinge trace | | | hi hinge trace | | |
|---|---|---|---|---|---|---|---|---|---|
|   | BaDep | GoSal | 3 | BaDep | GoSal | 3 | BaDep | GoSal | 3 |
| 1 | −2   | 398  | T | −15 | 342 |     | 0    | 463  |     |
| 2 | 13h  | 398  | h | 9   | 342 |     | 17h  | 463  |     |
| 3 | 35   | 398  | e | 31h | 342 |     | 43   | 463  |     |
| 4 | 63h  | 427  | 433 | 72 | 380 | 396 | 57 | 480 | 466 |
| 4 | 63h  | 438h | 433 | 72 | 412 | 396 | 57 | 452 | 466 |
| 5 | 95   | 507  | s | 100 | 474 |     | 85h  | 498h |     |
| 6 | 125  | 553  | a | 131 | 555 |     | 117  | 572  |     |
| 7 | 150  | 641  | m | 166 | 644 |     | 147  | 644  |     |
|   |      |      | e |     |     |     |      |      |     |

B) POSSIBLY INTERESTING STATES

|       | BaDep | GoSal |          |
|-------|-------|-------|----------|
| Ark   | 42    | 0  ⎫  |          |
| Tenn  | 79    | 267 ⎬ | outside  |
| NY    | 209   | 699 ⎭ |          |
| Alaska| −37   | 505 ⎫ |          |
| Wyo   | −15   | 301  |          |
| Nev   | −2    | 398  ⎬| slice 1  |
| Verm  | −0    | 398  |          |
| Ida   | 6     | 243 ⎭ |          |
| Tex   | 137   | 602 ⎫ |          |
| Penn  | 147   | 653  |          |
| Ill   | 150   | 653  ⎬| slice 7  |
| Cal   | 166   | 644  |          |
| (NY   | 209   | 699)⎭ |          |

|       | BaDep | GoSal |           |
|-------|-------|-------|-----------|
| Hawaii| 14    | 525 ⎫ | other     |
| Del   | 16    | 544  | high      |
| Iowa  | 76    | 477 ⎬ | in        |
| Ga    | 81    | 574  | slices    |
| Mich  | 128   | 602 ⎭ | 2 to 6    |
| SD    | 12    | 255 ⎫ | other     |
| ND    | 13    | 255  | low       |
| Ark   | 42    | 0    | in        |
| Nebr  | 49    | 255 ⎬ | slices    |
| Ind   | 80    | 398  | 2         |
| NJ    | 120   | 544  | to        |
| Mass  | 130   | 544 ⎭ | 6         |

P) PROBLEMS

27a) Can anything useful be said about the differences between "high" states and "low" states?

exhibit **28** of chapter 8: governor's salary

**Hinge and median traces--and fences--combining the two delineations (e.g., exhibits 23 and 26)**

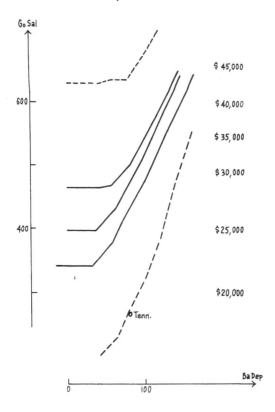

exhibit **29** of chapter 8: governor's salary

**Calculation of fences and spread**

**A) CALCULATION of FENCES**

|       |     |     |     |     |     |      |     | inner fences | |
|-------|-----|-----|------|------|------|------|------|------|-----|
| |BaDep| |loH| |hiH| |Hspr| |3| |step| |lo| |3| |hi| |3| |
| 0   | 358 | 463 | 111 |     |     | 116 | 186 |     | 629 |     |
| 20  | 352 | 463 | 111 |     |     | 166 | 186 |     | 629 |     |
| 40  | 362 | 463 | 101 | 102 |     | 153 | 209 |     | 616 | 629 |
| 60  | 382 | 484 | 102 | 101 |     | 152 | 230 |     | 636 |     |
| 80  | 412 | 498 | 86  | 92  |     | 138 | 274 |     | 636 |     |
| 100 | 450 | 542 | 92  | 86  |     | 129 | 321 |     | 671 |     |
| 120 | 508 | 592 | 84  |     |     | 126 | 382 |     | 718 |     |
| 140 | 570 | 635 | 65  |     |     | 98  | 472 |     | 733 |     |
| 160 | 628 | (675)| (47)|    |     | (71)| (557)|    | 746 |     |

→

exhibit **29** of chapter 8 (continued)

**B) HINGES and SPREAD in RAW (thousands of) DOLLARS**

| BaDep | loH | hiH | Hspread | 3 |
|---|---|---|---|---|
| 0 | 22.5 | 29.0 | 6.5 | 6.5 |
| 20 | 22.5 | 29.0 | 6.5 | 6.5 |
| 40 | 23.0 | 29.0 | 6.0 | 6.4 |
| 60 | 24.1 | 30.5 | 6.4 | 6.4 |
| 80 | 25.8 | 31.5 | 5.7 | 6.4 |
| 100 | 28.1 | 34.8 | 6.6 | 6.6 |
| 120 | 32.2 | 39.1 | 6.9 | 6.6 |
| 140 | 37.2 | 43.1 | 5.9 | 5.9 |
| 160 | 42.5 | (47.3) | (4.8) | (4.8) |

exhibit **30** of chapter 8: governor's salary

**The wandering schematic plot that results (based on exhibits 28 and 29)**

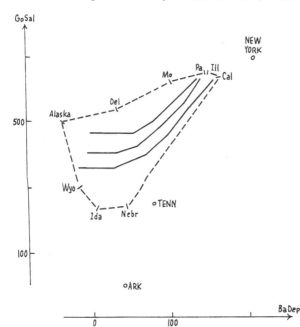

298    /8: Parallel and wandering schematic plots

### review questions

Will more care in making wandering schematic plots be needed for large batches? For small ones? What example did we consider in this section? How large a batch? Could it have been bigger? Why/why not? Did we need to take logs? Of the values of one variable? Of those of both variables? Why/why not? What did we do? What kind of slices did we take? What are particular choices for slice sizes? Why are they desirable?

What happened when we plotted hinge and median traces? Why might this have happened? What was the plausible way out? What did we do? How did it work? If we subtract before finding hinge and middle traces, what must we do afterwards? Is combining the result of two such analyses reasonable? Did we do it? How did we like the results?

Did we calculate fences? Were any states outside? Far out? How did the wandering schematic plot come out?

## 8G. Further questions/analysis in the example

We can--and should--take the example further in a number of ways.

### time changes

Of the five states on the upper boundary of the adjacent polygon, two (Alaska and Delaware) had raised their governor's salary in the previous year, while two more (Illinois and Missouri) had done this a year earlier. Of the four states on the lower boundary, three (California, Idaho, and Wyoming) raised their governor's salary in the next year, as did both states (Tennessee, New York) that were outside but not far out. Arkansas remained a far out exception.

If we had used smoothed values of governor's salaries, suitably smoothed over years, we would have had an appreciably tighter picture for our wandering schematic plot.

### vertical width

In addition to the wandering schematic plot, we also want to look at the behavior of the vertical width as BaDep changes. (With the very different slopes in different parts of the range, we dare not hope to "look" at exhibit 30 and see how the width changes.) Panel A of exhibit 29 shows a hinge-spread decreasing from 111 to 65 (or 47). This is in logarithmic units, and we wonder about behavior in raw dollars.

Panel B expresses the "hinges" in raw dollars. We find a quite constant hinge-spread, amounting to $6,500, over most of the range. (Since only eight states have BaDep's above 120, we do not have to pay too much attention to the apparent drop at the upper end.)

### slopes and levels

For small BaDep, the slope of all three traces is close to zero and the level of the middle trace is just about $25,000.

For large BaDep's, as exhibit 25 shows, the slope of GoSal on BaDep is close to 3. Taking account of units, this indicates that, for high BaDep, governor's salary varies about as the 0.3 power of bank deposits. A plausible multiplying constant turns out to be 31, so that, on the high side

governor's salary  about  31 (bank deposits)$^{0.3}$

is a reasonable fit.

### consolidated summary

We can now summarize the situation for governor's salary vs. bank deposits as follows:

◇ for low bank deposits, governor's salary seems to center near $25,000.

◇ for high bank deposits, governor's salary seems to center near 31(bank deposits)$^{0.3}$.

◇ the transition is moderately abrupt, and could be taken either as a dogleg or as somewhat smoothed out.

◇ in raw dollars, the vertical width seems quite constant, with an H-spread of $6,500.

◇ most states near the upper margin had just raised their governor's salaries.

◇ most states low outside or near the lower margin raised their governor's salary during the next year.

### moral

What we have learned from this example is that

◇ it is not safe to trust our middle-trace and hinge-trace calculations when these traces are too steep, where steepness relates to slice width. (If we had had 500 states, we would have had little trouble with steepness--except for the end slices.)

◇ we can avoid such difficulties by subtracting a simple function of $x$ to make matters less steep.

◇ we can combine results from different subtractions when the slope differs in different intervals.

We could have subtracted something that would work fairly well across the whole range, had we wished. Two candidates would be:

◇ the greater of 398 and 188 + 3(BaDep).

◇ the original middle trace interpolated between the final smoothed points (and extrapolated outside them) to give a definition everywhere it is needed.

We leave the exploration of these possibilities to the reader (see exhibit **31**). Exhibit **32** offers data for further problems.

exhibit **31** of chapter 8: Texas data

**Data on the male and female populations of the 254 counties of Texas**

MaPop = 100 $\log_{10}$ (male population in thousands)
FePop = 100 $\log_{10}$ (female population in thousands)

**A) EXAMPLES of CALCULATION**

| County | Male Population | | | Female Population | | |
|---|---|---|---|---|---|---|
| | raw | $\log_{10}$ | MaPop | raw | $\log_{10}$ | FePop |
| Anderson | 13,397 | 4.13 | 113 | 14,765 | 4.17 | 117 |
| Andrews | 6,887 | 3.84 | 84 | 6,563 | 3.82 | 82 |
| Angelina | 19,568 | 4.29 | 129 | 20,246 | 4.31 | 131 |

**B) The (MaPop, FePop) PAIRS, ORDERED on MaPop**

(−94, −95), (−45, −55), (−31, −40), (−27, −27), (−25, −29), (−23, 023 and −28), (−9, −11), (−6, −6), (1, −2), (3, 0 and 5), (5, −10), (9, 4), (10, 12), (12, 10 and 11), (14, 14), (15, 14), (16, 15 and 18), (17, 12 and 18), (19, 17 and 20), (21, 20), (24, 22), (25, 26), (26, 25 and 27), (27, 17 and 26 and 28), (28, 26), (29, 27 and 29 and 31), (30, 29), (31, 32), (32, 32 and 35), (34, 31 and 36), (35, 35), (36, 32), (37, 37), (39, 40 and 41), (40, 41), (41, 39 and 42), (43, 44 and 44 and 44), (45, 48), (46, 46 and 48), (47, 49), (48, 47 and 49), (49, 49 and 51), (50, 49 and 49 and 52), (51, 49 and 30), (52, 49 and 49 and 52), (54, 52), (55, 54 and 57), (56, 47 and 56 and 57 and 58 and 59), (58, 59 and 61), (59, 59 and 59 and 60 and 60 and 60 and 61 and 61), (60, 63), (61, 64 and 64), (62, 59 and 62), (63, 65 and 66), (64, 63 and 66), (66, 64 and 67 and 68), (67, 64), (68, 66), (69, 62 and 71 and 71), (70, 70), (71, 70 and 71), (72, 71 and 71), (73, 72 and 73 and 74), (74, 74 and 74 and 76 and 76), (75, 75), (76, 78), (77, 79), (80, 75 and 77 and 79), (81, 80), (82, 82), (83, 81 and 83 and 84), (84, 82 and 82 and 84), (85, 85 and 87), (86, 88), (87, 96 and 88 and 88), (88, 87 and 91), (89, 93), (90, 91), (91, 91 and 93 and 94), (92, 90 and 93 and 93 and 93 and 95), (93, 93 and 93 and 95), (94, 95 and 96 and 96), (96, 93 and 97 and 98), (97, 97 and 97 and 98 and 101 and 104), (98, 98 and 98 and 99), (99, 100 and 101 and 101 and 101), (100, 101 and 102 and 102), (101, 96 and 101 and 102 and 103 and 104), (104, 100 and 104 and 105 and 106 and 107), (106, 104 and 108 and 109), (107, 105 and 108 and 111), (109, 96 and 106 and 109), (110, 108), (111, 111), (113, 113 and 117), (114, 116), (115, 117), (116, 119), (118, 97), (119, 120), (120, 115 and 120), (121, 123), (122, 125 and 125), (123, 122 and 124 and 125 and 125), (124, 123), (125, 127), (127, 126), (128, 128 and 131), (129, 131), (131, 137 and 130), (133, 128), (135, 135), (136, 137 and 139), (138, 132), (146, 149), (149, 153), (152, 154), (153, 155), (160, 156), (162, 165), (166, 165), (170, 170), (171, 164), (178, 175), (180, 178), (184, 185), (187, 188 and 189), (189, 189), (195, 196), (202, 203), (204, 209), (208, 210), (220, 219), (242, 244), (253, 254), (266, 269), (279, 280).

**C) Some COUNTY NAMES and MAP REFERENCES**

| Name | Map | MaPop | FePop | Name | Map | MaPop | FePop |
|---|---|---|---|---|---|---|---|
| King | E9 | −45 | −55 | Kerr | J9 | 91 | 94 |
| Kenedy | N12 | −31 | −40 | Caldwell | J12 | 92 | 95 |
| Roberts | A8 | −27 | −27 | Beaver | G4 | 96 | 93 |
| Borden | F7 | −25 | −29 | Eastland | G10 | 97 | 101 |
| McMullen | L11 | −23 | −28 | Limestone | H13 | 97 | 104 |
| Glasscock | G7 | −22 | −29 | Houston | H15 | 101 | 96 |

➡

exhibit **31** of chapter 8 (panel C, continued)

| | | | | | | | |
|---|---|---|---|---|---|---|---|
| Jeff Davis | I3 | −9 | −11 | Falls | H13 | 101 | 104 |
| Real | H7 | 3 | 0 | Cooke | D12 | 104 | 107 |
| Oldham | B6 | 5 | −10 | Willacy | O12 | 104 | 100 |
| Edwards | J8 | 9 | 4 | Hill | I2 | 106 | 109 |
| Kinney | K8 | 11 | 6 | Brown | G10 | 107 | 111 |
| Culberson | G3 | 17 | 12 | Bee | L12 | 109 | 106 |
| Lipscomb | A9 | 24 | 22 | Walker | I14 | 109 | 96 |
| Hudspeth | G2 | 27 | 17 | Valverde | J7 | 110 | 108 |
| Ellis | F13 | 32 | 35 | Anderson | G14 | 113 | 117 |
| Crockett | I7 | 34 | 31 | Kaufmann | T14 | 116 | 119 |
| Zapata | N10 | 36 | 32 | Coryell | H12 | 118 | 97 |
| Delta | E14 | 45 | 48 | Kleberg | N12 | 120 | 115 |
| Dallam | A6 | 51 | 49 | Lamar | D14 | 122 | 125 |
| Brewster | I4 | 52 | 49 | Navarro | G13 | 122 | 125 |
| Cochran | E6 | 52 | 49 | Hunt | E14 | 128 | 131 |
| Orange | J17 | 56 | 47 | Fort Bend | J14 | 133 | 128 |
| Trinity | H15 | 56 | 59 | Harrison | F16 | 134 | 137 |
| Camp | E15 | 58 | 61 | Denton | E12 | 136 | 139 |
| Hardeman | D9 | 60 | 63 | Brazos | I14 | 138 | 132 |
| Childress | C9 | 61 | 69 | Bowie | E16 | 146 | 149 |
| Hamiton | H2 | 61 | 64 | Webb | M9 | 149 | 153 |
| Yoakum | E6 | 62 | 59 | Brazoria | K15 | 160 | 156 |
| Stephens | F10 | 63 | 66 | Smith | E15 | 162 | 165 |
| Bailey | D6 | 67 | 64 | Bell | H12 | 171 | 164 |
| Coleman | G10 | 78 | 81 | Potter | B7 | 178 | 175 |
| Pecos | H5 | 79 | 76 | Cameron | O12 | 187 | 189 |
| Gaines | F6 | 80 | 77 | Travis | J12 | 202 | 203 |
| Irion | H8 | 80 | 75 | Nueces | M12 | 204 | 209 |
| Zavala | L9 | 81 | 80 | Jefferson | J16 | 208 | 210 |
| Winkler | G5 | 84 | 82 | El Paso | G1 | 220 | 219 |
| Red River | D15 | 88 | 91 | Tarrant | F12 | 242 | 244 |
| Robertson | H14 | 89 | 83 | Dallas | F13 | 266 | 269 |
| | | | | Harris | J14 | 279 | 280 |

**P) PROBLEMS**

31a) Go through the analysis of GoSal against BaDep, subtracting max(398, 188 + 3(BaDep)) before finding middle and hinge traces. Compare your results with those in earlier exhibits.

31b) Go through the analysis of GoSal against BaDep, subtracting an inter(extra)polation of the final smooth of panel C of exhibit 22 before finding (new) middle and hinge traces. Compare your results with those in earlier exhibits.

31c) Analyze the data in panel B above, using the methods of this chapter. Discuss your results.

31d) Analyze the data in panel C above, using the methods of this chapter.

31cd2) Compare these analyses. Were the 77 values enough? Do you think 77 taken "at random" would be enough? Why/why not?

**S) SOURCE**

Census of Population: 1960; Volume I. Characteristics of the Population; Part 45. Texas, pages 45–181 to 45–244. U.S. Department of Commerce, Bureau of the Census, 1963.

exhibit **32** of chapter 8

**Time of high tide (24-hour clock), time since last high tide (in minutes over 12 hours) and height of tide in tenths of feet above mean low water) predicted for the 704 high tides at Portland, Maine 1975.**

A) EXAMPLES

For 24 Dec 75, high tides of 9.2 and 9.3 feet were forecast for 0248 (2:48 am) and 1506 (3:06 pm). For 25 Dec 75 a high tide of 9.3 feet was forecast for 0.343 (3:43 am). Thus (1506, 34, 93) and (−343, 37, 93) appear below, where 34 = 1506 − 1448 with 1448 = 0248 − 1200, and 37 = 0343 − 0306, with 0306 = 1506 − 1200.

B) DATA TRIPLES

| | | | |
|---|---|---|---|
| (0003, 32, 83) | (0004, 32, 84) | 0006, 13, 91) | (0006, 26, 108) |
| (0008, 8, 103) | (0009, 13, 90) | (0009, 23, 86) | (0010, 39, 92) |
| (0011, 27, 109) | (0012, 8, 94) | (0014, 12, 108) | 0014, 21, 88) |
| (0024, 20, 109) | (0024, 36, 88) | (0025, 28, 84) | (0025, 37, 97) |
| (0028, 7, 99) | (0033, 8, 83) | (0033, 33, 104) | (0037, 18, 109) |
| (0038, 30, 84) | (0041, 9, 101) | (0041, 14, 107) | (0041, 16, 89) |
| (0041, 21, 87) | (0041, 32, 91) | (0041, 33, 101) | (0043, 34, 84) |
| (0049, 37, 94) | (0050, 25, 86) | (0053, 39, 98) | (0056, 48, 103) |
| (0057, 39, 96) | (0059, 41, 94) | (0059, 29, 104) | (0101, 25, 108) |
| (0102, 11, 103) | (0103, 33, 83) | (0107, 34, 89) | (0110, 9, 90) |
| (0112, 7, 99) | (0113, 19, 87) | (0113, 29, 84) | (0115, 10, 90) |
| (0117, 18, 87) | (0117, 23, 105) | (0124, 31, 104) | (0125, 5, 96) |
| (0127, 28, 82) | (0127, 27, 84) | (0126, 41, 93) | (0130, 9, 94) |
| (0134, 19, 104) | (0137, 38, 96) | (0140, 10, 85) | (0141, 8, 91) |
| (0144, 51, 99) | (0145, 40, 87) | (0146, 35, 81) | (0147, 11, 101) |
| (0148, 12, 87) | (0149, 16, 87) | (0150, 29, 84) | (0152, 10, 98) |
| (0152, 23, 105) | (0155, 13, 90) | (0155, 33, 99) | (0155, 37, 91) |
| (0156, 20, 83) | (0201, 11, 97) | (0208, 29, 80) | (0211, 5, 91) |
| (0211, 43, 188) | (0213, 26, 100) | (0214, 40, 85) | (0218, 47, 103) |
| (0219, 12, 93) | (0219, 16, 100) | (0224, 9, 87) | (0227, 27, 83) |
| (0229, 18, 86) | (0229, 21, 100) | (0231, 16, 83) | (0232, 37, 80) |
| (0233, 38, 84) | (0235, 43, 90) | (0239, 24, 80) | (0240, 37, 98) |
| (0242, 16, 89) | (0243, 16, 98) | (0245, 11, 92) | (0247, 23, 101) |
| (0248, 37, 92) | (0253, 37, 92) | (0256, 13, 96) | (0256, 34, 78) |
| (0301, 10, 87) | (0306, 39, 88) | (0306, 42, 84) | (0309, 26, 82) |
| (0312, 14, 83) | (0312, 16, 92) | (0342, 19, 85) | (0312, 32, 94) |
| (0314, 37, 100) | (0315, 15, 94) | (0317, 21, 80) | (0324, 39, 82) |
| (0325, 40, 80) | (0327, 30, 77) | (0328, 24, 95) | (0333, 18, 88) |
| (0336, 45, 85) | (0338, 37, 97) | (0341, 14, 87) | (0343, 21, 93) |
| (0343, 37, 93) | (0346, 24, 95) | (0350, 38, 77) | (0356, 13, 83) |
| (0356, 18, 93) | (0356, 28, 82) | (0401, 16, 79) | (0404, 21, 85) |
| (0405, 43, 87) | (0405, 45, 81) | (0406, 25, 76) | (0412, 18, 91) |
| (0413, 30, 96) | (0415, 18, 88) | (0415, 37, 88) | (0415, 39, 80) |
| (0421, 34, 75) | (0426, 95, 82) | (0432, 22, 88) | (0432, 31, 90) |
| (0439, 35, 96) | (0441, 15, 83) | (0441, 47, 81) | (0442, 35, 94) |
| (0447, 25, 91) | (0447, 26, 82) | (0449, 23, 90) | (0150, 16, 79) |
| (0450, 42, 77) | (0455, 22, 76) | (0500, 21, 92) | (0502, 23, 86) |
| (0503, 32, 74) | (0505, 44, 79) | (0506, 41, 90) | (0506, 46, 82) |

→

exhibit **32** of chapter 8 (panel B, continued)

| | | | |
|---|---|---|---|
| (0510, 37, 80) | (0517, 21, 91) | (0518, 29, 92) | (0521, 41, 74) |
| (0523, 22, 84) | (0523, 41, 84) | (0530, 46, 85) | (0538, 23, 89) |
| (0539, 36, 86) | (0541, 18, 80) | (0541, 23, 84) | (0542, 32, 94) |
| (0544, 46, 79) | (0546, 34, 96) | (0547, 20, 77) | (0551, 27, 75) |
| (0553, 31, 89) | (0554, 21, 87) | (0555, 46, 80) | (0559, 35, 73) |
| (0603, 41, 80) | (0604, 22, 88) | (0605, 28, 80) | (0607, 26, 91) |
| (0607, 37, 94) | (0613, 45, 80) | (0623, 23, 92) | (0623, 24, 90) |
| (0623, 40, 76) | (0625, 19, 82) | (0631, 41, 90) | (0634, 45, 81) |
| (0639, 21, 79) | (0641, 23, 87) | (0644, 23, 92) | (0644, 25, 77) |
| (0644, 26, 94) | (0645, 40, 80) | (0647, 26, 98) | (0647, 32, 74) |
| (0647, 41, 84) | (0655, 26, 82) | (0655, 41, 85) | (0656, 34, 81) |
| (0649, 19, 85) | (0700, 36, 88) | (0700, 39, 75) | (0706, 29, 98) |
| (0707, 20, 92) | (0713, 30, 92) | (0717, 42, 80) | (0724, 18, 82) |
| (0725, 42, 81) | (0727, 21, 90) | (0728, 83, 96) | (0735, 21, 80) |
| (0737, 18, 91) | (0737, 29, 77) | (0737, 42, 81) | (0740, 34, 82) |
| (0743, 37, 76) | (0745, 21, 84) | (0745, 26, 84) | (0745, 31, 94) |
| (0748, 24, 97) | (0749, 20, 106) | (0752, 43, 83) | (0753, 36, 92) |
| (0756, 37, 78) | (0759, 18, 86) | (0803, 35, 88) | (0804, 24, 103) |
| (0809, 20, 97) | (0812, 37, 82) | (0815, 18, 83) | (0815, 34, 93) |
| (0820, 37, 87) | (0822, 23, 81 | (0823, 16, 90) | (0823, 26, 103) |
| (0826, 26, 85) | (0826, 32, 79) | (0828, 13, 87) | (0828, 19, 86) |
| (0830, 19, 86) | (0830, 29, 98) | (0833, 16, 97) | (0833, 37, 82) |
| (0833, 39, 78) | (0840, 15, 95) | (0846, 16, 102) | (0846, 44, 100) |
| (0847, 14, 101) | (0849, 15, 90) | (0849, 40, 84) | (0850, 41, 83) |
| (0859, 18, 106) | (0859, 30, 84) | (0902, 18, 85) | (0902, 40, 88) |
| (0905, 19, 103) | (0906, 20, 87) | (0907, 13, 89) | (0907, 26, 82) |
| (0910, 8, 90) | (0910, 32, 80) | (0911, 11, 91) | (0911, 33, 94) |
| (0913, 35, 94) | (0916, 22, 108) | (0921, 37, 82) | (0924, 35, 84) |
| (0926, 15, 103) | (0926, 29, 100) | (0931, 23, 95) | (0932, 14, 88) |
| (0938, 25, 106) | (0939, 13, 103) | (0939, 37, 89) | (0940, 25, 86) |
| (0942, 19, 86) | (0943, 40, 85) | (0943, 14, 89) | (0944, 17, 91) |
| (0944, 26, 105) | (0947, 39, 84) | (0950, 6, 93) | (0952, 15, 109) |
| (0955, 37, 82) | (0955, 40, 88) | (0957, 9, 91) | (1000, 20, 107) |
| (1002, 31, 101) | (1006, 37, 86) | (1006, 37, 94) | (1007, 17, 42) |
| (1007, 30, 85) | (1013, 15, 89) | (1014, 8, 95) | (1016, 20, 88) |
| (1019, 5, 93) | (1019, 13, 91) | (1019, 16, 107) | (1020, 22, 87) |
| (1021, 33, 101) | (1025, 29, 85) | (1027, 3, 95) | (1027, 21, 111) |
| (1027, 32, 95) | (1029, 37, 85) | (1036, 36, 85) | (1037, 10, 92) |
| (1037, 29, 106) | (1042, 13, 109) | (1043, 39, 88) | (1046, 27, 87) |
| (1048, 15, 90) | (1049, 17, 90) | (1051, 11, 92) | (1051, 26, 107) |
| (1053, 21, 109) | (1053, 31, 90) | (1054, 6, 95) | (1055, 4, 94) |
| (1055, 23, 87) | (1056, 37, 94) | (1057, 15, 113) | (1105, 32, 86) |
| (1107, 4, 98) | (1109, 17, 110) | (1110, 35, 86) | (1112, 8, 92) |
| (1112, 35, 100) | (1115, 32, 100) | (1117, 18, 113) | (1121, 14, 90) |
| (1121, 14, 91) | (1121, 25, 88) | (1122, 6, 93) | (1127, 29, 106) |
| (1127, 35, 88) | (1127, 35, 88) | (1127, 27, 88) | (1130, 25, 87) |
| (1131, 5, 94) | (1131, 10, 107) | (1132, 5, 95) | (1129, 36, 94) |
| (1140, 24, 110) | (1142, 35, 86) | (1144, 22, 109) | (1146, 9, 91) |
| (1146, 36, 92) | (1147, 31, 87) | (1148, 4, 99) | (1148, 14, 111) |
| (1153, 13, 91) | (1153, 22, 89) | (1156, 19, 89) | (1157, 7, 93) |
| (1200, 17, 110) | (1200, 37, 89) | (1202, 36, 98) | (1204, 27, 87) |
| (1204, 31, 104) | (1208, 5, 92) | (1208, 19, 113) | (1209, 5, 96) |

→

exhibit **32** of chapter 8 (panel B, continued)

| | | | |
|---|---|---|---|
| (1212, 37, 81) | (1219, 9, 104) | (1219, 32, 103) | (1220, 11, 90) |
| (1221, 34, 87) | (1225, 11, 91) | (1225, 19, 89) | (1225, 30, 87) |
| (1227, 36, 97) | (1228, 19, 88) | (1230, 5, 93) | (1230, 24, 111) |
| (1232, 38, 90) | (1233, 9, 99) | (1236, 25, 107) | (1237, 12, 107) |
| (1241, 29, 86) | (1244, 6, 90) | (1246, 38, 91) | (1250, 7, 96) |
| (1251, 37, 94) | (1253, 20, 108) | (1254, 13, 88) | (1254, 30, 106) |
| (1259, 9, 91) | (1259, 18, 89) | (1259, 18, 110) | (1301, 28, 87) |
| (1305, 8, 99) | (1305, 24, 86) | (1305, 37, 87) | (1307, 14, 106) |
| (1310, 33, 98) | (1311, 8, 92) | (1315, 34, 100) | (1318, 11, 98) |
| (1320, 39, 87) | (1321, 8, 88) | (1321, 32, 84) | (1322, 23, 109) |
| (1328, 10, 102) | (1329, 28, 102) | (1331, 18, 85) | (1333, 39, 85) |
| (1334, 7, 95) | (1336, 19, 88) | (1336, 26, 86) | (1336, 32, 86) |
| (1336, 40, 93) | (1342, 27, 83) | (1342, 40, 90) | (1347, 23, 103) |
| (1347, 30, 106) | (1350, 38, 87) | (1352, 15, 104) | (1355, 9, 91) |
| (1355, 10, 94) | (1400, 10, 85) | (1403, 17, 102) | (1403, 36, 92) |
| (1406, 41, 85) | (1407, 37, 83) | (1408, 34, 101) | (1411, 16, 96) |
| (1411, 22, 82) | (1415, 19, 87) | (1415, 27, 86) | (1415, 34, 84) |
| (1416, 21, 105) | (1421, 16, 96) | (1422, 14, 88) | (1424, 13, 96) |
| (1424, 32, 95) | (1426, 31, 81) | (1427, 13, 93) | (1427, 40, 94) |
| (1434, 42, 85) | (1440, 27, 104) | (1443, 16, 81) | (1443, 25, 97) |
| (1443, 42, 87) | (1445, 12, 88) | (1445, 13, 90) | (1451, 16, 98) |
| (1451, 40, 82) | (1453, 24, 79) | (1456, 25, 85) | (1456, 37, 82) |
| (1457, 18, 66) | (1500, 41, 86) | (1501, 21, 97) | (1504, 35, 100) |
| (1506, 18, 93) | (1512, 16, 87) | (1515, 33, 78) | (1516, 23, 100) |
| (1520, 14, 90) | (1522, 16, 91) | (1522, 35, 88) | (1522, 39, 95) |
| (1527, 42, 81) | (1528, 19, 77) | (1536, 12, 83) | (1538, 26, 101) |
| (1538, 31, 88) | (1541, 16, 89) | (1541, 24, 84) | (1543, 29, 89) |
| (1543, 31, 76) | (1543, 42, 81) | (1547, 20, 85) | (1554, 18, 92) |
| (1554, 42, 82) | (1557, 42, 81) | (1601, 33, 100) | (1604, 26, 91) |
| (1607, 24, 89) | (1608, 18, 86) | (1610, 37, 77) | (1620, 23, 95) |
| (1621, 16, 85) | (1621, 25, 74) | (1623, 39, 96) | (1625, 20, 89) |
| (1626, 40, 84) | (1626, 45, 79) | (1631, 25, 83) | (1633, 18, 78) |
| (1633, 32, 82) | (1634, 38, 80) | (1639, 25, 74) | (1639, 43, 90) |
| (1640, 19, 85) | (1642, 27, 91) | (1644, 18, 88) | (1649, 36, 83) |
| (1656, 44, 83) | (1658, 17, 88) | (1701, 46, 77) | (1703, 31, 98) |
| (1709, 19, 87) | (1710, 31, 86) | (1712, 30, 86) | (1715, 43, 78) |
| (1718, 31, 72) | (1722, 17, 81) | (1722, 35, 97) | (1723, 42, 78) |
| (1724, 21, 84) | (1724, 29, 83) | (1727, 32, 81) | (1728, 22, 91) |
| (1730, 24, 88) | (1733, 23, 76) | (1733, 44, 77) | (1741, 41, 93) |
| (1742, 40, 74) | (1747, 22, 86) | (1749, 26, 95) | (1750, 20, 90) |
| (1759, 41, 79) | (1800, 43, 87) | (1805, 21, 85) | (1800, 27, 98) |
| (1806, 43, 76) | (1814, 19, 90) | (1815, 24, 84) | (1818, 36, 82) |
| (1818, 37, 73) | (1818, 37, 79) | (1819, 32, 82) | (1821, 22, 86) |
| (1821, 35, 85) | (1821, 43, 81) | (1822, 19, 79) | (1824, 31, 99) |
| (1835, 23, 89) | (1837, 30, 89) | (1840, 46, 76) | (1843, 20, 90) |
| (1843, 36, 97) | (1847, 43, 77) | (1855, 21, 94) | (1855, 24, 92) |
| (1903, 40, 92) | (1906, 19, 86) | (1906, 21, 84) | (1906, 41, 77) |
| (1906, 43, 77) | (1909, 22, 98) | (1914, 35, 81) | (1917, 17, 90) |
| (1917, 22, 94) | (1919, 23, 79) | (1919, 38, 75) | (1924, 24, 100) |
| (1924, 29, 74) | (1924, 40, 81) | (1924, 40, 86) | (1929, 42, 85) |
| (1935, 18, 89) | (1940, 34, 90) | (1941, 28, 101) | (1941, 42, 76) |

exhibit **32** of chapter 8 (panel B, continued)

| | | | |
|---|---|---|---|
| (1943, 18, 95) | (1949, 42, 82) | (1954, 11, 90) | (1954, 17, 83) |
| (1956, 19, 94) | (1957, 29, 95) | (1957, 33, 79) | (1959, 24, 84) |
| (2000, 20, 85) | (2001, 43, 98) | (2007, 40, 77) | (2009, 13, 94) |
| (2009, 17, 98) | (2009, 24, 80) | (2015, 30, 75) | (2017, 24, 99) |
| (2017, 40, 79) | (2022, 19, 102) | (2023, 35, 93) | (2025, 40, 81) |
| (2029, 17, 90) | (2030, 41, 86) | (2034, 35, 78) | (2038, 12, 90) |
| (2038, 18, 100) | (2038, 23, 104) | (2041, 19, 87) | (2041, 37, 93) |
| (2044, 11, 94) | (2044, 29, 82) | (2046, 20, 85) | (2046, 37, 88) |
| (2049, 16, 94) | (2054, 26, 81) | (2054, 31, 98) | (2057, 27, 104) |
| (2100, 37, 78) | (2012, 12, 77) | (2102, 13, 98) | (2012, 13, 100) |
| (2012, 34, 71) | (2111, 38, 95) | (2113, 27, 103) | (2115, 13, 103) |
| (2115, 16, 90) | (2118, 8, 93) | (2118, 11, 89) | (2118, 29, 81) |
| (2118, 31, 100) | (2118, 38, 81) | (2123, 21, 85) | (2126, 40, 87) |
| (2124, 18, 98) | (2129, 16, 107) | (2129, 23, 86) | (21312, 20, 105) |
| (2136, 29, 81) | (2137, 13, 94) | (2137, 38, 94) | (2140, 35, 95) |
| (2144, 34, 79) | (2148, 22, 108) | (2148, 37, 80) | (2150, 34, 100) |
| (2152, 20, 98) | (2155, 16, 104) | (2156, 9, 92) | (2157, 7, 91) |
| (2158, 16, 88) | (2158, 26, 83) | (2200, 5, 96) | (2203, 31, 91) |
| (2204, 9, 103) | (2206, 23, 87) | (2206, 28, 106) | (2206, 35, 82) |
| (2208, 24, 106) | (2214, 30, 83) | (2216, 10, 102) | (2219, 13, 107) |
| (2219, 12, 94) | (2219, 13, 107) | (2219, 40, 89) | (2224, 34, 81) |
| (2225, 23, 108) | (2226, 10, 102) | (2227, 30, 82) | (2229, 37, 94) |
| (2232, 12, 90) | (2232, 15, 91) | (2232, 32, 102) | (2233, 8, 94) |
| (2233, 20, 86) | (2235, 6, 98) | (2237, 16, 110) | (2240, 4, 99) |
| (2240, 21, 87) | (2242, 35, 101) | (2243, 16, 108) | (2248, 32, 82) |
| (2251, 32, 83) | (2252, 9, 102) | (2252, 33, 97) | (2256, 10, 94) |
| (2258, 21, 110) | (2259, 32, 106) | (2303, 10, 105) | (2303, 36, 83) |
| (2304, 27, 95) | (2305, 10, 92) | (2306, 10, 107) | (2307, 2, 96) |
| (2307, 19, 80) | (2307, 19, 88) | (2316, 6, 97) | (2316, 25, 86) |
| (2316, 25, 109) | (2321, 39, 94) | (2322, 29, 106) | (2323, 6, 102) |
| (2326, 14, 110) | (2326, 32, 83) | (2327, 32, 84) | (2331, 10, 93) |
| (2333, 18, 109) | (2334, 37, 99) | (2335, 8, 100) | (2337, 7, 93) |
| (2337, 16, 89) | (2337, 25, 85) | (2340, 19, 89) | (2343, 34, 101) |
| (2344, 37, 85) | (2347, 5, 98) | (2347, 20, 111) | (2349, 32, 104) |
| (2350, 28, 86) | (2351, 12, 107) | (2354, 8, 104) | (2355, 8, 95) |

**P)** PROBLEMS (all large)

32a) Analyze the (time, interval) pairs, using the methods of this chapter. Discuss your results.

32b) Analyze the (time, height) pairs, using the methods of this chapter. Discuss your results.

32c/d/e/f/g/h) Analyze only those (time, height) pairs where the next-to-last digit of the time is 0/1/2/3/4/5.

32i/j/k/l/m/n/o/p/q/r) Analyze only those (time, interval) pairs where the last digit of the time is 0/1/2/3/4/5/6/7/8/9.

**S)** SOURCE

Tide Tables, 1975, East Coast of North and South America, including Greenland, pp. 32–33. U.S. Department of Commerce, National Oceanic and Atmospheric Administration. 1974.

#### review questions

What happened when we looked up changes in governor's salary? What did this suggest about smoothing over years? Did we dare look at the wandering schematic plot to "see" vertical width directly? Why/why not? How did vertical width seem to behave in logarithmic units? In raw dollars? What seemed to be the main conclusions about this example?

Do we believe it is generally safe to work with middle and hinge traces based directly on the data? When/when not? What do we do to avoid trouble? Can we plan to combine? What other sorts of things might we have subtracted in this example? Would we be free to do similar things generally?

### 8H. How far have we come?

This chapter started with parallel schematic plots, and progressed to wandering schematic plots. It is our first major stage in providing summaries for batches of $(x, y)$ points that can help us to understand:

◦ the broad pattern of distribution of most points,

◦ the existence and location of unusual points,

just as schematic plots did for us when dealing with batches of $x$-values (or of $y$-values).

We are now ready to:

◦ slice up an $(x, y)$ batch according to its $x$-values.

◦ do this using letter cuts, slice sizes selected by judgment, or slice sizes selected from such sequences as 5, 9, 17, 35,..., or 8, 16, 32,...

◦ find medians and hinges, within each slice, separately for both $x$-values and $y$-values.

◦ to replace, for a while, the $y$-hinges by letter-value differences and to smooth the results, obtaining a median trace and two letter-difference traces; and then, by addition and subtraction, two hinge traces.

◦ to be careful which $x$-hinge is matched with which $y$-hinge for each slice.

◦ to be prepared to go further, as in section 8F, when necessary.

◦ to go on, finding fences, outside and far-out points, and adjacent polygons, thus being ready to draw wandering schematic plots.

We now understand that it may be reasonable to:

◦ copy on in smoothing medians, but use the end-value rule in smoothing letter-differences or hinges.

◦ to subtract from $y$ known functions of $x$ before finding median and hinges, restoring what was subtracted in due course.

◇ to combine results from different subtractions before finalizing median and hinge traces.

If we have covered optional section 8I (below), we understand why:

◇ we expect to smooth $x$ whenever we smooth $y$.

### 8I. The need to smooth both coordinates (optional)

In section 8B, we came across--for the first time--the need to smooth a sequence of points in the plane. In such a case, we work with two matched sequences, one of the first coordinate, one of the second. If, as was the case in section 8C, the sequence of first coordinates comes to us in order, we might think it would be enough to smooth only the second coordinate.

How might we test such a view? Under what circumstances can we say "Any reasonable smooth of this pair of sequences must do thus-and-so!"? Surely if the initial points are on a straight line, it would be hard indeed to argue for smoothed points that do not fit on the same straight line.

Exhibit 33 shows arithmetically what can happen when the data all lie on a single straight line, here

$$y = 2x$$

and we smooth only the second, $y$, coordinate, using 3R'SSH or 3R'SSH3. (What would have happened with 3R or 3RSS?) Exhibit 34 shows the picture. Clearly, we go off the line! And, as a result, to most of us it would be clear that the sequence of points was rougher than before. (More nearly evenly spaced, perhaps, but rougher.)

There is no escape. When we smooth one coordinate, we are almost always wise to smooth the other coordinate in just the same way.

exhibit **33** of chapter 8: illustrative data

**Smoothing one of two coordinates**

| | | | y, the second coordinate | | | y minus 2x |
|---|---|---|---|---|---|---|
| x | raw | 3R' | 3R'SS | 3R'SSH | 3R'SSH3 | final smooth |
| 0   | 0   | T | T | 0   | T | 0 |
| 1   | 2   | h | h | 2   | h | 0 |
| 2   | 4   | e | e | 8h  | e | 4h |
| 12  | 24  |   |   | 20  |   | −4 |
| 14  | 28  | s | s | 28  | s | 0 |
| 16  | 32  | a | a | 36  | a | 4 |
| 26  | 52  | m | m | 62  | m | 10 |
| 56  | 112 | e | e | 119 | e | 7 |
| 100 | 200 |   |   | 200 |   | 0 |

308 exhibit 34/8: Schematic plots

**review questions**

Why is there a problem? What test can we impose on a smoother of points? What was our example like? What happened? What must we conclude?

exhibit **34** of chapter 8: illustrative only

**How smoothing only one coordinate can roughen a point sequence**

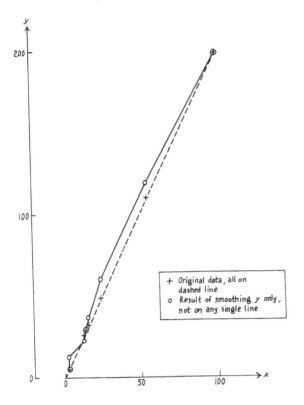

+ Original data, all on dashed line
o Result of smoothing y only, not on any single line

# Delineations of batches of points  9

chapter index on next page

If we want to be as effective in summarizing batches of points as schematic plots have made us in summarizing batches of numbers, we need to do rather more than we did in the last chapter. Wandering schematic plots tell us quite a lot about y, but we could be told more. And we need to be told about $x$ at the same time.

Actually, we will find that we already have all the tools we need. We can carry further some of the things we already do; we can display some of the things we usually calculate. So this will be a short chapter.

### review questions

Are wandering schematic plots effective enough in summarizing batches of points? Why/why not? What might we do? Will it take new tools? A long chapter?

### 9A. E-traces and D-traces

Our hinge, or H-traces were found, in section 8D, by starting with the identities

$$H_{lo} = M - (H\text{-to-}M)$$
$$H_{hi} = M + (M\text{-to-}H)$$

and then

◇ smoothing M to a middle trace.

◇ smoothing H-to-M--and, separately, M-to-H

◇ recombining (by addition or subtraction) to get smoothed H's, which we then take as H-traces.

If we want E-traces, we can start from

$$E_{lo} = H_{lo} - (E\text{-to-}H)$$
$$E_{hi} = H_{hi} + (H\text{-to-}E)$$

smooth E-to-H--and, separately, H-to-E--and combine the results with the H-traces we already had found. The results are naturally taken as E-traces.

Taking further steps going on to D-traces--thence to C-traces and beyond--introduces nothing new, just more arithmetic of the same kind. So let us plan to do this.

# index for chapter 9

review questions 309

**9A. E-traces and D-traces** 309
review questions 311

**9B. Simple delineation--Twin Rivers again** 311
*delineation* 311
an engineering example 312
review questions 313

**9C. Reduced and schematic delineations** 313
*reduced delineation* 314
*schematic (x, y)-plots* 315
*two-way schematic plot* 315
review questions 319

**9D. What our schematic plots and delineations have missed** 319
patterns of salaries 319
review questions 321

**9E. Three variables at once--or more** 321
handling more variables 321
index cards 322
how many data sets? 327

**9H. How far have we come?** 329

| EXHIBITS | | PAGES |
|---|---|---|
| 9A | | |
| 9B | | |
| 1 | | 311 |
| 2 | 9C | 312 |
| 3 | | 313 |
| 4 | | 314 |
| 5 | | 315 |
| 6 | | 316 |
| 7★ | 9D | 317 |
| 8★ | 9E | 320 |
| 9 | | 322 |
| 10★ | | 323 |
| 11★ | 9H | 328 |

**review questions**

Just how do we find an upper E-trace? A lower E-trace? An upper D-trace? How are these calculations related to one another? Do we have to learn anything new? Will we plan to go ahead?

## 9B. Simple delineation--Twin Rivers again

Let us go back to the Twin Rivers example of the last chapter, doing two things:

⋄ adding E-traces and D-traces over all that part of the data for which they are well defined.

⋄ adding the vertical letter cuts.

The result, shown in exhibit **1**, is appropriately called a full--or simple--

### delineation

of the batch. We now see the same things we saw (in exhibit 14 of chapter 8) earlier, but now even more clearly presented.

exhibit **1** of chapter 9: Twin Rivers

**A full delineation of the (ElUse, GaUse) scatter (GaUse = gas usage, ElUse = electric usage)**

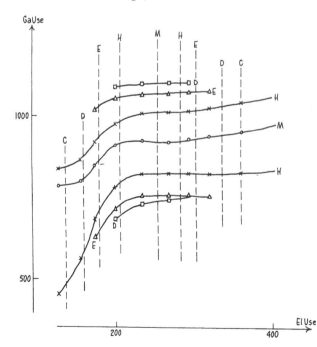

### an engineering example

In their book, Daniel and Wood (C. Daniel and F. S. Wood, 1971, *Fitting Equations to Data*, Wiley–Interscience) discuss quite carefully the fitting of suitable equations to 105 observed values of heat of solution for 14 cements. If we turn to their pages 255 and 256, we find 104 (fitted value, residual) pairs (one observation has been set aside as discrepant) which, when delineated, look as shown in exhibit **2.**

We see several things more or less clearly:

⋄ a squeezing-together of the fitted values at the low end (toward 500) (seen from the vertical lines at the letter cuts).

⋄ a wedging-together of the residuals--a decrease in spread--as the fitted values increase (seen from the H-traces).

⋄ perhaps a tendency to high residuals for the very lowest fitted values (seen from the H- and M-traces).

The first of these was to be expected from the nature of what was being done. As the cement samples aged, their heat of solution was known to decrease,

exhibit **2** of chapter 9: heats of solution

**Residuals delineated against fitted values for the heats of solution of 104 cement samples**

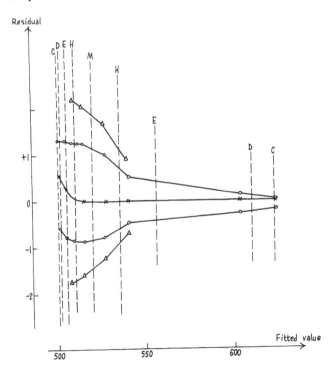

exhibit **4/9**: Delineations of batches of points

8.) Notice that we have used lighter marking, · · · ·, and ----, for the verticals than for the traces, ——— and – – – –. The reason for this is simple; straight lines, especially when parallel, seize the eye more effectively than do broken lines. If we are to give the verticals and the traces equal chance to seize the eye, we must favor the traces in how we show them.

Some, perhaps many, will find exhibit 3 somewhat too detailed. What can we do for them? What detail can we strip out?

Exhibit 4 shows the result of one such stripping out. The E-verticals and E-traces have been used to make a box. The M-vertical and M-trace have been shown dashed, ONLY WITHIN the E-box. The H-verticals and H-traces have been shown dashed, ONLY OUTSIDE the E-box. This

### reduced delineation

does a quite reasonable job of showing us, schematically, what the scatter of 50 (log bank deposits, Governor's salary) points are like. What more can we ask?

exhibit **4** of chapter 9: governor's salary

**Reduced delineation of governors' salary, 1969 against log bank deposits, 1968**

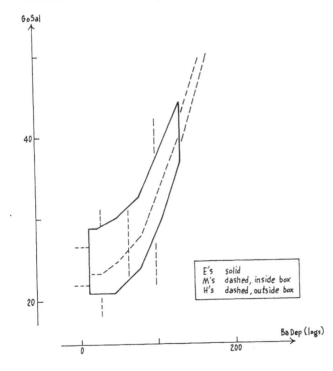

### schematic (x, y)-plots

Our schematic plots for $x$ alone used fences, outside and adjacent points. In the last chapter (8), we introduced $y$ fences and identified $y$-outside and adjacent points. We need, in addition, only look at the $x$-values--since the letter cuts are vertical lines, looking at $x$'s from $(x, y)$ pairs is now no different from looking at $x$'s alone. Thus providing $x$-fences is simple.

In this governor's salary example, we have chosen to take our human prerogative seriously, and treat Alaska distinctively. Exhibit **5** shows all nonoutside points, with $x$'s for those we are going to treat as adjacent. (Except for Alaska, which appropriately enough appears on the NW of this plot, this means exactly the points we would have taken as adjacent if Alaska were excluded.)

Exhibit **6** shows the completed

### two-way schematic plot

What about the states found outside?

◇Delaware is a geographically small state and probably does considerable banking in Philadelphia and Baltimore.

exhibit **5** of chapter 9: governor's salary

**All not-outside points, with the chosen rim points shown by x's**

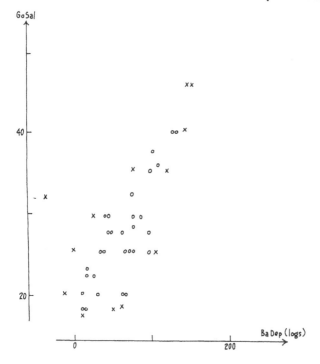

◇Hawaii and Alaska are noncontiguous--whether greater independence or mainland banking accounts for their location on this plot is questionable.

◇Arkansas, with a 10,000-dollar salary, clearly expected its governor to gain his living some other way.

◇New York was outside on bank deposits alone.

◇Georgia offers a problem--who knows the answer?

◇Massachusetts is formally outside, but very close to the E-box--probably an accident.

◇California is close to the fence, but below--for a state that large, perhaps a somewhat higher salary is to be expected. (It was raised the next year.)

(See also comments in Section 8G.)

Exhibit **7** sets some problems.

### exhibit **6** of chapter 9: governor's salary

**The completed schematic plot**

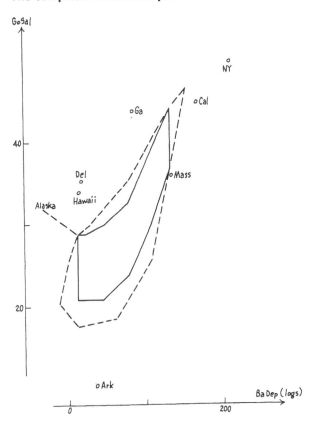

exhibit **7** of chapter 9: data and problems

**Some problems using schematic plots**

A) MARRIAGES and MOTOR VEHICLE DEATHS by STATE, 1958

|       | Marriages | Motor vehicle deaths | x   | y   |
|-------|-----------|----------------------|-----|-----|
| Ala   | 24,444    | 852                  | 39  | 93  |
| Ariz  | 9,917     | 508                  | 0   | 71  |
| Ark   | 15,574    | 444                  | 19  | 65  |
| Calif | 96,330    | 3510                 | 98  | 155 |
| Colo  | 14,688    | 396                  | 17  | 60  |
| Conn  | 16,879    | 251                  | 23  | 40  |
| Del   | 2,311     | 84                   | −64 | −8  |
| DC    | 8,094     | 62                   | −9  | −21 |
| Fla   | 35,243    | 1134                 | 55  | 105 |
| Ga    | 45,863    | 956                  | 66  | 98  |
| Ida   | 9,520     | 270                  | −2  | 43  |
| Ill   | 82,860    | 1886                 | 92  | 128 |
| Ind   | 41,226    | 1056                 | 62  | 102 |
| Iowa  | 25,101    | 598                  | 40  | 78  |
| Kans  | 15,481    | 554                  | 19  | 74  |
| Kent  | 26,095    | 789                  | 42  | 90  |
| La    | 21,447    | 844                  | 33  | 93  |
| Maine | 7,104     | 205                  | −15 | 31  |
| Maryl | 41,403    | 505                  | 62  | 70  |
| Mass  | 45,959    | 590                  | 66  | 77  |
| Mich  | 53,662    | 1375                 | 73  | 114 |
| Minn  | 23,410    | 707                  | 37  | 85  |
| Miss  | 36,198    | 548                  | 56  | 74  |
| Mo    | 33,976    | 975                  | 53  | 99  |
| Mont  | 6,160     | 193                  | −21 | 29  |
| Nebr  | 10,637    | 346                  | 3   | 54  |
| Nev   | 5,755     | 190                  | −24 | 28  |
| NH    | 7,078     | 101                  | −15 | 0   |
| NJ    | 39,113    | 754                  | 59  | 88  |
| NM    | 5,850     | 408                  | −23 | 61  |
| NY    | 124,513   | 2118                 | 110 | 133 |
| NC    | 27,228    | 1081                 | 43  | 103 |
| ND    | 4,306     | 155                  | −37 | 19  |
| Ohio  | 65,479    | 1812                 | 82  | 126 |
| Okla  | 33,444    | 667                  | 52  | 82  |
| Oreg  | 9,798     | 448                  | −1  | 65  |
| Penna | 64,529    | 1654                 | 81  | 122 |
| RI    | 5,653     | 74                   | −25 | −13 |
| SC    | 38,550    | 610                  | 59  | 79  |
| SD    | 5,662     | 240                  | −25 | 38  |
| Tenn  | 27,947    | 719                  | 45  | 86  |
| Tex   | 89,702    | 2342                 | 95  | 137 |
| Utah  | 6,741     | 191                  | −17 | 28  |
| Vt    | 3,376     | 77                   | −47 | −11 |
| Va    | 36,588    | 861                  | 56  | 94  |
| Wash  | 27,966    | 573                  | 45  | 76  |

→

exhibit 7 of chapter 9 (continued)

| | | | |
|---|---|---|---|
| WVa | 14,213 | 387 | 15 | 59 |
| Wisc | 25,073 | 822 | 40 | 91 |
| Wyo | 2,945 | 137 | −53 | 14 |

$x$ = −400 plus 100 times log of number of marriages, 1958.

$y$ = −200 plus 100 times log of number of motor vehicle deaths, 1958.

**B) NUMBERS OF STAMENS and PISTILS for 268 EARLY FLOWERS of *Ranunculus ficaria***

Number of stamens, $x$

| Number of pistils, $y$ | 18 | 20 | 22 | 24 | 26 | 28 | 30 | 32 | 34 | 36 | 37 | 38 | (Sum) |
|---|---|---|---|---|---|---|---|---|---|---|---|---|---|
| ≤6 | | | | | | | | | | | (37, 2) | | (1) |
|  | | | | | | 1 | | | | | | | (1) |
| 8 | | | | | | | | | | | | | |
|  | | | | | | 1 | | | | | | | (1) |
| 10 | | 1 | | | 1 | | | | | | | | (2) |
|  | | | 2 | | 1 | | | | | | | | (3) |
| 12 | | | 3 | 1 | 2 | 1 | 1 | 3 | 1 | 1 | | | (13) |
|  | | 1 | 1 | 1 | 4 | 1 | 1 | 1 | | 1 | 1 | | (12) |
| 14 | | 4 | 3 | | 1 | 2 | 4 | 1 | 2 | 3 | 1 | 1 | (22) |
|  | | | 1 | 2 | 4 | 3 | 7 | 4 | 4 | 5 | 2 | 1 | 1 1 | (35) |
| 16 | 1 | | | 2 | | | 1 | 5 | 3 | 5 | 4 | 5 | 3 2 | (31) |
|  | | | | 2 | 2 | 1 | 4 | 2 | 3 | 1 | 2 | 5 | 2 1 | (25) |
| 18 | | | 1 | | 2 | 4 | 3 | 1 | 7 | 1 | 3 | 2 | 1 2 | (27) |
|  | | | | | 1 | 2 | | 5 | 4 | 4 | 1 | 1 | 1 1 1 | (21) |
| 20 | | | | 2 | | 1 | 1 | 2 | 2 | | 1 | 3 | 4 2 1 | (19) |
|  | | | | 1 | | | | 2 | 2 | | 2 | 4 | 1 1 | (13) |
| 22 | | | | | | 1 | 2 | 3 | | 1 | 2 | 3 | 1 2 | (15) |
|  | | | | | | | 1 | 1 | | 2 | 1 | 1 2 | 2 | (10) |
| 24 | | | | | | | 1 | 1 | | 2 | | | (4) |
|  | | | | | | | | | | 1 | 2 | | **1** | (4) |
| 26 | | | | | | | | | 1 | 2 | | | (3) |
|  | | | | | | 1 | | 1 | | | | 2 | (4) |
| 28 | | | | | | | | | | | 1 | | (1) |
| 30 | | | | | | | | | | | | | |
|  | | | | | | | | 1 | | | | | (1) |
| 32 | | | | | | | | | | | | | |
| (Sum) | 1, 6, 8, 9, 16, 12, 22, 26, 26, 38, 14, 23, 20, 20, 13, 7, 1, 4, 1, 1 | | | | | | | | | | | | 268 → |

### review questions

For which familiar example did we present a full delineation? How did we differentiate the lines and traces? Why did we do this? What is a reduced delineation? How did it seem to work in this example? What were the essentials in a schematic plot (or a y-scatter)? How can we imitate a "middle half" box using parts of a full delineation? What is an E-box? How hard is it to find fences? x-wise? y-wise? How did we find the latter? What did we include in a finished schematic plot? Can we reasonably make modifications based on human judgment? Why/why not? What modifications did we make in our example? What did the schematic plot seem to tell us?

## 9D. What our schematic plots and delineations have missed

This section of chapter 9 is intended to show us something of what we have missed--missed because our summaries are designed to miss it/them.

### pattern of salaries

As a schematic device, we expect delineation to smooth out irregularities, and show us the smoothed-out picture. It does as we expect. Exhibit **8** shows the detailed distribution of Governor's salaries, one stem per thousand dollars; this points out to us what we already knew, that most salaries (30 out of 50) are in multiples of 5,000 dollars. (And of the 20 that remain, 8 are other multiples of 2,500 dollars and 8 other multiples of 2,000 dollars.)

exhibit 7 of chapter 9 (continued)

**P) PROBLEMS**

7a) Delineate x of panel A against y. Comment.

7b) Delineate y of panel A against x. Comment

7c) Delineate number of stamens of panel B against number of pistils. Comment.

7d) Delineate number of pistils of panel B against number of stamens. Comment.

7e) (uses a and b) Compare the results of 7a and 7b. Comment.

7f) (uses c and d) Compare the results of 7c and 7d. Comment.

**S) SOURCES**

For A: The World Almanac for 1960. Pages 306 and 308.

For B: L. H. C. Tippett, 1952. *The Methods of Statistics,* 4th edition. New York, John Wiley.

exhibit **8** of chapter 9: governor's salary
**The details of governors' salaries, 1969**

**A)** The EXPANDED STEM-and-LEAF        **B)** The THOUSANDS DIGIT

```
GoSal
10 | 0

17 | h
18 | 888h
19 |
20 | 00000
21 |
22 | 2
23 | q
24 |
25 | 5555555555
26 |
27 | hh
28 | xh
29 |
30 | 00000
31 |
32 | 2h
33 | h
34 |
35 | 55555
36 | 6
37 | h
38 |
39 |
40 | 000
41 |
42 | h
43 |
44 | d
45 | 55

50 | 0
```

```
(15)  0* | 000000000000000
      0. |
      1* |
      1. |
( 2)  2* | 22
( 2)  2. | hh
      3* |
(2)   3. | qh
      4* |
( 1)  4. | d
(17)  5* | 55555555555555555
      5. |
( 1)  6* | 6
      6. |
      7* |
( 4)  7. | hhhh
( 3)  8* | 888
( 3)  8. | hxh
      9* |
      9. |
```

Note that 32 of 50 end in 0 or 5
Note that 888h after **18|** means 18, 18, 18h, and so on.
(q = and a quarter, x = and .374, d = and .1)

**P)** PROBLEM

8a) Find the corresponding data for another year and make the corresponding analysis.

**S)** SOURCE

**The 1970 World Almanac, page 57. (Not for the same date as in the examples of chapter 8.)**

Any smooth-looking picture that shows us some aspects of our (BaDep, salary) swarm must--because it is smooth--fail to show us this lumpiness in governor's salaries. This may be good--and often is. It may also be bad--and sometimes is. It is an essential characteristic of any smooth schematic picture, whether we like it or not. We just need to know that it happens.

The same thing, of course, happens with a one-way schematic plot for these same 50 salary values, where the letter values are:

```
     50 salaries
  M25h|   27h    |              | 18 |
  H13 | 35    23 |12        f|53      5
  E  7| 40    20 |          F|71      —
                            xxx out xxx
                             50 adj 10
```

so that the one-way schematic plot is perfectly bland, with no indication of any granularity.

A similar thing would, of course, happen if our swarm had diagonal stripes instead of horizontal stripes--or if any of a variety of detailed patterns had occurred. We also need to know that such things happen. We also need to know that we shouldn't let their possibility worry us whenever it is reasonable for us to take a thoroughly schematic view.

### review questions

What do we expect delineation to show us? What do we want it to overlook? What about governors' salaries? Can any smooth summary tell about such lumps? In particular, what about an ordinary schematic plot for governors' salaries alone? Are there other kinds of lumping that a delineation should and would miss? What are some examples?

## 9E. Three variables at once--or more

### handling more variables

So long as we have only two variables--and want only to delineate one against the other--the extended stem-and-leaf techniques in which the leaves are "(trailing digits of $x$, all of $y$)" is quite reasonable. Once we have several variables, and may want to delineate any one against any other, the thought of copying down all the values many times is worse than just boring. We need another approach.

Our problem is to sort data-sets, at least as pairs, probably as wholes; to do this easily; to divide into slices; to sort each slice (on a second variable), and find its letter values. What is easy to sort, to divide into parts, and to re-sort? One obvious answer is a deck of index cards, 3" × 5" for modest numbers of variables, and 4" × 6" or 5" × 7" for larger numbers of variables.

### index cards

Exhibit **9** shows how this could be done, for instance with the 88 unincorporated places over 25,000 population found in the 1960 U.S. Census. We write the names and units of the quantities we want to work with in easily recognizable places on a master 3 × 5 card, and then put identification on each detail card, together with the values of the chosen variables, each in the place corresponding to its name on the master card.

We can now easily shuffle (unimportant), sort (in order according to any chosen one of the eight variables), and make piles of these cards, carrying out all the operations needed to form slices on any of the 8 variables and then to order any slice on any of the 7 variables remaining. Exhibit **10** offers some data and problems.

exhibit **9** of chapter 9: unincorporated places

**Illustration of master 3 × 5 card and two of the 88 unincorporated places in the 1962** *County* **and** *City Data Book* **(8 selected quantities)**

| U.S. rank (pop, 1960) | Land area (sq. mi) | % nonwhite 1960 |
|---|---|---|
| % housing units with automobile | MASTER | % foreign-born |
| % housing units with ≥ 1.01 persons/room | % completed less than 5 yrs school | % native with a foreign-born parent |

| 37 | 8.6 | 5.5 |
|---|---|---|
| 92.0 | Altadena, Calif. | 10.0 |
| 3.9 | 1.8 | 18.4 |

| 78 | 8.4 | 1.5 |
|---|---|---|
| 91.4 | Inglewood, Tenn. | 0.5 |
| 4.7 | 2.8 | 1.6 |

exhibit **10** of chapter 9: unincorporated places

**Data and problems**

**A)** The DATA---variables as in exhibit 9

| Place | U.S. rank | Land area (sq mi) | % non white | % foreign born | % native with for. parent | % <5 years school | % Housing units ≥1.01 pers/room | % Housing units using automobile |
|---|---|---|---|---|---|---|---|---|
| **CALIFORNIA** | | | | | | | | |
| Altadena | 37 | 8.6 | 5.5 | 10.0 | 18.4 | 1.8 | 3.9 | 92.0 |
| Arden-Arcade | 6 | 22.0 | 0.8 | 3.6 | 13.9 | 0.9 | 6.7 | 97.7 |
| Bell Gardens | 79 | 2.5 | 1.4 | 2.3 | 6.8 | 6.6 | 21.4 | 87.3 |
| Carson | 42 | 8.4 | 2.5 | 4.5 | 11.9 | 4.9 | 23.1 | 95.5 |
| Castro Valley | 44 | 6.4 | 0.7 | 5.7 | 18.8 | 3.0 | 6.5 | 94.2 |
| East Los Angeles | 1 | 7.9 | 3.8 | 20.3 | 34.0 | 19.3 | 22.8 | 74.6 |
| Florence-Graham | 41 | 3.0 | 45.8 | 8.4 | 13.6 | 14.4 | 23.8 | 71.3 |
| Lancaster | 84 | 11.3 | 0.6 | 2.8 | 11.2 | 1.5 | 8.3 | 93.7 |
| Lennox | 57 | 2.4 | 1.4 | 6.5 | 13.8 | 3.1 | 10.7 | 86.7 |
| South San Gabriel | 82 | 3.9 | 0.8 | 7.9 | 18.1 | 6.4 | 12.4 | 87.6 |
| Temple City | 54 | 5.1 | 0.4 | 7.3 | 17.7 | 2.1 | 3.9 | 91.9 |
| West Hollywood | 66 | 1.9 | 0.6 | 24.0 | 30.0 | 4.2 | 1.7 | 81.3 |
| **CONNECTICUT** | | | | | | | | |
| East Hartford | 31 | 18.2 | 0.8 | 7.8 | 27.3 | 3.9 | 10.7 | 82.0 |
| Enfield | 56 | 33.2 | 0.7 | 7.7 | 26.1 | 5.9 | 9.9 | 89.5 |
| Fairfield | 25 | 29.9 | 0.6 | 9.4 | 30.3 | 2.9 | 5.1 | 93.7 |
| Greenwich | 18 | 47.6 | 2.2 | 13.0 | 25.4 | 4.4 | 4.5 | 91.6 |
| Hamden | 36 | 24.8 | 1.5 | 9.7 | 32.1 | 4.0 | 3.3 | 90.8 |
| Manchester | 34 | 27.5 | 0.4 | 10.7 | 28.9 | 4.0 | 5.2 | 88.4 |
| Stratford | 27 | 18.5 | 3.0 | 8.9 | 31.6 | 4.8 | 6.5 | 93.1 |
| Wallingford | 62 | 41.6 | 0.1 | 9.1 | 26.7 | 4.5 | 9.6 | 90.3 |
| West Hartford | 10 | 21.2 | 0.4 | 11.7 | 31.1 | 2.9 | 1.8 | 91.5 |
| West Haven | 32 | 10.9 | 2.0 | 9.1 | 31.1 | 5.5 | 6.1 | 87.5 |

→

exhibit 10 of chapter 9 (panel A, continued)

| Place | U.S. rank | Land area (sq mi) | % non white | % foreign born | % native with for. parent | % <5 years school | % Housing units ≥1.01 pers/room | % Housing units using automobile |
|---|---|---|---|---|---|---|---|---|
| **FLORIDA** | | | | | | | | |
| Brownsville | 40 | 10.5 | 15.4 | 1.1 | 2.8 | 8.2 | 17.9 | 87.6 |
| **HAWAII** | | | | | | | | |
| Kailua-Lanikai | 87 | 5.4 | 40.1 | 3.6 | 17.6 | 3.7 | 18.9 | 94.3 |
| **MARYLAND** | | | | | | | | |
| Bethesda | 13 | 14.3 | 1.4 | 6.0 | 13.7 | 0.8 | 2.0 | 84.8 |
| Catonsville | 43 | 6.2 | 7.3 | 3.1 | 10.7 | 3.8 | 3.9 | 88.6 |
| Dundalk | 4 | 10.6 | 8.2 | 2.5 | 11.4 | 4.6 | 10.8 | 83.5 |
| Essex | 48 | 5.7 | 0.6 | 1.8 | 8.4 | 4.7 | 10.2 | 88.3 |
| Parkville-Carney | 74 | 6.1 | .. | 3.3 | 12.0 | 3.5 | 4.8 | 91.7 |
| Silver Spring | 7 | 9.3 | 1.7 | 5.7 | 18.2 | 1.2 | 3.3 | 94.4 |
| Wheaton | 15 | 9.2 | 1.4 | 3.3 | 13.5 | 1.4 | 9.3 | 98.6 |
| **MASSACHUSETTS** | | | | | | | | |
| Arlington | 23 | 5.1 | 0.3 | 12.1 | 32.1 | 2.8 | 4.4 | 87.2 |
| Belmont | 68 | 4.6 | 0.3 | 13.7 | 31.7 | 3.2 | 2.4 | 89.2 |
| Braintree | 58 | 14.3 | 0.2 | 8.0 | 28.3 | 1.9 | 6.8 | 95.0 |
| Brookline | 16 | 6.8 | 1.0 | 19.5 | 35.8 | 3.1 | 2.8 | 72.0 |
| Framingham | 30 | 24.1 | 0.6 | 8.5 | 25.0 | 3.8 | 6.2 | 84.8 |
| Lexington | 73 | 16.5 | 0.5 | 9.5 | 28.7 | 3.4 | 4.1 | 97.0 |
| Methuen | 71 | 22.5 | 0.3 | 12.7 | 34.8 | 7.0 | 6.1 | 85.6 |
| Milton | 81 | 12.9 | 0.1 | 11.3 | 32.1 | 1.9 | 1.9 | 93.0 |
| Natick | 67 | 15.1 | 0.5 | 8.2 | 26.1 | 2.5 | 6.2 | 89.9 |
| Needham | 86 | 12.5 | 0.3 | 8.2 | 25.9 | 1.9 | 1.9 | 95.4 |
| Watertown | 39 | 4.2 | 0.2 | 16.7 | 35.0 | 6.2 | 5.9 | 80.8 |
| Wellesley | 83 | 9.9 | 0.4 | 7.8 | 20.2 | 1.9 | 1.3 | 94.1 |
| Weymouth | 24 | 17.7 | 0.3 | 6.6 | 25.2 | 1.9 | 8.7 | 89.2 |

9E: three variables/exhibit 10(A)

| | | | | | | | |
|---|---|---|---|---|---|---|---|
| **NEW JERSEY** | | | | | | | |
| Cranford | 80 | 5.0 | 3.2 | 7.1 | 25.5 | 2.6 | 3.2 | 93.3 |
| Delaware | 55 | 24.4 | 1.9 | 4.6 | 19.2 | 3.2 | 4.2 | 96.4 |
| Edison | 29 | 31.0 | 3.3 | 7.9 | 26.4 | 4.5 | 9.0 | 93.3 |
| Ewing | 77 | 15.4 | 6.6 | 6.5 | 22.2 | 5.3 | 7.2 | 93.2 |
| Hamilton | 9 | 40.3 | 2.8 | 7.1 | 25.2 | 6.1 | 6.1 | 91.2 |
| Middletown | 38 | 54.9 | 2.6 | 5.8 | 19.2 | 3.0 | 7.3 | 95.4 |
| New Hanover | 69 | 24.4 | 9.2 | 3.6 | 15.3 | 0.7 | 18.2 | 85.9 |
| North Bergen | 33 | 5.6 | 0.3 | 17.7 | 34.5 | 6.8 | 6.5 | 75.4 |
| Parsippany-Troy Hills | 88 | 23.7 | 2.6 | 10.8 | 25.0 | 6.2 | 7.1 | 97.7 |
| Pennsauken | 51 | 10.4 | 4.5 | 4.9 | 21.0 | 4.6 | 4.1 | 89.4 |
| Teaneck | 35 | 6.1 | 4.2 | 11.8 | 33.0 | 2.8 | 2.8 | 90.2 |
| Union | 21 | 9.2 | 6.0 | 12.2 | 33.0 | 4.4 | 3.1 | 92.9 |
| Wayne | 64 | 25.3 | 0.3 | 7.5 | 25.9 | 3.0 | 5.4 | 97.4 |
| Woodbridge | 5 | 24.1 | 1.6 | 8.1 | 30.4 | 5.1 | 8.2 | 93.5 |
| **NEW YORK** | | | | | | | | |
| Baldwin | 60 | 4.1 | 0.6 | 8.0 | 27.2 | 2.0 | 2.6 | 90.7 |
| Cheektowaga-Northwest | 19 | 6.3 | 0.1 | 6.1 | 25.3 | 2.6 | 8.5 | 92.4 |
| East Meadow | 26 | 6.7 | 2.0 | 6.7 | 31.6 | 2.5 | 5.8 | 94.7 |
| Eggertsville | 28 | 10.9 | 0.2 | 6.6 | 23.2 | 1.2 | 3.1 | 84.7 |
| Elmont | 61 | 3.2 | 2.5 | 12.0 | 34.4 | 6.4 | 7.5 | 89.4 |
| Franklin Square | 53 | 2.8 | 0.2 | 11.5 | 34.0 | 4.2 | 3.4 | 93.1 |
| Hicksville | 22 | 6.7 | 0.4 | 6.7 | 27.6 | 2.9 | 7.1 | 95.6 |
| Levittown | 8 | 6.6 | 0.3 | 5.8 | 26.7 | 1.7 | 10.1 | 95.6 |
| Massapequa | 52 | 5.4 | 0.3 | 7.2 | 28.3 | 2.6 | 4.2 | 96.6 |
| Oceanside | 59 | 5.0 | 0.7 | 8.8 | 31.4 | 2.6 | 2.5 | 94.7 |
| Plainview | 72 | 3.6 | 0.4 | 6.3 | 33.7 | 1.7 | 2.6 | 98.5 |
| Tonawanda | 3 | 17.6 | 0.3 | 6.5 | 22.2 | 1.7 | 7.4 | 94.0 |
| Wantagh | 50 | 5.5 | 0.3 | 6.4 | 27.5 | 2.0 | 3.8 | 97.8 |
| **NORTH CAROLINA** | | | | | | | | |
| Kannapolis | 49 | 27.6 | 11.6 | 0.2 | 0.3 | 15.7 | 12.0 | 81.2 |
| **PENNSYLVANIA** | | | | | | | | |
| Abington | 14 | 16.3 | 4.8 | 5.7 | 20.2 | 2.3 | 2.7 | 92.7 |

exhibit 10(A–P)/9: Delineations

exhibit 10 of chapter 9 (panel A, continued)

| Place | U.S. rank | Land area (sq mi) | % non white | % foreign born | % native with for. parent | % <5 years school | % Housing units ≥1.01 pers/room | % Housing units using automobile |
|---|---|---|---|---|---|---|---|---|
| Bristol | 12 | 16.9 | 4.0 | 3.4 | 17.1 | 2.3 | 14.5 | 95.6 |
| Cheltenham | 45 | 9.0 | 2.4 | 7.8 | 27.7 | 2.6 | 1.5 | 90.4 |
| Falls | 65 | 27.2 | 0.7 | 2.5 | 16.3 | 1.8 | 13.3 | 97.7 |
| Haverford | 17 | 10.0 | 1.3 | 6.7 | 21.6 | 2.8 | 2.0 | 93.8 |
| Hempfield | 63 | 84.4 | 1.6 | 3.8 | 17.5 | 7.8 | 11.4 | 90.2 |
| Lower Merion | 11 | 24.0 | 4.9 | 7.3 | 19.7 | 2.9 | 1.4 | 89.2 |
| Middletown | 75 | 20.7 | 0.4 | 3.9 | 17.0 | 1.1 | 5.5 | 95.9 |
| Millcreek | 70 | 35.6 | 0.4 | 2.9 | 15.6 | 3.8 | 7.8 | 93.3 |
| Mount Lebanon | 47 | 6.4 | 0.2 | 3.7 | 18.3 | 1.1 | 1.3 | 89.3 |
| Penn Hills | 20 | 20.7 | 4.9 | 4.3 | 19.3 | 2.9 | 7.0 | 93.3 |
| Ridley | 46 | 5.5 | 1.7 | 4.0 | 18.6 | 3.9 | 6.5 | 91.8 |
| Ross | 85 | 13.8 | 1.4 | 4.0 | 18.9 | 1.8 | 5.5 | 94.4 |
| Springfield | 76 | 6.9 | 9.3 | 4.7 | 20.5 | 1.8 | 2.9 | 97.1 |
| Upper Darby | 2 | 8.5 | 0.2 | 6.7 | 23.3 | 2.8 | 2.7 | 64.8 |
| TENNESSEE | | | | | | | | |
| Inglewood | 78 | 8.4 | 1.5 | 0.5 | 1.6 | 2.8 | 4.7 | 91.4 |

**P) PROBLEMS**

10a) (Group) Share the 88 unincorporated urban places out among the group, so each does his/her share of making up index cards according to the pattern of exhibit 12 and his/her share of checking the correctness of cards made up by another. (It may help to make 8 copies of each card, so there can be 8 sets of 88.)

10a2) Divide into 8 subgroups. Let each subgroup take the 88 cards, sort in order according to their variable, and divide into slices.

10a3) Within each subgroup assign other variables to persons and let each person, in turn, sort each separate slice on his/her variable and record the corresponding letter values.

10a4) Let each person find, smooth and backadd (or subtract) the letter differences for his/her variable when in his/her group's slices.

10a5) Let each person make the corresponding delineation.

### how many data sets?

What happens if we want to work with more than two variables? The first question is: How many data sets will we need? To suggest an answer we turn back to batches of single values and batches of pairs.

We would not be too far out if we said that for batches of numbers:

⋄ 20 usually gives useful letter values.
⋄ 10 is likely to do so.
⋄ 50 does quite well.

(more may help more), while for batches of $(x, y)$ pairs:

⋄ 100 usually gives a useful delineation.
⋄ 50 is likely to do so.
⋄ 250 does quite well.

(more may help more). Thus, for batches of $(x, y, z)$ values, where we want to look at the three values together, we might expect that:

⋄ 500 usually gives a useful "three-way".
⋄ 250 is likely to do so.
⋄ 1250 does quite well.

If we are going to play with numbers this large by hand, even with only 3 variables, the index-card technique just described is surely called for--and our desire to have a computer do the work mounts rapidly.

And what will we do with our 500-odd cards when we have filled them out? Something strictly analogous to what we did for $(x, y)$ pairs, which can be described in these words: Arrange in order on $x$, letter-cut on $x$, treat each slice the way you would have treated a batch of $y$'s.

For our 500-odd $(x, y, z)$'s, what is the analog? Clearly: Arrange in order on $x$, letter-cut on $x$, treat each slice (of count at least $5 \times 5 = 25$) the way you would have treated a batch of $(y, z)$ pairs (this clearly means "delineate" them). The result of all this is a $(y, z)$-delineation for each $x$-slice.

How can we best look at these delineations? Side-by-side? Or one above another? Surely not--either of these arrangements tends to confuse the slicing variable $x$ with one or the other of the variables in the delineation. What does this leave? Two possibilities, both reasonable:

If we are willing to get some moderately stiff cards, and put one delineation on each, we can riffle the pack of cards, and watch the delineation move as we hasten from slice to slice. When we do this, we are reminded that we have smoothed the delineations within each slice, but we have as yet done no smoothing across slices.

So our next step is to take our delineations--more precisely to take the first cuts, and the smoothed cross-medians which define the delineations--and smooth them across slices. The result is a new set of points in each slice suitable for drawing its delineation. These delineations will now relate much more smoothly to their neighbors.

And what if we don't like low-grade imitations of movies--and refuse to riffle cards--how can we display our delineations? The best choice seems to be to put them on a diagonal. Perhaps we show one delineation for every slice, perhaps fewer. (One choice is to take medians between adjacent, nonoverlapping pairs of delineations.)

(If we were so rash as to try four variables, we could slice twice before delineating. We might use a NW-SE diagonal for our first variable and a NE-SW diagonal for the other.)

The brave may want to try the group problem in exhibit **11**.

exhibit **11** of chapter 9: problems

**Problem set for brave groups**

11a (group)) The 1960 U.S. Census found 676 cities of population 25,000 or more. The 1970 Census found 840. Pick out a few hopefully interesting variables (presumably from among those in some County and City Data Book) and make up a deck of index cards (as per exhibit 9) for either all 1960 cities or all 1970 cities.

11a2 (group)) Order and then slice the deck on one of these variables.

11a3 (subgroups)) Delineate a third variable on a second variable within each of these slices.

11a4 (different subgroups)) Smooth across slices.

11a5 (subgroups)) Construct the smoothed delineations.

11a6 (group)) Compare the resulting delineations, either by fitting or diagonally, or both.

11b (subgroups)) Take the data of exhibit 32 of chapter 8 and divide it into 8 three-hour sections--or 12 2-hour sections--starting at midnight. Delineate the height of tide on the interval since last high tide for each of these sections.

11b2 (different subgroups)) Smooth, around the clock, the results thus obtained.

11b3 (subgroups)) Construct the smoothed delineations.

11b4 (group)) Compare/combine the results and make any further analysis needed.

11c/11c2/11c3/11c4) Do the same, starting at 0100 instead of midnight.

11d (uses (11b4) and (11c4)) Compare the results of (11b4) and (11c4).

### 9H. How far have we come?

This chapter has been devoted to expanding the wandering schematic plots we learned to make in the last chapter to displays that tell us more about the joint behavior of $x$ and $y$ in an $(x, y)$-scatter.

We are now ready:

◇ to find E-traces, D-traces, and the like.

◇ to combine traces and cuts into a delineation.

◇ to convert a full delineation into a reduced one.

◇ to start from a full delineation and construct an $(x, y)$-schematic plot.

We now understand more clearly:

◇ what delineation can do for us.

◇ what delineation and schematic plots miss -- as they should.

◇ how we might go on to the analysis of three or more variables.

# 330 index for chapter 10

review questions 332

**10A. Two-way residuals; row-PLUS-Column analysis** 332
single and double lines 332
*double lines between things that add up* 332
effects, common, eff, all 335
row-PLUS-column 335
a comment 336
review questions 336

**10B. The row-PLUS-column fit** 337
have we learned more? 338
AN analysis--not THE analysis 340
review questions 341

**10C. Some points of technique** 343
use of bolding 343
zeros and decimal points 344
review questions 344

**10D. Row -TIMES-column analysis** 344
*row-TIMES-column* 348
review questions 348

**10E. Looking at row-PLUS-column fits and their residuals** 349
*row effect PLUS column fit* 349
the two-way plot 349
*a two-way plot* 350
TIMES fits 350
looking at residuals 351
*two-way plot of residuals* 351
other postponements 352
review questions 352

**10F. fitting one more constant** 352
*comparison values* 353
*diagnostic plot* 355
*row-PLUS-col-PLUS-one fit* 356
review questions 358

**10G. Converting PLUS to TIMES; re-expression** 358
consumption again 358
there could be others 358
review questions 360

**10H. How far have we come?** 350

| EXHIBITS | PAGES |
|---|---|
| 10A | |
| 1★ | 333 |
| 2★ | 336 |
| 10B | |
| 3★ | 338 |
| 4★ | 340 |
| 5★ | 342 |
| 10C | |
| 10D | |
| 6★ | 345 |
| 10E | |
| 7 | 349 |
| 10F | |
| 8 | 352 |
| 9★ | 354 |
| 10 | 356 |
| 11★ | 357 |
| 10G | |
| 12★ | 359 |
| 10H | |

flattening out to a constant value for each cement, these constant values being close to equality. With data taken in equal time steps, such squeezing together at small heats is only to be expected.

The strikingly larger residuals at the lower fitted values clearly deserve an explanation. We do not have one handy.

The upward kink is interesting, but it would be more so if we were surer that it was not an accident.

### review questions

What did we add? How symmetrically or unsymmetrically does the result trace $x$ and $y$? (Be specific.) What do we call it? How well did it work in the example?

### 9C. Reduced and schematic delineations

Exhibit 3 shows a full delineation of (raw) governor's salary, 1969, against (log) bank deposits, 1968. (Recall that we used log governor's salary in chapter

exhibit **3** of chapter 9: governor's salary

**A full delineation of governors' salary, 1969, against log bank deposits, 1968**

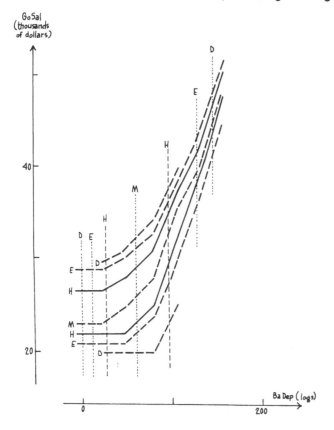

# Using two-way analyses   10

It is now high time for us to look at two-way analyses, both because of their importance, and because of the diversity and richness of analysis they introduce us to. Once we have two or more handles on the same set of data--as in a two-way table--there is much more flexibility in what we do. As a result, there can be many alternative approaches instead of one or perhaps two. A few of these will appear in this chapter.

The bases of a two-way table (implied "of responses") are:

◊ one kind of response.

◊ two kinds of circumstances, one of each kind occurring together for every observation.

If we have birth-rates, for example, for various combinations of age of mother with some social situation (country, date, relative income level, and number of living children are each possibilities), we have a two-way table of responses, perhaps with age of mother tagging columns while number of living children tags rows.

If we have yields of wheat, for example, for various combinations of named variety of wheat with amount and kind of fertilizer application, we have a two-way table of responses--with, say, variety tagging columns while fertilizer amount and kind tags rows. If we have hardnesses for metal specimens having various combinations of composition with heat treatment, we have a two-way table of responses--with, say, composition tagging columns while heat treatment tags rows, and so on.

This chapter and the next are organized together so that the rate of flow of new ideas:

◊ is moderately uniform.

◊ is slow enough to be bearable.

At the same time, repetition will help these new ideas to be more firmly grasped and understood. Many things that seem like loose ends in the present chapter will be tied up in the next. This means that we will not necessarily explain in the present chapter how to carry out the analyses we discuss. Some of that will have to be left for the next chapter.

On to the examples!

### review questions

Does it make a difference if we have two or more handles on a set of data? Why/why not? What are the essentials of a two-way table of responses? How does a "response" differ from a "circumstance"? What are some examples of two-way tables that you have made up for yourself? What were the three examples given above? Why is this chapter closely related to the next? How is this relationship unusual? What is hoped for the rate of new ideas for the two chapters? For the tying up of loose ends?

## 10A. Two-way residuals; row-PLUS-column analysis

Panel A of exhibit **1** shows the mean monthly temperatures for 7 months --from July to January--and 3 places in Arizona--Flagstaff, Phoenix, and Yuma. When we look at this table it is clear that:

◇ Yuma is somewhat warmer than Phoenix.

◇ both are much warmer than Flagstaff.

◇ it gets cooler from July to January.

The last surprises no one. The first two surprise no one who knows about Arizona, but may help the future visitor.

Is this all that such a table can tell us? Surely not.

We have already made good use of (one-way) residuals. How can we adapt the idea here?

Panel B shows one way to start--fitting a median value for each place and forming the corresponding residuals. With the 7 mean monthly temperatures at each place falling steadily, the easiest reasonable fit is the value for the middle month, October. This is exactly what has been used in Panel B.

The residuals, as always, have been found by subtraction, as in 19.1 = 65.2 − 46.1, 19.3 = 90.1 − 70.8, 18.2 = 94.6 − 76.4. We now see that the change of the seasons changes temperatures by about the same amount-- from month to month--in each of the three places. As simple a fact as this should have some use in letting us see deeper.

### single and double lines

In panel B we have begun a new standard:

### double lines between things that add up.

We will use single lines--or, if we are over-careful, dashed lines--to separate values from some kind of summary. Thus, the first row values of panel A and their median has

$$65.2 \quad 90.1 \quad 94.6 \;\big|\; 90.1$$

with a single line, while if we go on to residuals

65.2   90.1   94.6 | 90.1 ‖ −24.9   0.0   4.5

with a double line between the 90.1 and the residuals, things that add up (90.1 + (−24.9) = 65.2, 90.1 + 0.0 = 90.1, 90.1 + 4.5 = 94.6) in this case, give the original values.

The righthand side of panel B shows the medians for (i) the place fits and (ii) the residuals (after place fits) for each month. (If we were not going slowly, this single line and column of numbers would have been set up close to the main part of panel B.)

exhibit 1 of chapter 10: Arizona temperatures

**Two-way residuals (cooling down Arizona)**

A) The MEAN MONTHLY TEMPERATURES--in °F

|  | Flagstaff | Phoenix | Yuma |
|---|---|---|---|
| July | 65.2 | 90.1 | 94.6 |
| Aug | 63.4 | 88.3 | 93.7 |
| Sept | 57.0 | 82.7 | 88.3 |
| Oct | 46.1 | 70.8 | 76.4 |
| Nov | 35.8 | 58.4 | 64.2 |
| Dec | 28.4 | 52.1 | 57.1 |
| Jan | 25.3 | 49.7 | 55.3 |

B) FITTED PLACE VALUES, their RESIDUALS--and some MONTH MEDIANS

| fit | 46.1 | 70.8 | 76.4 | 70.8 |
|---|---|---|---|---|
| July | 19.1 | 19.3 | 18.2 | 19.1 |
| Aug | 17.3 | 17.5 | 17.3 | 17.3 |
| Sept | 10.9 | 11.9 | 11.9 | 11.9 |
| Oct | 0 | 0 | 0 | 0 |
| Nov | −10.3 | −12.4 | −12.2 | −12.2 |
| Dec | −17.7 | −18.7 | −19.3 | −18.7 |
| Jan | −20.8 | −21.1 | −21.1 | −21.1 |

C) The MEDIANS of panel B taken out; SECOND RESIDUALS

|  | eff 70.8 | −24.7 | 0 | 5.6 |
|---|---|---|---|---|
| July | 19.1 | 0 | .2 | −.9 |
| Aug | 17.3 | 0 | .2 | 0 |
| Sept | 11.9 | −1.0 | 0 | 0 |
| Oct | 0 | 0 | 0 | 0 |
| Nov | −12.2 | 1.9 | −.2 | 0 |
| Dec | −18.7 | 1.0 | 0 | −.6 |
| Jan | −21.1 | .3 | 0 | 0 |

Note the appearance of "eff" for "effect," used for values obtained from repeated fitting, whether or not these are fits to the original data. ➔

exhibit **1** of chapter 10 (continued)

**D) ONE set of FITS restored**

| eff | 70.8 | Flagstaff −24.7 | Phoenix 0 | Yuma 5.6 |
|---|---|---|---|---|
| July | 19.1 | 0 | .2 | −.9 |
| Aug | 17.3 | 0 | .2 | 0 |
| Sept | 11.9 | −1.0 | 0 | 0 |
| Oct | 0 | 0 | 0 | 0 |
| Nov | −12.2 | 1.9 | −.2 | 0 |
| Dec | −18.7 | 1.0 | 0 | −.6 |
| Jan | −21.1 | .3 | 0 | 0 |
| fit | 0 | 46.1 | 70.8 | 76.4 |

Examples: 
65.2 = 70.8 + 19.1 − 24.7 + 0
65.2 =   0  + 19.1 + 46.1 + 0

90.1 = 70.8 + 19.1 +  0  + .2
90.1 =   0  + 19.1 + 70.8 + .2

94.6 = 70.8 + 19.1 + 5.6 + (−.9)
94.6 =   0  + 19.1 + 76.4 + (−.9)

63.4 = 70.8 + 17.3 − 24.7 + 0
63.4 =   0  + 17.3 + 46.1 + 0

(The "0" appearing in the 2nd, 4th, ... lines is the "0" in the lower left, outside the double lines.)

**E) EXERCISES to be done by EVERY READER (replaces PROBLEMS)**

1a) Take medians out of the data in panel A in the other order, first by rows and then by columns. Did you get the same answer? Is the difference enough to matter?

1b) Instead of adding place fits, as in panel D, add month fits to the display of panel C. Where did you put the month fits? Why? What value did you put in the corner?

1c) Can you add BOTH place fits and month fits to panel C at the same time? If you can, what value could go in the lower right corner outside the double lines?

1d) Write out the expressions of these given values--52.1, 57.0, 57.1--in terms of residuals and effects.

1e) 88.3 occurs twice as a given value. Write out the two different expressions in terms of residuals and effects.

**S) SOURCE**
Climatography of the United States No. 60-2. Climates of the states. Arizona. U.S. Weather Bureau, Washington, D.C., September 1959.

**effects, common, eff, all**

Panel C goes on to use these medians as further fits, finding second residuals. Here "effect", usually abbreviated "eff" appears for the first time. We will use effect or eff routinely for the results of repeated fitting procedures. In panel C, −24.7 is a column effect, often abbreviated "col eff", while 19.1 is a "row eff".

Usually, as here, an effect is an allocation of part of each of several observed values to a circumstance or set of circumstances. It may be that what we have allocated really belongs to the circumstance, or it may not. But we can be sure that if what is allocated is larger than what is left over, we have made it easier to look at, and understand, what is going on.

The number taken out of each and every data value, here 70.8, is naturally called "common" and abbreviated "all". It is, if you like, an allocation to whatever circumstances are common to all the data--or to all but a few strays.

Thus the overall "verbal arithmetic" for panel C is

65.2 =   70.8    +   (−24.7)   +   19.1    +   0.
data = common  PLUS  col eff  PLUS  row eff  PLUS residual

Panel D goes on to show how we can include one set of fits along with all the effects, thus having a choice both in what we look at and in how we think of the given values as broken down. Taking this still further is placed on the reader's shoulders in panel E.

Notice that we have put the effects above and to the left of the residuals--and let the fits stay to the right and below. We have done this so that (i) the scanning eye is likely to see the effects first and (ii) the effects are as close to the tags as possible. Doing this betrays a bias--a belief that, on average, the effects will be more used--more often looked at and interpreted--than the fits. It is right to accept such a bias, if well-founded, as a general policy. It would be a mistake to feel that such a path should always be followed.

We routinely put effects first, but we feel free to

⋄ omit the fits.

⋄ omit the fits and put effects last.

Indeed, we even feel free to turn the display around--fits first, effects last--or to omit the effects and keep the fits. We can--and should--do whatever the special case seems to call for. The display, including its labels, should match the need that calls it forth. Exhibit 2 shows some alternative displays of the particular

**row-PLUS-column**

analysis of the 21 Arizona mean temperatures that we worked toward in exhibit 1.

### a comment

We are, of course, up to our old trick of seeking a simple partial description--a partial description that may be easier to comprehend--a partial description that, once subtracted away, will leave us able to look more effectively at what has so far not been described.

### review questions

What example did we start with? What did we find first, and what did we do with these values? What did we do next? What was the result? How do we use double lines? Single lines? What is an effect? What is a common term? What does "all" refer to? What do "col eff" and "row eff" refer to? What kind of residuals concern us here? What is the overall "verbal arithmetic"? How do we arrange the blocks in the all-purpose form? Of what kind of analysis is this a display? Can we use other forms of display of this analysis? Why/why not? What is the connection between a fit and a partial description?

exhibit **2** of chapter 10: Arizona temperatures

**Some alternative displays for a particular row-PLUS-column analysis (data from exhibit 1)**

**A)** The ALL-PURPOSE form

|       | eff   | Flagstaff | Phoenix | Yuma  | fit     |
|-------|-------|-----------|---------|-------|---------|
| eff   | 70.8  | −24.7     | 0       | 5.6   | 0       |
| July  | 19.1  | 0         | .2      | −.9   | 89.9    |
| Aug   | 17.3  | 0         | .2      | 0     | 88.1    |
| Sept  | 11.9  | −1.0      | 0       | 0     | 82.7    |
| Oct   | 0     | 0         | 0       | 0     | 70.8    |
| Nov   | −12.2 | 1.9       | −.2     | 0     | 58.6    |
| Dec   | −18.7 | 1.0       | 0       | −.6   | 52.1    |
| Jan   | −21.1 | .3        | 0       | 0     | 49.7    |
| fit   | 0     | 46.1      | 70.8    | 76.4  | (−70.8) |

**B)** The SAME with its PLUS nature EMPHASIZED

|       | eff   | Flagstaff | Phoenix | Yuma  | fit     |
|-------|-------|-----------|---------|-------|---------|
| eff   | 70.8  | −24.7     | 0       | 5.6   | 0       |
| July  | 19.1  | 0         | .2      | −.9   | 89.9    |
| Aug   | 17.3  | 0         | .2      | 0     | 88.1    |
| Sept  | 11.9  | −1.0      | 0       | 0     | 82.7    |
| Oct   | 0     | 0         | 0       | 0     | 70.8    |
| Nov   | −12.2 | 1.9       | −.2     | 0     | 58.6    |
| Dec   | −18.7 | 1.0       | 0       | −.6   | 52.1    |
| Jan   | −21.1 | .3        | 0       | 0     | 49.7    |
| fit   | 0     | 46.1      | 70.8    | 76.4  | (−70.8) |

## 10B. The row-PLUS-column fit

To get the fit we must leave out the residual. These words have different meanings when we want to find numbers for the fit and when we want to display its components. Exhibit 3 works with the same example. To get numbers for the fit, it is often easiest to use

$$\text{fit} = \text{given} \quad \text{MINUS} \quad \text{residual}$$

as has been done in panel A.

Panel B shows a check that the fit is actually of the form

$$\text{all} + \text{row eff} + \text{col eff}.$$

This check, which does NOT check that this is the fit we WANT--only that it is of "row-PLUS-column" FORM, that it is a fit of the kind we thought we were getting--is both important and useful.

Panel C shows the components of the fit. Notice (1) that the panel of residuals is left blank and (2) that this panel is now bordered with single lines

exhibit **2** of chapter 10 (continued)

**C) An EFFECTS-only FORM**

|       | eff    | Flagstaff | Phoenix | Yuma |
|-------|--------|-----------|---------|------|
| eff   | 70.8   | −24.7     | 0       | 5.6  |
| July  | 19.1   | 0         | .2      | −.9  |
| Aug   | 17.3   | 0         | .2      | 0    |
| Sept  | 11.9   | −1.0      | 0       | 0    |
| Oct   | 0      | 0         | 0       | 0    |
| Nov   | −12.2  | 1.9       | −.2     | 0    |
| Dec   | −18.7  | 1.0       | 0       | −.6  |
| Jan   | −21.1  | .3        | 0       | 0    |

**D) A FITS-only FORM**

|       | fit      | Flagstaff | Phoenix | Yuma |
|-------|----------|-----------|---------|------|
| fit   | (−70.8)  | 46.1      | 70.8    | 76.4 |
| July  | 89.9     | 0         | .2      | −.9  |
| Aug   | 88.1     | 0         | .2      | 0    |
| Sept  | 82.7     | −1.0      | 0       | 0    |
| Oct   | 70.8     | 0         | 0       | 0    |
| Nov   | 58.6     | 1.9       | −.2     | 0    |
| Dec   | 52.1     | 1.0       | 0       | −.6  |
| Jan   | 49.7     | .3        | 0       | 0    |

**P) PROBLEMS**

2a/b/c/d/e/f) Write down the same display in a way different from (i) all those above and (ii) those written down for earlier problems of this string, and discuss the pros and cons of the way chosen.

(double lines are left outside). Both of these indicate that we are NOT adding in the residuals.

Notice carefully how much--and how little--the check checks. Notice too that the difference of two row-PLUS-column fits is a row-PLUS-column fit. Thus we can check the difference of two fits in the same way.

### have we learned more?

It is natural to ask if this fit has taught us more than the three rather trivial statements we found after looking at panel A of exhibit 1. What more can we learn from the full two-way analysis? At least these things:

◊ the largest residual, 1.9, is small compared to differences in place effects and most differences between month effects. We can hardly fail to say that a "place-PLUS-month" fit has done a quite good job.

**exhibit 3** of chapter 10: Arizona temperatures

**Displaying and finding the fit (data from exhibit 1)**

**A) VALUES of the FIT--found as given MINUS residual**

|  | Flagstaff | Phoenix | Yuma |
|---|---|---|---|
| July | 65.2 | 89.9 | 95.5 |
| Aug | 63.4 | 88.1 | 93.7 |
| Sept | 58.0 | 82.7 | 88.3 |
| Oct | 46.1 | 70.8 | 76.4 |
| Nov | 33.7 | 58.6 | 64.2 |
| Dec | 27.4 | 52.1 | 57.7 |
| Jan | 25.0 | 49.7 | 55.3 |

**B) CHECKING the values of the FIT**

| Phoenix minus Flagstaff | Yuma minus Phoenix |
|---|---|
| 24.7 | 5.6 |
| 24.7 | 5.6 |
| 24.7 | 5.6 |
| 24.7 | 5.6 |
| 24.9* | 5.6 |
| 24.7 | 5.6 |
| 24.7 | 5.6 |

Note that this check is only a check of whether what we have written down is a possible fit--not that it is the fit we want.

* This value, unequal to the others, shows an error. Going back to exhibit 1, we find 35.8 MINUS 1.9 should be 33.9, not 33.7 (as in panel A above). We will make this correction in all that follows.

⋄ we now see explicitly that Flagstaff is about 25°F cooler than Phoenix, while Yuma is 5 or 6 degrees warmer than Phoenix.

⋄ we have a sequence of month effects--the rounded values are 19, 17, 12, 0, −12, −19, −21--that show a steady decrease from one month to the next--first slow, then fast, then slow--that is close to symmetric around October (changes from September to October and from October to November are almost twice as large as those between other pairs of adjacent months).

⋄ the large residuals are up for Flagstaff in November and December and down for Flagstaff in September and July.

Surely this is much more than we can gather easily from panel A. **We have lifted both lids**--the effects of season and the effects of place. Once both are lifted together we can **see quite a lot of what is left**--in this case, mainly that what is left is relatively small and, beyond that, which are the larger residuals, what is their sign, and roughly how large they are.

exhibit **3** of chapter 10 (continued)

C) the COMPONENTS of the FIT displayed

| eff | 70.8 ‖ | −24.7 | 0 | 5.6 ‖ | 0 |
|---|---|---|---|---|---|
| July | 19.1 | | | | 89.9 |
| Aug | 17.3 | | | | 88.1 |
| Sept | 11.9 | | | | 82.7 |
| Oct | 0 | | | | 70.8 |
| Nov | −12.2 | | | | 58.6 |
| Dec | −18.7 | | | | 52.1 |
| Jan | −21.1 | | | | 49.7 |
| fit | 0 ‖ | 46.1 | 70.6 | 76.4 ‖ | (−70.8) |

Some ways of finding the value of the fit in the upper left corner are:

$$70.8 - 24.7 + 19.1 = 65.2$$
$$0 + 46.1 + 19.1 = 65.2$$
$$0 - 24.7 + 89.9 = 65.2$$
$$(-70.8) + 46.1 + 89.9 = 65.2$$

**P) PROBLEMS**

3a/b/c) Write out the value of the fit in the upper-right/lower-left/lower-right corner in each of the four ways illustrated in panel C and also as "given MINUS residual".

3d) Do the same for a noncorner entry of your choice.

### AN analysis--not THE analysis

In dealing with row-PLUS-column fits we are deliberately flexible, just as we were earlier in dealing with

$$\text{value} = \text{fit} \quad \text{PLUS} \quad \text{residual}.$$

When "fit" was a typical value--or a straight line--we were prepared to have more than one alternative fit, each with its own set of residuals. Things are not different with row-PLUS-column fits.

Panels A and B of exhibit **4** show two (of the very many possible) alternatives to the row-PLUS-column analysis shown, in various forms, in exhibit 2. The differences in the three analyses are illustrated by the three breakdowns for July in Flagstaff.

$$65.2 = (70.8) + (19.3) + (-24.7) + (-.2) \quad \text{exhibit 2}$$
$$65.2 = (70.8) + (19.1) + (-24.7) + (0) \quad \text{A of exhibit 4}$$
$$65.2 = (70.8) + (18.7) + (-24.4) + (.1) \quad \text{B of exhibit 4}$$

The differences are small but do not vanish.

We will have, in the next chapter, some suggestions about which row-PLUS-column analyses to use in routine exploratory work. But these will only be suggestions. Anything that conforms to

$$\text{value} = \text{common} + \text{row eff} + \text{col eff} + \text{residual}$$

is a row-PLUS-column analysis. Whether it meets our needs in a particular case is up to us--after we look at it carefully and critically.

exhibit **4** of chapter 10: Arizona temperatures

**Two alternative row-PLUS-column analyses (in all-purpose form) of the same data (starts at exhibit 1)**

A) ANOTHER row-PLUS-column ANALYSIS--compare with panel D of exhibit 1

|       | eff    | Flagstaff | Phoenix | Yuma  | fit     |
|-------|--------|-----------|---------|-------|---------|
| eff   | 70.8   | -24.7     | 0       | 5.6   | 0       |
| July  | 19.1   | 0         | .2      | -.9   | 89.9    |
| Aug   | 17.3   | 0         | .2      | 0     | 88.1    |
| Sept  | 11.9   | -1.0      | 0       | 0     | 82.7    |
| Oct   | 0      | 0         | 0       | 0     | 70.8    |
| Nov   | -12.4  | 2.1       | 0       | .2    | 58.4    |
| Dec   | -18.7  | 1.0       | 0       | -.6   | 52.1    |
| Jan   | -21.1  | .3        | 0       | 0     | 49.7    |
| eff   | 0      | 46.1      | 70.8    | 76.4  | (-70.8) |

→

Thus there can be arguments about whether the analysis of panel B of exhibit 4 is as generally useful as the simpler-appearing analysis of exhibit 2, but there is no doubt that panel B shows clearly that the residuals can all be squeezed between $-0.8°F$ and $+0.8°F$, and one fit that does this. Is this what we want to do? If it is, fine. If not, down with this analysis--go back to exhibit 2 or try again!

### review questions

How do we find the fit? The values of the fit? How do we check the values of the fit? How often should we check the fit? What does/does not the check check? Did we learn more from our analysis than we learned from our first unguided look at the data? What form of display is nearest the all-purpose form in general usefulness? What did we learn from the analysis that we had not already learned? Should we be flexible in which fits we consider row-PLUS-column fits? In which fit we look at? Why/why not? How do we judge which fit to use in a specific instance? Do we necessarily prefer a fit whose residuals are squeezed together?

exhibit **4** of chapter 10 (continued)

**B) STILL ANOTHER row-PLUS-column analysis of the SAME DATA**

| eff  | 70.8  | −24.4 | −.2 | 5.4 | 0      |
|------|-------|-------|-----|-----|--------|
| July | 18.7  | .1    | .8  | −.3 | 89.5   |
| Aug  | 17.4  | −.4   | .3  | .1  | 88.2   |
| Sept | 11.4  | −.8   | .7  | .7  | 82.2   |
| Oct  | 0     | −.3   | .2  | .2  | 70.8   |
| Nov  | −11.4 | .8    | −.8 | −.6 | 59.4   |
| Dec  | −18.5 | .5    | 0   | −.6 | 52.3   |
| Jan  | −21.0 | −.1   | .1  | .1  | 49.8   |
| eff  | 0     | 46.4  | 70.6| 76.2| (−70.8)|

**E) EXERCISES to be done by ALL READERS (replaces PROBLEMS)**

4a) Write out a row-PLUS-column analysis of this same data different (in residuals, not just in form of display) from the three so far illustrated.

4b) Write out still another such.

4c) Comment on your feelings about choosing among the five row-PLUS-column analyses of this single set of data.

exhibit 5 of chapter 10: Arizona temperatures

**Alternating bolding applied to four forms of display (same analysis as panel A of exhibit 2; data in exhibit 1)**

**A) WITH "eff" and "fit" DRAWN TOGETHER—and alternating bolding**

|       | Flagstaff | Phoenix | Yuma | eff   | fit   |
|-------|-----------|---------|------|-------|-------|
| July  | −.2       | 0       | −1.1 | 19.3  | 90.1  |
| Aug   | −.2       | 0       | −.2  | 17.5  | 88.3  |
| Sept  | −1.0      | 0       | 0    | 11.9  | 82.7  |
| Oct   | 0         | 0       | 0    | 0     | 70.8  |
| Nov   | 2.1       | 0       | .2   | −12.4 | 58.4  |
| Dec   | 1.0       | 0       | −.6  | −18.7 | 52.1  |
| Jan   | .3        | 0       | 0    | −21.1 | 49.7  |
|       |           |         |      |       |       |
| eff   | −24.7     | 0       | 5.6  | 70.8  | 0     |
| fit   | 46.1      | 70.8    | 76.6 | 0     | −70.8 |

**B) OMITTING "fit" and MOVING "eff"**

|       | Flagstaff | Phoenix | Yuma | eff   |
|-------|-----------|---------|------|-------|
| July  | −.2       | 0       | −1.1 | 19.3  |
| Aug   | −.2       | 0       | −.2  | 17.5  |
| Sept  | −1.0      | 0       | 0    | 11.9  |
| Oct   | 0         | 0       | 0    | 0     |
| Nov   | 2.1       | 0       | .2   | −12.4 |
| Dec   | 1.0       | 0       | −.6  | −18.7 |
| Jan   | .3        | 0       | 0    | −21.1 |
|       |           |         |      |       |
| eff   | −24.7     | 0       | 5.6  | **70.8** |

**C) WITH ONE fit GIVEN but KEPT IN PARENTHESIS**

|       | Flagstaff | Phoenix | Yuma | (fit)   | eff    |
|-------|-----------|---------|------|---------|--------|
| July  | −.2       | 0       | −1.1 | (90.1)  | 19.3   |
| Aug   | −.2       | 0       | −.2  | (88.3)  | 17.5   |
| Sept  | −1.0      | 0       | 0    | (82.7)  | 11.9   |
| Oct   | 0         | 0       | 0    | (70.8)  | 0      |
| Nov   | 2.1       | 0       | .2   | (58.4)  | −12.3  |
| Dec   | 1.0       | 0       | .6   | (52.1)  | −18.7  |
| Jan   | .3        | 0       | 0    | (49.7)  | −21.1  |
|       |           |         |      |         |        |
| eff   | −24.7     | 0       | 5.6  | (0)     | **70.8** |

## 10C. Some points of technique

### use of bolding

Here is a good place to remind the reader that **bolding** is not something confined to the pages of books. We early recommended the use of multiple colors--or at least ink contrasted with pencil--by those preparing even the simplest displays by hand. Even those who stick to pencil--or to one color of ink--can always make a respectably clear impression on their lookers by encircling (individually or in blocks) those figures that might well be printed **bold**.

As our displays become a little more diversified and detailed, it becomes more and more important to use bolding--and to use it well. Exhibit 5 shows how--in any of a variety of forms of display--we can seize the eye a little, emphasizing the block structure by alternating bolding so that blocks of entries that are side by side or top by bottom are always different--one bold, one not.

exhibit **5** of chapter 10 (continued)

D) As B, but with BOLD and NONBOLD interchanged

|       | **Flagstaff** | Phoenix | Yuma | **eff** |
|-------|----------|---------|------|-------|
| **July**  | −.2      | 0       | −1.1 | 19.3  |
| **Aug**   | −.2      | 0       | −.2  | 17.5  |
| **Sept**  | −1.0     | 0       | 0    | 11.9  |
| **Oct**   | 0        | 0       | 0    | 0     |
| **Nov**   | 2.1      | 0       | .2   | −12.4 |
| **Dec**   | 1.0      | 0       | −.6  | −18.7 |
| **Jan**   | .3       | 0       | 0    | −21.1 |
| eff   | −24.7    | 0       | 5.6  | 70.8  |

P) PROBLEMS

5a/b/c/d) Imitate, respectively, panel A/B/C/D above for the analysis of panel A, exhibit 4.

5e/f/g/h) Copy, respectively, panel A/B/C/D above, using FOUR COLORS of ink or pencil for the numbers to emphasize the separateness of its parts.

5a2/b2/c2/d2) Do the same for the analysis of panel A of exhibit 4.

### zeros and decimal points

We have found it convenient, in our displays, to distinguish between exact zeros and near zeros. The difference between a very small residual and a residual that is exactly zero

◊ is entirely unimportant so far as the meaning of that residual--or the given value from whence it came--is concerned.

◊ is useful as a matter of technique, in showing us both something about how that particular analysis was done and something about the small-scale distortions of residuals associated with our method of fitting.

We have shown values with decimal points, contrary to our usual practice. So long as we do this, we can make the occurrence of exact zeros quite clear. All we need do is line the exact zeros up with the decimal points.

If, as often ought to be the case, we have no decimal points, it is easy to identify the exact zeros by some other character. Two choices are natural:

◊ the letter "z" for "zero,"

◊ the check-mark "√" for "the fit is exact."

We should feel free to use either.

### review questions

How do we use bolding? Rigidly or flexibly? What replaces bolding in hand work? How did we make exact zeros distinctive? For what purposes is an exact zero different from a near zero? For what purposes not? Do we have alternative markings for "zero"? What are they?

### 10D. Row -TIMES-column analysis

We learned earlier that it is often not wise to analyze our numbers in the form they came to us--whether from others or from our first writing of them down--roots, logs, and, occasionally, (negative) reciprocals ought by now to be not only commonplace but automatically considered. This is at least as true for row-PLUS-column analyses as for the simpler sorts of analyses we have looked at.

Panel A of exhibit **6** gives the expenditures for five kinds of personal consumption at five-year intervals from 1940 to 1960. Even a hasty look at the values shows that "row-PLUS-column" is not a good approach, at least so long as raw values are to be analyzed. Food and tobacco rose 64 billion--from 22 to 86--while private education and research rose only 3 billion--from 0.64 to 3.64.

Since both rises were by similar factors of about 4 or 5, however, we may expect to do better by taking logs. Panel B does just this, while panels C to E go on with a "rows-PLUS-columns analysis of LOGS." Looking at this analysis, and at the stem-and-leaf of its residuals in panel F, we see that

exhibit **6** of chapter 10: personal consumption

**Selected personal consumption expenditures (for the U.S.)**

A) The DATA--in billions of dollars

|  | 1940 | 1945 | 1950 | 1955 | 1960 |
|---|---|---|---|---|---|
| Food and tobacco | 22.2 | 44.5 | 59.6 | 73.2 | 86.8 |
| Household operation | 10.5 | 15.5 | 29.0 | 36.5 | 46.2 |
| Medical care and health expenses | 3.53 | 5.76 | 9.71 | 14.0 | 21.1 |
| Personal care | 1.04 | 1.98 | 2.45 | 3.40 | 5.40 |
| Private education and research | .641 | .974 | 1.80 | 2.60 | 3.64 |

B) LOGS of EXPENDITURES--log billions of dollars

| Food | 1.35 | 1.65 | 1.77 | 1.87 | 1.94 |
|---|---|---|---|---|---|
| Household | 1.02 | 1.19 | 1.77 | 1.87 | 1.94 |
| Medical | .55 | .76 | .99 | 1.15 | 1.32 |
| Personal | .02 | .30 | .39 | .53 | .73 |
| Private E/R | −.19 | −.01 | .26 | .41 | .56 |

C) A FIRST STAGE in the ANALYSIS of the LOGS in B--one fit removed

| Food | −.42 | −.12 | 0 | .10 | .17 || 1.77 |
|---|---|---|---|---|---|---|
| Household | −.44 | −.27 | 0 | .10 | .17 || 1.46 |
| Medical | −.44 | −.23 | 0 | .16 | .33 || .99 |
| Personal | −.37 | −.09 | 0 | .14 | .34 || .39 |
| Private E/R | −.45 | −.27 | 0 | .15 | .30 || .26 |

D) ONE row-PLUS-column ANALYSIS of the LOGS in B--using effects

| Food | .02 | .11 | 0 | −.04 | −.13 || .78 |
|---|---|---|---|---|---|---|
| Household | 0 | −.04 | 0 | −.04 | −.10 || .47 |
| Medical | 0 | 0 | 0 | .02 | .03 || 0 |
| Personal | .07 | .14 | 0 | 0 | .04 || −.60 |
| Private E/R | −.01 | −.04 | 0 | .01 | 0 || −.73 |
| eff | −.44 | −.23 | 0 | .14 | .30 || .99 |

E) ANOTHER row-PLUS-column ANALYSIS of the LOGS in B

|  | 1940 | 1945 | 1950 | 1955 | 1960 || eff |
|---|---|---|---|---|---|---|
| **Food** | .02 | .11 | 0 | −.04 | −.13 || .78 |
| **Household** | .04 | 0 | .04 | 0 | −.06 || .43 |
| **Medical** | 0 | 0 | 0 | .02 | .03 || 0 |
| **Personal** | .03 | .10 | −.04 | −.04 | 0 || −.56 |
| **Private E/R** | −.01 | −.04 | 0 | .01 | 0 || −.73 |
| eff | −.44 | −.23 | 0 | .14 | .30 || .99 |

exhibit **6** of chapter 10 (continued)

**F) STEM-and-LEAF of RESIDUALS in E**

```
      .1
 2|   .1* |10              25 residuals              |.04h|
      .0                                          f|-.05h  +.06h|
13|   .0* |24304002130     M13|    0    |           |one    two |
12|  -.0* |1004044400      M 7|-.01  .02|.03      F|-.10   +.11 |
 2|  -.0  |6
 1|  -.1* |3               adj = -.04 (four)      out = -.06 (1960 Household)
     -.1                   adj = +.04 (twice)     out = +.10 (1945 Personal)
                                                  out = +.11 (1945 Food)
                                                  far = -.13 (1960 Food)
```

**G) THE row-TIMES-column ANALYSIS corresponding to panel D**

| | | | | | | | |
|---|---|---|---|---|---|---|---|
| Food        | 1.05 | 1.29 | 1    | .91  | .74  | × | 6.00 |
| Household   | 1    | .91  | 1    | .91  | .80  |   | 2.94 |
| Medical     | 1    | 1    | 1.05 | 1.07 | 1    |   | 1    |
| Personal    | 1.17 | 1.38 | 1    | 1    | 1.10 |   | .25  |
| Private E/R | .98  | .91  | 1    | 1.02 | 1    |   | .186 |
|             | .36  | .59  | 1    | 1.38 | 2.00 | × | 9.80 |

MULTIPLY one entry of EACH of 4 KINDS to obtain each original VALUE

Examples:

(finding entries) log 1.05 = .02, log 1.29 = .11, log 6.00 = .78, log .36 = -.44,
   log .59 = -.23, log 9.80 = .99.
(combining entries) (1.05)(.36)(6.00)(9.80) = 22.2
   (1.29)(.59)(6.00)(9.80) = 44.5

**H) Row-TIMES-column ANALYSIS corresponding to panel E**

| | | | | | | | eff |
|---|---|---|---|---|---|---|---|
| Food        | 1.05 | 1.29 | 1 | .91 | .74 | × | 6.0  |
| Household   | 1    | .91  |   |     |     |   | 2.7  |
| Medical     |      |      |   |     |     |   | 1    |
| Personal    |      |      |   |     |     |   | .28  |
| Private E/R |      |      |   |     |     |   | .186 |
| eff         | .36  | .59  | 1 | 1.38 | 2.00 | × | 9.8 |

MULTIPLY one ENTRY of EACH of 4 KINDS to obtain each original VALUE.

**I) THE row-TIMES-column FIT of H—in billions of dollars**

|             | 1940 | 1945 | 1950 | 1955 | 1960 |
|---|---|---|---|---|---|
| Food        | 21.  | 35.  | 59.  | 81.  | 118. |
| Household   | 9.5  | 15.6 | 26.  | 36.  | 52.  |
| Medical     | 3.5  | 5.8  | 9.8  | 13.5 | 19.6 |
| Personal    | .98  | 1.6  | 2.7  | 3.7  | 5.4  |
| Private E/R | .65  | 1.07 | 1.8  | 2.5  | 3.6  |

◇ the fit is moderately good, only the residual for 1940 medical care and health expenses being outside (other largish residuals are food in 1945 and 1960 and personal care in 1945).

◇ the year effects increase fairly steadily, more rapidly between 1940 and 1950 than between 1950 and 1960.

◇ the category-of-expenditure effects are quite large.

◇ the three largest residuals are listed in panel F. When we review them in the context of the full analysis, we notice, first, that the residuals for food and tobacco fall steadily from 1945 to 1960--which might leave us wondering how much the recent rise in food prices is a return to a 1945 distribution of expenditures--and second, a much lower value for medical care and health expenses in 1940, a fact we might well want to understand.

exhibit **6** of chapter 10 (continued)

**J)** The RESIDUALS from H in %

| | | | | | |
|---|---|---|---|---|---|
| Food | 5% | 29% | √ | −9% | −26% |
| Household | | √ | −9% | | |
| Medical | | | | | |
| Personal | | | | | |
| Private E/R | | | | | |

**K)** The RESIDUALS from H in BILLIONS of DOLLARS--rounded to match H

| | | | | | |
|---|---|---|---|---|---|
| Food | 1 | 10 | √ | −8 | −31 |
| Household | 1 | −.1 | .3 | .5 | −5.8 |
| Medical | √ | √ | √ | .5 | 1.5 |
| Personal | .06 | .4 | −.3 | −.3 | √ |
| Private E/R | −.01 | −.10 | √ | .1 | .04 |

**P) PROBLEMS**

6a) Complete the entries in H.

6b) Make a third row-PLUS-column analysis of the logs in B, and convert it to a row-TIMES-column analysis of the values in A.

6c) Check that the entries of G reproduce at least three of the values of A. (Set out both what is to be multiplied and what multiplication gives.)

6d) Complete panel J.

6e) Discuss the meaning of the residuals in panel E.

6f) Discuss the meaning of the residuals in panel K.

6g) Compare your discussions in (6c) and (6d).

**S) SOURCE**

**1962 World Almanac and Book of Facts, page 756.**
Their source: Office of Business Economics, U.S. Department of Commerce.

If our audience is willing to be concerned with LOGS of EXPENDITURES, panels E and F present most of the story. But what if they do not? What if billions of dollars is the only form in which they can think?

All we can sensibly do is to undo the logarithms AFTER the ANALYSIS. Consider 1960 food as an example: The breakdown set out in panel E is

$$1.94 = (.99) + (.78) + (.30) + (-.13)$$

Now $.99 = \log 9.8$, $.78 = \log 6.0$; $.30 = \log 2.0$, and $-.13 = \log .74$, so the corresponding breakdown for billions of dollars is

$$86.8 \approx (9.8)(6.0)(2.0)(.74),$$

which we may wish to write as

$$86.8 \approx (9.8)(6.0)(2.0) \text{ down } 26\%.$$

The

row-TIMES-column

analyses corresponding to the row-PLUS-column analyses in panels D and E are set out in panels G and H. These tell the truth, but not strikingly enough to make an adequate effect on most people. How can we go further?

Panel I shows the actual numbers obtained by multiplying up the fit. For some, this will make a clearer impression than the year effects and component fits of panel G.

Panel J shows the residuals from this fit as percent. Some will find this presentation--which focuses on factors of change--quite satisfactory. Others will feel that it is essential to focus on amounts of change. For them, panel K is one way to set out the residuals effectively.

Once the raw data is better fitted by row-TIMES-column, the alternatives for displaying the answer are numerous--and the choice among them is often important, almost entirely as a matter of reaching the reader.

In a row-TIMES-column analysis, as the reader has probably already noticed, we can play the "exact 1" game when we have decimal points, in the same way we played the "exact 0" game for row-PLUS-column analyses. With trivial changes, the discussion at the end of the last section applies here also.

**review questions**

What example did we start with? Does a row-PLUS-column analysis of the given values seem sensible? What might we do to make a PLUS analysis reasonable? Did it work? Can we convert a PLUS analysis of logs to a TIMES analysis? Of what? How do we use triple lines? How do we distinguish TIMES analyses in how we display them? When we want to be particularly explicit? When we want to be particularly explicit about PLUS analyses? How can we show what is left over after a TIMES fit? Do we have a choice? (Be explicit.) What did we conclude about our example?

### 10E. Looking at row-PLUS-column fits and their residuals

Our next task is to learn to look more effectively--through pictures--first at row-PLUS-column fits, and then at their residuals. For the fits, we need a way to make a picture that shows them, for example in the form

$$\text{row effect} \quad \text{PLUS} \quad \text{column fit}$$

easily and clearly for each combination of row and column. As soon as there are more than a few each of both rows and columns, any attempt to do this along a single line will surely lead to crowding and confusion. We need to spread out the points that represent the different values of the fit "sidewise." As it happens, we can gain simplicity of picture at the same time. The only thing it really costs us is a very transient discomfort--one direction in our pictures will not have any meaning that we need (or want) to understand. We must learn to forget one direction in such pictures--something that is really quite easy to do.

#### the two-way plot

Exhibit 7 shows us the full picture corresponding to panel B of exhibit 4. We now have three columns--Flagstaff, Phoenix, and Yuma--and seven

**exhibit 7** of chapter 10: Arizona temperatures

**Another row-PLUS-column analysis (based on panel B of exhibit 4); behavior of the fit**

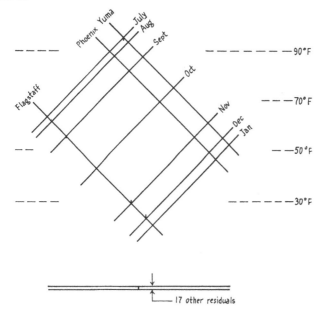

rows--the months from July to January. We can easily see that, at least so far as the fit goes, Yuma in January is warmer than Flagstaff in October--and that Flagstaff in July is close to Phoenix in October. The whole behavior of the fit is before us, set out clearly.

Even though this picture concentrates on the fit, we ought not to neglect the residuals. We need to know (1) something about how large the residuals are, generally, and (2) where and how big the largest ones are. At four of the intersections in exhibit 7--Flagstaff in September, November, and December, and Yuma in July--we have drawn short vertical lines. The lengths of these lines represent the sizes of the corresponding residuals, so that the heights of their far ends represent, not the fit, but

data = fit PLUS residual.

We have shown only a few residuals as explicitly as this for two reasons:

◇ if we try to show more than a few residuals, the picture tends to become so confused at to lose value instead of gaining it.

◇ in the case at hand, even the larger residuals are so small that the vertical lines we show are about as short as we can appreciate. Adding shorter lines would not help us.

Instead of showing the other 17 residuals individually, we have discharged our obligation to them by showing the stripe at the bottom of exhibit 7--and pointing out that all 17 starting at some center line (we would show the center time if the residuals were even a little larger) will be ended within the outer lines.

Notice also how the one-scale-forget-the-other character of this--or any other--two-way plot is emphasized by a "successive water-line" picturing of the vertical scale. This means dashed horizontal lines on both sides of the plot and **no** formal scale in either direction.

We will learn in the next chapter how we can make such a picture, which we will call

a two-way plot

for any rows-PLUS-columns fit.

**TIMES fits**

And what if we want a picture of a row-TIMES-column fit? There are two answers:

◇ make a picture of the corresponding row-PLUS-column fit to the logs-- you can put either both scales or only the unlogged scale on it. (Doesn't work if the fit has negative values.)

◇ turn to the early sections of chapter 12.

Either or both may meet your need.

### looking at residuals

We also want to look at residuals--especially after the largest residuals have been shown life-size, as in exhibit 7, however; only the rough size of each residual, rather than its precise value, concerns us. It will be enough to mark each one well enough to reveal which of the following it is:

**very** positive (#) or (⊞)
**large** positive (+)
**medium** positive (+)
**small** (either sign) (·)
**medium** negative (∘)
**large** negative (o)
**very** negative (⊚)

If our table is even moderately large, rules of thumb can help us choose what sizes of residual get which mark. In our present example, with 12 of the 21 residuals zero, we can do better by using judgment. Accordingly, we will call the residual of +2.1 very (positive)--those of −1.0, 1.0, and 1.1 large-- and the other 5 not zero medium.

The single most useful way to look at the residuals is to plot these coded characters "on top" of the intersections of the plot of fits. This leaves them "labelled" in many useful ways, including:

⋄ by the names of their row and column.
⋄ by the (rough) numerical values of their row and column.
⋄ by the (rough) value of the corresponding fit.
⋄ and by combinations of such.

This diversity of possible labels corresponds to a diversity of ways in which we can approach these coded residuals, thus making it easier for us to notice many of the kinds of structure that residuals not too infrequently show. Exhibit **8** shows what happens for the analysis of panel B of exhibit 4. We now can see (1) an apparent excess of negative residuals in July and August and (2) a clear tendency for Flagstaff to be cooler in summer and warmer in winter than the fit indicates. (We can see each of these if we study panel B of exhibit 4 carefully, but exhibit 8 thrusts them on our attention. We should not rely on careful examination, if we can avoid the need for it.)

All that we have left to the next chapter about making such a

### two-way plot of residuals

(given the two-way plot of fit) is the standard choice of which residuals to show with which symbol. Such a standard choice is often convenient, but ought not to be allowed to override good judgment.

**other postponements**

We also leave to the next chapter:

◊ the possible condensation of residuals, as when we decide that January and February are so close as to be best treated together.

**review questions**

Do we want to look at row-PLUS-column fits? Why? Can we use pictures where one direction is important but the other is to be forgotten? What do two families of parallel lines do? Why does this help? To what example did we return? What did the picture show us? What about large residuals? Other residuals? Have we yet learned to make a two-way plot for a fit? Why/why not? What about pictures for TIMES fits?

Do we want to look further at residuals? Why? In how much detail? What coding do we use? Where do we plot them? What kinds of labeling are thereby provided? What have we postponed to the next chapter?

**10F. Fitting one more constant**

Not infrequently we want to try to do just a little more than fit row-PLUS-column. Experience suggests that the most frequently useful fit of just a little

exhibit **8** of chapter 10: Arizona temperatures

**Residuals of panel B of exhibit 4 marked to show size and direction**

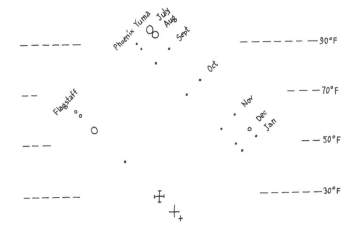

more is the fit of a constant times the product of the row effects and column effects already fitted. Two ways to write such a fit are

$$\text{all} + \text{row} + \text{col} + \frac{\text{row} \cdot \text{col}}{\text{div}}$$

and

$$\text{all} + \text{row} + \text{col} + k \cdot \frac{\text{row} \cdot \text{col}}{\text{all}}$$

where div (for "divisor") is a constant, as is $k$.

We will meet these last fractions

$$\frac{(\text{row})(\text{col})}{\text{all}}$$

often, almost always as a basis for some comparison, so we will call their values

**comparison values.**

One advantage of working with these values--rather than with the (row)(col) products alone--is that they have the same "dimension" as the effects. If we were to change from °F to °C in this example, all four kinds of quantities, "the fit", "row", "col", and "all", would have to be multiplied by 5/9 (after subtracting 32°F from both "the fit" and "all"). As a result, we could go on using the same numerical constant multiplying (row)(col)/all, which expression, like "the fit", would also have been multiplied by 5/9. (If we had chosen to work with "constant'(row)(col)", then, because (row)(col) would have been multiplied by $(5/9)^2$, we would have had to *divide* constant' by 5/9. Similar things, of course, happen when we change feet into inches--or make other changes of unit.

Exhibit **9** shows more monthly mean temperatures, this time warming on the East Coast (instead of cooling in Arizona), together with a row-PLUS-column analysis. Panel C of exhibit 9 sketches, in tabular form, what a two-way plot of the residuals would show, namely:

⋄whenever row effect and column effect are of the SAME sign, the corresponding residual is either small or NEGATIVE.

⋄whenever row effect and column effect are of OPPOSITE signs, the corresponding residual is either small or POSITIVE.

Having made the row-PLUS-column fit--and looked below it--we have seen a very clear and prominent behavior that is not yet described. It is up to us to go further--to do something more in this obvious case and to learn how to do something that we can try when matters are less obvious.

The simplest combination of row effect and column effect that behaves one way when the signs are the same--and the other when they are different--is the product

(row effect) TIMES (column effect).

**exhibit 9 of chapter 10: East Coast**

**Warming up on the East Coast**

**A)** The DATA--monthly mean temperatures in °F

|  | Laredo | Washington | Caribou |
|---|---|---|---|
| Jan | 57.6 | 36.2 | 8.7 |
| Feb | 61.9 | 37.1 | 9.8 |
| Mar | 68.4 | 45.3 | 21.7 |
| Apr | 75.9 | 54.4 | 34.7 |
| May | 81.2 | 64.7 | 48.5 |
| June | 85.8 | 73.4 | 58.4 |
| July | 87.7 | 77.3 | 64.0 |

**B)** ONE row-PLUS-column ANALYSIS of A--in 0.1°F

|  |  |  |  | eff |
|---|---|---|---|---|
| Jan | 0 | 1 | −77 | −183 |
| Feb | 33 | 0 | −76 | −173 |
| Mar | 16 | 0 | −39 | −91 |
| Apr | 0 | 0 | 0 | 0 |
| May | −50 | 0 | 35 | 103 |
| June | −91 | 0 | 47 | 190 |
| July | −111 | 0 | 64 | 229 |
|  | 215 | 0 | −193 | 544 |

**C)** The EFFECTS and RESIDUALS of B in CODED FORM

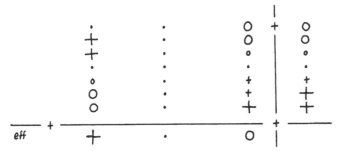

**D)** The COMPARISON VALUES

| Laredo | Washington | Caribou |
|---|---|---|
| −72 | 0 | 65 |
| −68 | 0 | 32 |
| −36 | 0 | 32 |
| 0 | 0 | 0 |
| 41 | 0 | −37 |
| 75 | 0 | −67 |
| 91 | 0 | −81 |

Examples: −72 = (215)(−183)/544 = (215/544)(−183),
 −68 = (215/544)(−171)

➤

Dividing every product by "all" will not change this. So our comparison values are appropriate tools or pointers.

Panel D sets out the comparison values, which clearly behave in the same general way as the residuals except for (1) a change of sign, (2) a possible difference in scale.

Clearly we ought to look at the possibility of extending our basic fit of

$$\text{common} + \text{row effect} + \text{column effect}$$

which we shall usually write as

$$\text{all} + \text{row} + \text{col}$$

to the form

$$\text{all} + \text{row} + \text{col} + k\frac{(\text{row})(\text{col})}{\text{all}}$$

where $k$ is a constant to be determined.

The natural way to look for a value of $k$ is to plot

$$\text{basic residual} = \text{data} \quad \text{MINUS} \quad (\text{all} + \text{row} + \text{col})$$

against

$$\text{comparison value} = \frac{(\text{row})(\text{col})}{\text{all}}.$$

Exhibit 10 does just this. The result is called a

$$\text{diagnostic plot.}$$

We see a very clear and strong dependence of basic residual on comparison value--with a suggestion of curvature, the negative slope appearing less pronounced on the left of zero than on the right. Trial with a transparent ruler or triangle shows us that

$$k = -1$$

exhibit **9** of chapter 10 (continued)

## P) PROBLEMS--longish and hardish

9a) Make up a table for which the residuals tend to have the same sign as the comparison values, and imitate all panels above.

9b) Do the same for a table in which the residuals are roughly −2 times the comparison values.

9c) Do the same for a table in which the residuals are roughly +3 times the comparison values.

9d) Find a set of real data, not used in this book, where the residuals are clearly related to the comparison values, and imitate all panels above.

## 356 / exhibit 10/10: Two-way analyses

is not a bad compromise. Accordingly, exhibit **11** shows the residuals after including

$$-1.0 \frac{(\text{row})(\text{col})}{\text{all}}$$

in the fit. Panel B of exhibit 11 shows how much the residuals have been drawn together by adding just one more constant to our fit.

Clearly, we have gained a surprising amount by going to an

$$\text{all-PLUS-row-PLUS-col-PLUS (constant)} \frac{(\text{row})(\text{col})}{\text{all}}$$

fit, which we will abbreviate as a

$$\text{row-PLUS-col-PLUS-one fit.}$$

It is safe to use this abbreviation, since no other way of adding a single constant to a row-PLUS-col fit is known that is nearly as often nearly as effective as the one we are learning about here.

exhibit **10** of chapter 10: East Coast

**Plotting basic residuals against comparison values (data from exhibit 9)**

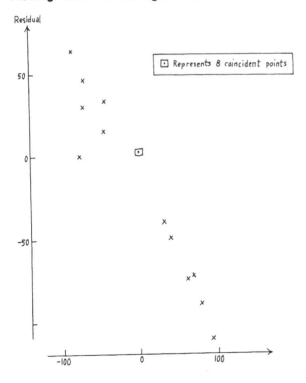

exhibit **11** of chapter 10: East Coast

**The analysis with one additional constant (monthly mean temperatures in 0.1°F, data in exhibit)**

A) The ANALYSIS

|  | |residuals| | | |values of −(row)(col)/544| | | |
|---|---|---|---|---|---|---|---|
|  | Laredo | Wash'n | Caribou | Laredo | Wash'n | Caribou | eff |
| Jan | −72 | 1 | −12 | 72 | 0 | −65 | −183 |
| Feb | −35 | 0 | −15 | 68 | 0 | −61 | −173 |
| Mar | −20 | 0 | −7 | 36 | 0 | −32 | −91 |
| Apr | 0 | 0 | 0 | 0 | 0 | 0 | 0 |
| May | 9 | 0 | −2 | 0 | 0 | 37 | 103 |
| June | −16 | 0 | −20 | −75 | 0 | 67 | 190 |
| July | −20 | 0 | −17 | −91 | 0 | 81 | 229 |
| eff | 215 | 0 | −193 |  |  |  | 544 |

Notes
1. The entries in the new panel are −1.0 times the comparison values found from the row, column, and common effects given. (Values in panel D of exhibit 9, for instance.)
2. The "PLUS-one" block can be omitted, and only the formula for its values given. Doing this saves space but decreases the viewer's understanding. Since there will almost always be room to put such a block above or below (if not to the left or right) of the residual block, it should almost always appear.

B) The RESIDUALS as a BATCH--before and after--rounded to 10's

```
  (B of 9)          (A of 11)
 0+|                    |
  s|6                   |
  f|5                   |
  t|323                 |
 0*|00100            0*|000000
-0*|0000            -0*|0111000
  t|4                t|222222
  f|5                f|4
  s|                 s|7
-0+|988
-1*|1
```

P) PROBLEMS

11a1) Look at the residuals of panel A. Can you see any structure? What seems to be going on?

11a2) (Teaser) Can you suggest a further fit that might help?

### review questions

If we want to fit just a little more--after a row-PLUS-column fit--what do we consider fitting? What is a comparison value? Why is it more convenient than just row effect times column effect? What example did we choose? What pattern did the row-PLUS-column residuals show? How was this related to the comparison values? How did we look? What is a diagnostic plot? What value of $k$ did we choose? Did this improve the analysis?

## 10G. Converting PLUS to TIMES; re-expression

We have just seen--in the previous section--an example where the natural PLUS-one fit seems to be close to
$$\text{all} + \text{row} + \text{col} - \frac{(\text{row})(\text{col})}{\text{all}}.$$
Much more frequent is the case where the natural PLUS-one fit seems to be close to
$$\text{all} + \text{row} + \text{col} + \frac{(\text{row})(\text{col})}{\text{all}}$$
where the additional term has the same size but the opposite sign. A hard look at this formula--and a pinch of elementary algebra--shows us that it is identical in value with
$$\text{all}\left(1 + \frac{\text{row}}{\text{all}}\right)\left(1 + \frac{\text{col}}{\text{all}}\right),$$
as any of us can easily verify by multiplying out the latter form.

Thus, PLUS-one fits with coefficient +1.0 are just another way to find row-TIMES-column fits.

### consumption again

It is now natural to return to the example of exhibit 6, where we skipped over a row-PLUS-column fit because it appeared so foolish. Exhibit **12** begins by setting out a row-PLUS-column analysis for this data--one in which the residuals are, of course, catastrophically large and systematic. Every reader should do the "everyone" problems (panel E), which will make him think hard. (*Hint.* compare the factors of increase from 1940 to 1960 for the various items of expenditure.)

A row-TIMES-column fit for a response--provided all row effects and all column effects are positive--is the same thing as a row-PLUS-column fit for the log of the response. (The log of a product is the sum of the logs, so we need only take logs of both sides.) Accordingly, finding a coefficient near +1.0 for a PLUS-one fit suggests, as one alternative, replacing the original reponse by its logarithm (unless there are negative values).

### there could be others

And we would not be surprised if other coefficients suggested other re-expressions, presumably--at least crudely--in accordance with the scale we learned about in chapter 3, namely

$$-1 \to \text{square}$$
$$-0.5 \to \text{take 3/2 power}$$
$$0 \to \text{leave alone}$$
$$0.5 \to \text{take square root}$$
$$1 \to \text{take log}$$
$$1.5 \to \text{take reciprocal square root}$$
$$2 \to \text{take reciprocal.}$$

exhibit **12** of chapter 10: personal consumption
**The data of exhibit 5 examined anew.**

**A)** A row-PLUS-column ANALYSIS of the data of **panel** A of **exhibit 5**

|  | 1940 | 1945 | 1950 | 1955 | 1960 | fit |
|---|---|---|---|---|---|---|
| Food | −31.2 | −11.1 | 0.0 | 9.3 | 15.8 | 59.6 |
| Household | −12.3 | −9.5 | 0.0 | 3.2 | 5.8 | 29.0 |
| Medical | .00 | .00 | .00 | .00 | .00 | 9.71 |
| Personal | 4.77 | 3.48 | .00 | −3.34 | −8.44 | 2.45 |
| Private | 5.02 | 3.12 | .00 | −3.49 | −9.55 | 1.80 |
| eff | −6.18 | −3.95 | .00 | 4.29 | 11.39 | |

*Note:* This is the analysis that was clearly so foolish that we skipped over it in exhibit 6 above

**E) PROBLEMS THAT EVERY READER SHOULD DO**

12a) Convert panel A into an analysis in terms of effects.

12b) Construct the table of comparison values.

12c) Plot the basic residuals against the comparison values. This plot indicates what value or values of A?

12d) Find the residuals for the corresponding row-PLUS-column-PLUS-one fit or fits.

12e) Find the residuals for the row-PLUS-column-PLUS-one fit with coefficient = +1.0. Discuss what you have found--do these residuals make sense? Why or why not?

12f) Write the analysis with coefficient = +1.0 in the rows-TIMES-columns form. Compare with the result of completing panel G of exhibit 6.

**P) FURTHER PROBLEMS**

12g) Take the row-TIMES-column analysis obtained by completing panel G of exhibit 6 and write it as a row-PLUS-column-PLUS-one analysis.

12deg) Compare the results of (12g) with those of A, (12d), and (12e) above.

This is, indeed, a pretty good guess--good enough to be a quite helpful start when we do need something other than raw values or log values.

**review questions**

What kind of PLUS-one fit is fairly common? What is "all + row + col + (row)(col)/all" equivalent to? What is a row-PLUS-column-PLUS-one fit with $k = +1.0$ equivalent to? Can we convert PLUS to TIMES? How do logs let us go back to PLUS? Can we interpret other values of $k$? How? How often do you expect to want to try?

### 10H. How far have we come?

This chapter is our introduction to the richness of analysis possible when our data have two or more handles. All the techniques in earlier chapters used at most one handle, beyond the spread inherent in all data. Even middling and delineation used only the values of $x$ to do something about the values of $y$. In this chapter we have dealt with the simplest case of one response and two families of circumstances, where all or nearly all combinations of a circumstance of one family with a circumstance of the other is represented by a response. This chapter needs to be paired with chapter 11, which follows, to give us a working set of tools.

We are now ready:

◇ to use double lines to separate blocks of values that fit together by addition--a technique that we can often use to our advantage in the calculations of chapters 7, 8 and 9.

◇ to check that a fit is really of the form row-PLUS-column by differencing both by rows and by columns, and then checking the differences for constancy in the other direction.

◇ to emphasize exact zeros without confusion. In addition to using "z" we are now prepared to use "$\sqrt{}$" and to line up unadorned zeros with nearby decimal points.

◇ to use triple lines to separate--and thus tie together--blocks of numbers that are to be combined by multiplication.

We are also ready:

◇ to look at row-PLUS-column analyses, given as a table, in terms of either effects or fits.

◇ to look at forget-one pictures (preferably with water-line scales) and focus our attention on (a) vertical position and (b) interlinking of points, giving no explicit attention to horizontal position.

◇ to look at pictures of coded residuals.

⋄to look at any row-TIMES-column fit, given as a table, and to wonder whether taking logs and reducing the table to a table for a row-PLUS-column fit to the log of the response is a good idea.

⋄to find comparison values, one for each cell, as row-effect TIMES column-effect *divided by* common term, and to plot residuals against comparison values in diagnostic plots.

⋄to fit a tilt to a diagnostic plot, thus converting our basic row-PLUS-column fit into a PLUS-one fit of the form

$$\text{all} + \text{row} + \text{col} + k \cdot \frac{(\text{row})(\text{col})}{(\text{all})}.$$

⋄to convert PLUS-one fits with a constant $k$ near $+1.0$ into TIMES fits.

⋄when a PLUS-one fit helps, to add an extra block to our table of results to show what has been fitted.

We now recognize:

⋄that there can be two-way residuals, and that finding them can be very useful.

⋄the nature and flexibility of two-way fits.

⋄the words "row", "column", "stripe" (for either row or column), "row effect", "column effect", "column fit", "row fit", and "common".

⋄that, besides row-PLUS-column analyses, we can have row-TIMES-column analyses and row-PLUS-column-PLUS-one analyses.

We find ourselves:

⋄well started on taking two-way tables apart.

⋄taking things apart into more than two terms, so as to learn much more about it.

We are still looking for better and more thorough ways:

⋄to find incomplete descriptions.

⋄to lift incomplete descriptions out of the way so we can see what lies under them.

Those that we have newly found are even more helpful, when they can be applied, than those we learned earlier.

# index for chapter 11

review questions 363

**11A. Taking medians out** 363
controlling errors 364
subtraction 365
back to the example 366
a rule of thumb 366
*four steps* 366
the fit 366
review questions 369

**11B. Alternative organizations of the arithmetic** 372
another pattern 372
review questions 374

**11C. Making the core of a two-way plot** 374
review questions 376

**11D. Going on with the residuals** 378
the residuals 378
review questions 382

**11E. Coding residuals; condensing fits and residuals** 382
condensing 384
review questions 387

**11F. We can combine!** 390
presidential Connecticut 390
a further opportunity 394
review questions 396

**11G. Guidance for expression** 396
amounts and counts 397
*diagnostic plot* 398
balances 398
counted fractions 398
a helpful thought 398
review questions 399

**11H. How far have we come?** 399

| EXHIBITS | PAGES |
|---|---|
| 11A | |
| 1★ | 364 |
| 2★ | 367 |
| 3★ | 368 |
| 4★ | 369 |
| 5★ | 370 |
| 11B | |
| 6★ | 372 |
| 11C | |
| 7★ | 375 |
| 8 | 377 |
| 9 | 378 |
| 11D | |
| 10★ | 379 |
| 11 | 380 |
| 12 | 381 |
| 13★ | 381 |
| 11E | |
| 14 | 383 |
| 15 | 384 |
| 16 | 385 |
| 17★ | 385 |
| 18★ | 386 |
| 19 | 387 |
| 20 | 388 |
| 21★ | 389 |
| 11F | |
| 22 | 390 |
| 23 | 391 |
| 24★ | 393 |
| 25 | 394 |
| 26 | 395 |
| 27★ | 396 |
| 11G | |
| 11H | |

# Making two-way analyses  11

As we pointed out in the opening to the last chapter (10, that is), these two chapters are side-to-side rather than end-to-end. (Thus, section 11A will have much more to do with section 10A than it will with 10E, 10F, and 10G, for example.) In this chapter we will usually concentrate on:

◇ the mechanisms of making our analyses, with some attention to
◇ the verbal arithmetic of what is going on.

As always, let us emphasize that we are doing exploratory data analysis, so that the techniques we illustrate are chosen to combine:

◇ relative ease and simplicity of use.
◇ adequate effectiveness (which may still leave them quite a way from the most stringent methods).
◇ consistency with the methods already learned for use in other situations.

We assert that the results these methods give are quite "good" enough for exploration--which need not mean that they are well chosen to settle confirmatory questions.

**review questions**

How is this chapter related to the last? What properties have we sought to combine in our techniques? Are these techniques being advertised for use in settling confirmatory questions?

## 11A. Taking medians out

We opened the discussion of two-way tables--at the beginning of the previous chapter (exhibit 1 of chapter 10)--by taking medians out first by column (by place) and then by row (by month) and then stopping. This gave us--in that example--a quite respectable row-PLUS-column analysis. It is too much to expect that things will always be quite so easy.

Exhibit **1** starts with counts of numbers of an insect, *Leptinotarsa decemlineata*, each the sum of the counts for two plots treated alike, for all combinations of 4 treatments and 6 areas of the field chosen to be likely to be relatively homogeneous. (Agricultural experimenters call these areas "blocks", but we need this word for blocks of figures in this section.)

Panel A shows the counts and, in parentheses, treatment (row) medians. Panel B adopts the single-line–double-line convention of the last chapter, taking out the row medians and stating the column medians of what is left. It also drops the headings and stub for the duration of the calculation.

### controlling errors

Panel C carries this two more steps. It also introduces an important practical device--circling the negative values. It is well known to those who do hand arithmetic (or hand algebra) that minus signs are not obtrusive enough. In the writer's experience, more errors in simple hand arithmetic can be traced to misreading--or insufficiently noticing--a sign than to all other causes com-

exhibit **1** of chapter 11: insect counts

**Repeated median removal, illustrated for numbers of *Leptinotarsa decemlineata* for two plots combined**

**A)** The DATA--and a beginning

| | Area | | | | | | |
|---|---|---|---|---|---|---|---|
| Treat | 1 | 2 | 3 | 4 | 5 | 6 | (median) |
| 1 | 492 | 410 | 475 | 895 | 401 | 330 | (442) |
| 2 | 111 | 67 | 233 | 218 | 28 | 18 | (89) |
| 3 | 58 | 267 | 283 | 279 | 392 | 141 | (273) |
| 4 | 4 | 1 | 53 | 14 | 138 | 11 | (12) |

**B)** ONE FULL STEP--and more medians

| | | | | | | | | | | | | |
|---|---|---|---|---|---|---|---|---|---|---|---|---|
| 492 | 410 | 475 | 895 | 401 | 330 | 442 | 50 | −32 | 33 | 453 | −41 | −112 |
| 111 | 67 | 233 | 218 | 28 | 18 | 89 | 22 | −22 | 144 | 129 | −61 | −71 |
| 58 | 267 | 283 | 279 | 392 | 141 | 273 | −215 | −6 | 10 | 6 | 119 | −132 |
| 4 | 1 | 53 | 14 | 138 | 11 | 12 | −8 | −11 | 41 | 2 | 126 | −1 |
| | | | | | | | 7 | −16 | 92 | 68 | 39 | −92 |

Example: 50 = 492 − 442

**C)** Taken TWO MORE STEPS--and negatives circled

| | | | | | | | | | | | | |
|---|---|---|---|---|---|---|---|---|---|---|---|---|
| 492 | 410 | 475 | 895 | 401 | 330 | 492 | 50 | (−32) | 33 | 453 | (−41) | (−112) |
| 111 | 67 | 233 | 218 | 28 | 18 | 89 | 22 | (−22) | 144 | 129 | (−61) | (−71) |
| 58 | 267 | 283 | 279 | 392 | 141 | 273 | (−215) | (−6) | 10 | 6 | 119 | (−132) |
| 4 | 1 | 53 | 14 | 138 | 11 | 12 | (−8) | (−11) | 41 | 2 | 126 | (−1) |
| | | | | | | | 7 | (−16) | 37 | 68 | 39 | (−92) |
| 53 | (−6) | 6 | 395 | (−70) | (−10) | (−10) | 43 | (−16) | (−4) | 385 | (−80) | (−20) |
| (−3) | (−24) | 89 | 43 | (−118) | 3 | 18 | 15 | (−6) | 107 | 61 | (−100) | 21 |
| (−189) | 43 | 6 | (−29) | 113 | (−7) | (−33) | (−222) | 10 | (−27) | (−62) | 80 | (−40) |
| (−19) | 1 | 0 | (−70) | 83 | 87 | 4 | (−15) | 5 | 4 | (−66) | 87 | 91 |
| −11 | −2 | 6 | 7 | 6 | −2 | | | | | | | |

→

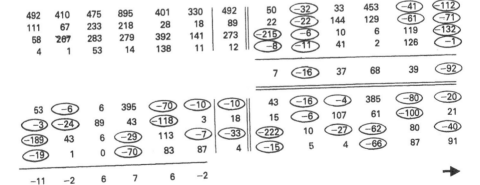

bined. Especially in the sort of calculations done in this section, it is often helpful to make the signs much more noticeable. The simple way to do it is to circle negative values, as we have done in panel C.

The careful reader will have noticed that the median in the third column of the second block of panel B is an error--92 = $\frac{1}{2}$(41 + 144) having been put where 37 = $\frac{1}{2}$(33 + 41) belongs. In panel C, when 92 was taken out, the result was

and since the three negative values were so noticeable, the error was detected and corrected. We must expect arithmetic errors when we do hand calculation. We need to arrange to (i) make fewer errors and (ii) catch the effects of most that we make, but it would be nonsense to have to work so carefully as not to make any.

Of course, if we can have a computer do our arithmetic for us, we can diminish errors greatly, as well as saving much effort. If the program has been carefully checked out--not always a simple task--we have only to worry about errors in getting the numbers in--and about the very infrequent printing error or machine breakdown.

### subtraction

With the circle-negatives rule, subtraction can be much safer. As we go from block 2(NE) to block 3(SE) of panel C, we need to take out 7 from the first column. This means lowering each value by 7, a process which sounds easy. The most difficult aspect of lowering is that this makes the size of positive

exhibit **1** of chapter 11 (continued)

**E) EXERCISES to be done by ALL READERS**

The logs (in 0.01) of the counts in panel A are:

| 269 | 261 | 268 | 295 | 260 | 252 |
| 204 | 183 | 237 | 234 | 145 | 126 |
| 176 | 243 | 245 | 245 | 259 | 215 |
| 60  | 0   | 172 | 115 | 214 | 104 |

1a) Imitate panel B, starting from the log counts.

1a2) Imitate panel C.

**S) SOURCE**

Geoffrey Beall, 1942, "The transformation of data from entomological field experiments so that the analysis of variance becomes applicable," Biometrika 32, 243–262. Table 4 on page 245.

numbers smaller, but the size of negative numbers larger. Here again the circled negatives help us keep our arithmetic straighter.

### back to the example

At the end of panel C, we have taken 3 steps--and are ready to take a fourth. The largest size of a median taken out has been 442 in step 1, 92 in step 2, 33 in step 3, and will be 11 in step 4. It would appear progress is being made. But it looks as though we might do well to go a little farther. (Some may like to stop here.)

Exhibit **2** takes repeated median removal as far as it can go with this example (unless we give up taking medians to the nearest integer and start using fractions.) The largest median taken out at the successive steps is now 442, 92, 33, 11, 6, 2, 2, 2, 1, 1. In view of the sizes of the residuals remaining --much, much larger than 1 or 2--it is quite unlikely that we needed to go "to the end". We could have stopped earlier.

### a rule of thumb

It is often both convenient and sparing of effort to take

### four steps.

(Some speak of this as "four steps of median polish"; others prefer to say "two cycles", since in four steps we have taken out row and column medians twice each.) This is a reasonable guideline, not an immutable law. If we take four steps, we need to at least find the medians we would use in the next step, to convince ourselves--if this is right--that we can stop with four steps in our particular instance.

### the fit

Having the residuals and the starting values, we can find the fit. Exhibit 3 of chapter 10 has already illustrated finding and checking the fit, but exhibit **3** here does this again to show how the checking can be laid out so as to be easier and more effective. Panel A finds what we hope is the fit. Panel B starts to check it by looking at column-to-column differences in each row. Between the first and second columns, to begin with, we have $414 - 428 = -14$, $89 - 103 = -14$, $222 - 236 = -14$, and $-2 - 12 = -14$, so all four of these differences are $-14$. We write $(-14)$ at the top of the space between these columns, and put a $\sqrt{}$ on every row where the difference is $-14$. Similarly with (61) from column 2 to column 3. When we go from column 3 to column 4, we find three differences of 32, and one that isn't. So we make only 3 $\sqrt{}$'s. And so on. The last check column, headed (90), has the differences from the last column to the **first** column.

Panel C does the same for differences between adjacent rows, as well as adjacent columns. It also puts the first-MINUS-last checks both before the first and after the last.

exhibit **2** of chapter 11: insect counts

**Repeated median removal taken further (initial stages from exhibit 1)**

**A) The CALCULATION**

| 492 | 410 | 475 | 895 | 401 | 330 | 442 |   | 50 | (-32) | 33 | 453 | (-41) | (-112) |
|-----|-----|-----|-----|-----|-----|-----|---|----|-------|----|-----|-------|--------|
| 111 | 67 | 233 | 218 | 28 | 18 | 89 |   | 22 | (-22) | 144 | 129 | (-61) | (-71) |
| 58 | 267 | 283 | 279 | 392 | 141 | 273 |   | (-215) | (-6) | 10 | 6 | 119 | (-131) |
| 4 | 1 | 53 | 14 | 138 | 11 | 12 |   | (-6) | (-11) | 41 | 2 | 126 | (-1) |

|   |   |   |   |   |   |   |   | 7 | (-16) | 37 | 68 | 39 | (-92) |
|---|---|---|---|---|---|---|---|---|-------|----|----|----|-------|

| 53 | (-6) | 6 | 395 | (-70) | (-10) | (-10) | 43 | (-16) | (-4) | 385 | (-80) | (-20) |
|----|------|---|-----|-------|-------|-------|----|-------|------|-----|-------|-------|
| (-3) | (-24) | 89 | 43 | (-118) | 3 | 18 | 15 | (-6) | 107 | 61 | (-100) | 21 |
| (-189) | 43 | 6 | (-29) | 113 | (-6) | (-33) | (-222) | 10 | (-27) | (-62) | 80 | (-39) |
| (-19) | 1 | 0 | (-70) | 83 | 87 | 4 | (-15) | 5 | 4 | (-66) | 87 | 91 |

| (-11) | (-2) | 6 | 7 | 6 | (-2) |
|-------|------|---|---|---|------|

| 64 | (-4) | 0 | 388 | (-76) | (-8) | (-2) | 66 | (-2) | 2 | 390 | (-74) | (-6) |
|----|------|---|-----|-------|------|------|----|------|---|-----|-------|------|
| 8 | (-22) | 83 | 36 | (-124) | 5 | 6 | 2 | (-28) | 77 | 30 | (-130) | (-1) |
| (-178) | 45 | 0 | (-36) | 107 | (-4) | (-2) | (-176) | 47 | 2 | (-34) | 109 | (-2) |
| (-8) | 3 | (-6) | (-63) | 77 | 89 | (-2) | (-6) | 5 | (-4) | (-61) | 79 | 91 |

|   |   |   |   |   |   |   | (-2) | 2 | 2 | (-2) | 2 | (-2) |
|---|---|---|---|---|---|---|------|---|---|------|---|------|

| 70 | (-2) | 2 | 394 | (-74) | (-2) | -2 | 68 | (-4) | 0 | 392 | (-76) | (-4) |
|----|------|---|-----|-------|------|----|----|------|---|-----|-------|------|
| 2 | (-32) | 73 | 30 | (-134) | (-1) | 2 | 4 | (-30) | 75 | 32 | (-132) | 1 |
| (-174) | 45 | 0 | (-32) | 107 | 0 | ✓ | (-174) | 45 | 0 | (-32) | 107 | 0 |
| (-4) | 3 | (-6) | (-59) | 77 | 93 | ✓ | (-4) | 3 | (-6) | (-59) | 77 | 93 |

| (-1) | ✓ | 1 | (-1) | 2 | ✓ |
|------|---|---|------|---|---|

| 71 | (-2) | 1 | 395 | (-76) | (-2) | ✓ | 71 | (-2) | 1 | 395 | (-76) | (-2) |
|----|------|---|-----|-------|------|---|----|------|---|-----|-------|------|
| 3 | (-32) | 72 | 31 | (-136) | (-1) | 1 | 2 | (-33) | 71 | 30 | (-137) | (-2) |
| (-173) | 45 | (-1) | (-31) | 105 | 0 | ✓ | (-173) | 45 | (-1) | (-31) | 105 | 0 |
| (-3) | 3 | (-7) | (-58) | 75 | 93 | ✓ | (-3) | 3 | (-7) | (-58) | 75 | 93 |

|   |   |   |   |   |   |   | ✓ | ✓ | ✓ | ✓ | ✓ | -1 |
|---|---|---|---|---|---|---|---|---|---|---|---|----|

| ✓ | 71 | (-2) | 1 | 395 | (-76) | (-1) |
|---|----|------|---|-----|-------|------|
| ✓ | 2 | (-33) | 71 | 30 | (-137) | (-1) |
| ✓ | (-173) | 45 | (-1) | (-31) | 105 | 1 |
| ✓ | (-3) | 3 | (-7) | (-58) | 75 | 94 |

**P) PROBLEMS**

2a) Draw a picture of this exhibit on a separate sheet of paper, and then draw a line showing how the calculation flows from block to block.

2b) Find a set of data of interest to you of about the same size and do the corresponding calculations.

**exhibit 3/11: Making two-way analyses**

In panel C it is easy to see that most of the values are tied firmly together with √'s. The two exceptions, 77 in the bottom row and 137 in the last column, stand out clearly. What does it seem that these should be? We have

$$59 + 32 = 91 \qquad 285 - 139 = 146$$
$$61 - (-30) \doteq 91 \qquad 13 + 133 = 146$$
$$315 - 224 = 91 \qquad -78 - (-224) = 146$$
$$507 - 416 = 91 \qquad 236 - 90 = 146$$

**exhibit 3 of chapter 11: insect counts**

**Finding and checking the fit**

**A) FINDING--after four steps**

```
     starting from                              residuals
492  410  475  895  401  330        64   -4    0   388  -76  -8
111   67  233  218   28   18         8  -22   83    36  -124  5
 58  267  283  279  392  141      -178   45    0   -36  107   4
  4    1   53   14  138   11        -8    3   -6   -63   77  89

          fit(?)
428  414  475  507  477  338     Examples: 428 = 492 - 64
103   89  150  182  152   13               414 = 410 - (-4)
236  222  283  315  285  137
 12   -2   59   77   61  -78
```

**B) CHECKING--one way, differences in parentheses**

```
(-14)      (61)      (32)      (-30)     (-139)     (90)
428  √  414  √  475  √  507  √  477  √  338  √
103  √   89  √  150  √  182  √  152  √   13  √
236  √  222  √  283  √  315  √  285       137
 12  √   -2  √   59       77        61  √  -78  √
```

**C) CHECKING--the second way also, just to be sure**

```
          (90)      (-14)      (61)      (32)      (-30)     (-139)     (90)
(416)             √         √         √                 √              √
           √  428  √  414  √  475  √  507  √  477  √  338  √
(-325)            √         √         √         √        √              √
           √  103  √   89  √  150  √  182  √  152  √   13  √
(133)             √         √         √         √        √
              236  √  222  √  283  √  315  √  285       137
(-224)            √         √         √                 √
           √   12  √   -2  √   59        77        61  √  -78  √
(416)             √         √         √                                 √
```

**P) PROBLEMS**

**3a)** (uses (2b)) Check the final step of problem (2b).

**3b)** Find another analysis of interest to yourself, and check it.

Whichever way we approach these values, we find that 77 should have been 91 and that 137 should have been 146.

Turning back to panel A, we find that we had reached 77 as

$$14 - (-63) = 77$$

and 137 as

$$141 - (-4) = \text{(should have made)}\ 145.$$

(That minus sign again!)

If we are to have a possible fit, we need to change the residual of −63 to one of −77 and the residual of −4 to one of −5. We will then have

$$14 - (-77) = 91$$
$$141 - (-5) = 146$$

and we will be as in panel A of exhibit **4**.

Panel B of exhibit 4 shows how the block values for the fit can be broken down, first into a row of column effects, and then into a column of row effects and a common term.

Exhibit **5** offers us a variety of sets of data to play with.

### review questions

How did we find residuals at the start of the last chapter? What example did we begin with here? What did we do to it? How did we arrange the arithmetic? What previously introduced convention did we make use of? What new convention did we add? Can you draw a picture showing the order of calculation of the blocks of numbers in exhibit 2? Why is circling negative numbers useful? In general? In thinking of taking out −5? How many steps might we ordinarily take? How many cycles? How do we calculate values for the fit? How do we check the results? How do we take a table of row-PLUS-column values apart?

exhibit **4** of chapter 11: insect counts

**Breaking down the fit (numbers from exhibit 3, corrected as in text)**

### A) RESIDUALS and FIT

| | | residuals | | | | | | fit | | | |
|---|---|---|---|---|---|---|---|---|---|---|---|
| 64 | −4 | 0 | 388 | −76 | −8 | 428 | 414 | 475 | 507 | 477 | 338 |
| 8 | −22 | 83 | 36 | −124 | 5 | 103 | 89 | 150 | 182 | 152 | 13 |
| −178 | 45 | 0 | −36 | 107 | −5 | 236 | 222 | 283 | 315 | 285 | 146 |
| −8 | 3 | −6 | −77 | 77 | 89 | 12 | −2 | 59 | 91 | 61 | −78 |

→

exhibit 4 of chapter 11 (continued)

## B) BREAKING IT DOWN

Taking the median out of one row:

12  −2  59  91  61  −78 | 36 ‖ −24  −38  23  55  25  −114

Taking one of these results out of the corresponding column

428           452                              258
103  less −24 gives  127  with median 194 giving  −67
236           260                               66
 12            36                             −158

Result

|  194 ‖ | −24 | −38 | 23 | 55 | 25 | −114 |
|---|---|---|---|---|---|---|
| 258 ‖ | | | | | | |
| −67 ‖ | | | | | | |
|  66 ‖ | | | | | | |
| −158 ‖ | | | | | | |

## C) The row-PLUS-column ANALYSIS

| Treat | | 1 | 2 | 3 | 4 | 5 | 6 |
|---|---|---|---|---|---|---|---|
| | −194 ‖ | −24 | −38 | 23 | 55 | 25 | −114 |
| 1 | 258 | 64 | −4 | 0 | 388 | −76 | −8 |
| 2 | −67 | 8 | −22 | 83 | 36 | −124 | 5 |
| 3 | 66 | −178 | 45 | 0 | −36 | 107 | −5 |
| 4 | −158 | −8 | 3 | −6 | −77 | 77 | 89 |

## P) PROBLEMS

4a) (uses (2b)) Break the fit of problem (2b) down in two different ways.

4b) Find another fit of interest to yourself; break it down.


exhibit 5 of chapter 11: data and problems

## Problems on repeated medians

### A) WHEAT DRYING DATA--Ratio of dry to wet weights for wheat grain

| | | NITROGEN APPLIED | | |
|---|---|---|---|---|
| Block | None | Early | Middle | Late |
| 1 | .718 | .732 | .734 | .792 |
| 2 | .725 | .781 | .725 | .716 |
| 3 | .704 | 1.035 | .763 | .758 |
| 4 | .726 | .765 | .738 | .761 |

→

exhibit **5** of chapter 11 (continued)

**B) A SMALL NEBRASKA VOTING EXAMPLE**--percent Democratic of major party vote

| Mark | '20 | '28 | '36 | '44 | '52 | '60 | County |
|------|-----|-----|-----|-----|-----|-----|--------|
| D0 | 46.7 | 59.0 | 54.6 | 34.5 | 18.9 | 23.6 | Grant |
| D1 | 33.7 | 23.7 | 39.9 | 23.8 | 14.9 | 13.8 | Hooker |
| D2 | 40.4 | 27.2 | 50.5 | 38.8 | 19.7 | 25.1 | Thomas |
| D4 | 34.9 | 26.6 | 51.6 | 40.4 | 23.0 | 21.2 | Blaine |
| D5 | 25.4 | 15.1 | 43.3 | 27.2 | 17.7 | 22.5 | Loup |
| D6 | 29.2 | 16.6 | 48.4 | 31.3 | 19.5 | 22.6 | Garfield |
| D7 | 31.9 | 35.4 | 57.5 | 44.2 | 33.7 | 36.0 | Wheeler |

Note: 7 counties in a row from west to east just north of the center of Nebraska.

**C) YIELDS OF SUGAR BEET**--tons

|  | Fertilizer Applied | | |
|---|---|---|---|
| Block | None | PO$_4$ only | PO$_4$ NO$_3$ |
| 1 | 2.45 | 6.71 | 6.48 |
| 2 | 2.25 | 5.44 | 7.11 |
| 3 | 4.38 | 4.92 | 5.88 |
| 4 | 4.35 | 5.23 | 7.54 |
| 5 | 3.42 | 6.74 | 6.61 |
| 6 | 3.27 | 4.74 | 8.86 |

**D) BIOLOGICAL VALUE OF PROTEINS**--as calculated from rat growth

| Pair of rats | Biological value | |
|---|---|---|
|  | Raw | Roasted |
| 1 | 61 | 55 |
| 2 | 60 | 54 |
| 3 | 56 | 47 |
| 4 | 63 | 59 |
| 5 | 56 | 51 |
| 6 | 63 | 61 |
| 7 | 59 | 57 |
| 8 | 56 | 54 |
| 9 | 44 | 62 |
| 10 | 61 | 58 |

**P) PROBLEMS**

5A) Take medians of the data of panel A repeatedly.

5b) The same for panel B.

5c) The same for panel C.

5d) The same for panel D.

**S) SOURCES**

Data of panel A: W. G. Cochran 1947. "Some consequences when the assumptions for the analysis of variance are not satisfied," *Biometrics* 3: 22–38, especially Table I at page 28.

Data of panel B: Richard M. Seammon, 1965. *America At the Polls*. A Handbook of American Presidential Election Statistics, 1920–1964. Pages 77–78.

Data of panel C: G. W. Snedecor 1946. *Statistical Methods*, 4th edition, Iowa State College Press, page 332.

Data of panel D: H. H. Mitchell, W. Burrough, J. R. Beadles, 1936, "The significance and accuracy of biological value of proteins computed from nitrogen metabolism data." *Journal of Nutrition* **11**: 257–274.

## 11B. Alternative organizations of the arithmetic

We have fixed one reasonable route to a row-PLUS-column fit:

◇ first repeatedly remove medians (usually at least four steps) to find residuals.

◇ from initial values and residuals, find the fit, and check that it is row-PLUS-column, correcting residuals as necessary. (If these corrections leave the residuals too far askew, some further steps of median removal may be necessary.)

◇ decompose the block of values of fit into column effects, row effects, and common.

This route is fairly nearly self-checking; moreover, errors can usually be corrected for without knowing where or how they occurred. This is a good safe procedure, and relatively easy to carry out.

Of course it is not the only reasonable procedure. Instances of procedures with different aims will come before us when we reach section 11J (in Chapter 11$^+$). One instance of a procedure with the same goals and results--one differing only in how we do the bookkeeping--is worth noting here.

### another pattern

Exhibit **6** goes through the successive steps, with a full effects-only analysis at each stage. One can always compute this way--sometimes (as we will see when we reach section 13A) we find this the procedure of choice.

The general pattern is simple:

◇ the median to be removed is ADDED to the corresponding main effect and SUBTRACTED from all the corresponding residuals.

Clearly, if we do the arithmetic correctly, such "sweepings out" do not alter the sum of effect and residual.

---

exhibit **6** of chapter 11: insect counts

**Repeated median removal in another format (initial numbers from exhibit 2, medians in parentheses)**

A) START

| 0 | 0 | 0 | 0 | 0 | 0 | 0 | 0 | (0) |
|---|---|---|---|---|---|---|---|---|
| 0 | 492 | 410 | 475 | 895 | 401 | 330 | (442) |
| 0 | 111 | 67 | 233 | 218 | 28 | 18 | (89) |
| 0 | 58 | 267 | 283 | 279 | 392 | 141 | (273) |
| 0 | 4 | 1 | 53 | 14 | 138 | 11 | (12) |

→

exhibit 6 of chapter 11 (continued)

**B) AFTER STEP 1**

| 0 | 0 | 0 | 0 | 0 | 0 | 0 | Examples: |
|---|---|---|---|---|---|---|---|
| 442 | 50 | (-32) | 33 | 453 | (-41) | (-112) | 442 = 0 + 442 |
| 89 | 22 | (-22) | 144 | 129 | (-61) | (-71) | 50 = 492 − 542 |
| 273 | (-215) | −6 | 10 | 6 | 119 | (-131) | 89 = 0 + 89 |
| 12 | (-8) | (-11) | 41 | 2 | 126 | (-1) | 22 = 111 − 89 |
| (181) | (7) | (−16) | (37) | (68) | (39) | (−92) | |

**C) AFTER STEP 2**

| 181 | 7 | −16 | 37 | 68 | 39 | −92 | (22) |
|---|---|---|---|---|---|---|---|
| 261 | 43 | (-16) | (-4) | 385 | (-80) | (-20) | (−10) |
| −92 | 15 | (-6) | 107 | 61 | (-100) | 21 | (18) |
| 92 | (-222) | 10 | (-27) | (-62) | 80 | (-39) | (−33) |
| −169 | (-15) | 5 | 4 | (-66) | 87 | 91 | (4) |

Examples:
181 = 0 + 181
261 = 442 − 181
−92 = 89 − 181
... ... ...
7 = 0 + 7
43 = 50 − 7
15 = 22 − 7
... ... ...

**D) AFTER STEP 3**

| 203 | −15 | −38 | 15 | 46 | 17 | −114 | Examples: |
|---|---|---|---|---|---|---|---|
| 251 | 53 | (-6) | 6 | 395 | (-70) | (-10) | 203 = 181 + 22 |
| −74 | (-3) | (-24) | 89 | 43 | (-118) | 3 | −15 = 7 − 22 |
| 59 | (-189) | 43 | 6 | (-29) | 113 | (-6) | −38 = −16 − 22 |
| −165 | (-19) | 1 | 0 | (-70) | 83 | 87 | ... ... ... |
| (−8) | (−11) | (−2) | (6) | (7) | (6) | (−2) | −114 = −92 − 22 |
| | | | | | | | 251 = 261 − 10 |
| | | | | | | | 53 = 43 + 10 |
| | | | | | | | ... ... ... |

**E) AFTER STEP 4**

| 195 | −26 | −40 | 21 | 53 | 23 | −116 | (−2) |
|---|---|---|---|---|---|---|---|
| 259 | 64 | (-4) | 0 | 338 | (-76) | (-8) | (−2) |
| −66 | 8 | (-22) | 83 | 36 | (-124) | 5 | (6) |
| 67 | (-178) | 45 | 0 | (-36) | 107 | (-4) | (−2) |
| −157 | (-8) | 3 | (-6) | (-77) | 77 | 89 | (−2) |

Examples: (Make your own)

→

**review questions**

Describe our first overall process for making a row-PLUS-column analysis. What is another pattern of calculation?

### 11C. Making the core of a two-way plot

Making the kind of two-way plot of a fit we found helpful in the last chapter is surprisingly easy. All we have to do is to write the **fit** as a sum of two parts, as in any of these examples:

$$\text{all + col} \quad \text{PLUS} \quad \text{row}$$
$$\text{all + row} \quad \text{PLUS} \quad \text{col}$$
$$(\tfrac{2}{3}\text{ all + col} \quad \text{PLUS} \quad \tfrac{1}{3}\text{ all + row})$$

exhibit **6** of chapter 11 (continued)

**F) AFTER STEP 5**

| 193 | -24 | -38 | 23 | 55 | 25 | -114 |
|---|---|---|---|---|---|---|
| 257 | 66 | -2 | 2 | 390 | -74 | -6 |
| -60 | 2 | -28 | 77 | 30 | -130 | -1 |
| 65 | -176 | 47 | 2 | -34 | 109 | -2 |
| -159 | -6 | 5 | -4 | -75 | 79 | 91 |
|  | (-2) | (2) | (2) | (-2) | (2) | (-2) |

Examples: (Make your own)

**P) PROBLEMS**

6a) Write down the analysis after Step 6.

6a2) After Step 7.

6a3) After Step 8.

6a4) After Step 9.

6b) Write out 7 further examples of the changes made in panel B, choosing values that involve every row and every column.

6c) Do the same with 7 further values for panel C.

6d) And for panel D.

6e/f) Write out 10 examples, for values involving every row and every column-- including the effect rows and columns--for panel E/F, respectively.

such that one part depends only on the column and the other only upon the row.

We then take these parts as (rectangular) coordinates. When we do this, the resulting points--one for each combination of a row and a column--make up a rectangular pattern. More precisely, they are represented by all the intersections of one family of horizontal lines and one family of vertical lines.

Exhibit **7** shows what happens numerically for the fit of panel B of exhibit 4. Exhibit **8** shows the resulting picture. The upper panel shows four vertical lines, corresponding to the four treatments, and six horizontal lines, corresponding to the six areas on which the experiment was repeated. The lower panel also shows a few of the lines whose equations are

one part PLUS the other part EQUALS constant.

exhibit **7** of chapter 11: insect counts

**Breaking the fit in two parts (initial numbers from exhibit 4)**

A) The FIT--in terms of common and effects

|     | A | B | C | D | E | F |
|-----|---|---|---|---|---|---|
| 194 | −24 | −38 | 23 | 55 | 25 | −114 |

| 1 | 258 |
| 2 | −67 |
| 3 | 66 |
| 4 | −158 |

B) ONE BREAKING IN TWO

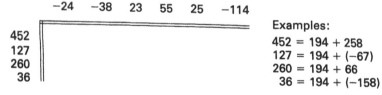

Examples:
452 = 194 + 258
127 = 194 + (−67)
260 = 194 + 66
36 = 194 + (−158)

C) ANOTHER

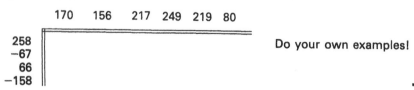

Do your own examples!

Since the fit is the sum of these parts, they are also lines

$$\text{sum} = \text{constant}$$

or

$$\text{fit} = \text{constant}.$$

We now have all the combinations of rows and columns plotted in such a way that the curves

$$\text{fit} = \text{constant}$$

are parallel straight lines.

To make a two-way plot instead, all we have to do is to twist the graph paper by 45° under the tracing paper before we make the tracing. Exhibit **9** shows the result. Here, as we will usually do, we have shown the level lines

$$\text{fit} = \text{constant}$$

as "water-levels" on both sides of the picture.

**review questions**

What is the first step in making a two-way plot? Do we have any choices? Why/why not? What do we plot on the graph paper? What does the main pattern look like? What do we add? How do we make a tracing? How do we show fit = constant?

exhibit **7** of chapter 11 (continued)

**D) YET ANOTHER**

|  | 70 | 56 | 117 | 149 | 119 | −20 |
|---|---|---|---|---|---|---|
| 358 |  |  |  |  |  |  |
| 33 |  |  |  |  |  |  |
| 166 |  |  |  |  |  |  |
| −58 |  |  |  |  |  |  |

Examples:
$358 = 258 + 100$
$33 = -67 + 100$
... ... ...
$70 = -24 + 94$
$56 = -38 + 94$
... ... ...

**P) PROBLEMS**

**7a)** Write out 3 further splittings into two parts.

**7b)** Imitate exhibit 8, starting from panel C. How does the result differ from exhibit 8?

**7c)** Same for panel D.

**7d)** (Uses (7a)) Same for one of the further splittings used as an answer to (7a).

exhibit **8** of chapter 11: insect counts

**Two stages in constructing a two-way plot of a fit (numbers from exhibit 7)**

**A)** The TWO PARTS plotted by LINES

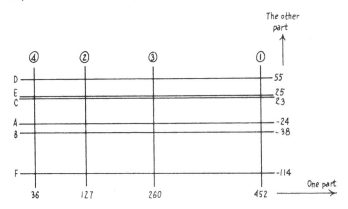

**B)** Some lines "ONE PART plus THE OTHER PART = CONSTANT" added

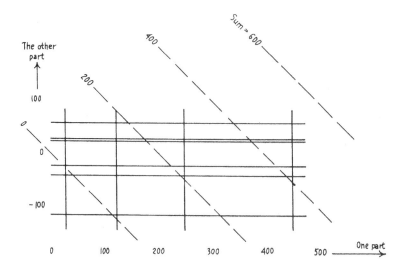

exhibit **9** of chapter 11: insect counts

**The bare two-way plot (numbers from exhibit 7)**

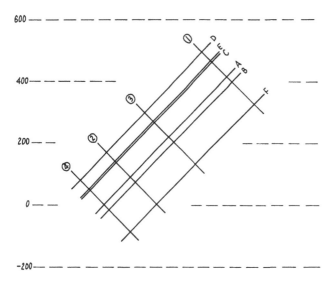

## 11D. Going on with the residuals

### the residuals

To complete this plot, we need to turn to the residuals. Exhibit **10** has stem-and-leaf and letter-value displays. We see that we have two outside negative and one far out positive. We now need to plot

$$\text{actual} = \text{fit} \quad \text{PLUS} \quad \text{residual}$$

for these three.

Just how we do this depends a little on how large the residuals are. Consider the residual of $-124$, which occurs at 2E, which is plotted as (127, 25). We want a point whose total is 124 lower. The easiest ones are (3, 25), found from $127 - 124 = 3$, and (127, $-99$), found from $25 - 124 = -99$. On the graph paper--before the 45° twist--these are directly to the left and directly below the fit. On the two-way plot, then, they will lie off at $\pm 45°$. We would rather go straight down, which we can do easily by subtracting $\frac{1}{2}(124) = 62$ from each original coordinate, finding (65, $-37$) from $127 - 62 = 65$ and $25 - 62 = -37$. Exhibit **11** shows how this looks on the graph paper.

For a big residual, like this one, it is easiest to begin by halving and applying one-half to each coordinate. For a little residual--as +5 would have been in this example--it is easier to go down each coordinate separately,

exhibit **10** of chapter 11: insect counts

**Summarization of residuals**

A) The RESIDUALS as they fall—rows and columns reordered

| Treat | D | E | C | A | B | F |
|---|---|---|---|---|---|---|
| 1 | 388 | −76 | 0 | 64 | −4 | −8 |
| 3 | −36 | 107 | 0 | −178 | 45 | −5 |
| 2 | 36 | −124 | 83 | 8 | −22 | 5 |
| 4 | −77 | 77 | −6 | −8 | 3 | 89 |

B) STEM-and-LEAF

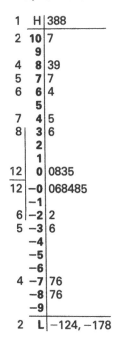

C) LETTER-VALUES

#24 residuals

| | M12h | 0 | |
|---|---|---|---|
| | H 6h | 54 | −15 |69
| | E 3h | 86 | −76 |
| | 2 | 107 | −124 |
| | 1 | 388 | −178 |

```
                  104
        f | 158   −119 |  out = −124, −178
          | xxx   two  |  far = 388
        F | 262   −223 |
          | one   xxx  |  adj: −77 and 107
```

D) For SPECIAL TREATMENT

| | Parts of fit | endpoints | actual |
|---|---|---|---|
| +388 at 1 D | (452, 55) → | (840, 55) or (452, 443) | or (646, 249) |
| −124 at 2 E | (127, 25) → | (3, 25) or (127, −99) | or (65, −37) |
| −178 at 3 A | (260, −24) → | (82, −24) or (260, −202) | or (171, −113) |

P) PROBLEMS

10a) (uses (5a)) Summarize the residuals with which you ended problem (5a).

10b) (uses (5b)) Same for (5b).

10c) (uses (5c)) Same for (5c).

10d) (uses (5d)) Same for (5d).

connecting the two points thus found by a line, and then--when making the tracing-paper version--drawing a line from the fit to the middle of this line.

Exhibit 11 also shows how the schematic plot for the other 21 residuals is made. We draw a diagonal line--at exactly 45°--for the center line. (It is easiest if this goes through the corners of the large squares.) Then we choose two or three convenient points along it, and step off--either horizontally or vertically--the adjacent and hinge values, here −76, −15, 54, and 107. We can then draw the ends of the box and the joins to the adjacent lines, and are ready to trace the result at the foot of the two-way plot.

Exhibit **12** is the result. We now see that the residuals are so large compared to what is described by the fit that the usefulness of this fit is close to being in serious doubt. (In part dealt with by one of the problems of exhibit **13**.)

exhibit **11** of chapter 11: insect counts

**Two constructions on the graph paper (numbers from exhibit 10)**

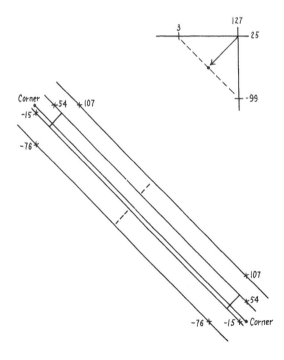

exhibit **12** of chapter 11: insect counts

**The completed two-way plot (see exhibits 9, 11, etc.)**

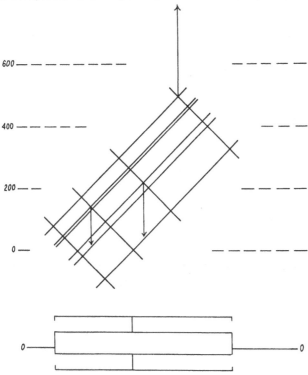

exhibit **13** of chapter 11: data and problems

**Problems on two-way plots**

A) AN ANALYSIS OF MEAN MONTHLY TEMPERATURES at SEVEN EAST COAST PLACES--in °F

|       | Laredo | Baton Rouge | Atlanta | Wash. | Boston | Portland | Caribou | eff   |
|-------|--------|-------------|---------|-------|--------|----------|---------|-------|
| Jan.  | 3.8    | 5.4         | 2.9     | 0.3   | 1.1    | −3.3     | −8.0    | −20.5 |
| Feb.  | 6.3    | 6.8         | 3.2     | −0.6  | −2.8   | −4.3     | −8.7    | −18.7 |
| Mar.  | 4.8    | 3.5         | 1.2     | −0.4  | −2.4   | −1.9     | −4.9    | −10.7 |
| Apr.  | 3.0    | 0.7         | 0.9     | −0.6  | −2.1   | −1.2     | −1.1    | −1.4  |
| May   | −1.0   | −1.8        | −0.1    | −0.4  | −0.8   | −0.1     | 3.4     | 7.9   |
| June  | −4.4   | −3.8        | −0.4    | 1.1   | 0.6    | 1.4      | 5.3     | 15.9  |
| July  | −6.2   | −6.0        | −2.3    | 1.3   | 1.9    | 3.7      | 7.2     | 19.6  |
| Aug.  | −5.0   | −5.3        | −2.1    | 0.5   | 2.3    | 3.4      | 6.1     | 18.5  |
| Sept. | −3.4   | −2.6        | −0.4    | 0.6   | 1.0    | 1.3      | 3.4     | 12.6  |
| Oct.  | −0.7   | −0.8        | −1.7    | −1.1  | 1.4    | 1.0      | 2.0     | 2.9   |
| Nov.  | −6.2   | −0.3        | −2.2    | −0.6  | 1.8    | 0.9      | 0.1     | −8.8  |
| Dec.  | 2.9    | 3.8         | 0.7     | −0.7  | −0.2   | −1.9     | −4.9    | −17.7 |
| eff   | 18.2   | 11.4        | 6.1     | 0.3   | −5.4   | −11.6    | −18.9   | 56.1  |

➔

#### review questions

What do we do with all residuals? Which do we show separately on the two-way plot? How do we find where to plot large residuals? What if those to be plotted are small? What kind of schematic plot do we make for the other residuals? How do we choose the spacings between lines? How do we plot them? What does the final two-way plot look like? In general? For our example? What does this lead us to think in the example?

### 11E. Coding residuals; condensing fits and residuals

For our two-way plots of residuals we use coding, using a standard set of seven symbols to indicate roughly the value of the residual. We have already seen--toward the end of section 10E--how we use these symbols, plotting them over an invisible two-way plot of the fit. We also saw how it can be quite reasonable to use judgment in deciding what range of values is converted into each symbol. We left to this chapter the statement of a standard coding--something to be used either when there is no urge to use judgment or as a start, as a basis from which judgment can begin.

Exhibit **14** sets out the standard cutting points, both generally and for the analysis we have been using as an example. In the example, we have 12 ·'s, as

---

exhibit **13** of chapter 11 (continued)

**P) PROBLEMS**

13a) Make the core of a two-way plot of the fit in panel A above.

13a2) Complete the two-way plot.

13b) (uses (5a)) Make the core of a two-way plot of the fit of problem (5a).

13b2) Complete the two-way plot.

13c/c2) (uses (5b)) Same for problem (5b).

13d/d2) (uses (5c)) Same for problem (5c).

13e/e2) (uses (5d)) Same for problem (5d).

**S) SOURCE--for panel A**

**Climatography of the United States, No.: #; Climatography of the states: ∗state∗; U.S. Weather Bureau**

| Place | Airport | # | ∗state∗ | Page |
|---|---|---|---|---|
| Laredo | Air Force Base | 60–41 | Texas | 17 |
| Baton Rouge | Ryan | 60–16 | Louisiana | 8 |
| Atlanta | Municipal | 60–9 | Georgia | 7 |
| Washington D.C.) | National | 60–44 | Maryland | 9 |
| Boston | Logan International | 60–19 | Massachusetts | 7 |
| Portland | City | 60–17 | Maine | 8 |
| Caribou | Municipal | 60–17 | Maine | 8 |

we might expect (12 = $\frac{1}{2}$(24) and the ·'s go from one hinge to the other). Of the six high, one is # and five are +. Of the six low, one is ⊙, one ○, and four ∘. These seem reasonable, as does the decoding into values set out in panel C.

exhibit **14** of chapter 11: general, and insect counts

**The standard cutting points, and their application to the residuals displayed in exhibit 10 (draws heavily on exhibit 10)**

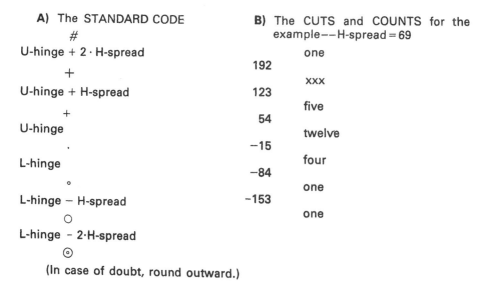

A) The STANDARD CODE
```
                        #
U-hinge + 2 · H-spread
                        +
U-hinge + H-spread
                        +
U-hinge
                        ·
L-hinge
                        ∘
L-hinge − H-spread
                        ○
L-hinge − 2·H-spread
                        ⊙
```
(In case of doubt, round outward.)

B) The CUTS and COUNTS for the example--H-spread = 69

```
            one
 192
            xxx
 123
            five
  54
            twelve
 −15
            four
 −84
            one
−153
            one
```

C) The ACTUAL RANGES used in the example

| code | values  |                | split in two |
|------|---------|----------------|--------------|
| #    | 388     |                | (194, 194)   |
| +    | xxx     |                |              |
| +    | 64 to 107 |              | (32, 32) to (53, 54) |
| ·    | −8 to 45 |               | (−4, −4) to (22, 23) |
| ∘    | −77 to −22 |             | (−38, −39) to (−11, −11) |
| ○    | −124    |                | (−62, −62)   |
| ⊙    | −178    |                | (−89, −89)   |

D) The CODED RESIDUALS arranged by ORDERED EFFECTS--as in panel A of exhibit 10

```
       D    E    C    A    B    F
  1    #    ∘    ·    +    ·    ·
  3    ∘    +    ·    ⊙    ·    ·
  2    ·    ○    +    ·    ∘    ·
  4    ∘    +    ·    ·    ·    +
```

Exhibit **15** shows the resulting plot. The scale at the right margin is constructed as in exhibit **16**. We again put half of each value in each coordinate. Exhibit **17** sets out some problems.

### condensing

When we look at exhibit 8, it is very hard to escape taking the distinction between area E, fitted at 25, and area C, fitted at 23, as useless and confusing. The distinction between A, fitted at −24 and B, fitted at −38, might be useful --so far as this exhibit goes. But when we look at the sizes of the residuals, we doubt the value of this also.

So far as the plot of fit goes, we gain by condensing E with C--and we can live with leaving A and B separate. The plot of coded residuals can usefully involve condensing both E with C and A with B. When we condense residuals, we replace the residuals involved by their median. (We do **not** recalculate residuals from the condensed fit, if any.) Exhibit **18** shows the residuals after condensation, and their coding. Exhibit **19** shows the two-way plot of residuals.

exhibit **15** of chapter 11: insect counts

**Plot of coded values (based on exhibit 14)**

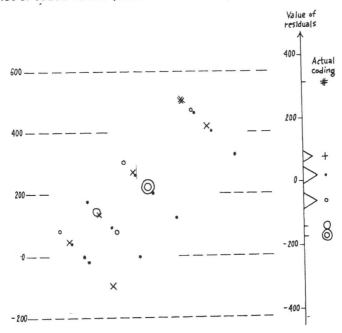

exhibit **16** of chapter 11: insect counts

**Construction of scale for exhibit 15**

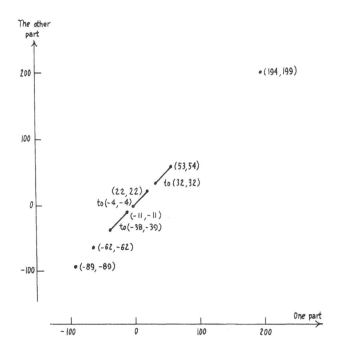

exhibit **17** of chapter 11: data and problems

**More problems on two-way plots**

A) AN ANALYSIS OF MEAN MONTHLY TEMPERATURES FOR FOUR ARIZONA AIRPORTS--in °F

|   | Flagstaff | Prescott | Phoenix | Yuma | eff |
|---|---|---|---|---|---|
| Jan | 0.3 | −0.3 | −0.1 | 0.2 | −19.6 |
| Feb | −0.5 | −0.5 | 0.2 | 0.7 | −15.3 |
| Mar | −0.4 | −0.6 | 0.3 | 0.8 | −9.8 |
| Apr | −0.3 | −0.1 | −0.1 | 0.3 | −2.2 |
| May | −0.3 | 0.0 | 0.4 | 0.0 | 5.7 |
| June | −0.4 | 0.8 | 0.4 | −1.0 | 14.1 |
| July | 0.1 | 0.3 | 0.1 | −0.6 | 20.5 |
| Aug | −0.1 | −0.2 | 0.0 | 0.1 | 18.9 |
| Sept | −0.7 | −0.1 | 0.2 | 0.5 | 13.1 |
| Oct | −0.1 | 0.2 | −0.2 | 0.1 | 1.6 |
| Nov | 1.2 | 0.4 | −1.0 | −0.5 | −10.0 |
| Dec | 0.9 | −0.3 | −0.2 | −0.5 | −17.1 |
| eff | −16.3 | −5.7 | 8.5 | 13.8 | 60.9 |

exhibit **17** of chapter 11 (continued)

**P) PROBLEMS**

17a) Stem-and-leaf the residuals of panel A, above, and code them.

17a2) Make a two-way plot of these coded residuals.

17b) Stem-and-leaf the residuals of panel A of exhibit 13 and code them.

17b2) (Should use (13a2)) Make a two-way plot of these coded residuals.

17c/17c2) (Should use (13b2)) Same for problem (5a).

17d/17d2) (Should use (13c2)) Same for problem (5b).

17e/17e2) Should use (13d2)) Same for problem (5c).

17f/17f2) (Should use (13e2)) Same for problem (5d).

**S) SOURCE**

Climatography of the United States No. 60–2. Climates of the states. Arizona. U.S. Weather Bureau, Washington, D.C., September 1959.

exhibit **18** of chapter 11: insect counts

**The condensed residuals for the analysis of exhibit 10**

A) The RESIDUALS

| Treat | D | EC | AB | F | #16 residuals | | | |
|---|---|---|---|---|---|---|---|---|
| 1 | 388 | −38 | 30 | −8 | M8h | −3h | | |
| 3 | −36 | 54 | −66 | −5 | H4h | 36 | −28h | 64h |
| 2 | 36 | −21 | −7 | 5 | E2h | 71 | −52 | |
| 4 | −77 | 36 | −2 | 89 | 1 | 388 | −77 | |

B) The CODING

| −157h | −93 | | −28h | | 36 | | 100h | | 165 | |
|---|---|---|---|---|---|---|---|---|---|---|
| ⊙ | ○ | | ○ | | · | | + | | + | # |
| xxx | xxx | | four | | seven | | four | | xxx | one |
| | | | −77 | | −21 | | 36 | | | 388 |
| | | | to | | to | | to | | | |
| | | | −36 | | 30 | | 89 | | | |

**P) PROBLEMS**

18a) Condense the analysis of panel A of exhibit 17.

18b) Condense the analysis of panel A of exhibit 13.

Exhibit **20** shows the bare-form condensed two-way plot of fit, where we have allowed the difference between E and C to be shown only in numbers (taking an opportunity by adding area effects and treatment fits). Note the pulling back of the "water-lines" to avoid both actual and indirect interference, with the plot and its surrounding tags.

Exhibit **21** offers some problems.

### review questions

How did we decide on coding in the previous chapter? Was that a reasonable thing to do? For what purposes can we use a standard code? What standard code do we use? Do we look to see what values get what code? Can we show this too? Do we? How do we plot the coded values? Should we consider condensing in either two-way plots of fit or two-way plots of residuals? Why/why not? In which plot(s)? What can we do to show condensation in the plot of fit? In the plot of residuals? How do we find condensed residuals?

exhibit **19** of chapter 11: insect counts

**Condensed two-way plot of residuals based on exhibit 18**

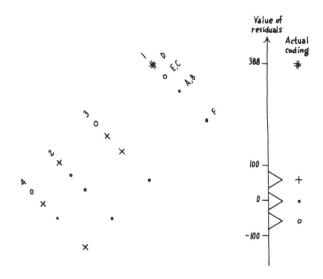

exhibit **20** of chapter 11: insect counts

## Condensed two-way plot

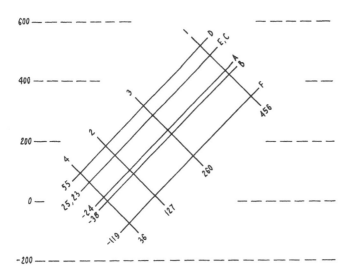

exhibit **21** of chapter 11: data and problems

## Problems on condensed analyses

**A) AN ANALYSIS OF A LARGER NEBRASKA PRESIDENTIAL VOTING EXAMPLE--percent Democratic among major parties**

| County location | '20 | '24 | '28 | '32 | '36 | '40 | Election '44 | '48 | '52 | '56 | '6) | 64 | eff |
|---|---|---|---|---|---|---|---|---|---|---|---|---|---|
| I2 | 3.2 | −0.3 | −0.5 | −1.6 | 4.2 | 3.6 | 1.3 | 0.8 | −1.1 | −2.6 | 1.6 | 5.7 | 1.6 |
| H3 | 0.1 | −2.1 | 0.0 | 2.8 | 2.6 | −0.4 | −4.0 | 0.3 | 1.3 | 2.3 | −1.8 | −2.6 | −4.8 |
| G4 | 2.4 | 3.1 | −2.0 | 2.4 | 1.3 | −0.1 | −1.6 | −0.6 | −1.1 | −2.0 | −1.2 | −0.7 | −5.0 |
| G6 | 0.8 | −1.2 | −3.5 | −1.6 | 1.3 | 1.1 | 2.3 | 1.3 | −2.3 | 0.1 | 2.1 | 0.4 | −1.3 |
| H7 | −4.2 | −2.9 | −3.4 | −5.1 | 3.1 | 0.1 | 1.2 | 1.6 | 2.7 | 2.0 | 2.2 | 2.4 | 0.2 |
| G7 | −6.1 | −4.3 | 0.6 | −1.6 | −2.0 | −0.8 | 0.2 | 0.4 | 3.8 | 2.3 | 1.7 | 5.4 | 0.4 |
| E7 | −3.1 | 4.2 | 3.9 | 1.4 | −1.7 | 0.7 | 1.9 | 1.7 | −2.7 | −2.0 | −2.7 | −1.6 | 17.1 |
| G8 | 3.3 | 3.3 | 2.8 | 0.2 | −3.2 | −1.2 | −3.2 | −1.0 | 1.9 | −1.6 | −1.5 | 0.6 | −4.3 |
| E8 | 1.9 | 0.7 | 1.2 | 2.4 | −2.9 | −2.1 | −0.4 | −2.6 | −0.4 | −0.3 | 1.2 | 1.5 | −3.3 |
| E9 | 2.2 | −0.7 | 1.0 | 1.0 | −3.2 | −0.4 | 2.0 | −2.4 | −2.5 | 11.1 | 0.6 | 1.4 | −0.6 |
| eff | −7.7 | −4.7 | −8.1 | 23.6 | 16.4 | −1.1 | −2.4 | 3.6 | −13.8 | −9.3 | −7.1 | 10.8 | 40.2 |

**P) PROBLEM**

21a) Condense the analysis of panel A above.

21b/c/d/e) (uses (5a/5b/5c/5d)) Condense the analysis of problems (5a/5b/5c/5d).

21f) Condense the analysis of panel B of exhibit 22 (following).

**S) SOURCE**

Richard M. Scammon, 1965, *America at the Polls. A Handbook of American Presidential Election Statistics, 1920–1964.* Pages 278–284.

## 11F. We can combine!

The last big idea we learned before starting on two-way tables was smoothing. What about combining the two?

### presidential Connecticut

Exhibit 22 shows the percent Democratic of the major party vote for President in each of the eight counties of Connecticut in every presidential election from 1920 to 1964. Panel B shows a row-PLUS-column analysis. Clearly this analysis has done a fair amount of good--the residuals are even smaller than the county effects, which are smaller than the election effects. (The H-spreads are, respectively, 16, 60 and 116. A useless analysis, of a table with so many rows and columns, would, by contrast, have effects noticeably smaller than residuals.) If we look hard, however, we see that not everything has been pulled out that might be.

When we look at the election effects, we see that they are far from irregular. If we were to smooth them simply, say with 3RSS (see Chapter 7), the standard breakdown

$$\text{input} = \text{smooth} \quad \text{PLUS} \quad \text{rough}$$

would give a lot of smooth. (In fact, the results would be:

$$\text{smooth} = -152, -152, -65, -3, 37, 37, 26, 3, -43, -43, 35, 200,$$

and

$$\text{rough} = 0, 19, 0, 0, 26, 0, 0, 0, -15, -77, 0, 0.)$$

exhibit 22 of chapter 11: presidential Connecticut

**The data, and the PLUS analysis (continued in exhibit 23)**

**A) The DATA**

| Election | County | | | | | | | |
|---|---|---|---|---|---|---|---|---|
| | Lich | Fairf | Mddx | Toll | NL | NH | Windh | Hartf |
| 1920 | 32.5 | 30.9 | 33.1 | 31.0 | 34.6 | 36.5 | 37.1 | 35.9 |
| 24 | 30.0 | 24.5 | 29.9 | 30.3 | 32.1 | 34.4 | 36.6 | 31.4 |
| 28 | 36.0 | 43.7 | 39.7 | 39.6 | 43.3 | 50.5 | 48.5 | 46.4 |
| 1932 | 41.9 | 47.1 | 46.3 | 46.0 | 49.8 | 52.4 | 53.1 | 49.9 |
| 36 | 48.1 | 56.3 | 52.9 | 52.8 | 53.9 | 60.5 | 52.4 | 61.2 |
| 1940 | 46.0 | 50.7 | 49.2 | 50.5 | 54.7 | 55.0 | 55.4 | 56.5 |
| 44 | 44.4 | 48.9 | 48.6 | 48.5 | 54.8 | 53.9 | 55.3 | 57.3 |
| 48 | 41.0 | 43.3 | 47.5 | 46.9 | 51.8 | 50.2 | 53.0 | 54.3 |
| 1952 | 36.1 | 38.9 | 41.5 | 41.2 | 45.1 | 45.1 | 46.4 | 49.4 |
| 56 | 30.1 | 29.8 | 35.2 | 36.5 | 38.6 | 37.0 | 40.4 | 41.9 |
| 1960 | 46.1 | 46.6 | 50.1 | 48.6 | 51.6 | 58.0 | 57.0 | 58.9 |
| 64 | 65.8 | 60.8 | 67.5 | 69.0 | 69.1 | 69.1 | 73.5 | 73.0 |

→

Once we have noticed this, we naturally ask what would happen if we were also to smooth the eight sequences of residuals, one for each county.

Panel C--which appears in exhibit 23--shows the result of such 3RSS smoothing. We see that there are clear trends over time in several counties. Thus, the smooth for Fairfield falls from 7 (and then 20) to −56, while those

exhibit **22** of chapter 11 (continued)

**B)** A county-PLUS-election ANALYSIS--IN 0.1% Democratic

| 484 | −62 | −30 | −14 | −18 | 15 | 31 | 42 | 50 |
|---|---|---|---|---|---|---|---|---|
| −152 | 55 | 7 | 13 | −4 | −1 | 2 | −3 | −23 |
| −171 | 49 | −38 | 0 | 8 | −7 | 0 | 11 | −49 |
| −65 | 3 | 48 | −8 | −5 | −1 | 55 | 24 | −5 |
| −3 | 0 | 20 | −4 | −3 | 2 | 12 | 8 | −32 |
| 63 | −4 | 46 | −4 | −1 | −23 | 27 | −65 | 15 |
| 37 | 1 | 16 | −15 | 2 | 11 | −2 | −9 | −6 |
| 26 | −4 | 9 | −10 | −7 | 23 | −2 | 1 | 13 |
| 3 | −15 | −24 | 2 | 0 | 16 | −16 | 1 | 6 |
| −58 | −3 | −7 | 3 | 4 | 10 | −6 | −4 | 18 |
| −120 | −1 | −36 | 2 | 19 | 7 | −25 | −2 | 5 |
| 35 | 4 | −23 | −4 | −15 | −18 | 30 | 9 | 20 |
| 200 | 36 | −46 | 5 | 24 | −8 | −24 | 9 | −4 |

**S) SOURCE**

Richard M. Scammon, 1965, *America At the Polls*. A Handbook of American Presidential Election Statistics, 1920–1964. Pages 77–78.

exhibit **23** of chapter 11: presidential Connecticut

**Continuation of exhibit 22; the smoothing, etc.**

**C)** The 3RSS SMOOTHS over YEARS--for each county separately

| eff | Litch | Fairf | Mddx | Toll | NL | NH | Windh | Hartf | #96 | smooths | |
|---|---|---|---|---|---|---|---|---|---|---|---|
| −152 | 55 | 7 | 8 | −4 | −1 | 2 | 11 | −23 | M48h | 1h | |
| −152 | 49 | 7 | 0 | −4 | −1 | 2 | 11 | −23 | H24h | 8 | −4 |
| −65 | 3 | 20 | 20 | −3 | −1 | 12 | 11 | −23 | E12h | 12 | −7h |
| −3 | 0 | 20 | −4 | −3 | −1 | 12 | 8 | −6 | D 6h | 19h | −23 |
| 37 | 0 | 20 | −4 | −1 | 2 | 12 | 2 | −5 | | | |
| 37 | | | −4 | −1 | 11 | −2 | 1 | 6 | *#93h | *smooths | |
| 26 | −4 | 9 | 2 | 0 | 16 | −2 | 1 | 6 | *M47 | 2 | |
| 3 | −4 | −7 | 2 | 0 | 16 | −6 | 1 | 6 | *H24 | 8 | −4 |
| −58 | −3 | −23 | 2 | 4 | 10 | −6 | 1 | 6 | *E12h | 12 | −7h |
| −58 | −1 | −24 | 2 | 4 | 7 | −16 | 9 | 6 | *D6h | 19h | −23 |
| 35 | 4 | −36 | 2 | 19 | −8 | −24 | 9 | 5 | | | |
| 200 | 14 | −46 | 2 | 24 | −8 | −24 | 9 | 3 | | | |

➡

exhibit 23 of chapter 11 (continued)

**D) The CORRESPONDING ROUGHS--where smooth PLUS rough ≡ residual**

```
  0    0    0    5    0    0    0  -14    0    #96 roughs
-19    0  -45    0   12   -6   -2    0  -26    M48h |  0  |
  0    0   28   -4   -2    0   43   13   18    H24h |  0  -4 |
  0    0    0    0    0    3    0    0  -26    E12h |13  -11|
 26   -4   26    0    0  -25   15  -67   20    D 6h |19  -21|
  0    5    0  -11    3    0    0  -10  -12
  0    0    0  -12   -7    7    0    0    7    *#73 roughs
  0  -11  -17    0    0    0  -16    0    0    *M37 |  0  |
  0    0   16    1    0    0    0   -5   12    *H19 |  5   -7 | 12
-62    0  -12    0   15    0   -9    1   -1    : E10 |15  -12|
  0    0   13   -6  -34  -10   54    0   15    *D 5h|21  -25h|
  0   22    0    0    0    0    0    0   -7
```

**E) A CODING**

|  | −24h |  | −14h |  | −4h |  | 4h |  | 14h |  | 24h |  |  |
|---|---|---|---|---|---|---|---|---|---|---|---|---|---|
| ⊙ |  | ○ |  | ∘ |  | · |  | + |  | + |  | # |  |
| two |  | eight |  | seven |  | (45) |  | (24) |  | eight |  | two | (smooths) |
| −36 |  | −16 |  | −5 |  | −4 |  | 5 |  | 16 |  | 49 |  |
| to |  | to |  | to |  | to |  | to |  | to |  | to |  |
| −46 |  | −24 |  | 8 |  | 4 |  | 14 |  | 24 |  | 55 |  |
|  |  |  |  |  |  |  |  |  |  |  |  |  |  |
| six |  | one |  | sixteen |  | (54) |  | eight |  | seven |  | four | (roughs) |
| −25 |  | −17 |  | −5 |  | −4 |  | 5 |  | 15 |  | 26 |  |
| to |  | to |  | to |  | to |  | to |  | to |  | to |  |
| −67 |  | −17 |  | −8 |  | 4 |  | 13 |  | 22 |  | 54 |  |

**F) SEMI-GRAPHICAL EXAMPLES--Fairfield and Hartford counties**

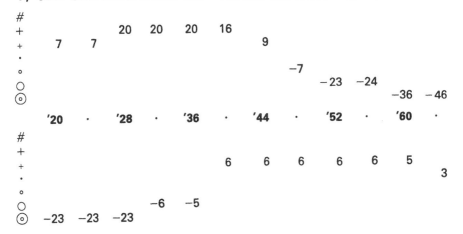

for Litchfield, Windham, and Hartford are lower--or higher--in the middle than at the ends. Panel D has the corresponding roughs. Panel E provides a--judgment-based--coding that seems reasonably well adapted to both smooth and rough. Panel F illustrates its application to semigraphical display of the smooths. Panel G--appearing in exhibit **24**--shows an analysis into (i) common, (ii) county effects, (iii) election effects, (iv) rough election effects, (v)

exhibit **24** of chapter 11: presidential Connecticut

**Continuing exhibits 22 and 23; the analysis**

G) An ANALYSIS--smooth above and rough below

|  | 484 | Litch −62 | Fairf −23 | Toll −18 | Mddx −12 | NL 15 | NH 29 | Windh 50 | Hartf 54 |
|---|---|---|---|---|---|---|---|---|---|
| 1920 | −152 | 55 | 0 | −4 | 6 | −1 | 4 | 3 | −27 |
| 24 | −152 | 49 | 0 | −4 | −2 | −1 | 4 | 3 | −27 |
| 28 | −65 | 3 | 13 | −3 | −6 | −1 | 14 | 3 | −27 |
| 32 | −3 | 0 | 13 | −3 | −6 | −1 | 14 | 0 | −10 |
| 36 | 37 | 0 | 13 | −1 | −6 | 2 | 14 | −6 | −9 |
| 1940 | 37 | −4 | 9 | −1 | −6 | 11 | 0 | −7 | 2 |
| 44 | 26 | −4 | 2 | 0 | 0 | 16 | 0 | −7 | 2 |
| 48 | 3 | −4 | −14 | 0 | 0 | 16 | −4 | −7 | 2 |
| 52 | −43 | −3 | −30 | 4 | 0 | 10 | −4 | −7 | 2 |
| 56 | −43 | −1 | −31 | 4 | 0 | 7 | −14 | 1 | 2 |
| 1960 | 35 | 4 | −43 | 19 | 0 | −8 | −22 | 1 | 1 |
| 64 | 200 | 14 | −53 | 24 | 0 | −8 | −22 | 1 | −1 |
| 1920 | ✓ | 0 | 0 | 0 | 5 | 0 | 0 | −14 | 0 |
| 24 | −25 | 0 | −45 | 12 | 0 | −6 | −2 | 0 | −26 |
| 28 | ✓ | 0 | 28 | −2 | −4 | 0 | 43 | 13 | 18 |
| 32 | ✓ | 0 | 0 | 0 | 0 | 3 | 0 | 0 | −26 |
| 36 | 26 | −4 | 26 | 0 | 0 | −25 | 15 | −66 | 20 |
| 1940 | ✓ | 5 | 0 | 3 | −11 | 0 | 0 | −10 | −12 |
| 44 | ✓ | 0 | 0 | −7 | −12 | 7 | 0 | 0 | 7 |
| 48 | ✓ | −11 | −17 | 0 | 0 | 0 | −10 | 0 | −0 |
| 52 | −15 | 0 | 16 | 0 | 1 | 0 | 0 | −5 | 12 |
| 56 | −77 | 0 | −12 | 15 | 0 | 0 | −9 | −11 | −1 |
| 1960 | ✓ | 0 | 13 | −34 | −6 | −10 | 54 | 0 | 15 |
| 64 | ✓ | 22 | 0 | 0 | 0 | 0 | 0 | 0 | −7 |

P) PROBLEMS

24a) Write down an analysis where the county effects are combined with the smooths.

24b) Do the same including the common.

24b2) Plot the results of 24b.

smooths of residuals, and (vi) roughs of residuals. In this analysis, we took out the county medians of the smooths, transferring them to the county effects. (Transferring election medians would be a little complicated; we leave this to those intrigued by the challenge.)

Exhibit **25** shows the smooths in coded form. The trends are now somewhat clearer than when given in "mournful numbers". The vertical coordinate has been taken to be the fitted "county effect", which seems as likely to be relevant as any numbers or ordering that we have. (Certainly much better than alphabetic order.) A corresponding plot for the roughs is left to the reader. No clear structure appeared in our plot of the roughs, so we may be near the bottom of this data.

### a further opportunity

To see what is happening graphically, we would naturally think of plotting the percent Democratic against date for each of the eight counties. How well would this have done? The election effects differ among themselves by (20.0) −

exhibit **25** of chapter 11: presidential Connecticut

**Smooths of exhibit 23 in coded form**

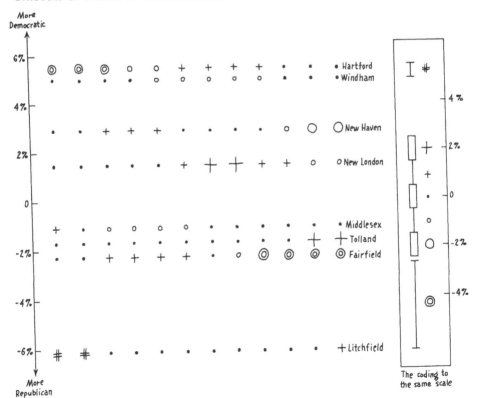

(−17.7) = 37.7%, as compared with (55) − (−65) = 12.0% for the residuals. Had we made eight separate plots, each would have shown the same general pattern (actually national in scope). Had we tried to plot the eight counties on top of one another, the actual votes would have been just irregular enough to be confusing. If we want something like a set of superposed "curves", we need to do two things: (i) take out an election effect (so that we can see smaller differences) and (ii) get rid of some of the irregularities in the behavior of the separate counties (so we can put eight "curves" on the same picture without too much confusion). In our analysis (exhibit 24), we have all the pieces we need to get a good set of curves. We need only add the county effects back to the election-wise smoothed residuals, getting a smoothed representation of the displacement of each county from a statewide election effect.

The results are plotted in exhibit 26, which does a rather good job of telling us about the smooth parts of the county idiosyncrasies. (The three counties labelled in capital letters produced between 300 and 325 thousand votes each in the 1964 election; the **total** for the other five combined was

exhibit **26** of chapter 11: presidential Connecticut

**Smooth relation of counties to state election effects (combination of two smooths)**

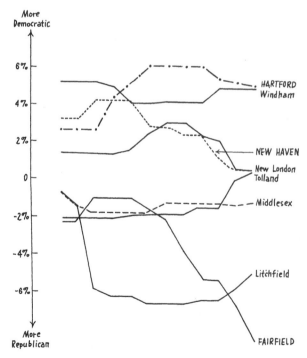

somewhat less than 300 thousand.) The most striking things in this picture are probably:

◊ the decline (in % Dem.) in *Litchfield County* (NW Conn.) in the twenties.

◊ the decline in *Fairfield County* (N.Y.C. suburbs, etc.) in the forties, fifties and sixties.

◊ lesser changes in *Hartford* (rise in the thirties) and *New Haven* (fall in the forties and fifties).

Thus, combining smoothing and two-way analysis can lead us to useful pictures accessible by neither one alone.

Exhibit **27** sets some problems.

### review questions

What might we combine? What did we take for an example? What did a PLUS analysis show? What did we do next? Roughly what was the result? Did we code the result(s)? How many kinds of results were coded? How did the resulting plots appear? Did we reach a new analysis?

### 11G. Guidance for expression

Given some data, how should we express it for row-PLUS-column analysis? The answer for any specific set of data comes mainly by trial and

---

exhibit **27** of chapter 11: problems

**Some combined analyses to be done (all rather large)**

27a) (uses (5b)) Smooth the analysis of problem (5b) from W to E for each election. Display result.

27b) (uses (5b)) Smooth the analysis of problem (5b) over time for each county. Display result.

27c) Smooth the analysis of panel A of exhibit 13 over the cycle of the months for each airport. Display the result.

27d) Smooth the analysis of panel A of exhibit 17 likewise. Display the result.

27e) Smooth the analysis of panel A of exhibit 21 over time for each county. Display the result.

27f) Could we have smoothed this example over counties?

27g) What happens when we can smooth both ways, and do?

27k) What is the difference between smoothing "over time" or "W to E" and smoothing "over the cycle of the months"?

error. Rules of thumb about where to begin, however, can be very helpful, as can suggestions about how to decide where to go next.

We learned a little about re-expression in two-way tables at the close of the previous chapter (in section 10G). It is time to think a little more about such questions.

Good rules seem to depend most heavily upon the nature of the response, and how widely it varies. As always, we can do well by bringing in other bodies of sufficiently similar data, for:

◇ it is a rare thing that a specific body of data tells us as clearly as we would wish how it itself should be analyzed.

### amounts and counts

Amounts and counts are--by definition--nonnegative. Accordingly, we can re-express them by logs or reciprocals. Zeros may seem to offer a difficulty, but there are two ways out:

◇ the easy way, where log 0 is L and $-1/0$ is L, where L is taken to be lower than any other value involved. (The fitting technique with which this chapter opened works quite well in the presence of a few L's.)

◇ the careful way, where the given values are "started" by 0.1, 0.25, 0.5, 1, or more, before logging or reciprocating.

Often either works well. Sometimes the easy way is a little crude, and the careful way is worth the trouble.

If the ratio

largest amount (or count)/smallest amount (or count)

is close to 1, it will not make any appreciable difference whether we take logs (or reciprocals) or not. So, if this ratio is close to unity, we think about what we might expect to work well **if** the ratio were large, and start our analysis with that expression.

If the ratio is large, we will probably have to go to

◇ logs fitted row-PLUS-column, or

◇ raw values fitted row-TIMES-column.

(These two tap equivalent collections of possibilities, but using them will usually give somewhat different fits.) Hand arithmetic is usually easier when we re-express the given values as logs--and the fit is often better, too--so we usually start there.

If the ratio is intermediate in size, we ask ourselves whether we have a reasonable belief as to what ought to be preferred (in a reasonable world). If we have such a belief, we start from there--and let the diagnostic plot tell us which way we ought to move. If we have no belief, we do a row-PLUS-column fit for the raw-values first, carrying through at least as far as a

### diagnostic plot

--a plot of residuals against comparison values. Now we look to see if a slope of 1 is better than a slope of 0. It it is, we can convert our row-PLUS-column fit into a row-TIMES-column fit--but we usually do better by going to logs. If 0 is much better than 1, we stand pat. In between, we think--and look--hard.

Rarely, we will have a place for a re-expression other than logs or reciprocals. The need for this will usually be shown when two adjacent choices--from (square), raw, log, reciprocal, (reciprocal square)--show diagnostic plots with opposite tilts. Exploration and, particularly, thought are then in order.

### balances

If the values in front of us include some that are negative, we cannot take logs. We can start with a row-PLUS-column fit, nevertheless, and if the diagnostic plot favors a coefficient near +1.0, we can convert this "PLUS" into a row-TIMES-column fit with ease. (We have already found the comparison values in order to make the diagnostic plot--multiplying them by +1.0 is easy. If something other than row-PLUS-column or row-TIMES-column fits is needed, exploration and thought are again the proper medicine. (See optional section 11L for an extreme example, and optional section 11I for a less extreme one.)

If we have effects that are well-balanced in sign, the row-PLUS-column fit may not be a good enough start for a row-TIMES-column fit. (Sometimes it fails miserably.) If all we really want is the TIMES fit, we can change the signs both in as many entire rows as we wish and, separately, in as many entire columns as we wish, and then make first a PLUS fit and than a TIMES fit to the new table. We can then put back the signs we took out, changing the signs of residuals accordingly.

### counted fractions

We will learn some ways to re-express counted fractions when we come to chapter 15. No more here.

### a helpful thougnt

We now see even more use for comparison values and diagnostic plots then we did before. It is important that PLUS-one offers us conversion from row-PLUS-column to row-TIMES-column. We ought to take essentially every fit as far as a diagnostic plot. We then always have the opportunity to choose to:

⋄ask if TIMES seems better than PLUS.

◇fit one more constant.

◇or do neither, if neither seems helpful/reasonable.

**review questions**

Must we rely on trial-and-error in choosing an expression for row-PLUS-column analysis? Can there be guidance? Why/why not? Does any body of data usually tell us clearly about just how it ought to be analyzed? What of amounts and counts? Should we worry about logs or reciprocals of zeros? Why/why not? What if largest/smallest is close to 1? Quite large? In a middle range? Do we expect to take most PLUS analyses as far as a diagnostic plot? What two slopes play a particular role? What of balances? Counted fractions?

**11H. How far have we come?**

This chapter is basically a how-to-do-it chapter, filling in the techniques for making all the analyses and displays we met in chapter 10.

We are now ready:

◇to make row-PLUS-column fits by repeatedly "shifting out" medians.

◇and to call this process "median polish", organizing its calculations in either of two ways.

◇to lessen the chance of arithmetic error by circling negative values in our working calculations.

◇to arrange in a better way the check that what we have is really a row-PLUS-column fit.

◇to make a two-way plot of any row-PLUS-column fit, starting from any of various ways of dividing its values into two suitable parts, drawing horizontal and vertical lines, and turning through 45°.

◇to identify the residuals deserving of appearance in the plot of the fit--those outside--and to include them in the fit (mechanically, by adding half of each such residual to each of the two parts of the fit).

◇and to make a suitable schematic plot for the remaining residuals.

◇to use the standard code for residuals in a two-way plot of residuals.

◇and to prepare a display of the actual sizes of residuals that receive each code.

◇to use reasonable judgment followed by a diagnostic plot in choosing a re-expression of a response to be analyzed in a two-way table.

Moreover, we have seen how:

◇to condense wither a two-way plot of fit or a two-way plot of residuals.

◇to fit smoothing inside some two-way tables.

◇ and to display appropriately the extra piece (or pieces) such an analysis produces.

◇ to use L, taken as less than any number, for the log or reciprocal of zero, so that we have an alternative to "starting" when analyzing two-way tables with a few zeros. (And, even, occasionally, to go farther, using L for logs of negative numbers, too.)

◇ to open the way to a row-TIMES-column fit by way of a row-PLUS-column one, through changing signs in several steps, each step either by entire rows or by entire columns.

We now understand much more clearly:

◇ that the general maxim--it is a rare thing that a specific body of data tells us clearly enough how it itself should be analyzed--applies to choice of re-expression for two-way analysis.

◇ that it is wise to take almost every two-way fit as far as a diagnostic plot.

We are now equipped to do, with relative ease, a reasonable variety of things in connection with two-way fit. The things we learned as possibilities in chapter 10 are now all things **we can do**. As we find diverse opportunities to try them out--whether or not closely related to this book--we will come to recognize more and more what they can do for us.

# Optional sections for chapters 10 and 11    11⁺

chapter index on next page

Even though we separated "What we can do!" and "How to do it!" into two chapters, there was still more related material than seemed wise to pack into those chapters. The present chapter, then, contains a mixed grill of topics that belong close to, rather than in, chapters 10 and 11.

## 11I. Exploring beyond PLUS-one (extends for chapter 10)

Exhibit **28** offers, for those who wish to play, an expedition into one part of the wilderness of row-PLUS-col-PLUS-two fits.

One thing learned by those who go on this expedition is that, once we fit a PLUS-one supplement to our row-PLUS-column fit, there may be still more to gain from a further, supplementary, row-PLUS-column fit. Suppose we try this with the simple row-PLUS-column-PLUS-one fit of exhibit 11 of chapter 10.

Exhibit **29** shows both what can be gained, and something of where it leaves us. Clearly, we are likely to engage in back-and-forth fitting and refitting between "row-PLUS-column" and "(new constant)(new comparison value)" if we plan to follow this road. One step that we can take without complicating the calculation is to fit a supplementary value for the common term. Panel C of exhibit 29 shows the result.

exhibit **28** of chapter 11⁺: East Coast

**Optional problems taking the analysis further**

O) OPTIONAL PROBLEMS--all teasers

28a1) Plot the residuals of panel A of exhibit 11 of chapter 10 against the comparison values of panel D of exhibit 9 of chapter 10. What sort of very simple additional fit comes to mind?

28a2) What plot would make fitting one more constant (beyond the -PLUS-one constant) easy? Make the plot. Choose this constant. (Now you have ONE example of ONE kind of a row-PLUS-col-PLUS-two fit.)

28a3) Find the residuals for this fit. Go back and readjust each of (i) row fits and column fi•s, (ii) the constant $k$ multiplying comparison values, to get a better row-PLUS-col-PLUS-two fit. Display its residuals. How do they compare with the residuals of exhibit 10 of chapter 10?

## index for chapter 11+

**11I. Exploring beyond PLUS-one (extends for chapter 10)** 401
    review questions 404

**11J. Taking out any summary** 404
    repeated midextreme removal 404
    the general case 404
    repeated mean removal 405
    review questions 408

**11K. An example of re-expression--city killings** 408
    review questions 414

**11L. An unusual fit** 415
    coastwise distances 415
    review questions 418

**11M. How much may we have learned?** 419

| EXHIBITS | PAGES |
|---|---|
| **11I** | |
| 28★ | 401 |
| 29★ | 403 |
| **11J** | |
| 30★ | 405 |
| 31★ | 407 |
| **11K** | |
| 32★ | 409 |
| 33 | 410 |
| 34 | 411 |
| 35 | 412 |
| 36 | 413 |
| 37 | 413 |
| 38★ | 414 |
| **11L** | |
| 39 | 415 |
| 40★ | 417 |
| **11M** | |

exhibit **29** of chapter 11[+]: East coast

**A supplementary row-PLUS-column fit after a row-PLUS-column-PLUS-one fit (starts from exhibit 11 of chapter 10)**

A) If we allow ourselves a SUPPLEMENTARY row-PLUS-column FIT

| The data[†] | | | One row-PLUS-column fit | | | | |
|---|---|---|---|---|---|---|---|
| −72 | 1 | −12 | −53 | 0 | 0 | ‖ | 1 |
| −35 | 0 | −15 | −13 | 2 | 0 | ‖ | −2 |
| −20 | 0 | −7 | 0 | 0 | 6 | ‖ | 0 |
| 0 | 0 | 0 | 7 | −13 | 0 | ‖ | 13 |
| −9 | 0 | −2 | 0 | 11 | 0 | ‖ | 11 |
| −16 | 0 | −21 | 4 | 0 | −8 | ‖ | 0 |
| −20 | 0 | −17 | 0 | 0 | −4 | ‖ | 0 |
|  |  |  | −7 | 13 | 0 | ‖ | −13 |

[†]Actually, the residuals from a PLUS-one fit.

B) WHERE ARE WE?--and where might we go?

We have now fitted

$$\text{all} + \text{row} + \text{col} - \frac{(\text{row})(\text{col})}{544} + \text{all}^* + \text{row}^* + \text{col}^*$$

where the *'s denote the supplementary fit of panel A. This is equivalent to

$$(\text{all} + \text{all}^*) + (\text{row} + \text{row}^*) + (\text{col} + \text{col}^*) - \frac{(\text{row})(\text{col})}{544}.$$

One way to neaten things up would be to find $k^*$ in

$$(\text{all} + \text{all}^*) + (\text{row} + \text{row}^*) + (\text{col} + \text{col}^*) + k^* \frac{(\text{row} + \text{row}^*)(\text{col} + \text{col}^*)}{(\text{all} + \text{all}^*)}$$

to do as well we can. (This we leave to panel P.)

C) The LEAST we CAN DO--two versions of a restricted (common only) fit to the residuals of exhibit 11

| −65 | 8 | −5 | ‖ | 0 |
|---|---|---|---|---|
| −28 | 7 | −8 | ‖ | 0 |
| −13 | 7 | 0 | ‖ | 0 |
| 7 | 7 | 7 | ‖ | 0 |
| −2 | 7 | 5 | ‖ | 0 |
| −9 | 7 | −14 | ‖ | 0 |
| −13 | 7 | −10 | ‖ | 0 |
| 0 | 0 | 0 | ‖ | −7 |

| −70 | 3 | −10 | ‖ | 0 |
|---|---|---|---|---|
| −33 | 2 | −13 | ‖ | 0 |
| −18 | 2 | −5 | ‖ | 0 |
| 2 | 2 | 2 | ‖ | 0 |
| −7 | 2 | 0 | ‖ | 0 |
| −14 | 2 | −19 | ‖ | 0 |
| −18 | 2 | −15 | ‖ | 0 |
| 0 | 0 | 0 | ‖ | −2 |

P) PROBLEMS

29a) Find another set of data that deserves such treatment, and do the PLUS-one fit.

29a2) Continue, parallel to panel A.

29a3) Is a continuation parallel to panel C justified? Why? (Why not?) If justified, do it.

### review questions

Might we go beyond-PLUS-one? What might we learn? What is the least we might do? What does this do to our example?

### 11J. Taking out any summary

Panel C of exhibit 1 of chapter 10 offered a row-PLUS-column analysis of the Arizona temperatures. When we check it out, we find that further steps of median removal would have no effect.

It offers a convenient starting place, however, for other kinds of removal --as would, in most other examples, the results of four steps of median removal One kind of removal that is easy to illustrate, because the arithmetic is easy, is the removal of **midextremes.** Because it is so sensitive to single exotic values, we are not likely to use it to analyze real data. (It has a place, of course, when we are constructing simple approximations to error-free tables.) We shall use it here only as an illustration--an illustration of something just about as far from the median as we can find. (If it works out all right, anything else we are likely to think of trying will work out all right, also.)

#### repeated midextreme removal

So let us try removing the midextreme of every row and the midextreme of every column. (Some will want to recall that the midextreme has often been called the midrange.) Exhibit **30** shows the calculation, which we have put in the same pattern that we used in section 11A for repeated median removal. The result depends a little upon where and how we start, but the three 3 × 7 tables that we reach, namely (in 0.1°F),

| −1 | 6 | −5 | −2 | 6 | −6 | −1 | 6 | −6 |
|---|---|---|---|---|---|---|---|---|
| −4 | −3 | 1 | −4 | 4 | 1 | −3 | 4 | 0 |
| −8 | 7 | 7 | −8 | 8 | 7 | −7 | 8 | 7 |
| −3 | 2 | 2 | −3 | 3 | 2 | −2 | 3 | 2 |
| 8 | −8 | −6 | 7 | −8 | −7 | 8 | −8 | −7 |
| 5 | 0 | −6 | 5 | 1 | −6 | 6 | 1 | −6 |
| −1 | 1 | 1 | −1 | 2 | 1 | −1 | 1 | 0 |
| (panel A) | | | (panel B) | | | (panel C) | | |

all (happen to) have the same maximum size of residual--0.8. (If we want to use half-tenths, we might get this down to 0.7h.)

#### the general case

We can do this sort of calculation, removing some chosen summary value repeatedly, using any summary value we like.

**repeated mean removal**

Those who like lots of arithmetic can even use the mean, as in exhibit **31**. This time we take the calculations only to two steps. (They would have stopped exactly, had we used fractions of tenths.) The results obtained by calculating to one decimal are not identical. (They would have been, had we gone to fractions.) In fact, they are, in units of 0.1°F,

|  |  |  |  |  |  |  |  |  |
|---|---|---|---|---|---|---|---|---|
| −1 | 4 | −4 | 0 | 5 | −4 | 0 | 4 | −4 |
| −4 | 1 | 2 | −3 | 2 | 2 | −3 | 1 | 2 |
| −10 | 3 | 6 | −9 | 4 | 6 | −9 | 3 | 6 |
| −3 | 0 | 3 | −3 | 0 | 2 | −2 | 0 | 3 |
| 10 | −8 | −3 | 11 | −7 | −3 | 11 | −8 | −3 |
| 6 | −1 | −4 | 6 | −1 | −5 | 6 | −2 | −5 |
| −1 | −1 | 2 | −1 | −1 | 1 | 0 | −1 | 2 |
| (panel A) | | | (panel B) | | | (panel C) | | |

At no location do they differ by more than a single tenth.

exhibit **30** of chapter 11[+]: Arizona temperatures

**Taking out midextremes (extreme illustration!)**

A) STARTING FROM panel C of exhibit 1 of chapter 10

| 0 | .2 | −.9 | −.4 | .4 | .6 | −.5 | Examples: |
|---|---|---|---|---|---|---|---|
| 0 | .2 | 0 | .1 | −.1 | .1 | −.1 | |
| −1.0 | 0 | 0 | −.5 | −.5 | .5 | .5 | $-.4 = \frac{1}{2}(.2 + .9)$ |
| 0 | 0 | 0 | 0 | 0 | 0 | 0 | |
| 1.9 | −.2 | 0 | .8 | 1.1 | −1.0 | −.8 | $.1 = \frac{1}{2}(0 + .2)$ |
| 1.0 | 0 | −.6 | .2 | .8 | −.2 | −.8 | |
| .3 | 0 | 0 | .2 | .1 | −.2 | −.2 | $-.5 = \frac{1}{2}(-1.0 + 0)$ |
| | | | | .3 | −.2 | −.2 | ... |

| −.1 | .6 | −.5 | .2 | .1 | .8 | −.3 | |
|---|---|---|---|---|---|---|---|
| −.4 | .3 | .1 | √ | −.4 | .3 | .1 | $.4 = 0 - (-.4)$ |
| −.8 | .7 | .7 | √ | −.8 | .7 | .7 | $.6 = .2 - (-.4)$ |
| −.3 | .2 | .2 | √ | −.3 | .2 | .2 | $-.1 = 0 - (.1)$ |
| .8 | −.8 | −.6 | √ | .8 | −.8 | −.6 | |
| .5 | 0 | −.6 | √ | .5 | 0 | −.6 | |
| −.1 | .1 | .1 | −.1 | −.2 | 0 | 0 | |
| √ | √ | √ | | | | | |

➡

exhibit 30(B–P)/11+: Optional sections

exhibit 30 of chapter 11+ (continued)

**B) STARTING from the SAME place the OTHER way**

```
   0    .2   -.9
   0    .2    0
 -1.0    0    0
  1.9   -.2   0
  1.0    0   -.6
   .3    0    0
         √   -.4
```

```
 -.4   -.5    -.2    .4   -.3
 -.4    .2   -.4    .2    .4
 -1.4   .4   -.9    .5    .9
 -.4    0   -.4     0    .4
  1.5  -.2    .9   -.8   -.2
  .6    0    .4   -.2   -.4
 -.1    .4   -.3   -.2    .2
              √           .2
```

```
 -.2    .6   -.2    .6   -.5
 -.4    .4   -.4    .6    .2
 -.8    .8   -.9    .4    .7
 -.3    .3   -.4    .7    .2
  .7   -.8    .9    .2   -.4
  .5    .1    .4   -.6   -.4
 -.1    .2   -.4    0    .6
                           0
```

```
  .1    √    √    √
  .6    √    √    √
 -.2    .6         √
 -.4    .4         √
 -.8    .8         √
 -.3    .3         √
  .7   -.8         √
 -.7
  .5    .1
 -.1    .2
```

**C) STARTING from the RAW DATA**

```
 65.2   90.1   94.6   79.9 | -14.7   10.2   14.7
 63.4   88.3   93.7   78.6 | -15.2    9.7   15.1
 57.0   82.7   88.3   72.6 | -15.6   10.1   15.7
 46.1   70.8   76.4   61.2 | -15.1    9.6   15.2
 35.8   58.4   64.2   50.0 | -14.2    8.4   14.2
 28.4   52.1   57.1   42.8 | -14.4    9.3   14.3
 25.3   49.7   55.3   40.3 | -15.0    9.4   15.0

                                   -14.9    9.3   15.0
```

```
 -.1    .6   -.6    .3          .2    .9   -.3
 -.3    .4    .1   √           -.3    .4    .1
 -.7    .8    .7   √           -.7    .8    .7
 -.2    .3    .2   √           -.2    .3    .2
  .8   -.8   -.7  -.1           .7   -.3   -.8
  .6    .1   -.6  -.1           .5    0   -.7
 -.1    .1    0   √            -.1    .1    0
       √
```

**P) PROBLEM**

30a) Start from the raw data and go the other way. Compare your result with those of panels A, B, and C.

exhibit **31** of chapter 11+: Arizona temperatures

**Taking out means instead**

**A) STARTING FROM panel C of exhibit 1 of chapter 10**

| | | | | | | |
|---|---|---|---|---|---|---|
| 0 | .2 | -.9 | -.2 | .2 | .4 | -.7 |
| 0 | .2 | 0 | .1 | -.1 | .1 | -.1 |
| -1.0 | 0 | 0 | -.3 | -.7 | .3 | .3 |
| 0 | 0 | 0 | 0 | 0 | 0 | 0 |
| 1.9 | -.2 | 0 | .6 | 1.3 | -.8 | -.6 |
| 1.0 | 0 | -.6 | .1 | .9 | -.1 | -.7 |
| .3 | 0 | 0 | .1 | .2 | -.1 | -.1 |
| | .3 | √ | | .3 | √ | -.3 |

| | |
|---|---|
| √ | -.1 .4 .1 -.4 |
| √ | -.4 .1 .2 .2 |
| √ | -1.0 .3 .6 .3 |
| √ | -.3 0 -.8 -.3 |
| √ | 1.0 -.8 -.1 -.4 |
| √ | .6 -.1 -.1 .2 |
| √ | -.1 -.1 -.1 |

**B) STARTING from the SAME place the OTHER way** ↑

| | | | | | | |
|---|---|---|---|---|---|---|
| 0 | .2 | -.9 | -.3 | 0 | .5 | -.4 |
| 0 | .2 | 0 | √ | -.3 | .2 | .2 |
| -1.0 | 0 | 0 | -.4 | -.9 | .4 | .6 |
| 0 | 0 | 0 | 0 | -.3 | 0 | .2 |
| 1.9 | -.2 | 0 | .5 | 1.1 | -.7 | -.3 |
| 1.0 | 0 | -.6 | .1 | .6 | -.1 | -.5 |
| .3 | 0 | 0 | .1 | -.1 | -.1 | .1 |
| | .3 | √ | -.2 | | | √ |

| | | | | | | |
|---|---|---|---|---|---|---|
| -.3 | -.7 | .2 | -.3 | | | |
| -.3 | .2 | .2 | √ | | | |
| -1.3 | .2 | 0 | -.4 | | | |
| -.3 | .2 | 0 | .5 | | | |
| 1.6 | -.2 | 0 | .1 | | | |
| .7 | 0 | -.4 | .1 | | | |
| 0 | 0 | .2 | | | | |

408    exhibit 31(C–P)/11⁺: Optional sections

You, too, could invent your own summary, and take it out repeatedly to find yet another fit, yet another array of residuals.

### review questions

What example did we choose? What did we try taking out first? What would a single very stray value have done to us? (Work an example, if need be.) How many ways did we take midextremes out of our example? How did the results compare? What did we try taking out next? What would a single **very,** very stray value have done to us? (Work an example, if need be.) How well did the results of various calculations agree? Can you take out other summaries? Which ones? In what way? Even if you invent a new summary?

### 11K. An example of re-expression--city killings

A simple example illustrates many of the points just stated. Exhibit **32** shows the data, a row-PLUS-column analysis, and the corresponding comparison values. Exhibit **33** shows the resulting diagnostic plot. Exhibit **34** shows an

exhibit **31** of chapter 11⁺ (continued)

**C) STARTING from the RAW DATA--get out the adding machine!**

| 65.2 | 90.1 | 94.6 | 83.3 | −18.1 | 6.8 | 11.3 |
| 63.4 | 88.3 | 93.7 | 81.8 | −18.4 | 6.5 | 11.9 |
| 57.0 | 82.7 | 88.3 | 76.0 | −19.0 | 6.7 | 12.3 |
| 46.1 | 70.8 | 76.4 | 64.4 | −18.3 | 6.4 | 12.0 |
| 35.8 | 58.4 | 64.2 | 52.8 | −17.0 | 5.6 | 11.4 |
| 28.4 | 52.1 | 57.1 | 45.9 | −17.5 | 6.2 | 11.2 |
| 25.3 | 49.7 | 55.3 | 43.4 | −18.1 | 6.3 | 11.9 |
| | | | | −18.1 | 6.4 | 11.7 |
| | | | ✓ | 0    | .4  | −.4 |
| | | | ✓ | −.3  | .1  | .2  |
| | | | ✓ | −.9  | .3  | .6  |
| | | | ✓ | −.2  | 0   | .3  |
| | | | ✓ | 1.1  | −.8 | −.3 |
| | | | ✓ | .6   | −.2 | −.5 |
| | | | ✓ | 0    | −.1 | .2  |

**P) PROBLEM**

**31a)** Start from the raw data, and go the other way. Compare your result with that of panels A, B, and C.

exhibit **32** of chapter 11+: city killings

**Numbers of murders and nonnegligent manslaughters (combined) for 1961, 1964, and 1966 in 18 large cities**

| A) The DATA | | | | B) A PLUS ANALYSIS | | | | C) COMPARISON VALUES | | |
|---|---|---|---|---|---|---|---|---|---|---|
| |City| | 1961 | 1964 | 1966 | |eff| | 1961 | 1964 | 1966 | 1961 | 1964 | 1966 |
| New York | 482 | 636 | 653 | 528 | −130 | 0 | 11 | −117.3 | 0 | 29.3 |
| Chicago | 365 | 398 | 510 | 290 | −9 | 0 | 106 | −64.4 | 0 | 16.1 |
| Los Angeles | 159 | 177 | 226 | 75 | 0 | −6 | 37 | −16.7 | 0 | 4.2 |
| Detroit | 136 | 125 | 214 | 52 | 0 | −35 | 48 | −11.6 | 0 | 2.9 |
| Philadelphia | 144 | 188 | 178 | 64 | −4 | 16 | 0 | −14.2 | 0 | 3.6 |
| Baltimore | 89 | 144 | 175 | 36 | −31 | 0 | 25 | −8.0 | 0 | 2.0 |
| Washington | 88 | 132 | 141 | 24 | −20 | 0 | 3 | −5.3 | 0 | 1.3 |
| Cleveland | 80 | 116 | 139 | 8 | −12 | 0 | 17 | −1.8 | 0 | .4 |
| Dallas | 99 | 149 | 120 | 15 | 0 | 26 | −9 | −3.3 | 0 | .8 |
| New Orleans | 52 | 82 | 113 | −26 | −6 | 0 | 25 | 5.8 | 0 | −1.4 |
| St. Louis | 77 | 120 | 106 | −7 | 0 | 19 | −1 | 1.6 | 0 | −.4 |
| Akron | 16 | 12 | 17 | −96 | 28 | 0 | −1 | 21.3 | 0 | −5.3 |
| Boston | 26 | 52 | 58 | −56 | −2 | 0 | 0 | 12.4 | 0 | −3.1 |
| Buffalo | 19 | 21 | 24 | −87 | 22 | 0 | −3 | 19.3 | 0 | −4.8 |
| Denver | 32 | 33 | 39 | −75 | 23 | 0 | 0 | 16.7 | 0 | −4.2 |
| Jersey City | 11 | 17 | 16 | −91 | 18 | 0 | −7 | 20.2 | 0 | −5.1 |
| Kansas City | 49 | 48 | 59 | −55 | 20 | −5 | 0 | 12.2 | 0 | −3.1 |
| Long Beach | 10 | 17 | 20 | −91 | 17 | 0 | −3 | 20.2 | 0 | −5.1 |
| eff | | | | 108 | −24 | 0 | 6 | | | |

P) PROBLEMS

**32a)** Select five comparison values that are not zero, and check that they do correspond to the fit of panel B. Show your work.

**32b)** Find a 3RSS middle trace for residuals against comparison values and compare with exhibit 17. Show your work.

**32c)** Find the numerical values of a TIMES fit to the data based on the PLUS analysis of panel B and its comparison values. Write out the fit.

**32c2)** Find the residuals corresponding to the fit just found, make a stem-and-leaf, and compare with the stem-and-leaf given in exhibit 38, below.

**32c3)** Make a diagnostic plot of these residuals against the original comparison values. Discuss the results.

**32c4)** Make a 3RSS middle trace for these 54 points.

S) SOURCE

**World Almanac and Book of Facts for: 1963, page 310; 1966, page 307; 1968, page 903.**

enlarged view of the center of exhibit 33, showing a 3RSS middle trace with the points suppressed and the two comparison lines

residual = zero

residual = comparison value

added. Clearly the 3RSS middle trace is closer to the later, although the dissymmetry between left and right sides is a possible worry. Next, in view of this judgment, we should go either to a row-TIMES-column fit of the counts or to a row-PLUS-column for the logs of the counts. We leave the row-TIMES-column fit based on the row-PLUS-column fit of counts to the reader (see problems (32c1/2/3)) and proceed to the logs. Exhibit 35 has log counts, a row-PLUS-column analysis, and its comparison values. Exhibit 36 shows the 3RSS middle trace for the diagnostic plot. It seems pretty quiet and uninteresting, and we can believe that we may have a satisfactory analysis.

exhibit **33** of chapter 11⁺: city killings

**Diagnostic plot based on exhibit 32**

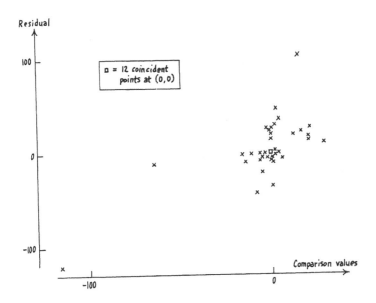

We next want to return from log counts to raw counts, in order to make a clear comparison of residuals. Exhibit **37** has the antilogs of the PLUS fit to logs and the corresponding residuals. (These antilogs are, of course, a TIMES fit to the raw counts.) Exhibit **38** shows both stem-and-leaf and letter-value displays for three sets of residuals,

⋄ the row-PLUS-column fit to counts.
⋄ the row-TIMES-column fit based on same.
⋄ the row-TIMES-column fit from the PLUS fit to logs.

We see a steady decrease in the size of the residuals, and conclude that

⋄ it was worthwhile going from PLUS to TIMES.
⋄ the TIMES fit via the logs was even closer.

exhibit **34** of chapter 11⁺: city killings

**Enlarged view of center of exhibit 33**

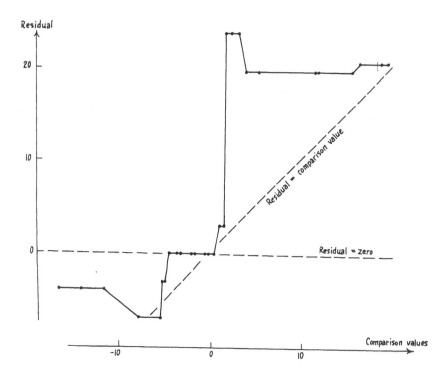

This last point suggests that

⋄ the TIMES fits to raws (here counts) that is based upon a PLUS fit to raws is likely to be a good beginning.

⋄ doing it by logs instead may be worth the effort.

These considerations are, of course, already reflected in the attitudes set out in section10G.

exhibit **35** of chapter 11⁺: city killings

**Analysis of the logs of counts**

| City | Given values (logs*) | | | eff | Row-PLUS-column | | | Comparison values | | |
|---|---|---|---|---|---|---|---|---|---|---|
| | 1961 | 1964 | 1966 | | 1961 | 1964 | 1966 | 1961 | 1964 | 1966 |
| New York | 268 | 280 | 281 | 74 | 0 | 4 | −1 | −2.93 | 0 | 2.20 |
| Chicago | 256 | 260 | 271 | 62 | 0 | −4 | 1 | −2.46 | 0 | 1.84 |
| Los Angeles | 220 | 225 | 235 | 26 | 0 | −3 | 1 | −1.03 | 0 | .77 |
| Detroit | 213 | 210 | 233 | 10 | 0 | −11 | 6 | −.75 | 0 | .56 |
| Philadelphia | 216 | 227 | 225 | 22 | 0 | 3 | −5 | −.87 | 0 | .65 |
| Baltimore | 195 | 216 | 224 | 14 | −13 | 0 | 2 | −.55 | 0 | .42 |
| Washington | 194 | 212 | 215 | 7 | −7 | 3 | 0 | −.28 | 0 | .21 |
| Cleveland | 190 | 206 | 214 | 4 | −8 | 0 | 2 | −.16 | 0 | .12 |
| Dallas | 200 | 217 | 208 | 6 | 0 | 9 | −6 | −.24 | 0 | .18 |
| New Orleans | 172 | 191 | 205 | −11 | −11 | 0 | 8 | .44 | 0 | −.33 |
| St. Louis | 189 | 208 | 203 | −5 | 0 | 11 | 0 | .20 | 0 | −.15 |
| Akron | 120 | 108 | 123 | −85 | 11 | −9 | 0 | 3.37 | 0 | −2.52 |
| Boston | 142 | 172 | 176 | −32 | −20 | 2 | 0 | 1.27 | 0 | −.95 |
| Buffalo | 128 | 132 | 138 | −70 | 4 | 0 | 0 | 2.77 | 0 | −2.08 |
| Denver | 151 | 152 | 159 | −49 | 6 | −1 | 0 | 1.94 | 0 | −1.46 |
| Jersey City | 104 | 123 | 120 | −88 | −2 | 9 | 0 | 3.48 | 0 | −2.61 |
| Kansas City | 169 | 168 | 177 | −31 | 6 | −3 | 0 | 1.23 | 0 | −.92 |
| Long Beach | 100 | 123 | 130 | −79 | −15 | 0 | 1 | 3.13 | 0 | −2.35 |
| eff | | | | 202 | −8 | 0 | 6 | | | |

* Unit = 0.01

exhibit **36** of chapter 11[+]: city killings

**3RSS middle trace for diagnostic plot of exhibit 35**

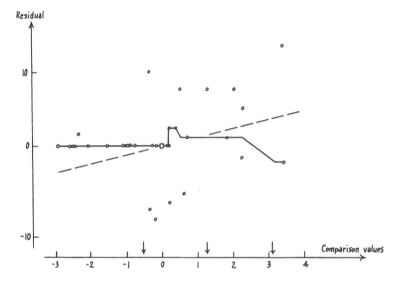

exhibit **37** of chapter 11[+]: city killings

**Antilogs of the fit of exhibit 35--which is a rows-TIMES-column fit to counts--and the corresponding residuals**

A) FIT and RESIDUALS

| City | fit (in counts) | | | residuals | | |
|---|---|---|---|---|---|---|
| | 1961 | 1964 | 1966 | 1961 | 1964 | 1966 |
| New York | 479 | 575 | 661 | 3 | 61 | −8 |
| Chicago | 363 | 436 | 501 | 2 | −38 | 9 |
| Los Angeles | 158 | 191 | 219 | 1 | −14 | 7 |
| Detroit | 135 | 162 | 186 | 1 | −37 | 28 |
| Philadelphia | 145 | 174 | 200 | −1 | 14 | −22 |
| Baltimore | 120 | 145 | 166 | −31 | −1 | 9 |
| Washington | 102 | 123 | 141 | −14 | 9 | 0 |
| Cleveland | 96 | 115 | 132 | −16 | 1 | 7 |
| Dallas | 100 | 120 | 138 | −1 | 29 | −18 |
| New Orleans | 68 | 81 | 93 | −1 | 27 | −1 |
| St Louis | 78 | 93 | 107 | −1 | 27 | −1 |
| Akron | 12 | 15 | 17 | 4 | −3 | 0 |
| Boston | 42 | 50 | 58 | −16 | 2 | 0 |
| Buffalo | 19 | 21 | 24 | 1 | 0 | 0 |
| Denver | 32 | 32 | 39 | 0 | 1 | 0 |
| Jersey City | 11 | 14 | 16 | 0 | 3 | 0 |
| Kansas City | 43 | 51 | 59 | 6 | −3 | 0 |
| Long Beach | 15 | 17 | 20 | −5 | 0 | 0 |

#### review questions

What example did we try? How does the diagnostic plot look? How did we do better than just looking at its points? Where did it tell us to go next? How did this work? Could we come back? To what kind of a fit? How many fits did we compare? How did they compare? What tentative conclusions did this suggest?

**exhibit 38** of chapter 11⁺: city killings

**Three sets of residuals for three analyses of the counts of murders plus nonnegligent manslaughters**

**A)** the STEM-and-LEAF DISPLAYS

```
      |PLUS on counts|   |TIMES from ←|   |TIMES from logs|
  H   |106,              55, 90            61
  ----|---------        
  4*  |8                 5                 
  3*  |7                 3                 
  2*  |0235568           366               0789
  1*  |1667789           026779            4
   .  |                  678               6799
  0*  |3                 222333344         1111122334
   Z  |⊠ ⊠               ⊠ · ·             ⊠ : ·
  -0* |112334            122234            1111331
   .  |566799            56                58
  -1* |2                 0023458           446668
  -2* |0                 3                 2
  -3* |15                5                 178
  -4* |
  ----|----
  L   |-130,
```

Note: The ±1's in the righthand display, together with most ±2's and 3's, would have been zero had the log analysis been taken to a third decimal, something that would not have been worth the trouble. (Recall that "⊠" stands for a count of 10.)

**B)** The LETTER-VALUE DISPLAYS

```
#54  PLUS on counts           TIMES from ←              TIMES from logs
M27h|        0        |            0            |            0
H14 | 17       -3| 20 |  8        -2| 10        |  4        -3|  7
E 7h| 24       -8| 32 | 21       -11| 32        |  9       -16| 25
D 4 | 28      -20| 48 | 33       -15| 48        | 27       -22| 49
  3 | 37      -31| 68 | 45       -18| 63        | 28       -31| 59
  2 | 48      -35| 83 | 55       -23| 78        | 29       -37| 66
  1 |106     -130|236 | 90       -35|125        | 61       -38| 99
                                                                    →
```

## 11L. An unusual fit

Before we close this chapter we ought to show that row-PLUS-column and row-TIMES-column--either bare or -PLUS-one--are not all the reasonable possibilities for fitting two-way tables. We may want to start in quite a different way--or we may want to start in one of these ways and go further. (One example here; others will surface in Chapter 12.)

### coastwise distances

One kind of two-way table that sometimes appears is a table of distances by sea between any two of a list of ports. Exhibit **39** sketches the simplest

---

exhibit **38** of chapter 11⁺ (continued)

### C) Some SPREADS JUXTAPOSED

|  | PLUS on counts | | TIMES from ← | | TIMES from logs |
|---|---|---|---|---|---|
| H-spread | 20 | * | 10 | * | 7 |
| E-spread | 32 |   | 32 | * | 25 |
| D-spread | 48 |   | 48 |   | 49 |
| 3-spread | 68 | * | 63 | * | 59 |
| 2-spread | 83 | * | 78 | * | 66 |
| range | 236 | * | 125 | * | 99 |

Note: "*" marks a decrease of the next column right, as compared to the next column left.

### P) PROBLEMS

38A) Collect similar data for other years (preferably both earlier and later) and a similar set of cities; do an analysis parallel to exhibit 32.

38a2/a3/a4/a5/a6/a7) Continue parallel to exhibits 33/34/35/36/37/38.

---

exhibit **39** of chapter 11⁺: illustrative

### The simplest situations in sea distances

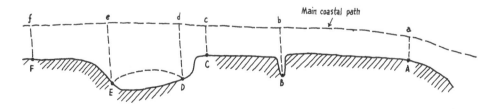

situations. Capital letters--A, B, C, ..., F--are ports, while small letters--a, b, c, ..., f--are points on the main path of coastal traffic, where shipping for the corresponding port turns off. (In practice, things are rarely this simple.)

The sea distances between adjacent ports mainly fall into three parts, as follows:

| | |
|---|---|
| **A and B** | ⟨A to a⟩ + ⟨a to b⟩ + ⟨b to B⟩ |
| **B and C** | ⟨B to b⟩ + ⟨b to c⟩ + ⟨c to C⟩ |
| **C and D** | ⟨C to c⟩ + ⟨c to d⟩ + d to D⟩ |
| **D to E** | (see below) |
| **E to F** | ⟨E to e⟩ + ⟨e to f⟩ + ⟨f to F⟩ |

where the ⟨ ⟩, for this example only, indicate distances. If we are going to fit not only distances between adjacent ports but distances among other pairs, we need to do the fitting in terms of (for A and D, for example)

local A + path a to d + local D

where, for instance, local A more or less corresponds to the aA distance. If we can find, somehow, reasonable values for the "local terms", then we ought to be able to fit

distance   MINUS   local for row   MINUS   local for column

by

|path location for row   MINUS   path location for column|

where the vertical bars--|···|--mean "take the absolute value of"--that is, "take the size of the difference with a + sign no matter what sign the difference has".

There will inevitably be residuals. Some may have simple interpretations; others may not. In the sketch of exhibit 39, the fitted distance between D and E will be

⟨D to d⟩ + ⟨d to e⟩ + e to E⟩

but the sketch shows a much shorter direct path from D to E. This surely ought to make the residual for D and E negative. Exhibit **40** gives the result of such a fitting process (but not how it was done) for eight ports on the West Coast of South America. Of the 28 pairs of ports, the residuals are no larger than 1 mile in all but eight cases. Of these eight, three suggest shortcuts from Coquimbo to ports on the Punta Arenas side, two more suggest shortcuts between adjacent ports, one is −2, and the last--+5 for Lota-Punta Arenas--is quite unexplained.

The most important morals of this example are these:

◇**while it almost always is wise to start trying a row-PLUS-column approach,** there are two-way tables that can be fitted well but only by another approach--**there is no panacea for treating all two-way tables.**

◊ if we understand, even imprecisely, what is really happening in a two-way table--or in any other data structure--it is inexcusable not to use that understanding to guide our approach to its analysis. **It is wrong to expect to learn again, by making unguided approaches to each new body of data, all that we have already learned.** Rather we should use all we knew

exhibit **40** of chapter 11⁺: port distances

**Shiproute distances between South American ports (a different kind of fit)**

A) The DATA--distances in sea miles

|  | Ant | Arr | Cal | Coq | Iqu | Lota | P A | Val |
|---|---|---|---|---|---|---|---|---|
| Antofagista | x | 325 | 215 | 396 | 224 | 828 | 1996 | 576 |
| Arrea | 325 | x | 522 | 702 | 110 | 1134 | 2301 | 8828 |
| Caldera | 215 | 522 | x | 196 | 420 | 628 | 1795 | 396 |
| Coquimbo | 396 | 702 | 196 | x | 602 | 455 | 1623 | 203 |
| Iquique | 224 | 110 | 420 | 602 | x | 1033 | 2201 | 782 |
| Lota | 828 | 1134 | 628 | 455 | 1033 | x | 1191 | 268 |
| Punta Arenas | 1996 | 2301 | 1795 | 1623 | 2201 | 1191 | x | 1432 |
| Valparaiso | 576 | 882 | 376 | 203 | 782 | 268 | 1432 | x |

B) The DATA adjusted for LOCAL DISTANCES--row and column

| (local distance) | (9) | (12) | (8) | 14) | (9) | (16) | (0) | (6) |
|---|---|---|---|---|---|---|---|---|
| Antofagista | x | 304 | 198 | 373 | 206 | 803 | 1987 | 561 |
| Arrea | 304 | x | 502 | 676 | 89 | 1106 | 2289 | 864 |
| Caldera | 198 | 502 | x | 174 | 403 | 604 | 1787 | 362 |
| Coquimbo | 373 | 676 | 174 | x | 579 | 425 | 1609 | 183 |
| Iquique | 206 | 89 | 403 | 579 | x | 1008 | 2192 | 767 |
| Lota | 803 | 1106 | 604 | 425 | 1008 | x | 1175 | 246 |
| Punta Arenas | 1987 | 2289 | 1787 | 1609 | 2192 | 1175 | x | 1426 |
| Valparaiso | 561 | 864 | 362 | 183 | 767 | 246 | 1426 | x |

Examples: 304 = 325 − 9 − 12; 198 = 215 − 9 − 8; 502 = 522 − 12 − 8,.

C) The FIT and the PATH LOCATIONS

|  | Ant | Arr | Cal | Coq | Iqu | Lota | P A | Val |
|---|---|---|---|---|---|---|---|---|
| Antofagista | x | 303 | 200 | 373 | 205 | 803 | 1987 | 562 |
| Arrea | 303 | x | 503 | 676 | 98 | 1106 | 2290 | 865 |
| Caldera | 200 | 503 | x | 173 | 405 | 603 | 1787 | 362 |
| Coquimbo | 373 | 676 | 173 | x | 578 | 430 | 1614 | 189 |
| Iquique | 205 | 98 | 405 | 578 | x | 1008 | 2192 | 767 |
| Lota | 803 | 1106 | 603 | 430 | 1008 | x | 1184 | 241 |
| Punta Arenas | 1987 | 2290 | 1787 | 1614 | 2192 | 1184 | x | 1425 |
| Valparaiso | 562 | 865 | 362 | 189 | 767 | 241 | 1425 | x |
| path | 1994 | 2297 | 1794 | 1621 | 2199 | 1191 | 7 | 1432 |

Examples: 303 = |1994 − 2297|; 200 = |1994 − 1794| (Here "||" indicates "absolute value of" = "size of".)

➡

before to both guide and structure our analyses. (It is **wrong**--not quite so wrong but wrong enough--**to trust our understanding to the point where we never try approaches that our understanding tells us will not work.** Some of our understandings are wrong--we will not find out about their failure unless we probe data, at least occasionally, in ways they suggest are not useful.)

### review questions

What example did we start with? What seemed to be a natural way to begin an analysis? What picture was in our mind? How might we expect the "path" terms to be further broken down? What happened when we tried this? What are the morals of this example?

### exhibit 40 of chapter 11⁺ (continued)

#### D) The RESIDUALS

| | | | | | | | | |
|---|---|---|---|---|---|---|---|---|
| Antofagista | x | 1 | -2 | 0 | 1 | 0 | -1 | -1 |
| Arrea | 1 | x | -1 | 0 | -9 | 0 | -1 | -1 |
| Caldera | -2 | -1 | x | 1 | -2 | 1 | 0 | 0 |
| Coquimbo | 0 | 0 | 1 | x | 1 | -5 | -5 | -6 |
| Iquique | 1 | -9 | -2 | 1 | x | 0 | 0 | 0 |
| Lota | 0 | 0 | 1 | -5 | 0 | x | -9 | 5 |
| Punta Arenas | 0 | -1 | 0 | -5 | 0 | -9 | x | 1 |
| Valparaiso | -1 | -1 | 0 | -6 | 0 | 5 | 1 | x |

Note: Of 28 residuals, twenty are ±1, 0, or −1; two each are −2, −5, −9; one each is 5 and −6.

#### S) SOURCE

1963 World Almanac and Book of Facts, page 684.

#### P) PROBLEMS

40a) Point out a few specific numbers that show that a row-PLUS-column fit cannot fit this data well.

40b) Do the same for a row-TIMES-column fit.

#### T) TEASERS

*Note:* A teaser is a problem, methods for whose solution are not found in this book.

40c) Construct a way to fit the local distances and path locations.

40d) Would you expect a reasonable fit in this form to distances among West Indies, Caribbean, and Gulf ports? (These were in the World Almanac, too!)

## 11M. How much may we have learned?

This chapter has been concerned with four quite diverse topics, each of which can be read and grappled with separately:

⋄ exploring the use of a second row-PLUS-column fit after a PLUS-one fit has been made, and removed.

⋄ using summary polish to fit row-PLUS-column for any summary -- including means and midextremes -- and not merely for median.

⋄ an example where re-expression lets us deal more effectively with data on numbers of metropolitan killings.

⋄ an example, involving coastwise distances, where a quite different kind of supplement to a row-PLUS-column fit (really a row-MINUS-column fit) is desirable and effective.

**The thrust of most of these topics is simple: What we have learned in chapters 10 and 11 is only the root -- a root on which there can be many branches. And any of us may need to invent still another branch.**

# index for chapter 12

review questions 421

**12A. PLUS-one fits** **421**
row-TIMES-column-PLUS-one-fit 421
PLUS-one fit 422
the special response 422
review questions 424

**12B. Pictures for "-PLUS-one" fits** **424**
another example 425
PLUS-one plot 425
-PLUS-one plot of residuals 427
review questions 427

**12C. Making those pictures** **428**
review questions 430

**12D. Sometimes we can have parallel-line plots, still** **431**
but not always 432
review questions 433

**12E. More extended fits** **433**
telephones by continent 433
extended row-PLUS-column fits 436
interpretation of the example 437
an interpretation 438
review questions 438

**12F. Simplification is sometimes possible** **438**
other possibilities 440
the moral 440
review questions 441

**12H. How far have we come?** **441**

| EXHIBITS | PAGE |
|---|---|
| **12A** | |
| 1★ | 423 |
| **12B** | |
| 2 | 424 |
| 3 | 426 |
| 4 | 426 |
| 5 | 427 |
| **12C** | |
| 6★ | 428 |
| 7 | 429 |
| 8 | 430 |
| **12D** | |
| 9 | 431 |
| 10 | 432 |
| **12E** | |
| 11 | 433 |
| 12 | 435 |
| 13★ | 435 |
| **12F** | |
| 14★ | 439 |
| **12F** | |

# Advanced fits 12

This chapter is to cap the climax so far as two-way fits go in this volume. We learned about "-PLUS-one" fits in chapter 10--but we did not learn a variety of things that can help us understand and interpret them. Such questions are discussed here, together with what two-way plots look like for such fits and how such plots can be made.

We saw one extended two-way fit at the end of chapter 11. In this chapter we will learn more about some natural extensions of row-PLUS-column fits.

### review questions

Where did we learn about "-PLUS-one" fits? How much did we learn? Where will we learn more? Are there natural extensions of row-PLUS-column fits? Where will they be discussed?

### 12A. PLUS-one fits

When we introduced row-PLUS-column-PLUS-one fits in the form

$$(*) \qquad \text{all} + \text{row} + \text{col} + \frac{(\text{row})(\text{col})}{\text{constant}}$$

we noticed that when

$$\text{constant} = \text{all}$$

this becomes a row-TIMES-column fit. What happens in general?

A careful eye and a little algebra show that the expression $(*)$ can be written as:

$$(\text{all} - \text{constant}) + \text{constant}\left(1 + \frac{\text{row}}{\text{constant}}\right)\left(1 + \frac{\text{column}}{\text{constant}}\right)$$

which, if we rearrange it to

$$\text{constant}\left(1 + \frac{\text{row}}{\text{constant}}\right)\left(1 + \frac{\text{col}}{\text{constant}}\right) + (\text{all} - \text{constant}),$$

is easily seen to be a

row-TIMES-column-PLUS-one fit.

**We thus see that**

**row-PLUS-column-PLUS-one**

and

**row-TIMES-column-PLUS-one,**

which can be written more explicitly as, respectively,

**row-PLUS-column-PLUS constant times comparison value**

and

**row-TIMES-column-PLUS-another constant,**

**cover exactly the same range of possibilities.** Since these two classes are **identical,** it is natural to speak--or write--simply of a

PLUS-one fit.

It will often be true, as we will see illustrated shortly, that the

row-TIMES-column-PLUS-constant

form is the easy one to understand.

**the special response**

In the case of no added constant, of the

row-TIMES-column

fit, the response "zero" is special. For if EITHER factor is zero, so too is the product. Thus if there are rows that go far enough, there is a row all of whose fitted responses are zero--and if there are columns that go far enough, there is a column all of whose fitted responses are zero. In a row-TIMES-column fit, this

⋄ must happen for zero.

⋄ cannot happen for any other value.

In general, if there is an added constant, given from the additive side, as we know, by

all   MINUS   divisor of (row) × (col),

then this is a special value in exactly the same sense: If the rows go far enough, there is a row all of whose fitted values are the same, equal to this special value--and columns similarly. Again, this

⋄ must happen for the special value.

⋄ cannot happen for any other value.

Let us turn back to the example of warming up the East Coast. There we fitted

$$54.4°F + \text{row} + \text{col} - 1.0 \frac{(\text{row})(\text{col})}{54.4°F},$$

which is the same as

$$54.4°F + \text{row} + \text{col} + \frac{(\text{row})(\text{col})}{-54.4°F}.$$

Thus the special response is

$$54.4°F - (-54.4°F) = 109°F.$$

If our form for the fit were to continue to apply to very extreme cases, the following would have to happen:

◇ if there were a month so hot that any one place had a mean temperature of 109°F, all places would have (up to small residuals) a temperature of 109°F in that month.

◇ if there were a place so hot that any one (ordinary) month had a mean temperature of 109°F there, all months would (up to small residuals) have a temperature of 109°F at that place.

Clearly, there is no such special month--and aren't we on the East Coast glad! There almost might be such a special place--if we went farther and farther south we would expect higher temperatures and less difference between winter and summer. (Somewhere near the equator, things have to cross over, so that Southern-hemisphere summer can match Northern-hemisphere winter, and vice versa).

This is probably the easy way to think about the PLUS-one fit to the East Coast cooling down:

◇ an "ideal equatorial" temperature of 109°F.

◇ depression of mean monthly temperature below this by the product of a month factor and a place factor.

Exhibit 1 gives one set of these factors--and its problems ask you for several more alternative sets.

### exhibit 1 of chapter 12: East Coast

**The fit of exhibit 11 of chapter 10 in TIMES form**

A) FIT

FIT = 109°F − (month factor)(place factor)

B) The FACTORS--one choice

| Month | Temp | Place | Factor |
|---|---|---|---|
| Jan | 72.7°F | Laredo | 0.60 |
| Feb | 71.5°F | Washington | 1.00 |
| Mar | 63.5°F | Caribou | 1.35 |
| Apr | 54.4°F | | |
| May | 44.1°F | | |
| June | 35.4°F | | |
| July | 31.5°F | | |

P) PROBLEMS

1a) Find an equivalent set of factors with Laredo's factor being 1.00.
1b) Do the same for Caribou's factor = 1.00.
1c) Find an equivalent set of factors with April's factor = 1.00.
1d) Do the same with January's factor = 1.00.
1e) Do the same with July's factor = 1.00.
1f) (Teaser) Can you describe by a formula the general pair of sets of factors that reproduce the given fit?

**review questions**

Can we reformulate row-PLUS-column-PLUS-one analyses with $k$ different from +1? Why/why not? What is a row-TIMES-column-PLUS-one analysis? How is it related to a row-PLUS-column-PLUS-one analysis? Why would we want to call anything just a "PLUS-one fit"? What is a special response? What is the special response for a row-TIMES-column analysis? For a PLUS-one analysis? To what example did we apply these ideas? What value was the special response? How might this be interpreted? In how many ways? Which did/didn't make sense? What form of display for the elements of the fit did we reach? Are there simple alternatives?

## 12B. Pictures for "-PLUS-one" fits

We are fortunate that we can make pictures of "-PLUS-one" fits that are almost as simple as those for "row-PLUS-column" fits. Instead of two sets of parallel lines, we will have:

◇ one set of parallel lines.

◇ one set of lines with a common intersection.

We need something of this sort to take care of the special value.

Exhibit 2 shows such a picture, which is purely illustrative. (We will write down the numbers in a later exhibit.) The line $L$ (from the parallel lines -- here, by chance, vertical) and the line $C$ (from the intersecting lines) correspond to the special value, which happens here to be 100.

**exhibit 2 of chapter 12: Illustrative**

**A PLUS-one fit whose special value is clearly visible**

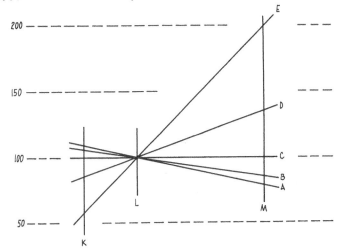

It does not matter whether you choose *A, B, C, D* or *E*--or another line of the same family--the intersection with *L* will come at the same height, 100--because the intersection will be at their common point.

It does not matter whether you choose *K* or *L* or *M*--or another line of the same family--its intersection with *C* will come at the same height, 100--because *C* is horizontal.

This example is unusual in two ways:

◇ the parallel lines are vertical.

◇ the special value is in the picture.

Usually, the parallel lines are slanted. Often, perhaps usually, the intersection point is outside the picture--either because the special value is far from the values actually found, or because the rows actually examined are far from the special row, or because the columns are far from the special column.

### another example

Let us look, then, at the East Coast temperatures again. All the real months are far from the hypothetical special month when all these places would have an equal mean monthly temperature (108.8°F!). All the places are far from the hypothetical place (near the equator, presumably) where the mean monthly temperature would be the same in every month (108.8°F again). So the intersection point will be off the picture.

Exhibit **3** shows the

### PLUS-one plot

of the fit given in exhibit 13 of chapter 10, which was of the form

$$544 + \text{row} + \text{col} - \frac{(\text{row})(\text{col})}{544}$$

which is exactly the same as

$$1088 - \frac{1}{544}(544 - \text{row})(544 - \text{col}),$$

so that the point of intersection is at 1088 (in 0.1°F, hence at 108.8°F).

In particular, we see that the fitted July temperature in Caribou almost matches that for February in Laredo, and that the fitted April temperature in Caribou almost matches that for January in Washington. (The former near 65°F, the latter near 35°F.)

In exhibit 3 we chose to put places on the intersecting lines, and months on the parallel lines. This was in no way compulsory. In exhibit **4**, these choices have been interchanged. When you check the heights of corresponding points in the two exhibits, you will find them exactly the same (to the accuracy of the construction of the pictures). They do not look quite alike--indeed they stress different information, but they **contain** the same information.

426    exhibits 3 and 4/12: Advanced fits

exhibit 3 of chapter 12: East Coast

**PLUS-one plot of the fit in exhibit 13 of chapter 10**

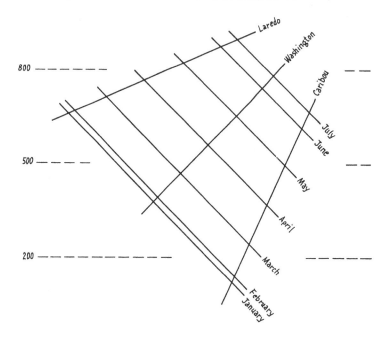

exhibit 4 of chapter 12: East Coast

**PLUS-one plot of the same fit interchanging months and places**

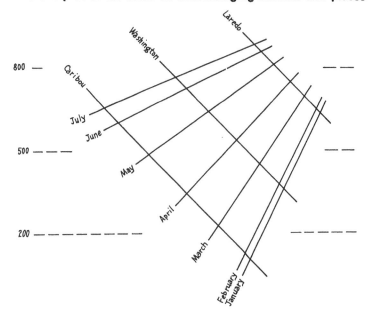

So far as stress goes, exhibit 3 urges more strongly that Caribou and Laredo differ less in temperature in July than in January, while exhibit 4 emphasizes more clearly that the January–July increase in temperature is less in Laredo than Caribou. Both facts can, of course, be read off either picture--we speak of what is stressed while both are shown. In another example, the choice of which set of lines is used for rows--and which for columns--might well be settled by which phenomenon most deserved emphasis.

### -PLUS-one plot of residuals

If we wish to give more attention to the residuals--especially to their pattern--we can follow the path we took in section 9E--plotting the coded values of the residuals over the grid points and omitting the grid entirely. Exhibit 5 shows the result of doing this, starting from exhibit 4. Essentially all the information about the lack of fit is preserved.

### review questions

What is the essential in learning to picture PLUS-one fits? What from elementary geometry helps us do this? How? What does a PLUS-one plot involve? For what example did we draw such? Is there a choice between

exhibit **5** of chapter 12: East Coast

**Plot of coded values overlying exhibit 4**

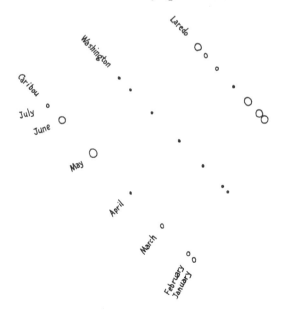

428 exhibit 6/12: Advanced fits

distinct forms? Why/why not? Can we make a corresponding plot of coded residuals? What, if anything, is new at this stage? Which forms of PLUS-one plots of fits have which advantages?

### 12C. Making those pictures

Let us turn back, yet again, to the PLUS-one fit of exhibit 13 of chapter 10. The fit takes the form

$$\text{all} + \text{row} + \text{col} - \frac{(\text{row})(\text{col})}{544}$$

which we can rearrange and bracket as

$$(\text{all} + \text{col}) \quad \text{PLUS} \quad \left(\text{row} - \frac{(\text{row})(\text{col})}{544}\right).$$

Exhibit 6 shows these two terms written out.

We are going to do the same thing here that we did for row-PLUS-column plots:

⋄ break the fit into two parts.

⋄ take these parts as coordinates.

⋄ notice that

$$\text{fit} = \text{sum of coordinates}$$

and that

$$\text{sum of coordinates} = \text{constant}$$

occurs along 45° lines.

exhibit 6 of chapter 12: East Coast

**A two-term form of the analysis of exhibit 11 of chapter 10**

A) The ANALYSIS--incomplete, of course

|      | Laredo | Wash | Caribou |                          |
|------|--------|------|---------|--------------------------|
| Jan  | −111   | −183 | −248    |                          |
| Feb  | −103   | −171 | −232    |                          |
| Mar  | −55    | −91  | −123    |                          |
| Apr  | 0      | 0    | 0       | row                      |
| May  | 62     | 103  | 140     | MINUS                    |
| June | 115    | 190  | 251     | (row)(col)/544           |
| July | 138    | 229  | 310     |                          |
|      | 759    | 544  | 351 }   | all + col                |

P) PROBLEMS

6a) Write out the alternative two-term analysis using (al! + row) and ("what it takes").

6a2) Make the analog of exhibit 7 that is based on the answer to (6a1).

○ rotate the picture so that these lines, which are also

$$\text{fit} = \text{constant}$$

are horizontal.

If "(all + col)" is one coordinate, say the horizontal one, then to each column there corresponds a vertical line. What about the other coordinate? For any fixed row

$$\text{row} - \frac{(\text{row})(\text{col})}{544}$$

reduces to

$$A + B(\text{col})$$

for suitable $A$ and $B$. When plotted against "col" it has to give a straight line.

Thus the columns turn up as vertical lines, while the rows turn up as slanting lines. If we know one point on each vertical line, and two points on each slanting line, we can fill in the whole plot.

Exhibit 7 shows the result of plotting the terms of exhibit 6 as rectangular coordinates, with place lines and month lines drawn in. Clearly, we need have

exhibit **7** of chapter 12: East Coast

**Plotting exhibit 6 on rectangular coordinates (first step toward a two-way-PLUS-one plot)**

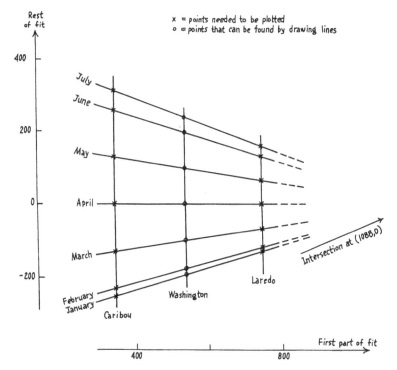

## 12D. Sometimes we can have parallel-line plots, still

In the example at hand, where the fit can be written

$$1088 - \text{data} = \frac{1}{544}(544 - \text{row})(544 - \text{column}),$$

we can take logs, finding

$$\log(1088 - \text{data}) = -\log 544 + \log(544 - \text{row}) + \log(544 - \text{col}),$$

which is in row-PLUS-column form, and can thus be plotted in terms of two sets of parallel lines. Exhibit **9** shows the result, when we preserve "up" by measuring $\log(1088 - \text{data})$ from above downward.

As we learned above, making this picture, especially when a few residuals have to be shown, is not easier than making exhibit 4. Moreover, the things noted in the last section as stressed by exhibits 3 and 4 are only to be extracted from exhibit 9 by very careful consideration of the nature of the vertical scale.

exhibit **9** of chapter 12: East Coast

**Two-way plot using log (1088 − temperature) as a response**

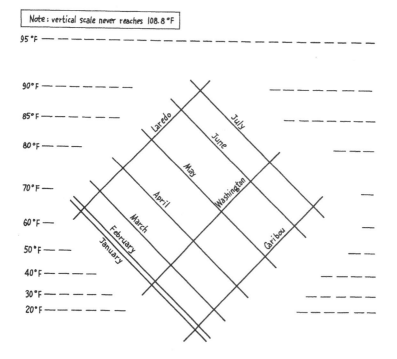

If we had no intuition for--and no interest in--a temperature scale in which a change of 1°F is everywhere an equal change, we would probably find exhibit 6 the display of choice and would have to give careful attention to its scale--emphasizing, perhaps, that the difference between 20°F and 40°F appears, on this scale, closely equal to the difference between 80°F and 85°F. (Such a choice seems unlikely for this example, but possible for many others.) If, on the other hand, we are attached--for some good reason--to the °F scale, we would probably find exhibit 4 (or 3) the display of choice.

### but not always

Whether or not we have a choice about turning our row-TIMES-column fit back to row-PLUS-column depends on the data. Sometimes we have the choice, but not always.

Exhibit 10 shows a simple contrived example of this. One of the simplest plus-ONE plots looks like our earlier exhibit 2, in which all five of AL, BL,..., EL plot as a single point. Not a good prospect for a rectangular grid! (We should be happy with what exhibit 2 does for us.)

exhibit **10** of chapter 12: arbitrary example

**A row-TIMES-column fit where logs do not help**

**A) The DATA**

|   | A | B | C | D | E |
|---|---|---|---|---|---|
| K | 109 | 106 | 100 | 85 | 58 |
| L | 100 | 100 | 100 | 100 | 100 |
| M | 79 | 86 | 100 | 135 | 198 |

**B) A PERFECT FIT to DATA-MINUS-100**

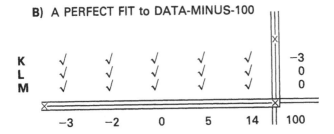

**C) REMARKS**

$$(\text{data} - 100) \equiv (\text{ROW effect}) \times (\text{COL effect})$$

but we cannot take logs because some effects are NEGATIVE.

**P) PROBLEM**

**10)A)** Invent and analyze a similar but larger example.

## review questions

Can we make a two-sets-of-parallel-lines picture of a PLUS-one fit? Why/why not? When do we get in trouble taking logs? What then?

## 12E. More extended fits

### telephones by continent

A rather different kind of example is set out in exhibit **11**. The basic numbers are the numbers of telephones in use on 1 January of seven years in each of seven continents. (*Note.* Here Oceania includes Australia and New

exhibit **11** of chapter 12: world's telephones

**The world's telephones by continent**

**A)** The RAW COUNTS--in thousands

|  | 1951 | 1956 | 1957 | 1958 | 1959 | 1960 | 1961 |
|---|---|---|---|---|---|---|---|
| N. Amer | 45,939 | 60,423 | 64,721 | 68,484 | 71,799 | 76,036 | 79,831 |
| Eur | 21,574 | 29,990 | 32,510 | 35,218 | 37,598 | 40,341 | 43,173 |
| Asia | 2,876 | 4,708 | 5,230 | 6,062 | 6,856 | 8,220 | 9,053 |
| S. Amer | 1,815 | 2,568 | 2,695 | 2,845 | 3,000 | 3,145 | 3,338 |
| Oceania | 1,646 | 2,366 | 2,526 | 2,691 | 2,868 | 3,054 | 3,224 |
| Africa | 895 | 1,411 | 1,546 | 1,663 | 1,769 | 1,905 | 2,005 |
| MidAmer. | 555 | 733 | 773 | 836 | 911 | 1,008 | 1,076 |

**B)** The LOG COUNTS--in 0.001 of log count in millions

| N. Amer | 1662 | 1781 | 1811 | 1836 | 1856 | 1881 | 1902 |
|---|---|---|---|---|---|---|---|
| Eur | 1334 | 1477 | 1512 | 1547 | 1575 | 1606 | 1635 |
| Asia | 459 | 673 | 718 | 783 | 836 | 915 | 957 |
| S. Amer | 259 | 410 | 431 | 454 | 477 | 498 | 523 |
| Oceania | 216 | 374 | 402 | 430 | 458 | 485 | 508 |
| Africa | −48 | 150 | 189 | 221 | 248 | 280 | 302 |
| MidAmer. | −256 | −135 | −112 | 78 | −40 | 4 | 32 |

**C)** ONE row-PLUS-column ANALYSIS of the LOGS of B

|  |  |  |  |  |  |  |  | eff | fit |
|---|---|---|---|---|---|---|---|---|---|
| N. Amer | 39 | 12 | 7 | 0 | −7 | −14 | −15 | 1382 | 1836 |
| Eur | 0 | −3 | −3 | 0 | 1 | 0 | 7 | 1093 | 1547 |
| Asia | −111 | −43 | −33 | 0 | 26 | 67 | 93 | 329 | 783 |
| S. Amer | 18 | 2 | 9 | 0 | −4 | −15 | −12 | 0 | 454 |
| Oceania | −1 | 11 | 4 | 0 | 1 | −4 | −3 | −24 | 430 |
| Africa | −56 | −4 | 0 | 0 | −0 | 0 | 0 | −233 | 221 |
| MidAmer. | 25 | 0 | −12 | −10 | 1 | 13 | 19 | −522 | −68 |
| eff | −213 | −67 | −32 | 0 | 27 | 59 | 81 | 454 | |

**430  exhibit 8/12: Advanced fits**

plotted only the points on the two outermost place lines--no matter how many places there were--since all the points for a given month lie along a straight line.

Exhibit **8** shows the line pattern with a few

$$\text{sum of coordinates} = \text{constant}$$

lines drawn in. Since the sum of coordinates is exactly the fit, we have only to rotate exhibit 8 through 45° in order to obtain exhibit 4.

**review questions**

What sorts of two-term forms of PLUS-one analysis can we use to help make PLUS-one plots? How? What example did we use?

exhibit **8** of chapter 12: East Coast

**The lines of exhibit 7 with some lines "ONE PART plus THE OTHER PART = CONSTANT" indicated**

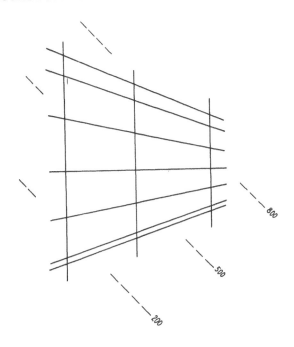

Zealand.) Even a glance at the raw counts indicates that we should start with logs. Where the original logs ranged from −256 to 1902, the residuals range only from −111 to 93. We have accounted for 9/10 (some would say 99/100) of the variation by our continent-and-year effects. They clearly tell most of the story.

Because we have lifted these effects out of the way, we can now see the residuals quite clearly. They now ask for us to go further. Look at the rows for North America and Asia, for instance. They show quite substantial and steady trends from year to year. Something needs to be done to take account of this. What? And how much?

At a rough glance, the steady trends of the residuals remind us of the steady trend of column effects. To look harder at this, it is natural to plot one against the other. This is done in exhibit **12**. Here we see that the North American residuals are rather well fitted by a straight line, as are those for South America.

If we look at the residuals for Asia, they fit a straight line fairly well for the five years 1957 to 1961, but not so well for 1951 and 1956. The Middle America residuals behave somewhat similarly.

If we draw eye-fitted lines on exhibit 12 and find their slopes, the corresponding multiple of the column effects of panel C of exhibit 11 are as set out in panel D. Comparing these values with the residuals of panel C shows that we can gain considerably by such an additional fit.

Panel A of exhibit **13** shows the new residuals, obtained by subtracting these additional fits, and indicates some further adjustments to row effects that now seem worthwhile. Panel B shows the analysis after this further adjustment. The fit is now of the form

$$\text{all} + \text{row eff} + \text{col eff} + (\text{row extra}) \times (\text{col eff})$$

where we have used "row extra" as a harmless name for the additional constant we are using for each row. This fit, as we will shortly see, has a reasonably simple interpretation.

exhibit **11** of chapter 12 (continued)

**D) ADDITIONAL FITS as MULTIPLES of COLUMN EFFECT--from exhibit 12**

|         |      |     |     |   |    |     |     | multiple |
|---------|------|-----|-----|---|----|-----|-----|----------|
| N. Amer | 43   | 13  | 6   | 0 | −5 | −12 | −16 | −.2      |
| Asia    | −234 | −74 | −35 | 0 | 30 | 65  | 89  | 1.1*     |
| S. Amer | 21   | 7   | 3   | 0 | −3 | −6  | −8  | −.1      |
| MidAmer | −85  | −27 | −13 | 0 | 11 | 24  | 32  | .4*      |

* Fits well only for later years.

**S) SOURCE**

The World's Telephones 1961, American Telephone and Telegraph Company, pages 2 and 3.

exhibit **12** of chapter 12: world's telephones

**Plots of residual against column effect for four continents, showing four different slopes (at least for later years)**

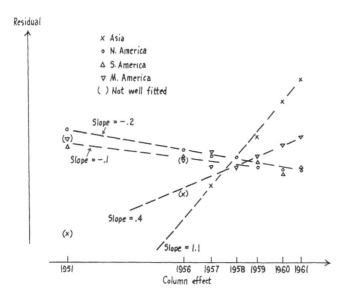

exhibit **13** of chapter 12: world's telephones

**Continuation of exhibit 11**

A) NEW RESIDUALS--from panels C and D of exhibit 11

|  | 1951 | 1956 | 1957 | 1958 | 1959 | 1960 | 1961 | Possible* |
|---|---|---|---|---|---|---|---|---|
| N. Amer | −4 | −1 | 1 | 0 | −2 | −2 | 1 | −1 |
| Eur | 0 | −3 | −3 | 0 | 1 | 0 | 7 | √ |
| Asia | 123 | 31 | 2 | 0 | −4 | 8 | 4 | 2 |
| S. Amer | −3 | 16 | 6 | 0 | −1 | −9 | −4 | −3 |
| Oceania | −1 | 11 | 4 | 0 | 1 | −4 | −3 | √ |
| Africa | −56 | −4 | 0 | 0 | −0 | 0 | 0 | √ |
| M. Amer | 110 | 27 | 1 | −10 | 0 | −11 | −13 | −10 |

*Possible change in row effect, to reduce residuals further.
Example: −4 = 39 − 43, −1 = 12 − 13, 0 = 0 unchanged, −3 = −3 unchanged.

**exhibit 13(B–P)/12: Advanced fits**

**extended row-PLUS-column fits**

The kind of fit we have just made is clearly one that we can try once we have made any row-PLUS-column fit of the form

all + row eff + col eff.

Indeed, we have our choice of adding

(row extra)  TIMES  (col eff)

or

(col extra)  TIMES  (row eff)

or both.

exhibit **13** of chapter 12 (continued)

**B) NEW ANALYSIS--with adjusted row effects**

|          |      |     |     |    |    |    |    | eff   | extra* |
|----------|------|-----|-----|----|----|----|----|-------|--------|
| N. Amer. | −3   | 0   | 2   | 1  | −1 | −1 | 3  | 1381  | −.2    |
| Eur      | 0    | −3  | −3  | 0  | 1  | 0  | 7  | 1093  | .0     |
| Asia     | 121  | 29  | 0   | −2 | −6 | 6  | 2  | 331   | 1.1    |
| S. Amer  | 0    | 19  | 9   | 3  | 2  | −6 | −1 | 2     | −.1    |
| Oceania  | −1   | 11  | 4   | 0  | 1  | −4 | −3 | −24   | .0     |
| Africa   | −56  | −4  | 0   | 0  | 0  | 0  | 0  | −233  | .0     |
| M. Amer  | 120  | 37  | 11  | 0  | −1 | −1 | 0  | −532  | .4     |
|          | −213 | −67 | −32 | 0  | 27 | 59 | 81 | 454   |        |

\* The fit now has the form:

454 + row eff + col eff + (row extra)(col eff).

**C) LETTER-VALUE displays of RESIDUAL of B**

Those before                Those after

#11 residuals              #58 residuals

```
M 6 | 11           M19h| 0
H 3h|32  −4h|28h   ·H10 |1  −1|2
E 2 |120  −1       E 5h |2h −3
```

With hinges at −1 and 1, and extremes of −3 and 2h, the 38 residuals after the marks are small enough to exhibit a very good fit.

**P) PROBLEMS**

13a) Try a "row extra" of −.05 (instead of zero) for Oceania, and one of −.15 (instead of −.1) for South America, and complete the corresponding analysis.

13b) Display the residuals for the analysis of (13a) as in panel C above.

We can regard making a PLUS-one fit by adding

$$\frac{(\text{row})(\text{col})}{\text{div}} = \frac{(\text{row eff})(\text{col eff})}{\text{divisor}}$$

as a special case of any of these, since, if we put

$$\text{row extra} = \frac{\text{row eff}}{\text{divisor}},$$

this last addition becomes

$$(\text{row extra})(\text{col eff}),$$

while if we put

$$\text{col extra} = \frac{\text{col eff}}{\text{divisor}},$$

it becomes

$$(\text{col extra})(\text{row eff}).$$

The additions we are now considering are substantial generalizations of such "PLUS-one" additions. They leave "row extra" or "col extra" wholly at our choice, instead of restricting them to multiples of "row effect" or "col effect", respectively.

A look at panel B (of exhibit 13) shows how little "row extra" looks like a multiple of "row effect" in our case. These generalizations are not always helpful, but they can be tried--with any two-way table.

They are more likely to be useful when (1) there are several rows [or columns] which differ substantially or (2) the data is unusually well behaved. (Our example comes under (2) since we are getting a good fit to three-decimal logarithms, as panel C of exhibit 11 shows.)

### interpretation of the example

How are we to interpret the "row extra" values in our example? Were they NOT there, we might have said;

"In logarithmic form the data are rather well fitted by a steady increase from year to year, the same increase for each continent."

With the "row extra" included, we have an increase of the same shape for each continent, the amount differing.

If we write the fit as

$$\text{all} + \text{row eff} + (1 + \text{row extra}) \times (\text{col eff})$$

or, equivalently, as

$$\text{row fit} + (1 + \text{row extra}) \times (\text{col eff}),$$

we see that the fitted rates of increase for the different continents are in the ratios of the values of

$$1 + \text{row extra}$$

(in particular, 2.1 for Asia, where we fitted a row extra of 1.1, and 0.8 for North America, where we fitted a row extra of $-0.2$).

We can therefore describe what we have fitted in these words.

"With the exceptions noted, the fitted increases--in the log count --for the various continents follow similar paths, the relative rates of increase being, roughly, 2.1 for Asia (after 1957); 1.4 for Middle America (after 1958); 1.0 for Europe, for Oceania (after 1957), and for Africa (after 1956); 0.9 for South America (after 1958); and 0.8 for North America."

So described, the additional fit seems both understandable and meaningful.

Note that the usual construction gives a plot with one family of parallel lines and one set of skew lines.

### an interpretation

One way to look at such fits is as a mixture of PLUS and TIMES fitting. The "(1 + row extra) × (col eff)" term is clearly a TIMES fit. To this we have added "all + row eff", which is surely a PLUS fit. Further thoughts--and generalizations--are left to the reader.

### review questions

What example? What expression? Why? How did a PLUS analysis work? How did we look farther? What additional fit did this suggest? How much did this help? What kind of fit do we now have? How generally can we try such a fit? How many kinds are there? How related to PLUS-one fits? How did we reformulate our fit? Was it harder/easier to understand?

### 12F. Simplification is sometimes possible

This same example can be used to illustrate how row-PLUS-column fits--and their extensions--can sometimes be simplified. If we set the year effects of panel B (of exhibit 13) against the years, as follows, we see a clear parallelism:

| Year | 1951 | 1956 | 1957 | 1958 | 1959 | 1960 | 1961 |
|---|---|---|---|---|---|---|---|
| Year − 1958 | −7 | −2 | −1 | 0 | 1 | 2 | 3 |
| eff | −213 | −67 | −32 | 0 | 27 | 59 | 81 |

The effect is close to [30 times (year − 1958)].

Without losing much, then, we could simplify our fit by replacing

$$454 + \text{row effect} + \text{col effect}$$

by

$$454 + \text{row effect} + 30(\text{year} - 1958),$$

or, starting with the extended fit, we could replace

$$454 + \text{row effect} + (1 + \text{row extra}) \cdot (\text{col effect})$$

by

$$454 + \text{row effect} + (1 + \text{row extra})(30)(\text{year} - 1958),$$

which is equivalent to

$$\text{row fit} + (\text{row slope}) \times (\text{year} - 1958),$$

where

$$\text{row slope} = 30(1 + \text{row extra}).$$

Exhibit **14** shows, in its panel A, the analysis that the last replacement leads to. While not quite as good as panel B of exhibit 13, it does quite well for a simple fit.

exhibit **14** of chapter 12: world's telephones

**Some reduced fits to the telephone data (data from exhibit 11)**

A) The REDUCED FORM of the EXTENDED FIT--based on panel B of exhibit 12

|  | residuals | | | | | | | row fit | row slope |
|---|---|---|---|---|---|---|---|---|---|
|  | 1951 | 1956 | 1957 | 1958 | 1959 | 1960 | 1961 | | |
| N. Amer | −5 | −6 | 0 | 1 | −3 | 2 | 5 | 1835 | 24 |
| Eur | −3 | −10 | −5 | 0 | −2 | −1 | −2 | 1547 | 30 |
| Asia | 115 | 14 | −4 | −2 | −12 | 4 | −17 | 785 | 63 |
| S. Amer | −8 | 8 | 2 | −2 | −6 | −12 | −14 | 456 | 27 |
| Oceania | −4 | 4 | 2 | 0 | −2 | −5 | −12 | 430 | 30 |
| Afr | −59 | −11 | −2 | 0 | −3 | −1 | −9 | 221 | 30 |
| M. Amer | 116 | 27 | 8 | 0 | −4 | −2 | −16 | −78 | 42 |
| (year−1958) | (−7) | (−2) | (−1) | (0) | (1) | (2) | (3) | | |

Fit = (row fit) + (row slope) × (year − 1958)

P) PROBLEMS

14a/b/c/d/e/f/g) Fit a straight line against (year − 1958) by the methods of chapter 6, to the log counts for North America, South America, ..., Middle America, respectively, being careful to omit counts for early years (if any) where fit is poor. (Usefully spread through the students of a class.)

14h) Combine the results of (14a) to (14g) into a new analysis of the form A. How much improvement do you think has been made?

→

**other possibilities**

We clearly were not confined to

$$(\text{constant})(\text{year} - 1958)$$

as a replacement for

column effect

in this example. We could, for instance, have tried

$$(\text{constant})(\text{year} - 1958) + (\text{another constant}) \times (\text{year} - 1958)^2$$

or

$$(\text{constant})(\log \text{year} - \log 1958),$$

to say nothing of many other possibilities. Where the column [or row] labels are essentially numerical--or are intimately associated with a set of numbers outside the data--we can always explore any of a variety of simplifications-- either in the basic row-PLUS-column fit or in one of its extensions.

**the moral**

Such analyses bring to the data in each row [or column] the sort of analysis we entered into in chapters 5 and 6. They are the most obvious example of stacking modes of analysis one upon another in this book. We need to recognize--and to emphasize to ourselves--that stacking one analytical approach onto another--sometimes for several layers--is the only way we know to do a respectable job of data analysis when the data can be grasped by more than one handle.

For us, the essential novelty of two-way tables is the existence of at least two handles--rows and columns. Many sets of data have three, four, or five handles; others have ten--or even a hundred.

> Learning to work with two handles is an essential part of an education in exploratory data analysis. It opens our eyes to some of the things we can do with more than one handle.

exhibit **14** of chapter 12 (continued)

14i) Compare the analysis of (14h) with that of panel B of exhibit 13. Do you think that reducing the fit (was/was not) going too far?

14j) Produce a reduced fit analysis corresponding to panel C of exhibit 11. How do you feel it compares with that of that panel?

14k) Write out the computations leading to the first row of residuals in A above. Explain in detail.

14l) Do the same for another row of your choice.

The possibility of simplification is just a case of more handles. In our telephone example we have a choice of three handles:

⋄ the name of the continent,
⋄ the name of the year,
⋄ the number of the year,

or, if we prefer, several handles:

⋄ the name of the continent,
⋄ the name of the year,
⋄ the number of the year,
⋄ the number of the year, and its square,
⋄ the logarithm of the number of the year,
⋄ and so on.

In deciding whether to simplify or nor, we are choosing between one set of two handles and another.

**We need to recognize what handles are naturally available in a problem -- and try to learn which handle or set of handles exposes the behavior of the data to us most usefully.**

### review questions

How did we simplify our fit? Was the simplified fit fairly satisfactory? To what earlier chapters is this technique related? How does its existence illustrate the potentialities of data with two or more handles? What handles did we have in the example?

### 12H. How far have we come?

This chapter introduces us to various extensions of the PLUS and TIMES fits, including a deeper look at PLUS-one fit. It gives us a broader hold on what we can do when we need it. Making a row-PLUS-column fit appears more and more a matter of fitting an incomplete description, one whose fitting is quite likely to disclose the need of going further.

We are now ready:

⋄ to throw back any PLUS-one fit from the form

$$\text{all} + \text{row} + \text{col} + \text{constant} \times (\text{comparison value})$$

to the form

$$(\text{constant}') + (\text{row}') \times (\text{col}')$$

(thus showing that PLUS-one fits are equally closely related to TIMES fits as to PLUS fits),

◊ and to find the value of the special response, the response such that, if it occurs in one cell of a fit, it also occurs both in every cell in that column and in every cell in that row.

◊ to make pictures involving one set of parallel lines and one set of radiating lines for either PLUS-one fits or their residuals, using the initial expression of the response,

◊ and, by taking logs of the differences between actual response and special response, to make parallel-line pictures of such fits.

We have also seen how to go well beyond PLUS-one fits, fitting a straight line for each stripe (all rows **or** all columns) of residuals against the corresponding effects. The result can be written in various forms, and thought of, in particular, as the sum of a special PLUS fit and a general TIMES fit, thus opening the way to further generalizations.

We now understand more clearly that numerical labels for our rows-- once we have fitted either row effects alone or both row effects and row extras--offer us a further opportunity, one of fitting row effects--or row extras--usefully well by some function of row labels. (Clearly we can change "row" to "column" throughout what we have just said.)

We now recognize--and can make use of--the relation of PLUS-one fits to both PLUS and TIMES fits. The same is true for generalizations combining PLUS and TIMES.

**We have reached a natural terrace in our ascent of the mountain of what can be done to data once it has two simple handles. We are high enough on the shoulder of this mountain to look back and see our earlier steps in clearer perspective. We can also see that the mountain keeps on rising.**

# Three-way fits    13

chapter index on next page
___

The chapters spent on two-way fits are time and space well spent. We have learned to deal with two-way data quite well, and in a variety of ways. The arithmetic has not always been easy, but it has been feasible by hand, especially since there were comfortably usable checking techniques.

Those who were fortunate enough to have access to a suitably programmed computer could relax--those who, out in the real world, will need to get at least rough answers with pen or pencil and scratch paper can have learned to do just this.

We learned, at the end of chapter 12, a little of how more than one handle on our data makes a variety of attacks and explorations possible.

In the present short chapter, we will find an introduction to three- and more-way tables. In simple cases we really can do the arithmetic by hand, as an example will show. We do have to be more careful, working harder to get it right to begin with, since checking and salvage are not as easy.

As we deal with more more-way tables, the need for a computer goes up (and the chance that our computer is all ready to do what we want goes down). The short account here is intended:

⋄ to show what sorts of pieces turn up in such analyses.

⋄ to show that the simplest cases can be done by hand.

⋄ to encourage readers who have or are presented with three- or more-way data to analyze it simply and effectively.

### review questions

What about the chapters on two-way analyses? And their arithmetic? Was there a place for a computer? To what is this chapter an introduction? What can we say about the feasibility of the required arithmetic? Is there a place for the computer? What are the aims of this chapter?

### 13A. Three- and more-way analyses: arrangement and tagging

The same sort of things we have done for two-way tables can be done for three- and more-way tables. Fortunately, we do not need three- or four-way sheets of paper. We can push everything into a two-way layout, provided we are willing to use more blocks.

# index for chapter 13

review questions 443

**13A. Three- and more-way analyses: arrangement and tagging** 443
a psychological experiment 445
a tag example 446
review questions 448

**13B. An analysis of the psychological example** 448
review questions 451

**13C. Making three-way analyses** 452
review questions 457

**13D. Three-way re-expression** 458
review questions 461

**13E. More about the example** 462
re-formulated responses 464
*log (time of detection)* 464
review questions 465

**13H. How far have we come?** 465

| EXHIBITS | PAGE |
|---|---|
| **13A** | |
| 1★ | 446 |
| **13B** | |
| 2★ | 449 |
| 3 | 451 |
| **13C** | |
| 4 | 453 |
| 5 | 454 |
| 6★ | 455 |
| 7★ | 456 |
| **13D** | |
| 8 | 459 |
| 9 | 460 |
| 10★ | 461 |
| **13E** | |
| 11★ | 462 |
| 12★ | 463 |

### a psychological experiment

In 1944, Johnson and Tsao published an account of an experiment for measuring the sensitivity of individuals to changes in pull on a ring held by one finger. Measurements were made with different initial steady pulls (here called weight) and different rates of increase of the pull once increase was begun (here called rate). The persons tested included four who were blind and four who could see. The measurements recorded were the increase in pull before the subjects reported feeling a change.

We discuss only the part of the experiment involving:

◇three persons, all blind, called here K, L, and M [IB(2), IIB(1), and IIB(2) in Johnson's notation].

◇three initial pulls (weights), called here 1, 4, and 7 (as in Johnson 1949) and amounting to 100, 250, and 400 grams.

◇four rates of increase, called here $a$, $b$, $c$, and $d$ (as in Johnson 1949), amounting to 100, 200, 300, and 400 grams per minute.

This small part of the Johnson–Tsao data provides an example of convenient size, one that can show us how two-way methods can extend to three-way data. Exhibit **1** gives the sums, over two experimental sessions, called "dates" by Johnson and Tsao, of the mean of five responses on each date. (The values given have been rounded to whole numbers for simplicity; they are expressed in grams of weight.)

This table, as is almost necessary for a table put on a page, is arranged in rows and columns. The columns correspond to one variable (weight), while the rows correspond to two variables (subject and rate). In any situation involving all combinations of certain versions of three or more variables, we can think of the numbers laid out similarly, with certain variables combined into columns and others combined into rows. In Johnson's table of all 448 numbers (his table 87) for example:

◇sex, sight, date, and individual are combined into columns, of which there are $2 \times 2 \times 2 \times 2 = 16$.

◇weight and rate are combined into rows, of which there are $7 \times 4 = 28$.

Arrangements of this kind are the only natural way to put the numbers down. They also provide, as we shall see, a convenient way to begin their analysis.

Where we have 3, 4, ..., 10, ... factors, the problem of tagging them can be somewhat troublesome. Surely "row" and "column" are no longer useful. Two choices are natural:

◇when we are analyzing "for real," tags linked to the way we think of the factor--such as "rate" for rate, "init" for initial pull, "subj" for person who is the subject--are likely to be best.

## exhibit 1(A–C)/13: Three-way fits

◇ when we are analyzing to learn about analysis, it often helps to link the tag for a factor to the tags for its versions--as when KLM is used to tag the factor whose versions are K, L, or M.

### a tag example

Panel B shows the first column of panel A as a two-way table, for which panel C shows a row-PLUS-column analysis. In panel D we have taken the analysis of panel C and stretched it out into a single column again. If this were

### exhibit 1 of chapter 13: finger limens

**The 3 × 3 × 4 = 36 basic observations**

**A) The DATA**

| Rate & subject | Initial pull (weight) | | |
|---|---|---|---|
| | 1 | 4 | 7 |
| aK | 39 | 31 | 26 |
| aL | 16 | 12 | 12 |
| aM | 18 | 13 | 14 |
| bK | 85 | 55 | 56 |
| bL | 32 | 22 | 20 |
| bM | 31 | 26 | 26 |
| cK | 101 | 84 | 70 |
| cL | 43 | 33 | 30 |
| cM | 42 | 38 | 40 |
| dK | 151 | 124 | 98 |
| dL | 63 | 51 | 37 |
| dM | 58 | 55 | 46 |

**B) ONE-THIRD of ABOVE, RE-ARRANGED--data for pull 1**

| Subject | Rate | | | |
|---|---|---|---|---|
| | a | b | c | d |
| K | 39 | 85 | 101 | 151 |
| L | 16 | 32 | 43 | 63 |
| M | 18 | 31 | 42 | 58 |

**C) The SAME, ANALYZED**

| | a | b | c | d | eff |
|---|---|---|---|---|---|
| K | −32 | −2 | 2 | 33 | 55 |
| L | 0 | 0 | −1 | 0 | 0 |
| M | 4 | 1 | 0 | −3 | −2 |
| | −22 | −6 | 6 | 25 | 38 |

→

all the data, putting it in one-column form would be an empty exercise. There is more data, however. To set it all out in rows and columns, we have to divide the three factors into two groups--one for rows and one for columns. One group has to include at least two factors. (Since it is easier to have many rows than many columns, we can be almost sure that rows will have to accommodate two factors.) This means each column will need the kind of analysis we have just illustrated.

exhibit **1** of chapter 13 (continued)

**D)** The ANALYSIS, RE-ARRANGED

|      | 1   |
|------|-----|
| aK   | −32 |
| aL   | 0   |
| aM   | 4   |
| bK   | −2  |
| bL   | 0   |
| bM   | 1   |
| cK   | 2   |
| cL   | −1  |
| cM   | 0   |
| dK   | 33  |
| dL   | 0   |
| dM   | −3  |
| aØ   | −22 |
| bØ   | −6  |
| cØ   | 6   |
| dØ   | 25  |
| ØK   | 55  |
| ØL   | 0   |
| ØM   | −2  |
| ØØ   | 38  |

**P) PROBLEMS**

**1a)** Imitate panels B, C and D with the second column of panel A.

**1b)** Do the same for the third column.

**S) SOURCE**

Palmer O. Johnson, 1949. **Statistical Methods in Research,** Table 87 on page 290.

Note how we have tagged the lines in panel D. The (two-way) residuals are tagged by their full tag--rate and subject. The subject effects are tagged with a "ø" for rate (to say that rate isn't there any more) and the tag of the subject. Similarly the rate effects are tagged with the tag of the rate and an "Ø" for the subject (to say that subject isn't there any more). Finally the common term is tagged "øØ" (to say that neither rate nor subject is there any more). Here "Ø" is a Scandinavian vowel, pronounced roughly somewhere between "uh" and "urh" and used by modern mathematicians for empty sets--sets that contain no element. We will use this same approach to tag both rows and columns in the three-way case.

### review questions

Can we do for three-way tables what we have done for two-way tables? For more-way tables? What example? What three factors or kinds of circumstances? What factors in the full experiment? What example did we use to illustrate how we tag rows and columns? How did we push it around? What is ø or Ø or 0 used to mark? What expression did we begin the main analysis with?

### 13B. An analysis of the psychological example

Let us now turn to the three-way analysis of our $3 \times 3 \times 4$ of difference limens. We will learn below that it is good to analyze logs of weight changes rather than weight changes themselves. Accordingly, exhibit **2** opens with the logs of the values in panel A of exhibit 1. The analysis in panel B has eight blocks ($2^3$ here, not $2^2 = 4$, as in the two-way case) and ALL EIGHT have to be added up to get the given values. (Notice that we are using an "effects only" analysis. We could include the fits, but this might be confusing.)

For 66, the given value for **dM7**, the eight terms in the breakdown are

0, the residual at **dM7**

−2, the effect at **dM0**

−5, the effect at **dØ7**

5, the effect at **øM7**

22, the effect at **dØ0**

0, the effect at **øM0**

−4, the effect at **øØ7**

50, the common term at **øØ0**

These do sum to 66, as you can easily check.

exhibit **2** of chapter 13: finger limens

**The log-difference limens and a three-way analysis (entries in panel A are $-100 + 100 \log_{10}$ of entries in panel A of exhibit 1)**

A) The DATA (logs)

|  | 1 | 4 | 7 |
|---|---|---|---|
| aK | 59 | 49 | 42 |
| aL | 20 | 8 | 8 |
| aM | 26 | 11 | 15 |
| bK | 93 | 74 | 75 |
| bL | 51 | 34 | 30 |
| bM | 49 | 42 | 42 |
| cK | 100 | 92 | 85 |
| cL | 63 | 52 | 48 |
| cM | 62 | 58 | 60 |
| dK | 118 | 109 | 99 |
| dL | 80 | 71 | 57 |
| dM | 76 | 74 | 66 |

B) One THREE-Way ANALYSIS -- ($\checkmark$ means 0)

|  | 1 | 4 | 7 | 0 |
|---|---|---|---|---|
| aK | $\checkmark$ | 2 | −1 | −1 |
| aL | −2 | $\checkmark$ | 4 | $\checkmark$ |
| aM | 6 | −3 | $\checkmark$ | $\checkmark$ |
| bK | $\checkmark$ | −5 | $\checkmark$ | 4 |
| bL | 1 | $\checkmark$ | $\checkmark$ | −1 |
| bM | $\checkmark$ | 1 | $\checkmark$ | $\checkmark$ |
| cK | $\checkmark$ | 2 | −1 | −2 |
| cL | −1 | $\checkmark$ | $\checkmark$ | $\checkmark$ |
| cM | $\checkmark$ | $\checkmark$ | 1 | $\checkmark$ |
| dK | $\checkmark$ | −1 | 2 | $\checkmark$ |
| dL | $\checkmark$ | 1 | $\checkmark$ | $\checkmark$ |
| dM | $\checkmark$ | $\checkmark$ | $\checkmark$ | −2 |
| aø | $\checkmark$ | $\checkmark$ | $\checkmark$ | −36 |
| bø | 2 | $\checkmark$ | $\checkmark$ | −9 |
| cø | −2 | $\checkmark$ | $\checkmark$ | 8 |
| cø | $\checkmark$ | 4 | −5 | 22 |
| øK | $\checkmark$ | $\checkmark$ | $\checkmark$ | 34 |
| øL | 2 | $\checkmark$ | $\checkmark$ | −6 |
| øM | −6 | $\checkmark$ | 5 | $\checkmark$ |
| øø | 12 | $\checkmark$ | −4 | 50 |

Note that we have tried to keep our fonts straight, using ø for "none of abcd", Ø for "none of KLM", and **0** (zero) for "none of 147". Be careful to pronounce "147" as "one-four-seven"; it is NOT "one forty-seven".

C) Some EXAMPLES of the BREAKDOWN -- with row labels

$59 = \underbrace{0 + (-1)}_{aK} + \underbrace{0 + (-36)}_{a\emptyset} + \underbrace{0 + 34}_{\emptyset K} + \underbrace{12 + 50}_{\emptyset\emptyset}$

$49 = \underbrace{2 + (-1)}_{aK} + \underbrace{0 + (-36)}_{a\emptyset} + \underbrace{0 + 34}_{\emptyset K} + \underbrace{0 + 50}_{\emptyset\emptyset}$

$20 = \underbrace{-2 + 0}_{aL} + \underbrace{0 + (-36)}_{a\emptyset} + \underbrace{2 + (-6)}_{\emptyset L} + \underbrace{12 + 50}_{\emptyset\emptyset}$

$93 = \underbrace{0 + 4}_{bK} + \underbrace{2 + (-9)}_{b\emptyset} + \underbrace{0 + 34}_{\emptyset K} + \underbrace{12 + 50}_{\emptyset\emptyset}$

We now have terms in our breakdown for, taking the above eight terms from below upwards,

◇ a common term, here 50 at ∅∅0.

◇ an initial pull effect (1, 4, or 7; often called an initial pull main effect), in particular 4 at ∅∅7.

◇ a subject effect (K, L, or M, often called a subject main effect), in particular 0 at ∅M0.

◇ a rate effect (a, b, c, or d; often called a rate main effect), in particular 22 at d∅0.

◇ a subject-initial pull effect (K1, K4,..., M4, or M7; often called a subject-initial pull interaction), in particular 5 at ∅M7.

◇ a rate-initial pull effect (a1, a4,..., d4, or d7, often called a rate-initial pull interaction), in particular −5 at d∅7.

◇ a rate-subject effect (aK, aL,..., dL, or dM, often called a rate-subject interaction), in particular, −2 at **dM0.**

◇ a (three-way) residual (aK1 to dM7), in particular, 0 at **dM7.**

In the case of **dM7,** whose breakdown we have just looked at, the large terms are the common and the rate effect (50 and 22), followed by two interactions (−5 and 5).

Looking at panel B of exhibit 2 as a whole, we see that the common and the three main effects are large, while the two-factor interactions do not appear much larger than the three-way residuals.

It is easy to put together detailed plots for the larger effects with box-and-whiskers (or even schematic) plots for the smaller effects. Exhibit **3** shows one such. Clearly, the three big things are:

◇ the effect of rate (more weight change the faster the rate).

exhibit **2** of chapter 13 (continued)

P) PROBLEMS

2a) Write out the breakdowns for 8, 34, and 71.

2b) Is there any breakdown involving both −36 and 23? Both −6 and −6? both the −5 in line **bK** and the −5 in line **d∅**? Explain.

2c) Pick out seven entries not so far used as examples or problems and write out the breakdown, tagging as in panel C.

2d) Copy the breakdowns of panel C, tagging each value with its full identification.

2e) What kind of an effect is a∅0? ∅L0?

◇ the distinction between subject K (slower to respond) and subjects L and M.

◇ the effect of initial pull (slower when lower).

After these come some individual two-factor effects and residuals. (Among these latter, the countervailing nature of **M1** and *a***M1** and of *b***K** and *b***K4** may deserve notice. Can you explain or discuss?)

**review questions**

What example did we take? What expression was analyzed? How many blocks do we get from a full three-way analysis? How are initial values expressed in terms of effects and residuals? How can we summarize the components of the breakdown? How many large effects did we finally have? What were they (describe)? What came next? What seemed interesting about them?

exhibit **3** of chapter 13: finger limens

**Summary of the effects and residuals of the analysis of exhibit 2**

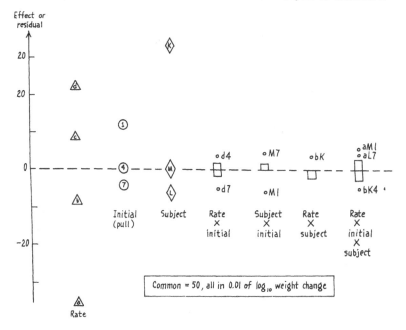

### 13C. Making three-way analyses

Systematically analyzing a three-way set of data is straightforward. Repeated removal--in particular repeated median removal--works both well and easily. Exhibit **4** follows the details step-by-step for the example of exhibit 2.

In panel A, the separate box well below the data shows the *abcd*-medians. Their box is cut up into four parts, three of which are empty (at this stage). The value 96, tagged **K1**, is the median of 59, 93, 100, and 118, which are tagged *a***K1**, *b***K1**, *c***K1**, and *d***K1**, respectively. In the separate box, the blank tagged **Ø1**, similarly, is the median of the four blanks tagged *a***Ø1**, *b***Ø1**, *c***Ø1**, and *d***Ø1**, respectively, and so on.

Were we in a two-way situation, we would have written

| | | |
|---|---|---|
| 59 | | −37 |
| 93 | and could have | −3 |
| 100 | converted this to | 4 |
| 118 | | 22 |
| 96 | | 96 |

(In the two-way case, we dropped the 96, and recovered the fit later. In three- and more-way cases, it does not pay to do this. This means that we need to be more careful with our arithmetic.) In panel B, we find that this is exactly what has happened to these five entries. Step 1 consists of taking the *abcd*-medians out of the values medianed and putting them into the corresponding effects.

In panel B, as well as showing our state after step 1, we show the 147-medians (the 1, 4, 7 medians, not the "one forty-seven" medians). Here the value −37, in the *a***K0** line, is the median of −37, −35, and −38 (tagged, respectively *a***K1**, *a***K4**, and *a***K7**). Similarly 84, tagged **ØK**, is the median of 96, 84 and 80 (tagged, respectively **ØK1**, **ØK4** and **ØK7**). Of which three entries is the blank following 22, 23, 20 in the column of 147-medians the median?

In a two-way table, we would have had

$$-37 \quad -35 \quad -38 \quad | \quad -37$$

and would have replaced this by

$$0 \quad 2 \quad -1 \quad \| \quad -37$$

Again it is just this, and its analog for the other fourteen nonblank 147-medians ("one-four-seven" not "one forty-seven"), that we do.

Panel C shows where we stand after the second step of median removal. The upper lefthand block (12 rows and 3 columns) has been shrunk till no value is larger in size than 6. Again we also show, below the main table, the medians to be used in the next step. This time all four kinds are nonblank. The −5, tagged *d***7**, is the median of −3, −5, and −5 (tagged *d***K7**, *d***L7**, and *d***M7**, respectively). The −36, tagged *a***ø**, is the median of −37, −35, and −36 (tagged *a***Kø**, *a***Lø**, and *a***Mø**, respectively). The 12, tagged **ø**, is the median of 12, 14 and 5 (tagged **øK1**, **øL1** and **øM1**, respectively). The 51, tagged **ø0**, is the median of 84, 43 and 51 (tagged **øK0**, **øL0**, and **øM0**, respectively).

Now the machinery is running at full diversity, so that the main display of panel D (found in exhibit **5**) has no blanks. (It does, of course, have some zeros.) We have now finished one cycle of median removal. This analysis is, in fact, relatively satisfactory. Since this is not always so, and since we are doing a leading example, we will continue for a second cycle of median removal.

exhibit **4** of chapter 13: finger limens (panels A, B and C)

**Polishing by medians (note that blanks can be read as zeros)**

A) The DATA

|     | 1   | 4   | 7  | 0 |
|-----|-----|-----|----|---|
| aK  | 59  | 49  | 42 |   |
| aL  | 20  | 8   | 8  |   |
| aM  | 26  | 11  | 15 |   |
| bK  | 93  | 74  | 75 |   |
| bL  | 51  | 34  | 30 |   |
| bM  | 49  | 42  | 42 |   |
| cK  | 110 | 92  | 85 |   |
| cL  | 63  | 52  | 48 |   |
| cM  | 62  | 58  | 60 |   |
| dK  | 118 | 109 | 99 |   |
| dL  | 80  | 71  | 57 |   |
| dM  | 76  | 74  | 66 |   |
| aØ  |     |     |    |   |
| bØ  |     |     |    |   |
| cØ  |     |     |    |   |
| dØ  |     |     |    |   |
| ØK  |     |     |    |   |
| ØL  |     |     |    |   |
| ØM  |     |     |    |   |
| ØØ  |     |     |    |   |

B) After STEP 1

|     | 1   | 4   | 7   | 0 |
|-----|-----|-----|-----|---|
| aK  | −37 | −35 | −38 |   |
| aL  | −37 | −35 | −31 |   |
| aM  | −30 | −39 | −36 |   |
| bK  | −3  | −10 | −5  |   |
| bL  | −6  | −9  | −9  |   |
| bM  | −7  | −8  | −9  |   |
| cK  | 4   | 8   | 5   |   |
| cL  | 6   | 9   | 9   |   |
| cM  | 6   | 8   | 9   |   |
| dK  | 22  | 25  | 19  |   |
| dL  | 23  | 28  | 18  |   |
| dM  | 20  | 24  | 15  |   |
| ØK  | 96  | 84  | 80  |   |
| ØL  | 57  | 43  | 39  |   |
| ØM  | 56  | 50  | 51  |   |

C) After STEP 2

|     | 147-medians | 1  | 4  | 7  | 0   |
|-----|-------------|----|----|----|-----|
| aK  | −37         | 0  | 2  | −1 | −37 |
| aL  | −35         | −2 | 0  | 4  | −35 |
| aM  | −36         | 6  | −3 | 0  | −36 |
| bK  | −5          | 2  | −5 | 0  | −5  |
| bL  | −9          | 3  | 0  | 0  | −9  |
| bM  | −8          | 1  | 0  | −1 | −8  |
| cK  | 5           | −1 | 3  | 0  | 5   |
| cL  | 9           | −3 | 0  | 0  | 9   |
| cM  | 8           | −2 | 0  | 1  | 8   |
| dK  | 22          | 0  | 3  | −3 | 22  |
| dL  | 23          | 0  | 5  | −5 | 23  |
| dM  | 20          | 0  | 4  | −5 | 20  |
| ØK  | 84          | 12 | 0  | −4 | 84  |
| ØL  | 43          | 14 | 0  | −4 | 43  |
| ØM  | 51          | 5  | −1 | 0  | 51  |

abcd-medians

|   | 1  | 4  | 7  |
|---|----|----|----|
| K | 96 | 84 | 80 |
| L | 57 | 43 | 39 |
| M | 56 | 50 | 51 |

Ø

KLM-medians

|   | 1  | 4 | 7  | 0   |
|---|----|---|----|-----|
| a | 0  | 0 | 0  | −36 |
| b | 2  | 0 | 0  | −8  |
| c | −2 | 0 | 0  | 8   |
| d | 0  | 4 | −5 | 22  |
| Ø | 12 | 0 | −4 | 51  |

454    exhibit 5(D–F)/13: Three-way fits

Accordingly, panel D (of exhibit 5) shows *abcd*-medians again. This time fifteen are zero and one is +1. Thus we have little to change in taking step 4 and going to panel E. This panel shows 147-medians (one-four-seven; also for the second time) only two of which are nonzero.

Taking step 5, and going to panel F is also easy. This panel shows **KLM** medians (again for the second time) of which two are nonzero. An easy step 7 takes us to panel G (shown in exhibit **6**).

exhibit **5** of chapter 13: finger limens

**Polishing by medians continued (panel lettering continued from exhibit 4)**

D) After STEP 3

|     | 1  | 4  | 7  | 0  |
|-----|----|----|----|----|
| aK  | 0  | 2  | −1 | −1 |
| aL  | −2 | 0  | 4  | 1  |
| aM  | 6  | −3 | 0  | 0  |
| bK  | 0  | −5 | 0  | 3  |
| bL  | 1  | 0  | 0  | −1 |
| bM  | −1 | 0  | −1 | 0  |
| cK  | 1  | 3  | 0  | −3 |
| cL  | −1 | 0  | 0  | 1  |
| cM  | 0  | 0  | 1  | 0  |
| dK  | 0  | −1 | 2  | 0  |
| dL  | 0  | 1  | 0  | 1  |
| dM  | 0  | 0  | 0  | −2 |
| aØ  | 0  | 0  | 0  | −36 |
| bØ  | 2  | 0  | 0  | −8 |
| cØ  | −2 | 0  | 0  | 8  |
| dØ  | 0  | 4  | −5 | 22 |
| ØK  | 0  | 0  | 0  | 33 |
| ØL  | 2  | 0  | 0  | −8 |
| ØM  | −7 | −1 | 4  | 0  |
| ØØ  | 12 | 0  | −4 | 51 |

E) After STEP 4

|     | 1  | 4  | 7  | 0  | 147-medians |
|-----|----|----|----|----|-------------|
| aK  | 0  | 2  | −1 | −1 | 0 |
| aL  | −2 | 0  | 4  | 0  | 0 |
| aM  | 6  | −3 | 0  | 0  | 0 |
| bK  | 0  | −5 | 0  | 3  | 0 |
| bL  | 1  | 0  | 0  | −2 | 0 |
| bM  | −1 | 0  | −1 | 0  | −1 |
| cK  | 1  | 3  | 0  | −3 | 1 |
| cL  | −1 | 0  | 0  | 0  | 0 |
| cM  | 0  | 0  | 1  | 0  | 0 |
| dK  | 0  | −1 | 2  | 0  | 0 |
| dL  | 0  | 1  | 0  | 0  | 0 |
| dM  | 0  | 0  | 0  | −2 | 0 |
| aØ  | 0  | 0  | 0  | −36 | 0 |
| bØ  | 2  | 0  | 0  | −8 | 0 |
| cØ  | −2 | 0  | 0  | 8  | 0 |
| dØ  | 0  | 4  | −5 | 22 | 0 |
| ØK  | 0  | 0  | 0  | 33 | 0 |
| ØL  | 2  | 0  | 0  | −7 | 0 |
| ØM  | −7 | −1 | 4  | 0  | −1 |
| ØØ  | 12 | 0  | −4 | 51 | 0 |

F) After STEP 5

|     | 1  | 4  | 7  | 0  |
|-----|----|----|----|----|
| aK  | 0  | 2  | −1 | −1 |
| aL  | −2 | 0  | 4  | 0  |
| aM  | 6  | −3 | 0  | 0  |
| bK  | 0  | −5 | 0  | 3  |
| bL  | 1  | 0  | 0  | −2 |
| bM  | 0  | 1  | 0  | −1 |
| cK  | 0  | 2  | −1 | −2 |
| cL  | −1 | 0  | 0  | 0  |
| cM  | 0  | 0  | 1  | 0  |
| dK  | 0  | −1 | 2  | 0  |
| dL  | 0  | 1  | 0  | 0  |
| dM  | 0  | 0  | 0  | −2 |
| aØ  | 0  | 0  | 0  | −36 |
| bØ  | 2  | 0  | 0  | −8 |
| cØ  | −2 | 0  | 0  | 8  |
| dØ  | 0  | 4  | −5 | 22 |
| ØK  | 0  | 0  | 0  | 33 |
| ØL  | 2  | 0  | 0  | −7 |
| ØM  | −6 | 0  | 5  | −1 |
| ØØ  | 12 | 0  | −4 | 51 |

*abcd*-medians

|   | 0 | 0 | 0 | 0 |
|---|---|---|---|---|
| K | 0 | 0 | 0 | 0 |
| L | 0 | 0 | 0 | 1 |
| M | 0 | 0 | 0 | 0 |
| Ø | 0 | 0 | 0 | 0 |

**KLM**-medians

|   | 0 | 0 | 0 | 0 |
|---|---|---|---|---|
| a | 0 | 0 | 0 | 0 |
| b | 0 | 0 | 0 | −1 |
| c | 0 | 0 | 0 | 0 |
| d | 0 | 0 | 0 | 0 |
| ø | 0 | 0 | 0 | −1 |

exhibit **6** of chapter 13: finger limens

**Polishing by medians concluded (panel lettering continued from exhibits 4 and 5)**

G) After STEP 6

|     | 1 | 4 | 7 | 0 | medians |
|-----|---|---|---|---|---------|
| aK  | 0 | 2 | -1 | -1 | 0 |
| aL  | -2 | 0 | 4 | 0 | 0 |
| aM  | 6 | -3 | 0 | 0 | 0 |
| bK  | 0 | -5 | 0 | 4 | 0 |
| bL  | 1 | 0 | 0 | -1 | 0 |
| bM  | 0 | 1 | 0 | 0 | 0 |
| cK  | 0 | 2 | -1 | -2 | 0 |
| cL  | -1 | 0 | 0 | 0 | 0 |
| cM  | 0 | 0 | 1 | 0 | 0 |
| dK  | 0 | -1 | 2 | 0 | 0 |
| dL  | 0 | 1 | 0 | 0 | 0 |
| dM  | 0 | 0 | 0 | -2 | 0 |
| aØ  | 0 | 0 | 0 | -36 | 0 |
| bØ  | 2 | 0 | 0 | -9 | 0 |
| cØ  | -2 | 0 | 0 | 8 | 0 |
| dØ  | 0 | 4 | -5 | 22 | 0 |
| ØK  | 0 | 0 | 0 | 34 | 0 |
| ØL  | 2 | 0 | 0 | -6 | 0 |
| ØM  | -6 | 0 | 5 | 0 | 0 |
| ØØ  | 12 | 0 | -4 | 50 | 0 |

H) SAME--zeroes omitted

|     | 1 | 4 | 7 | 0 |
|-----|---|---|---|---|
| aK  |   | 2 | -1 | -1 |
| aL  | -2 |   | 4 |   |
| aM  | 6 | -3 |   |   |
| bK  |   | -5 |   | 4 |
| bL  | 1 |   |   | -1 |
| bM  |   | 1 |   |   |
| cK  |   | 2 | -1 | -2 |
| cL  | -1 |   |   |   |
| cM  |   |   | 1 |   |
| dK  |   | -1 | 2 |   |
| dL  |   | 1 |   |   |
| dM  |   |   |   | -2 |
| aØ  |   |   |   | -36 |
| bØ  | 2 |   |   | -9 |
| cØ  | -2 |   |   | 8 |
| dØ  |   | 4 | -5 | 22 |
| ØK  |   |   |   | 34 |
| ØL  | 2 |   |   | -6 |
| ØM  | -6 |   | 5 |   |
| ØØ  | 12 |   | -4 | 50 |

I) SAME--±1 also omitted

|     | 1 | 4 | 7 | 0 |
|-----|---|---|---|---|
| aK  |   | 2 |   |   |
| aL  | -2 |   | 4 |   |
| aM  | 6 | -3 |   |   |
| bK  |   | 5 |   | 4 |
| bL  |   |   |   |   |
| bM  |   |   |   |   |
| cK  |   | 2 |   | -2 |
| cL  |   |   |   |   |
| cM  |   |   |   |   |
| dK  |   |   | 2 |   |
| dL  |   |   |   |   |
| dM  |   |   |   | -2 |
| aØ  |   |   |   | -36 |
| bØ  | 2 |   |   | -9 |
| cØ  | -2 |   |   | 8 |
| dØ  |   | 4 | -5 | 22 |
| ØK  |   |   |   | 34 |
| ØL  | 2 |   |   | -6 |
| ØM  | -6 |   | 5 |   |
| ØØ  | 12 |   | -4 | 50 |

**abcd**-medians

| 0 | 0 | 0 | 0 |
|---|---|---|---|
| 0 | 0 | 0 | 0 |
| 0 | 0 | 0 | 0 |
| 0 | 0 | 0 | 0 |

P) PROBLEMS

6a) Take the as-initially-expressed data of panel A of exhibit 1 and carry out steps 1, 2, and 3.

6a) Continue with steps 4, 5, and 6.

6a3) Do you feel a need to go further? Make a picture analogous to exhibit 7. Summarize in words what it seems to want to tell us.

6b) Go on beyond panel (I), omitting ±2's also. Which of the three--this, panel (H), panel (I)--seems to you to convey the clearest and most informative picture?

6c) (teaser) Can we tell whether or not we also took the five blank medians out of the corresponding entries as part of step 2?

This panel shows not only 147-medians--all twenty of which are zero--but also *abcd*-medians--all sixteen of which are zero. Since we have just taken out all 20 KLM-medians, we know that all 20 KLM-medians of panel G will also be zero. We can be happy--at least as doers of arithmetic--since we have come to the end of our process of median removal. Unless we want to change our analytical process, we are done. We have, in particular, reached the analysis set out in detail in panel B of exhibit 2 and pictured in exhibit 3.

In this case, two cycles of median removal have done all that median removal could do. In general, this need not happen. Most often, we think two cycles of median removal will be enough--whether or not we are at the end of the process. "Two are probably enough" is good advice, but does not absolve the analyst from looking hard to see what one or more additional cycles might do.

The remaining panels of exhibit 6--panels H and I--show the result of omitting first zeros and then also ±1's (and, in a problem, ±2's). These offer one simple way to try to show more clearly what seems to be going on in such a moderately complex situation--**taking the smallest values out of the way.**

Exhibit 7 sets some problems using other sets of data.

exhibit 7 of chapter 13: data and problems

**Some three-way problems**

### A) CLEANING ABILITY OF DETERGENT COMBINATIONS

|  | $A_1B_1$ | $A_1B_2$ | $A_1B_3$ | $A_2B_1$ | $A_2B_2$ | $A_2B_3$ | $A_3B_1$ | $A_3B_2$ | $A_3B_3$ |
|---|---|---|---|---|---|---|---|---|---|
| $C_1$ | 106 | 197 | 223 | 198 | 329 | 320 | 270 | 361 | 321 |
| $C_2$ | 149 | 255 | 294 | 243 | 364 | 410 | 315 | 390 | 415 |
| $C_3$ | 182 | 259 | 297 | 232 | 389 | 416 | 340 | 406 | 387 |

$A_1, A_2, A_3$ = three concentrations of a detergent;
$B_1, B_2, B_3$ = three concentrations of sodium carbonate,
$C_1, C_2, C_3$ = three concentrations of sodium carboxy-methyl cellulose.

Larger values correspond to better cleaning.

### B) POROSITIES IN PACKED TUBES

|  | LC | LY | BC | BY | SC | SY | GC | GY |
|---|---|---|---|---|---|---|---|---|
| 0 | 473 | 440 | 490 | 457 | 478 | 450 | 479 | 424 |
| 3 | 366 | 342 | 378 | 331 | 392 | 291 | 390 | 298 |
| 9 | 373 | 366 | 402 | 373 | 406 | 375 | 385 | 349 |

L, B, S, G = Lead, phosphor-Bronze, polyStyrene, Glass
C, Y = Cascaded, dropped 1 Yard
0, 3, 9 = distance from wall in tenths of an inch (in 3-inch diameter tube)
Entries are porosities in percent (in thousandths).

**review questions**

How can we tackle analyzing a three-way table? How can we lay out three-way data? Initially? After analysis? What kind of repeated removal did we try? What change did we make in the details of our calculation? On what example? How many cycles did we go through? Why did we stop there? Might we have stopped somewhere else? Why/why not?

exhibit **7** of chapter 13 (continued)

**C) STRENGTHS OF CEMENT**

|   | 3p | 3P | 7p | 7P | 8p | 8P | 9p | 9P |
|---|---|---|---|---|---|---|---|---|
| m | 15.959 | 16.151 | 16.443 | 16.636 | 16.780 | 16.859 | 17.086 | 17.309 |
| M | 16.028 | 16.221 | 16.451 | 16.607 | 16.788 | 16.948 | 17.090 | 17.318 |

3, 7, 8, 9 = cure for 3 days, 7 days, 28 days, and 28 days under special conditions
p, P = old press-sheet, new press-sheet
m, M = old mould, new mould
Entries are sums of 6 logarithms of compressive strengths in $kg/cm^2$.

**PROBLEMS (longish)**

7a) Set up the data from panel A in a convenient way, and remove medians once in each of the three possible ways.

7a2) Continue the removal for a further round.

7a3) Complete the analysis.

7b) Do the same as (7a) for the data of panel B.

7b2) Continue as in (7a2)

7b3) Complete as in (7a3).

7c) Do the same as (7a) for the data of panel C.

7c2) Continue as in (7a2).

7c3) Complete as in (7a3).

**S) SOURCES**

(For panel A) A. J. Feuell and R. E. Wagg, 1949, "Statistical Methods in Detergency Investigations" **Research** 2: 334. (Used by R. L. Wine, 1964, **Statistics for Scientists and Engineers,** Prentice Hall, at page 492).

(For Panel B) J. C. MacRae and W. A. Gray, 1961, "Significance of the properties of materials in the packing of real spherical particles," **British Journal of Applied Physics,** 12: 164–72 (from their table 6, used by Wine at page 493).

(For panel C) A. Hald, 1952. **Statistical Theory with Engineering Applications,** John Wiley (see page 481).

## 13D. Three-way re-expression

While dealing with three-way analysis, we should take at least a brief look at re-expression--in particular at how we can try to detect a leaning toward a

$$\text{row-TIMES-column-TIMES-layer}$$

analysis once we have made a

$$\text{row-PLUS-column-PLUS-layer}$$

analysis.

Suppose that we have fitted

$$\text{all} + \text{row} + \text{col} + \text{lay}$$

where "lay" stands for "layer," a convenient word to use even if, as is the case, we are not going to show our numbers as layers of rows crossing columns. (Even if we do not plan to show them in such a way, we may do well to think about them thus.) These are the leading terms in

$$(\text{all})\left(1 + \frac{\text{row}}{\text{all}}\right)\left(1 + \frac{\text{col}}{\text{all}}\right)\left(1 + \frac{\text{lay}}{\text{all}}\right),$$

the remaining terms of which (when we multiply out) being

$$+ \frac{(\text{row})(\text{col})}{\text{all}} + \frac{(\text{row})(\text{lay})}{\text{all}} + \frac{(\text{col})(\text{lay})}{\text{all}} + \frac{(\text{row})(\text{col})(\text{lay})}{\text{all}^2}.$$

Thus we might expect to take as comparison values:

◊ for each two-factor interaction, a quantity formed from main effects just the way we would do for a two-way analysis.

◊ for a three-way residual, a quantity formed by multiplying together all three of the corresponding main effects and then dividing through by the square of the common term.

We thus can look separately at each of four parts--or we can look at all of them together.

Exhibit **8** shows a three-way PLUS analysis of the original Johnson and Tsao data (from exhibit 1) and the four sets of comparison values. Exhibit **9** shows the four diagnostic plots. For each of these, if we had to choose between slope zero (residual = 0) and slope one (residual = comparison value), we

exhibit **8** of chapter 13: finger limens

**A rate-PLUS-subject-PLUS-initial analysis of the Johnson and Tsao data, and the corresponding comparison values (data in exhibit 1)**

A) The ANALYSIS         B) Some COMPARISON VALUES

|     | 1  | 4  | 7  | 0   | 1    | 4 | 7    |       |
|-----|----|----|----|-----|------|---|------|-------|
| aK  | −7 | 0  | 2  | −16 | −6.4 | 0 | 1.3  | −19.7 |
| aL  | 0  | 0  | 0  | 3   | .7   | 0 | −.1  | 2.3   |
| aM  | 7  | 0  | −3 | 0   | 0    | 0 | 0    | 0     |
| bK  | 0  | −9 | 1  | 4   | −1.8 | 0 | .4   | −5.5  |
| bL  | 0  | 0  | 0  | 0   | .2   | 0 | 0    | .6    |
| bM  | 1  | 0  | −2 | 0   | 0    | 0 | 0    | 0     |
| cK  | 0  | 4  | 0  | −9  | 2.1  | 0 | −.4  | 6.6   |
| cL  | 0  | 0  | 0  | 0   | −.2  | 0 | .1   | −.8   |
| cM  | 0  | 0  | 1  | 1   | 0    | 0 | 0    | 0     |
| dK  | 4  | 0  | −5 | 35  | 8.5  | 0 | −1.7 | 26.3  |
| dL  | 0  | 0  | 0  | 0   | −1.0 | 0 | .2   | −3.1  |
| dM  | −3 | 0  | 1  | 0   | 0    | 0 | 0    | 0     |
| aØ  | −6 | 0  | 2  | −18 | −5.8 | 0 | 1.2  |       |
| bØ  | 0  | 0  | 0  | −5  | −1.6 | 0 | .3   |       |
| cØ  | 0  | 0  | −1 | 6   | 1.9  | 0 | −.4  |       |
| dØ  | 2  | 0  | −12| 24  | 7.7  | 0 | −1.6 |       |
| Ø   | 11 | 0  | −7 | 34  | 11.0 | 0 | −2.2 |       |
| ØL  | 0  | 0  | 0  | −4  | −1.3 | 0 | .3   |       |
| ØM  | −6 | 0  | 4  | 0   | 0    | 0 | 0    |       |
| ØØ  | 10 | 0  | −2 | 31  |      |   |      |       |

Examples: $-6.4 = (-18)(34)(10)/(31)^2$
$-5.8 = (-18)(10)/31$
$11.0 = (34)(10)/31$
$-19.7 = (-18)(34)/31$

would choose the latter. (The certainty with which we would make the choice does vary appreciably.) In this clear case, all four voices speak out for re-expression by logs.

We may also want to deal with less clear instances. In such circumstances we might like to persuade all the voices to speak together. Exhibit **10** has a stem-and-leaf combining all four sets of comparison values, and a listing of those--with near ties combined--where the comparison value seems large enough to be worth attention. Those who do problem (10a) will learn how strongly the four diagnostic plots speak out when united.

exhibit **9** of chapter 13: finger limens

**The four separate diagnostic plots (to the same scale numbers from exhibit 22)**

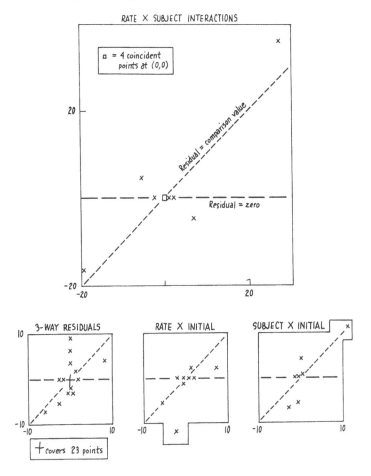

Exhibit **11** shows a second analysis of the same data--obtained in fact by mean removal--and asks, as problems, for its diagnosis.

**review questions**

What is the relation of TIMES to PLUS in the three-way case? (Can you forecast what happens in the four-way case?) How many kinds of comparison values are there in the three-way case? How many kinds of diagnostic plots? What example? What did the diagnostic plots indicate? Might we want to combine diagnostic plots? How might we do this? (Is there a place for 3RSS middling?)

**exhibit 10** of chapter 13: finger limens

**Combining and focussing the four diagnostic plots**

**A) STEM-and-LEAF--**
**FOR ALL COMPARISON**
**VALUES**

```
H | 11.0, 26.3
8 | 5
7 | 7
6 | 6
5 |
4 |
3 |
2 | 13
1 | 239
0 | 12233467
Z | ⊠⊠⊠ ∴
-0 | 12448
-1 | 036678
-2 | 2
-3 | 1
-5 | 58
-6 | 4
L | -19.7,
```

**B) Those of SIZE ≥ 1.5 ORDERED**
**WITH THE CORRESPONDING**
**RESIDUALS**

| comparison value(s) | residual(s) |
|---|---|
| 26.3 | 35 |
| 11.0 | 11 |
| 8.5 | 4 |
| 7.7 | 2 |
| 6.6 | 9 |
| 2.1, 2.3 | 0,3 |
| 1.9 | 0 |
| -1.6, -1.6, -1.7, -1.8 | 0, 0, -5, -12 |
| -2.2 | -7 |
| -3.1 | 0 |
| -5.5 | 4 |
| -5.8 | -6 |
| -6.4 | -7 |
| -19.7 | -16 |

**P) PROBLEMS**

**10a)** Plot the seventeen (comparison value, residual) pairs of panel B in comparison with the lines "residual = zero", and "residual = comparison value".

### 13E. More about the example

We have learned that the Johnson and Tsao example deserves either a TIMES analysis of the weight changes or a PLUS analysis of their logs. Can we go further?

**exhibit 11 of chapter 13: finger limens**

**Another analysis of the same data**

A) The ANALYSIS

|     | 1    | 4    | 7    | ø     |
|-----|------|------|------|-------|
| aK  | -.9  | 2.5  | -1.5 | -1.1  |
| aL  | -2.4 | 0    | 2.5  | .4    |
| aM  | 3.4  | -2.4 | -.9  | .8    |
| bK  | 1.1  | -2.5 | 1.5  | 1.5   |
| bL  | .9   | .3   | -1.2 | -1.3  |
| bM  | -2.0 | 2.2  | -.3  | -.2   |
| cK  | .0   | .4   | -1.3 | -1.0  |
| cL  | .8   | -.8  | 0    | .2    |
| cM  | -.7  | -.5  | 1.3  | .9    |
| dK  | 0    | -1.3 | 1.4  | .6    |
| dL  | .8   | .5   | -1.3 | .8    |
| dM  | -.7  | .8   | .0   | -1.5  |
| aø  | .4   | -1.8 | 1.2  | -29.8 |
| bø  | 1.8  | -2.4 | .6   | -1.8  |
| cø  | -2.1 | .7   | 1.4  | 12.7  |
| dø  | .1   | 3.4  | -3.3 | 27.1  |
| øK  | 1.4  | .3   | -1.7 | 24.7  |
| øL  | 1.9  | -.2  | -1.7 | -14.8 |
| øM  | -3.3 | -.2  | 3.3  | -9.8  |
| øø  | 8.1  | -2.0 | -6.0 | 56.2  |

P) PROBLEMS

11a) Find the comparison values corresponding to the analysis of panel A.

11a2) Make the four diagnostic plots using these comparison values.

11a3) Combine the comparison values as in exhibit 10.

11a4) Make a unified and focussed diagnostic plot.

11b) Do (11a) for the analysis of exhibit 10.

11b2) Do (11a2) for this.

11b3) Do (11a3) for this.

11b4) Do (11a4) for this.

11c/c2/c3/c4) Do (11a/a2/a3/a4) for some other analysis as specified.

Exhibit 12 looks further at the main effects.

Panel A treats the effect of rate of change of pull. A careful look shows us that the fitted term varies quite closely with the log of the rate, the differences "fitted minus log" ranging only from −36 to −40. Let us take this for an observed fact, and return to it in a moment.

Panel B deals with the three observers present in our limited set of data. Here it looks as though sex might be relevant--though three cases are hardly enough to make any strong suggestion.

Panel C deals with the factor or initial pull, which is found to have some effect, though not a large one. Here the effect is crudely like −1/4 log(initial pull), the differences "fitted minus 1/4 log" ranging only from 10 to 12.

exhibit 12 of chapter 13: finger limens

**The main effects made explicit (fitted terms from exhibit 6)**

A) THE abcd OR "RATE" FACTOR

| version | Rate (gm/min) | log rate* | fitted term | diff |
|---|---|---|---|---|
| a | 100 | 0.0 | −36 | −36 |
| b | 200 | 30.1 | −9 | −39 |
| c | 300 | 47.7 | 8 | −40 |
| d | 400 | 60.2 | 22 | −38 |

* These columns are −200 + 100 log$_{10}$ (previous column).

B) THE KLM OR "OBSERVER" FACTOR:

| Version | Sex | fitted term |
|---|---|---|
| K | Male | 34 |
| L | Female | −6 |
| M | Female | 0 |

C) THE 147 OR "INITIAL PULL" FACTOR:

| Version | Initial pull (grams) | log | fitted term | ** |
|---|---|---|---|---|
| 1 | 100 | 0.0 | 12 | 12 |
| 4 | 250 | 39.8 | 0 | 10 |
| 7 | 400 | 60.2 | −4 | 11 |

** Fitted term PLUS (1/4)(log initial pull)

P) PROBLEMS

12a) For the modified response discussed in the text, modify exhibit 3 accordingly.

12a2) Describe in words what this new picture tells us.

Thus some discussion--interesting and possibly useful--can be made of each of the three main-effect terms.

In order to understand the first of these three it is time to note what we have been analyzing. In fact:

$$y = \text{change in pull before a change was detected,}$$
$$\log y = -100 + 100 \cdot \log_{10}(\text{change in grams}).$$

Since the latter seems to move parallel to:

$$\log \text{rate} = -200 + 100 \cdot \log_{10}(\text{grams per minute}),$$

we ought to think about the--little-changing--difference:

$$\log y - \log \text{rate} = 100 + 100 \log_{10} \frac{\text{change in grams}}{\text{grams per minute}}.$$

When we do this, we realize that

$$\frac{\text{change in grams}}{\text{grams per minute}} = \text{delay in minutes,}$$

and that what we have observed can be very simply put in words as follows:

◇ the time taken to detect a beginning of a change in pull is much more constant than the change in pull taken to detect such a beginning.

## re-formulated responses

Once we know this, there seems little reason not to take account of this from the beginning of our analysis. We ought to do better by analyzing

### log(time of detection)

rather than either the weight change for detection or the log of this weight change. Doing this alters all initial entries by an amount that is uniform for each version of *abcd*. This can mean that the residuals are unaffected, while the fit is changed in the exactly corresponding way.

Combining the common term and the rate term we have:

| Version | Old fit* | Change | New fit |
|---------|----------|--------|---------|
| a | 14 | 0.0 | 14 |
| b | 41 | 30.1 | 11 |
| c | 58 | 47.7 | 10 |
| d | 72 | 60.2 | 12 |

which leads to a common term of, say, 12 and a much smaller rate term, whose values are 2, −1, −2 and 0.

---

* Panel G of exhibit 6.

### review questions

What happened when we looked harder at the main effects? Did this make us change the way we thought about the experiment and its data? Why/why not? Where did we come out?

### 13H. How far have we come?

This chapter shows us the way to generalize our analyses of two-way tables to the analysis of three-way tables, leaving the extensions to more-way tables open to those who wish to go that way. Again it is mainly a how-to-do-it chapter.

We are now ready:

◇ to array three- (or more-)way data on the page, so that we can both look at and work with it effectively. As a part of this we have to (a) think of how we can effectively label rows and columns with suitable tags, using different fonts (capitals, lower case, digits, etc.) where that can help, and (b) choose a suitable symbol (ø, ∅, or zero) for the factor ( = set of circumstances) that is no longer there.

◇ to "median polish" such tables by "shifting out" values from all the blocks that are susceptible to taking medians over a given set of circumstances. We have to do this for each factor ( = set of circumstances) at least once, often twice, and sometimes more than twice.

◇ to plot a summary of all the kinds of effects--and the residuals. We do not have all the nice properties that we have become used to in the two-way case, but we can get a reasonably general picture (and perhaps be told which two-way plots will help the most).

◇ to define, in the three-way case, not just a single kind of comparison value (as in the two-way case) but four kinds,

◇ and to plot different blocks of the analysis against each kind, both separately and together. (We then have a choice between a PLUS-four analysis, and re-expression, among many things.)

We now understand that the obligation to simplify and interpret the blocks of our analysis, particularly (of course) the one-factor effects, is, if anything, stronger for three- and more-way tables than for two-way ones.

We now see that we can extend all that we have learned to do with two-way tables in a new direction. We see great opportunities--and we cannot help being reluctant to do all the computations by hand. Our desire--and need--for effective computer programs has to be great.

These are practical matters--often of crucial importance--but the matters of principle are untouched by their presence. We can now believe that **when data has three or more handles**--and we know how to grasp them two at a time--**there will be a way,** whether or not we see it at first, **to grasp all of them together.**

# index for chapter 14

review questions 467

**14A. Coordinates and level traces** 467
  *level lines* 468
  *level curves* 468
  *trace* 468
  *level traces* 468
  review questions 470

**14B. Different middle traces for the same slices** 470
  changing the traced coordinate 470
  review questions 475

**14C. An explanation** 475
  curved traces 476
  review questions 476

**14D. Changing the slicing coordinate** 476
  review questions 479

**14E. What matters?** 481
  some morals 481
  review questions 481

**14F. Rematching and strength of relationship** 482
  dependency ratios 485
  *dependency ratio* 486
  *dependency trace* 486
  reconstituted traces 490
  a comment 490
  *actual middle* 490
  review questions 490

**14H. How far have we come?** 491

**14I. The ubiquity of medians (optional section)**
  odd count 492
  even count 493
  review questions 493

| EXHIBIT | PAGE |
|---|---|
| 14A | |
| 1 | 469 |
| 14B | |
| 2★ | 471 |
| 3 | 473 |
| 4 | 474 |
| 5 | 474 |
| 14C | |
| 6 | 477 |
| 14D | |
| 7 | 478 |
| 8 | 479 |
| 9 | 480 |
| 10★ | 480 |
| 14E | |
| 14F | |
| 11★ | 482 |
| 12 | 485 |
| 13 | 486 |
| 14★ | 487 |
| 15 | 489 |
| 16 | 489 |
| 14H | |
| 14I | |

# Looking in two or more ways at batches of points  14

In chapters 8 and 9 we looked at batches of $(x, y)$ points in a single rather dull way: First slice on $x$ and then trace on $y$. Doing this produced quite helpful pictures; we learned to grasp such batches and coax useful displays out of them. The time has come for us to go on, to ask such natural questions as: Why not interchange $x$ with $y$? What if we use $x + y$? Or $x - y$, or still other choices?

Sometimes using $x$ to cut seems almost inevitable, but by no means always. If, for example, one is height and the other is weight, it is hard to say that either variable "must" come first, that it "must" be used for cutting. The present chapter explores the consequence of change, of looking in two or more ways at the same batch of $(x, y)$ points.

**review questions**

What did we do earlier? In which chapters? What did this do for us? What might we now do? Are we going to do it? Is it inevitable that we use $x$ to cut? Give an example. What are we now to do?

**14A. Coordinates and level traces**

When we talk about $(x, y)$ pairs, we are thinking about objects or situations involving two numbers. Say

$$x = \text{height}$$

and

$$y = \text{weight}.$$

We naturally think of them as plotted in ordinary rectangular coordinates. And how do--or should--we think of these coordinates? Once we have this clear, we can think much more easily about as many possible coordinates, of as many different kinds, as may help us in a specific situation.

Saving of pencil and ink--avoiding confusion of what we plot: both often lead us to use "axes" or "scales" to tell us about rectangular coordinates in a finished picture. No such picture begins life with only axes or scales, however

--rather it begins on a piece of graph paper. What tells us about the $x$-coordinates of our points are all the vertical lines on our graph paper--both those that are there and those that could be drawn between the ones that are there. Similarly, all possible horizontal lines tell us about our y-coordinates.

The verticals that tell us about $x$ are lines $x$ = constant. Because $x$ is constant, the points on any one vertical line are at the same level as far as $x$ goes, so we should not be surprised that such lines are called

level lines

of $x$. What really tells us about $x$ is the sequence of all its level lines. On graph paper we give a dense enough set of level lines for us to behave as if they were all there.

Some coordinates are constant on curves, not lines. They have

level curves

not level lines. To a mathematician, a line is a special case of curve, so he uses "level curve" as the general term. So might we, were we doing mathematics--but we're not. In the common language, a line is straight and a curve is curved, and never the twain shall meet.

We need a word to cover a straight line, a broken line, a curve, etc. We will use

trace

Thus every coordinate has its 
level traces.

Not all level traces are straight lines. Not all level lines are parallel. Much surveying is done by measuring the direction of interesting points from each of two base stations. The level traces for the directions from any one point are straight lines radiating from that point. The second strip of exhibit **1** shows how these families of level lines look, alone and together.

Instead of two directions, we could try two **distances,** which will give us circles as level curves. The third strip of exhibit 1 does this for the U.S. and distances from New York and Los Angeles. We would be in a little trouble here, since there are now pairs of points both of whose coordinates are the same. But if we use distances from Sept Iles, Quebec, and Vancouver, British Columbia, as the lowest (fourth) strip shows, we can get away from this difficulty, so long as we know we are staying in the U.S.

This shows us that we can be quite free with what we use for a coordinate, even if we want to think of the points as plotted on a very definite and fixed map. All we have to do is choose two sets of level traces that, together, meet our needs, and we are practically in business.

We do have to choose what values to assign to each level trace, but the difference between two sets of choices is just a matter of re-expression. Each family of level traces by themselves corresponds to a variable that we can name, but not yet both measure and express. A family of level traces, each with a value, corresponds to something we have named, measured, and expressed.

Thus, if we are concerned with the total bank deposits in a state, as we were in chapter 8, the same sets of level lines (or level curves) work for

◇ bank deposits in dollars.
◇ the square root of bank deposits in dollars.
◇ the log of bank deposits in dollars.
◇ and, of course, bank deposits in cents--or in millions of dollars.

Only the numerical labels on the level lines (or level curves) are different.

exhibit **1** of chapter 14

**Level traces (several examples involving one and two sets)**

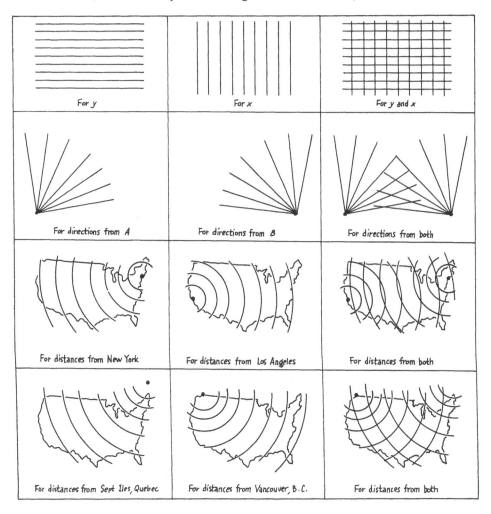

#### review questions

What tells us about the *x*-coordinate of a point on graph paper? The *y*-coordinate? What is a level line? A level curve? A level trace? Why did we choose the term "trace"? Can we use circles as level traces? How many sets? Given a fixed map, how much is our choice of coordinates restricted? How much do we know when we know all about a family of level traces except the value taken on each? What does knowing the values add? What effect does re-expression have on the family of level traces?

## 14B. Different middle traces for the same slices

It is now time to look at different middle traces of the same point scatter, asking which ones are much alike and which very different.

#### changing the traced coordinate

The 1960 Census reported on a total of 88 unincorporated places with 25,000 or more inhabitants. We shall look at only two of the 67 columns of information provided in the 1962 *County and City Data Book*:

⋄ the % of occupied housing units with ≥1.01 persons per room (column 255, called *x* or "% crowding" below).

⋄ the % of occupied housing units with one or more automobiles (column 267, called *y* or "% automobiles" below).

Exhibit **2** sets out the data on slices according to *x*, using an alternate method, and gives corresponding medians (raw and smoothed), all this for *x*, for *y*, for *x* + *y*, and for *y* − *x*. It then gives the corresponding *y* values for the smooths of *y*, *x* + *y*, and *y* − *x*. This allows us to compare the middle traces for:

$$y \text{ against } x$$

$$x + y \text{ against } x$$

$$y - x \text{ against } x$$

which we do in exhibit **3**. We see a very similar picture for each of these three middle traces:

⋄ 90% automobiles for very low % of crowding.

⋄ 92 to 93% automobiles for just below and at median % crowding.

⋄ a fall in % automobiles for high % of crowding (to about 75% for the highest piece--3 of 88 places).

We need to ask **both** "Why were these traces so much alike?" **and** "Why were they different at all?"

exhibit 2 of chapter 14: unincorporated places

**Sliced data and medians for $x =$ % housing units with 1.01 persons per room, for $y =$ % housing units with one or more automobiles, for $x + y$, and for $y - x$, (each data parenthesis contains ($x$, $y$, $x + y$, $y - x$), in that order)**

### A) The DATA

| slice | | Data |
|---|---|---|
| #, count | range | |
| (#1, 3) | 13 and 14: | (13, 941, 954, 928)   (14, 892, 906, 878)<br>(13, 893, 906, 880) |
| (#2, 5) | 15 to 19: | (17, 813, 830, 796)   (18, 915, 933, 897)<br>(19, 930, 949, 911)   (19, 954, 973, 935)<br>(15, 904, 919, 889) |
| (#3, 10) | 20 to 28: | (20, 948, 968, 928)   (24, 892, 916, 868)<br>(26, 907, 933, 881)   (25, 947, 972, 922)<br>(26, 985, 1011, 959)  (27, 927, 954, 900)<br>(20, 938, 958, 918)   (27, 648, 675, 621)<br>(28, 720, 748, 692)   (28, 902, 930, 874) |
| (#4, 17) | 29 to 45 | (39, 920, 959, 881)   (39, 919, 958, 880)<br>(33, 908, 941, 875)   (39, 886, 925, 847)<br>(33, 944, 977, 911)   (41, 970, 1011, 929)<br>(32, 933, 965, 901)   (42, 964, 1006, 922)<br>(41, 894, 935, 853)   (31, 929, 960, 898)<br>(31, 947, 978, 916)   (42, 966, 1008, 924)<br>(38, 978, 1016, 940) (29, 971, 1000, 942)<br>(44, 872, 916, 828)   (45, 916, 961, 871)<br>(34, 931, 965, 897) |
| (#5, 19) | 47 to 67: | (65, 942, 1007, 879) (51, 937, 988, 886)<br>(52, 884, 936, 832)   (65, 931, 996, 866)<br>(61, 875, 936, 814)   (48, 917, 965, 869)<br>(62, 848, 910, 786)   (61, 856, 917, 795)<br>(62, 899, 961, 837)   (59, 808, 867, 749)<br>(61, 912, 973, 851)   (65, 754, 819, 689)<br>(54, 974, 1028, 920) (58, 947, 1005, 889)<br>(55, 959, 1011, 904) (65, 918, 983, 853)<br>(55, 944, 999, 889)   (67, 977, 1044, 910)<br>(47, 914, 961, 867) |
| (#6, 17) | 68 to 99: | (83, 937, 1020, 854) (99, 895, 994, 796)<br>(96, 903, 999, 807)   (93, 986, 1079, 893)<br>(68, 950, 1018, 882) (87, 892, 979, 805)<br>(72, 932, 1004, 860) (73, 954, 1027, 881)<br>(71, 977, 1048, 906) (82, 935, 1017, 853)<br>(85, 924, 1009, 839) (75, 894, 969, 819)<br>(71, 956, 1027, 885) (74, 940, 1014, 866)<br>(70, 933, 1003, 863) (90, 933, 1023, 843)<br>(78, 933, 1011, 855) |

→

Exhibit **4** shows, at an enlarged scale, the actual points and the four **median lines** for slice #8. On $x$, the median of these five was New Hanover, New Jersey. On $y$--and on $y - x$--the median of these 5 was Brownsville, Florida. On $y + x$ the median of these 5 was Bell Gardens, California. (Notice that Bristol, Pennsylvania, and Kailua-Lanikai, Hawaii, escaped being medians for these particular coordinates. It would, however, be easy to find other simple coordinates for which either of them is a median.)

With nothing to make as many as three of the median lines meet in a point, we should not be surprised that no three median lines do this. Accord-

exhibit **2** of chapter 14 (continued)

(#7, 9)  101 to 133:  (107, 867, 974, 760)     (124, 876, 1000, 752)
                      (107, 889, 996, 782)     (108, 835, 943, 727)
                      (102, 883, 985, 781)     (101, 956, 1057, 855)
                      (120, 812, 932, 692)     (133, 977, 1110, 844)
                      (114, 902, 1016, 788)

(#8, 5)  145 to 214:  (214, 873, 1087, 659)    (179, 876, 1055, 697)
                      (189, 943, 1132, 754)    (182, 859, 1041, 677)
                      (145, 956, 1101, 811)

(#9, 3)  228 to 238:  (231, 995, 1186, 724)    (228, 746, 974, 518)
                      (238, 713, 951, 475)

B) The MEDIANS

| slice | Raw medians for | | | | 3R'SSH3 smooths for | | | | y values from smooth of | | |
|---|---|---|---|---|---|---|---|---|---|---|---|
|  | x | y | x+y | y−x | x | y | x+y | y−x | y | y+x | y−x |
| #1 | 13 | 896 | 906 | 880 | 13 | 893 | 906 | 880 | 893 | 893 | 893 |
| #2 | 18 | 915 | 933 | 897 | 19 | 910 | 928 | 888 | 910 | 909 | 907 |
| #3 | 26 | 917 | 942 | 890 | 27 | 916 | 946 | 888 | 916 | 919 | 915 |
| #4 | 39 | 931 | 965 | 898 | 42 | 917 | 962 | 884 | 917 | 920 | 926 |
| #5 | 61 | 917 | 973 | 866 | 60 | 917 | 976 | 869 | 917 | 916 | 929 |
| #6 | 78 | 935 | 1014 | 855 | 81 | 908 | 980 | 840 | 908 | 909 | 921 |
| #7 | 108 | 883 | 996 | 781 | 119 | 890 | 990 | 778 | 890 | 871 | 897 |
| #8 | 182 | 876 | 1087 | 697 | 176 | 845 | 990 | 674 | 845 | 814 | 850 |
| #9 | 231 | 746 | 974 | 518 | 321 | 746 | 974 | 518 | 746 | 743 | 749 |

Examples: 893 = 906 − 13, 893 = 880 + 13, 909 = 928 − 19, 909 = 888 + 19.

P) PROBLEMS (largish)

2a) Slice the data in slices of respective sizes 5, 9, 17, 26, 17, 9, 5 and calculate the analog of panel B.

2b) Do the same for slices of 9, 9, 9, 9, 16, 9, 9, 9, 9.

ingly, the point which enters the smoothing process from this slice is not the same for y, y + x, and y − x, as exhibit 5 emphasizes.

Not the same, but not too different. No three of the four median lines intersect, but all four come reasonably close together. (Indeed, a hard look at panel B of exhibit 2 suggests that slice #8 was chosen as the piece where the failure to come together exactly was largest.)

exhibit **3** of chapter 14: unincorporated places

**Middle traces for y, y + x, and y − x, all against x, plotted in (x, y) coordinates (numbers in exhibit 2)**

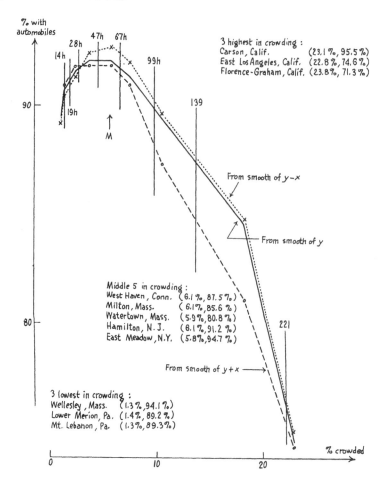

exhibit **4** of chapter 14: unincorporated places

**Data points and various medians for slice #8 (numbers in exhibit 2)**

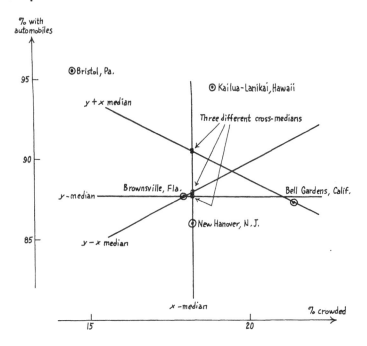

exhibit **5** of chapter 14: unincorporated places

**Detail of the center of exhibit 4, showing the three alternate summary points**

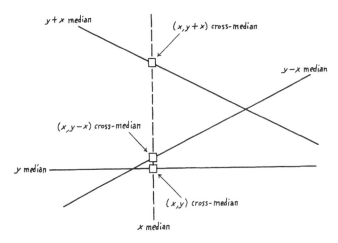

Thus there is no reason to expect that what we have seen here--"some difference in the middle traces, but not much"--is anything other than what we will usually see. This is fortunate, for it tells us that we will probably get away with showing only

◇ the middle trace.
◇ some first cuts.

If the choice of the coordinate traced against the slicing coordinate had really mattered, we would have had to show what it was like also, and this could have been a little too much for a clear picture.

### review questions

Can a single point scatter have several middle traces? Why/why not? What example did we choose to illustrate the behavior of several middle traces? What did we take as the slicing variable? How did we slice? How many different tracing variables did we try? How did the three traces behave? What two questions were then important? How did we try to look at what goes on? If we draw three or more median LINES for a point scatter, will they always/never have a point in common? -Why/why not? Are they likely to come close to having a point in common? What might bring them closer to this?

### 14C. An explanation

Why does this happen? If we look at--or think about--a piece with rather more points, the situation will become clearer. Consider slice #5, with 19 points and the following letter-value displays:

| #19 | x-values |     |     |     |
|-----|----------|-----|-----|-----|
| M10 |          | 61  |     |     |
| H5h | 63h      |     | 54h | 11  |
| E3  | 65       |     | 51  | 14  |

| #19 | y-values |     |     |     |
|-----|----------|-----|-----|-----|
| M10 |          | 917 |     |     |
| H 5h | 943     |     | 880 | 63  |
| E 3 | 959      |     | 848 | 99  |

The spread of $y$ is 5 to 7 times as great as that of $x$. This makes it almost certain that
$$\text{median } (y + x)$$
and
$$\text{median } (y + \text{median } x) = \text{median } y + \text{median } x$$

will be reasonably close to each other, since both (i) the order of the values of $y + x$ will resemble that of the values of $y + $ median $x$, and (ii) the values of $y + x$ will resemble the values of $y + $ median $x$.

A similar argument holds for

$$\text{median } (y - x)$$

and

$$\text{median } (y - \text{median } x) = \text{median } y - \text{median } x.$$

The narrower the slices, the less it can matter whether we take $y$, $y + x$, or $y - x$ against $x$. So, also, the larger the batch of $(x, y)$ pairs, the less will it matter.

Some would like these words backed up with a picture. Exhibit **6** shows how the points in a slice shift if we move them along $y + x = $ constant (left panel) or $y - x = $ constant (center and right panels). The righthand panel shows what sort of special pattern it takes to move things "a lot", which is still not large compared to the spread of the $y$'s in the slice.

This picture may also help to remind us that, with luck, such shifts of median (as we change the traced coordinate) will be quite irregular from slice to slice. Thus the smoothing will greatly reduce their effect on the final middle traces.

### curved traces

The explanation just given works just about as well if we bring in a halving coordinate whose level traces are not straight lines. All we need is that the slopes of the level curves are moderate everywhere. (Draw yourself an analog of some of the panels of exhibit 6 using level CURVES for halving--and even one set of curves for slicing and another for halving!

Of course this argument, like most approximate arguments, is helpful only over a limited range. If, in our example, we had used $y + 25x$ or $y - 25x$, where $25x$ is now much more variable than $y$, we could not be as certain of what would happen.

### review questions

What explanation did we consider? How well did it work? To what picture did we turn? What did it help us see? Would things be similar for halving traces that are not straight lines?

### 14D. Changing the slicing coordinate

We are now ready to see what happens when we change the slicing coordinate. How does middling against $x$ compare with middling against $y$? What about the comparison of both of these with such alternatives as middling against $x + y$ and middling against $x - y$?

Exhibit **7** shows, in four separate panels, middle traces for the unincorporated-places example against $y + x$, $x$, $y - x$, and $y$, respectively. The differences appear more striking than the similarities. All four are shown superposed in exhibit **8**, where we can now see a tendency toward concurrence of all four middle traces near a point at about 5 or 6% crowding and 92 or 93% automobiles.

exhibit **6** of chapter 14: purely illustrative

**Examples of sliding the points of a slice onto the sliced median**

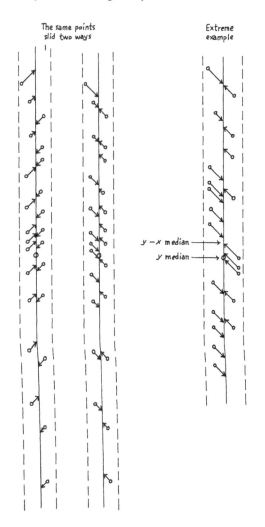

**exhibit 7 of chapter 14: unincorporated places**

**Four different middle traces against y + x, x, y − x and y (two "of y" and two "of x")**

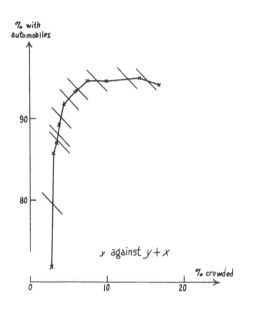
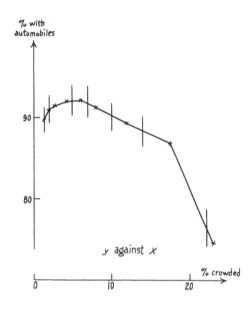
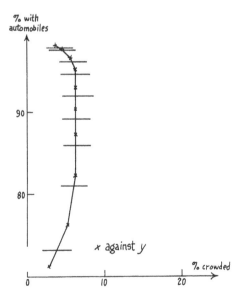
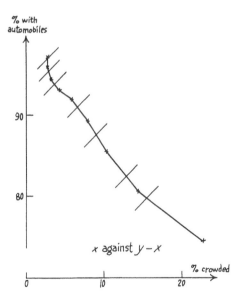

This concurrence will perhaps make more sense when we look at exhibit **9**, which shows the medians of $y + x$, $x$, $y - x$, and $y$, all four of which come close to a point at about the same location. This is a reasonable facsimile of a center for our particular point scatter. If there is an "almost center", it is surely not surprising that all "middle traces" run somewhere near it.

Exhibit **10** presents some problems.

### review questions

What happens when we change the slicing coordinate? Is the middle trace little affected? Why/why not? Do we expect near concurrence at a point? Why/why not?

exhibit **8** of chapter 14: unincorporated places

**The four middle traces of exhibit 7 superposed**

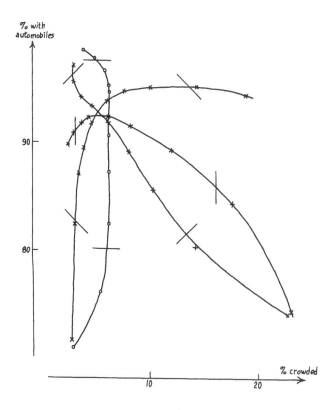

exhibit **9** of chapter 14: unincorporated places

**Four medians also nearly meet at a point**

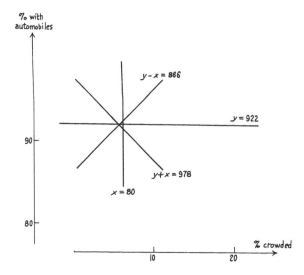

exhibit **10** of chapter 14: problems

**Some further problems on changing the sliced coordinate**

10a) Take the (BaDep, GoSal) data from section 8F and slice it on (i) BaDep + GoSal, (ii) BaDep − GoSal; then, for each, find the middle traces, middling the unsliced one of these two new coordinates. Plot them.

10a2) Add to the plot the results of slicing on BaDep (exhibit 21 of chapter 8) and on GoSal (problem 21 of chapter 8). How nearly do all four coincide?

10b) Plot the results of (i) slicing the Twin Rivers data on ElUse and middling on GaUse (exhibit 9 of chapter 8 and (ii) slicing on GaUse and middling on ElUse.

10c) Slice the Twin Rivers data on each of GaUse + ElUse and GaUse − ElUse, middling on the other.

10bc) Combine all four middlings on the same picture. How nearly do all four coincide?

10e) Do separate medians on BaDep, GoSal, BaDep + GoSal, BaDep − GoSal for the governor's salary data (exhibit 21 of chapter 8). Plot the four lines. How nearly do they coincide?

10e2) Do the same also for BaDep ± 1/2 GoSal. Plot all six lines. How closely do they coincide?

10f) Do the same for GaUse, ElUse, GaUse + ElUse, GaUse − ElUse for the Twin Rivers data (exhibit 20 of chapter 8).

## 14E. What matters?

It is not hard to see that, given any point, there is a coordinate of the form

$$cx + dy$$

for some $c$ and $d$ not both zero, or, if you prefer, of the form

$$x \cos \theta - y \sin \theta$$

for some $\theta$, such that the corresponding median goes through the given point. (Those who wish to look at a not-quite-easy proof may refer to section 14I.) It is not clear that, given a reasonable point, there is a middle trace that goes exactly through that point, but it is clear that there will be one that goes close.

If we use our first cuts to fit straight lines, instead of finding middle traces, we will find the same thing: Given a point, there will be a family of first cuts whose fitted line goes close to the given point.

### some morals

We have now learned that:

⋄ it does NOT matter very much if we change the traced coordinate--the coordinate whose middle trace we are looking at--so long as (i) we keep the set of first cuts the same, and (ii) we stick to reasonable combinations of the coordinate we started with.

⋄ it matters VERY MUCH indeed which coordinate we take a middle trace against--take as the slicing coordinate; by taking the right first cuts, we can make the middle trace come close to any pre-chosen point.

⋄ similar results hold for fitted lines.

The morals follow:

⋄ **the important thing to know about a middle trace is the family of first cuts that lead to it. Which coordinate was traced against this important slicing coordinate usually matters relatively little.**

⋄ thus we should continue to accompany each middle trace with a sufficient collection of first cuts to show us what it came from.

⋄ **the same thing is likely to be true for any fitted line.**

It's an old saying that "it's baloney, no matter how you slice it!", but we have now learned that "How you slice the point scatter can make all the difference in the world to your middle trace--or your fitted line!".

### review questions

Given a point scatter, can you find a new point (anywhere in the plane) such that no median line goes through that point? Why/why not? Which is more important to know about a median trace--the family of slicing cuts or the family of tracing cuts? Why? What family of traces thus best supports a middle trace?

## 14F. Rematching and strength of relationship

Suppose, in the Twin Rivers example (exhibit 1 to 19 of chapter 8), that we had been given--not the 152 (ElUse, GaUse) pairs--but only the two stem-and-leaf displays, one for each variable. We would not have known about the actual relationship, but we could have decided on what the strongest relationship--in either direction--could have been that would be consistent with these two stem-and-leafs, one for 152 electric usages, the other for 152 gas usages.

### exhibit 11 of chapter 14: energy usages

**Some numbers for ElUse and GaUse separately (based on exhibit 1 of chapter 8)**

#### A) MEDIANS of LETTER SEGMENTS

| Depth* | | Tag | | ElUse | | |
|---|---|---|---|---|---|---|
| Point | Segment | Point | Segment | Value | Median | (**) |
|  | (1) |  | 1 |  | 89 | (424) |
| 1 |  | 1 |  | 89 |  |  |
|  | (2) |  | 1A |  | 108 | (388) |
| 2 |  | A |  | 108 |  |  |
|  | (3) |  | aB |  | 119 | (778) |
| 3 |  | B |  | 119 |  |  |
|  | (5) |  | bC |  | 134 | (832) |
| 5–6 |  | C |  | 134 |  |  |
|  | (8) |  | CD |  | 152h | (828) |
| 10 |  | D |  | 156 |  |  |
|  | (15) |  | dE |  | 170 | (766) |
| 19h |  | E |  | 177h |  |  |
|  | (29) |  | eH |  | 192 | (958) |
| 38h |  | H |  | 207 |  |  |
|  | (58) |  | hm |  | 235 | (922) |
| 76h |  | M |  | 253 |  |  |
|  | (57h) |  | mh |  | 267h | (891) |
| 38h |  | H |  | 283h |  |  |
|  | (29) |  | He |  | 292 | (942) |
| 19h |  | E |  | 301h |  |  |
|  | (15) |  | Ed |  | 316 | (924) |
| 10 |  | D |  | 333 |  |  |
|  | (8) |  | Dc |  | 347 | (946) |
| 5h |  | C |  | 359h |  |  |
|  | (4h) |  | Cb |  | 369 | (925) |
| 3 |  | B |  | 422 |  |  |
|  | (3) |  | Ba |  | 422 | (1222) |
| 2 |  | A |  | 429 |  |  |
|  | (2) |  | A1 |  | 429 | (1002) |
| 1 |  | 1 |  | 435 |  |  |
|  | (1) |  | 1 |  | 435 | (860) |

→

Clearly, the **strongest positive relation** would have come about when the highest gas usage was paired with the highest electrical usage, the 2nd highest with the 2nd highest, and so on to the lowest with the lowest. Equally, the **strongest negative relationship** would have paired the highest with the lowest, the 2nd highest with the 2nd lowest, and so on, to the lowest with the highest. The corresponding middle traces would have their ends as extreme as possible.

The neutral relationship--where knowing about one set of values offers no clue about the other--would correspond to a horizontal middle trace, one located at the median of the quantity taken as a response.

exhibit **11** of chapter 14 (continued)

| Depth* | | Tag | | GaUse | | |
|---|---|---|---|---|---|---|
| Point | Segment | Point | Segment | Value | Median | (**) |
|  | (1) |  | 1 |  | 388 | (108) |
| 1 |  | 1 |  | 388 |  |  |
|  | (2) |  | 1A |  | 424 | (89) |
| 2 |  | A |  | 424 |  |  |
|  | (3) |  | aB |  | 530 | (175) |
| 3 |  | B |  | 530 |  |  |
|  | (4h) |  | bc |  | 576 | (198) |
| 5h |  | C |  | 597 |  |  |
|  | (8) |  | cD |  | 612 | (211) |
| 10 |  | D |  | 658 |  |  |
|  | (15) |  | dE |  | 684 | (247) |
| 19h |  | E |  | 726 |  |  |
|  | (29) |  | eH |  | 762 | (266) |
| 38h |  | H |  | 789 |  |  |
|  | (57h) |  | hm |  | 839 | (255) |
| 76h |  | M |  | 910 |  |  |
|  | (57h) |  | mh |  | 963 | (263h) |
| 38h |  | H |  | 1016 |  |  |
|  | (29) |  | He |  | 1058 | (252) |
| 19h |  | E |  | 1068 |  |  |
|  | (15) |  | Ed |  | 1096 | (272) |
| 10 |  | D |  | 1144 |  |  |
|  | (8) |  | De |  | 1150 | (231) |
| 5h |  | C |  | 1171 |  |  |
|  | (4h) |  | Cb |  | 1185 | (202) |
| 3 |  | B |  | 1214 |  |  |
|  | (3) |  | Ba |  | 1214 | (299) |
| 2 |  | A |  | 1222 |  |  |
|  | (2) |  | A1 |  | 1222 | (422) |
| 1 |  | 1 |  | 1298 |  |  |
|  | (1) |  | 1 |  | 1298 | (196) |

*Ties are shown in the "point" column where meaningful, for example 5–6 at upper left, and allowed for in the segment column.
** Values in parenthesis are medians--for the other energy usage--for that segment of townhouses.

→

When we ask how much relationship there is between the two values, we can be asking one of two quite different questions (see opener of chapter 5 for "response" and circumstance"):

◊ how fast does the response shift as the circumstance shifts? (A question to be answered by the tilt of a fitted line or by the, probably changing, slope of a middle trace, or by such displays as those to come next.)

◊ how much of the possible relationship is present? (A question partly answered by the relation of the actual middle trace to what might have happened, given the values of each variable separately--without their matching. Displays that show this more clearly will follow.)

Exhibit 11 has some arithmetic details. Exhibit 12 shows the actual GaUse middle trace (using a 3R smooth) against a background of three possible middle traces, most positive, neutral, and most negative. Exhibit 13 shows the same picture for the actual ElUse middle trace. To make the two exhibits more easily comparable in exhibit 13, we have preserved our original choice--ElUse as horizontal--and GaUse as vertical--though it would often seem more natural to interchange them, since we are now cutting on GaUse and halving on ElUse.

We see in both exhibits, actual close to most positive on the extreme lower left, but actual closer to neutral elsewhere.

**dependency ratios**

A natural way to take things further is to look at the ratios

$$\frac{\text{actual MINUS median}}{\text{most positive MINUS median}}$$

exhibit **11** of chapter 14 (continued)

**P) PROBLEMS**

11a/b    Imitate panel A for upper/lower hinges instead of medians.

11a2/b2) Make a plot similar to exhibit 12 for upper/lower hinge traces instead of median traces.

11a3/b3) Make a plot similar to exhibit 13 for upper/lower hinge traces instead of median traces.

11a4/b4) Discuss the relationship of what you have just done to exhibits 12 and 13--and to itself.

which compares actual amount up with largest possible amount up (or, if this is negative, at

$$\frac{\text{actual MINUS median}}{\text{median MINUS most negative}},$$

which compares actual down with largest possible down, instead).

At a point where this ratio is close to +1, the actual middle trace is close to the most positive trace and the relationship is strongly positive. Where it is

exhibit **12** of chapter 14: energy usages

**Actual and three potential middle traces for GaUse against ElUse (from exhibit 11)**

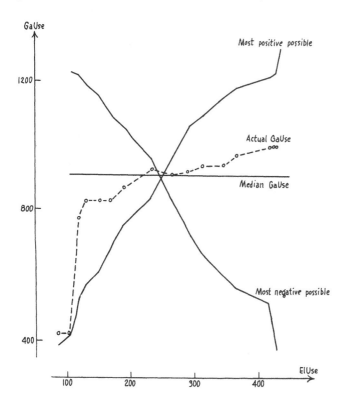

close to $-1$, the relation is strongly negative. Where it is close to zero, the actual trace is close to the median and there is little sign of dependence. Hence it is natural to call each individual value a

### dependency ratio,

and to call a trace reflecting its several values a

### dependency trace.

Exhibit **14** finds these ratios, and then smooths them with 3R. Panel B gives the median ratios (omitting the extreme values found in the first two lines), both unweighted and weighted. We see that, if we weight, the apparent relationship is quite feeble (possibly zero) except for those two first slices.

exhibit **13** of chapter 14: energy usages

**Actual and three potential middle traces for ElUse against GaUse from exhibit 11 (coordinates as in exhibit 12 rather than as natural)**

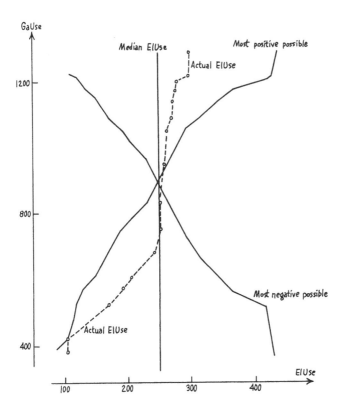

exhibit 14 of chapter 14: energy usages

**Finding and smoothing dependency ratios, and making reconstituted traces (from exhibit 11)**

A) CALCULATION and SMOOTHING of RATIOS

| | for ElUse against GaUse | | | | |
|---|---|---|---|---|---|
| | value − median | | Ratio = Act/Pot | | |
| slice | Potent. | Actual | Raw | 3R | (wts) |
| 1 | −164 | −145 | .88 | .88 | (1) |
| 1A | −145 | −164 | 1.13 | .88 | (1) |
| AB | −134 | −78 | .58 | .58 | (1) |
| BC | −119 | −55 | .46 | .46 | (2) |
| CD | −101 | −42 | .42 | .42 | (5) |
| DE | −83 | −6 | .07 | .07 | (9) |
| EH | −61(39) | 13 | −.22(−.33) | −.19(−.15) | (19) |
| H1 | −18(13) | 2 | −.19(−.15) | −.19(−.15) | (38) |
| MH | 13 | 10h | .81 | −.03 | (38) |
| HE | 39(−61) | −1 | −.28(−.02) | .23 | (19) |
| ED | 63 | 19 | .30 | .23 | (9) |
| DC | 94 | −22 | .23 | .23 | (5) |
| CB | 116(−119) | −51 | −44(−.43) | .23 | (2) |
| BA | 169 | 46 | .27 | .27 | (1) |
| A1 | 176 | 169 | .96 | .27 | (1) |
| 1 | 182(−164) | −57 | −31(−.35) | .27 | (1) |

| | for GaUse against ElUse | | | | |
|---|---|---|---|---|---|
| | value − median | | Ratio = Act/Pot | | |
| slice | Potent. | Actual | Raw | 3R | (wts) |
| 1 | −522 | −486 | .93 | .93 | (1) |
| 1A | −486 | −522 | 1.07 | .93 | (1) |
| AB | −380 | −132 | .35 | .35 | (1) |
| BC | −334 | −78 | .23 | .31 | (3) |
| CD | −298 | −82 | .28 | .31 | (4) |
| DE | −226 | −144 | .64 | .28 | (10) |
| EH | −148(148) | 48 | −.32(−.32) | −.17(−.23) | (19) |
| HM | −71(53) | 12 | −.17(−.23) | −.17(−.23) | (38) |
| MH | 53(−153) | −19 | −.36(−.36) | −.17(−.23) | (37) |
| HE | 148 | 32 | .22 | .08 | (11) |
| ED | 186 | 14 | .08 | .08 | (10) |
| DC | 240 | 36 | .15 | .15 | (5) |
| CR | 275 | 15 | .05 | .15 | (2) |
| BA | 304 | 312 | 1.03 | .29 | (1) |
| A1 | 312 | 92 | 29 | 29 | (1) |
| 1 | 388(522) | −50 | −13(−.10) | .29 | (1) |

Examples: −164 = 89 − 253; −145 = 108 − 253; .88 = (−145)/(−164);
−522 = 388 − 910; −486 = 424 − 910; .93 = (−486)/(−522);
−145 = 108 − 253; −164 = 89 − 253; 1.13 =(−164)/(−145).

Note: Values in parenthesis from--or figured from--other extreme.

→

Exhibits **15** and **16** plot the resulting

dependency traces,

which do a good job of answering the second question above.

exhibit **14** of chapter 14 (continued)

**B) MEDIAN RATIOS--omitting two first lines**

|      | ElUse Ratio (smoothed) | GaUse Ratio (smoothed) |
|------|------------------------|------------------------|
| Raw  | .23                    | .22                    |
| Wtd. | −.02                   | .17                    |

**C) RECONSTITUTED TRACES**

| slice | ElUse against GaUse Smoothed value − median | | | Value recon. | GaUse against ElUse Smoothed value − median | | | Value recon. |
|-------|-------|----------|--------|-------|-------|--------|--------|-------|
|       | ratio | Potent.  | Recon. |       | ratio | Potent | Recon. |       |
| 1     | .88   | −164     | −144   | 109   | .93   | −522   | −485   | 425   |
| 1A    | .88   | −145     | −128   | 125   | .93   | −486   | −452   | 458   |
| aB    | .58   | −134     | −78    | 175   | .35   | −380   | −133   | 777   |
| bC    | .46   | −119     | −55    | 198   | .31   | −334   | −104   | 806   |
| cD    | .42   | −101     | −42    | 211   | .31   | −298   | −92    | 818   |
| dE    | .07   | −83      | −6     | 247   | .28   | −226   | −63    | 847   |
| eH    | −.15  | (39)     | −6     | 247   | −.17  | −148   | 25     | 935   |
| hm    | −.15  | (13)     | 0      | 253   | −.17  | −71    | 12     | 922   |
| mh    | −.03  | (13)     | 0      | 253   | −.17  | 53     | −9     | 901   |
| He    | .23   | 39       | 9      | 262   | .08   | 148    | 12     | 922   |
| Ed    | .23   | 63       | 14     | 267   | .08   | 186    | 15     | 925   |
| Dc    | .23   | 94       | 22     | 275   | .15   | 240    | 36     | 946   |
| Cb    | .23   | 116      | 27     | 280   | .15   | 275    | 41     | 951   |
| Ba    | .27   | 169      | 46     | 299   | .29   | 304    | 88     | 998   |
| A1    | .27   | 176      | 48     | 301   | .29   | 312    | 90     | 1000  |
| 1     | .27   | 182      | 49     | 302   | .29   | 388    | 113    | 1023  |

**P) PROBLEMS**

14a/b/c/d) Make calculations analogous to panels A, B, C for the data of (21a/b/c/d), respectively, in chapter 8.

14a2/b2/c2/d2) Make a plot like exhibit 15 for this data.

14a3/b3/c3/d3) Make a plot like exhibit 16 for this data.

14e) Make a plot of the two reconstituted traces in panel C in comparison with most positive and most negative traces.

14f) Smooth the reconstituted traces, and then repeat the ratioing and smoothing processes. Compare the result with panel A. Discuss the changes.

exhibit **15** of chapter 14: energy usage

**Dependency ratio for GaUse against ElUse (from exhibit 14, 3R smoothed)**

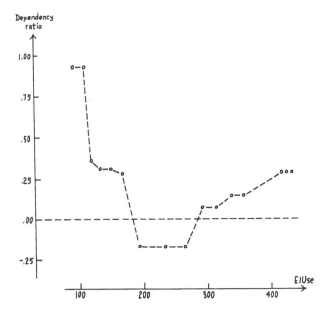

exhibit **16** of chapter 14: energy usages

**Dependency ratio for ElUse against GaUse (from exhibit 14)**

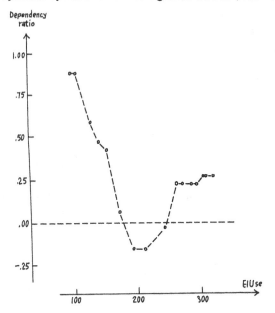

### reconstituted traces

Panel C (of exhibit 14) illustrates the reconstitution of traces by multiplying the smoothed dependency ratios by the potential differences. This may sometimes be useful (see also problem (14f).)

### a comment

If we had not already seen, in Section 14D, how greatly the choice of the slicing coordinate affects both the middle trace and the fitted line, we would have **had** to stress here **the difference between y middled against x and x middled against y.** The point is so important, however, that we are going to stress it anyway.

In most cases, we expect to find the ends of the

### actual middle

"somewhere between"--either between the most positive trace and the horizontal median OR between the horizontal median and the most negative trace. Similarly, we expect the other

### actual middle

found by interchanging variables to also lie "somewhere between"--but not in the SAME somewhere between. Rather, with the axes preserved, as in exhibit 13, we expect it either between the most positive trace and the vertical median or between the vertical median and the most negative trace.

Thus we expect the two middle traces to have different tilts, the former less than the tilt of the most positive possible, the latter greater. Whether we fit lines or smooth middle traces, we always expect two different tilts, etc. If we are dealing with (Adata, Bdata) pairs we expect TWO tilts:

⋄ one for Adata against Bdata.

⋄ the other for Bdata against Adata.

### review questions

What could we do with our $x$'s and $y$'s if we did not know which $x$ went with which $y$? What does rematching mean? What do we have to know? What do we do? How many potential middle traces do we find? What are they? Can we use such results as a background? Why/why not? What might "how much relationship" mean? What two questions have we emphasized? How do they differ? Have we dealt with one meaning already? Why/why not ? Can we compare actual relationship with possible relationship? Graphically? Numerically? What number or numbers might be a partial substitute? What is a dependency ratio? Can its value be greater than 1? Why/why not? Less than $-1$? Why/why not? Against what do we usually plot dependency ratios? How would we interpret a dependency ratio of −0.97? Of +0.82? Of −0.05? Of +0.32? How did the dependency ratio behave in our example? What is a reconstituted trace? Do we expect "$y$ middled against $x$" and "$x$ middled against $x$" to be quite similar? Why/why not? Will they ever cross? Why/why not? Where do we expect to find the ends of middle traces? What if we look at lines, instead, for "$y$ fitted to $x$" and "$x$ fitted to $y$"? Will they agree with each other? Why/why not? Will they be parallel? Why/why not? Will they cross? Why/why not?

### 14H. How far have we come?

In this chapter we have looked at any $(x, y)$ batch in two or more ways--changing either the slicing variable or the tracing variable used to lead us to a middle trace--or swapping $x$ and $y$ in fitting lines. We have also compared middle traces with what the actual values of $x$ and of $y$, taken separately, could provide.

We are now ready:

◇ to insist on supporting our middle traces with adequate indications of what cuts were used to make the slices.

◇ to rematch $x$ and $y$ values, finding the most positive and most negative relationships that could have happened, given the $x$'s and given the $y$'s.

◇ to assess the strength of relationship actually present by comparison with these extremes, either graphically or numerically.

We now understand more clearly:

◇ what a coordinate can be.

◇ what to expect when we keep the slicing variable fixed and change the tracing variable.

◇ what to expect when we do the converse.

◇ why the slicing variable is MORE important, whether we are finding middle traces or fitting straight lines.

◇ that the straight lines for "$y$ fitted to $x$" and "$x$ fitted to $y$" are expected to cross and often to be quite different.

◇ that the same is true for the traces: "$y$ middled on $x$" and "$x$ middled on $y$".

### 14I. The ubiquity of medians (optional section)

Medians are rather different for batches of odd count or even count. We will want to take these cases separately.

#### odd count

Suppose we take some arbitrary point $(x_{arb}, y_{arb})$ and a batch of $n$ points $(x_1, y_1), (x_2, y_2), \ldots, (x_n, x_n)$, where $n$ is an odd number. Can we show that there is a coordinate of the simple form

$$z = cx + dy$$

with $c$, $d$ not both zero, whose median passes through $(x_{arb}, y_{arb})$?

Note, first, that the problem depends ONLY on the directions of the points $(x_i, y_i)$ from $(x_{arb}, y_{arb})$. For whether $z = cx_{arb} + dy_{arb}$ is a median of an **odd** number of points or not depends only on on what side of this line, which goes through $(x_{arb}, y_{arb})$, the various points lie. So we may as well draw a circle around $(x_{arb}, y_{arb})$--any circle with this center will do--and project the $n$ points on the circle.

This gives us a simpler-seeming problem: Given $n$ points on a circle, is there a diameter that halves them? This is easy to deal with if we work with directed diameters, with diameters with an arrow--something to give them left and right sides.

Once we do this we can write down:

$$\text{discrepancy} = \begin{pmatrix} \text{number of points} \\ \text{on left} \end{pmatrix} \text{ MINUS } \begin{pmatrix} \text{number of points} \\ \text{on right} \end{pmatrix}$$

for any directed diameter. Several properties of this are easy to see, including:

⋄ if we turn a directed diameter through 180 degrees, the new discrepancy is MINUS the original (we have interchanged left and right!).

⋄ if we turn the directed diameter smoothly, its discrepancy does not change unless one or both ends of the diameter hit one of the $n$ points.

⋄ accordingly, discrepancies are constant on not more than $n$ intervals, with a point of change linking each interval to the next.

⋄ if the discrepancy changes sign as we pass through some direction--some change point--that direction is a median. (The count of $(x_i, y_i)$ points strictly to the left of the directed diameter in that direction is the lesser of the counts for the adjacent segments and is thus strictly less than $n/2$. The same holds on the right. There must be enough points on this directed diameter to make the count "left or on" $\geq n/2$. The same has to be true for "right or on" and the result is proved.)

⋄ a segment of constant discrepancy cannot have discrepancy zero. (Some direction in such a segment goes through NO point. With an odd number of points, the discrepancy must now itself be an odd number.)

These facts are enough to settle matters: There are segments of constancy. None have discrepancy zero. For a segment of one sign there is an exactly opposite segment of the other sign. Somewhere a PLUS segment must touch a MINUS segment. The linking point fixes a direction that is a median.

### even count

The case of an even number of points is a little more difficult and is left to the reader as a teaser. (*Hint.* Don't try to move the points to a circle. Define for each direction, as shown by its diameter:

$$\text{lateral discrepancy} = \begin{pmatrix} \text{smallest PLUS devi-} \\ \text{ation (from diameter)} \end{pmatrix} - \begin{pmatrix} \text{least MINUS devi-} \\ \text{ation (from diameter)} \end{pmatrix}.$$

You CAN go on to show that there is still a median through the arbitrary point.

### review questions

What question concerned us? How could we reduce it? To what form? To what indicating quantity did we then turn? What properties did this have? How did they combine to settle the question?

# index for chapter 15

review questions 496

**15A. Started counts and counted fractions 496**
*split count* 496
*ss-count* 496
ss-fractions 497
review questions 497

**15B. Three matched scales for counted fractions 498**
review questions 502

**15C. Quicker calculation 502**
review questions 508

**15D. Examples where careful expression clearly pays off 508**
a 2 × 2 opinion example 508
two-way tables of counts 510
review questions 512

**15E. Double folding--the 2 × 2 case 513**
*a double flog* 516
*fflog* 516
review questions 516

**15F. Double folding--larger cases 516**
review questions 520

**15G. Easy froots and flogs with a slide rule (optional) 520**
review questions 521

**15H. How far have we come? 522**

| EXHIBIT | PAGE |
|---|---|
| 15A | |
| 1★ | 497 |
| 15B | |
| 2★ | 499 |
| 3 | 502 |
| 15C | |
| 4★ | 503 |
| 5★ | 506 |
| 15D | |
| 6★ | 510 |
| 7★ | 511 |
| 8★ | 512 |
| 15E | |
| 9★ | 517 |
| 15F | |
| 10★ | 518 |
| 15G | |
| 15H | |

# Counted Fractions    15

We need to deal with counted fractions, like the 17/23 of 23 marbles that were blue and the 244/59,085,182 of 59,085,182 U.S. 1958 incomes that were over a million dollars. The second example shows a cutting value--we will shorten this to cut--a value that divides "below"--in our example up to one million dollars, from "above"--in our example from one million and one dollars up. We need to be concerned with counted fractions both for their own sake--we may want to point out values, compare, or otherwise analyze counted fractions--and as an essential element in describing how something is distributed.

In very many instances it is natural to have a sequence of counted fractions referring to the same information. If we can count incomes over a million, we might have counted incomes (i) over two million, (ii) over half a million, (iii) over each of a fairly long list of cuts. Something similar can happen for the 23 marbles; there may be a sequence of cuts "above" which fall:

  20 marbles that are at least bluish,

  17 marbles that are "blue",

  13 marbles that are "very blue".

We cannot be surprised by any observed counted fraction appearing as one of a sequence of counted fractions, defined by a sequence of cutting values. (If we know the fraction of 1958 incomes above, say, each of 24 well-chosen cuts, we know a lot about how 1958 income was distributed.)

Whether we deal with counted fractions separately or as part of a sequence of such, it pays us to face up to two considerations:

⋄there are likely to be observed values that are "on the fence"--that EXACTLY equal a cutting value.

⋄where we have observed "none", it may be reasonable that further repeated observations may find an occasional instance, so that using "zero" for the count seems too drastic.

Having dealt with these two considerations, it is natural to go on and be sure that we are treating the fraction "in" and the fraction "out"--often the fraction "below" and the fraction "above"--symmetrically.

We can extend all these ideas to looking at cuts on two different variables --say, sex and occupation--at the same time.

**review questions**

What is a counted fraction? Do counted fractions always come alone? Why/why not? What two considerations have we to face? Should we treat the fractions "in" and "out" differently? What about cuts on two different bases at once?

### 15A. Started counts and counted fractions

Suppose we have 27 observations; what fraction do we assign to:

◇ the next-to-lowest observation?

◇ a cutting value such that two of the observations are below it and none equal to it?

◇ a cutting value such that two of the observations are below it and three equal to it?

◇ a cutting value so low that NO observation is below it?

These are only a few of the possibilities with which we must deal.

It is a long-used statistical practice to treat any observed value exactly equal to a cutting value as if it were half below and half above. We shall continue this tradition. Thus we will treat "two below and three equal to" as $2 + \frac{1}{2}(3) = 3.5$ below. We refer to such a value as 3.5 as a

split count

when a specific term is needed.

The desirability of treating "none seen below" as something other than zero is less clear, but also important. Here practice has varied, and a number of different usages exist, some with excuses for their being and others without. The one we here recommend does have an excuse, but since this excuse (i) is indirect and (ii) involves more sophisticated considerations, we shall say no more about it. What we recommend is adding 1/6 to all split counts, thus "starting" them. We would use "zero" only for the "count" below a cutting value that was such that NO value could be below it (and even one equal to it is essentially out of the question). Since we are **starting** a **split** count when we use

(the count below)  PLUS  (1/2 the count equal)  PLUS  1/6

or

(the count above)  PLUS  (1/2 the count equal)  PLUS  1/6,

it is not unreasonable to distinguish either of these as an

ss-count.

### ss-fractions

Notice that

ss-count below PLUS ss-count above = total count PLUS 1/3

no matter where the cut falls. Thus, if we define

$$\text{ss-fraction below} = \frac{\text{ss-count below}}{\text{ss-count below PLUS ss-count above}},$$

we can write

$$\text{ss-fraction below} = \frac{\text{ss-count below}}{\text{total count PLUS 1/3}},$$

which, when we put in all the details for "ss-count below", can be written

$$\text{ss-fraction below} = \frac{\text{count below PLUS 1/2 count equal PLUS 1/6}}{\text{total count PLUS 1/3}}.$$

Exhibit 1 gives some examples and problems for the use of ss-fractions.

Whenever we say "fraction" in the rest of this chapter, we will mean "ss-fraction". (When counts are large, we can, of course, neglect the starts -- and when there are no observations exactly equal to our cutting values, we shall not need to split any counts.)

### review questions

What is the simplest way to count ordered observations? And the empty segments between them? Why may this not be a good choice? What is another way? What are some of its difficulties? What is a reasonable compromise? What is an "ss-count"? An "ss-fraction"?

exhibit 1 of chapter 15: examples and problems

**The use of ss-fractions**

**A) EXAMPLES**

| observations | cut | # below | # equal | ss-count | total # PLUS $\frac{1}{3}$ | ss-fraction |
|---|---|---|---|---|---|---|
| 0, 1, 2, 2, 3, 7 | $\sqrt{2}$ | 2 | 0 | $2\frac{1}{6}$ | $6\frac{1}{3}$ | .342 |
| 0, 1, 2, 2, 3, 7 | 2 | 2 | 2 | $3\frac{1}{6}$ | $6\frac{1}{3}$ | .500 |
| 0, 1, 2, 2, 3, 7 | 3 | 4 | 1 | $4\frac{2}{3}$ | $6\frac{1}{3}$ | .737 |
| 0, 1, 2, 2, 3, 7 | 4 | 5 | 0 | $5\frac{1}{6}$ | $6\frac{1}{3}$ | .816 |
| .123, .456, .654, .789 | .2 | 1 | 0 | $1\frac{1}{6}$ | $4\frac{1}{3}$ | .269 |
| .123, .456, .654, .789 | .5 | 2 | 0 | $2\frac{1}{6}$ | $4\frac{1}{3}$ | .500 |
| .123, .456, .654, .789 | .654 | 2 | 1 | $2\frac{2}{3}$ | $4\frac{1}{3}$ | .615 |

➜

## 15B. Three matched scales for counted fractions

Now we have "started" our counted fractions, how shall we express them?

Treating "those who are" and "those who aren't" symmetrically is often useful and important. This means two things:

o choosing scales with a value of 0 at 50%;

o arranging matters so that a swap of $f$ with $1 - f$, where $f$ is the fraction concerned, will change the sign, but not the size, of our expression of $f$.

While we are at it, since $f = \frac{1}{2}$ is to be scored as 0, we may as well plan to have a variety of choices that agree, not only exactly for $f = \frac{1}{2}$ *but quite closely for values around* $f = \frac{1}{2}$. This is a matter of convenience. It also allows ever-repeated emphasis on how similar all the expressions are "in the middle"—on how little it matters which we choose if all our fractions are near $\frac{1}{2}$. (If this is mysterious enough to bother you, turn ahead to exhibit 2, where three modes of expression are matched near 50%, though they differ widely at extreme %'s.)

Such matching costs us nothing, because whether we analyze a given set of numbers or a set of numbers each exactly three times as large as those given is almost certain to make no essential difference: When complete, the results can, in any case, be translated to a common, generally understandable scale. Trivial re-expressions are indeed trivial.

With these guideposts, what can we do that will be simple and effective? The simplest measure that re-expresses $f = \frac{1}{2}$ as 0 is the "plurality"

$$(f) - (1 - f)$$

which is the difference between the fraction of those that are and the fraction of those that aren't. If we take $f$ and $1 - f$ in %, we call the result "folded percents", abbreviated "fcents".

exhibit **1** of chapter 15 (continued)

### P) PROBLEMS

| observations | cut | ss-fraction |
|---|---|---|
| 1a) 13, 27, 29, 35, 47, 53 | 29 | ? |
| 1b) 7, 9, 9, 9, 11, 15, 25 | 9 | ? |
| 1c) 13, 27, 29, 35, 47, 53 | 34 | ? |
| 1d) 7, 9, 9, 9, 11, 15, 25 | 5 | ? |
| 1e) 13, 27, 29, 35, 47, 53 | 204 | ? |

1f) Out of 323 earthquakes, only 3 were stronger than a given cut-off. If none were at the cut-off, what is the ss-fraction "to the right" of the cut-off?

1g) As (1f) but with 2 **at** the cut-off.

1h) As (1f) but with only one beyond.

1i) As (1f) but with only one beyond and 2 **at** the cutoff.

exhibit **2** of chapter 15: reference table

**Pluralities, folded roots, folded logarithms--alternative expressions for counted fractions (take sign of answer from head of column giving %)**

**A) MAIN TABLE**

| + | Plur. | froot | flog | − | + | Plur. | froot | flog | − |
|---|---|---|---|---|---|---|---|---|---|
| 50% | use | .00 | use | 50% | 85% | .70 | .76 | .87 | 15% |
| 51 | → | .02 | ← | 49 | 86 | .72 | .78 | .91 | 14 |
| 52 |  | .04 |  | 48 | 87 | .74 | .81 | .95 | 13 |
| 53 |  | .06 |  | 47 | 88 | .76 | .84 | 1.00 | 12 |
| 54 |  | .08 |  | 46 | 89 | .78 | .87 | 1.05 | 11 |
| 55% | use | .10 | use | 45% | 90.0% | .80 | .89 | 1.10 | 10.0% |
| 56 | → | .12 | ← | 44 | 90.5 | .81 | .91 | 1.13 | 9.5* |
| 57 |  | .14 |  | 43 | 91 | .82 | .92 | 1.16 | 9 |
| 58 |  | .16 |  | 42 | 91.5 | .83 | .94 | 1.19 | 8.5 |
| 59 |  | .18 |  | 41 | 92 | .84 | .96 | 1.22 | 8 |
| 60% | use | .20 | use | 40% | 92.5% | .85 | .97 | 1.26 | 7.5% |
| 61 | → | .22 | ← | 39 | 93 | .86 | .99 | 1.29 | 7 * |
| 62 |  | .24 |  | 38 | 93.5 | .87 | 1.01 | 1.33 | 6.5 |
| 63 | .26 | .26 | .27 | 37 | 94 | .88 | 1.02 | 1.37 | 6 |
| 64 | .28 | .28 | .29 | 36 | 94.5 | .89 | 1.04 | 1.42 | 5.5 |
| 65% | .30 | .30 | .31 | 35% | 95.0% | .90 | 1.06 | 1.47 | 5.0% |
| 66 | .32 | .32 | .33 | 34 | 95.5 | .91 | 1.08 | 1.53 | 4.5 |
| 67 | .34 | .35 | .35 | 33 | 96 | .92 | 1.10 | 1.59 | 4 |
| 68 | .36 | .37 | .38 | 32 | 96.5 | .93 | 1.12 | 1.65 | 3.5 |
| 69 | .38 | .39 | .40 | 31 | 97 | .94 | 1.15 | 1.74 | 3 |
| 70% | .40 | .41 | .42 | 30% | 97.2% | .94 | 1.16 | 1.77 | 2.8% |
| 71 | .42 | .43 | .45 | 29 | 97.4 | .95 | 1.17 | 1.81 | 2.6 |
| 72 | .44 | .45 | .47 | 28 | 97.6 | .95 | 1.18 | 1.85 | 2.4 |
| 73 | .46 | .47 | .50 | 27 | 97.8 | .96 | 1.19 | 1.90 | 2.2 |
| 74 | .48 | .50 | .52 | 26 | 98.0 | .96 | 1.20 | 1.95 | 2.0 |
| 75% | .50 | .52 | .55 | 25% | 98.2% | .96 | 1.21 | 2.00 | 1.8% |
| 76 | .52 | .54 | .58 | 24 | 98.4 | .97 | 1.22 | 2.06 | 1.6 |
| 77 | .54 | .56 | .60 | 23 | 98.6 | .97 | 1.24 | 2.13 | 1.4 |
| 78 | .56 | .59 | .63 | 22 | 98.8 | .98 | 1.25 | 2.21 | 1.2 |
| 79 | .58 | .61 | .66 | 21 | 99.0 | .98 | 1.27 | 2.30 | 1.0 |
| 80% | .60 | .63 | .69 | 20% | 99.2% | .99 | 1.28 | 2.41 | 0.8% |
| 81 | .62 | .66 | .72 | 19 | 99.4 | .99 | 1.30 | 2.55 | 0.6 |
| 82 | .64 | .68 | .76 | 18 | 99.6 | .99 | 1.32 | 2.76 | 0.4 |
| 83 | .66 | .71 | .79 | 17 | 99.8 | 1.00 | 1.35 | 3.11 | 0.2 |
| 84 | .68 | .73 | .83 | 16 | 100.0% | 1.00 | 1.41 | * | 0.0 |

Supplementary Table ➡

Another reasonable name for this expression would be folded fraction. Clearly, if we fold any other expression of a fraction--by subtracting the given function of those who aren't from the same function of those who are--we will force $f = \frac{1}{2}$ to have score zero, and will make the score of 63% the negative of the score of 37%, whatever expression we chose to begin with. Indeed, if we are to have the sort of symmetry around 50% we seek, we can confine our attention to folded expressions. (Proof?)

exhibit **2** of chapter 15 (continued)

**B) SUPPLEMENTARY TABLE**--for flogs of fractions beyond 1% or 99%

| +      | \|flog\| | −    | +      | \|flog\| | −    |
|--------|----------|------|--------|----------|------|
| 99.0%  | 2.30     | 1.0% | 99.80% | 3.11     | .20% |
| .1     | 2.35     | .9   | .82    | 3.16     | .18  |
| .2     | 2.41     | .8   | .84    | 3.22     | .16  |
| .3     | 2.48     | .7   | .86    | 3.28     | .14  |
| .4     | 2.55     | .6   | .88    | 3.36     | .12  |
|        |          |      |        |          |      |
| 99.50  | 2.65     | .50  | 99.90  | 3.45     | .10  |
| .52    | 2.67     | .48  | .91    | 3.51     | .09  |
| .54    | 2.69     | .46  | .92    | 3.57     | .08  |
| .56    | 2.71     | .44  | .93    | 3.63     | .07  |
| .58    | 2.73     | .42  | .94    | 3.71     | .06  |
|        |          |      |        |          |      |
| 99.60  | 2.76     | .40  | 99.95  | 3.80     | .05  |
| .62    | 2.78     | .38  | .96    | 3.91     | .04  |
| .64    | 2.81     | .36  | .97    | 4.06     | .03  |
| .66    | 2.84     | .34  | .98    | 4.26     | .02  |
| .68    | 2.87     | .32  | .99    | 4.61     | .01  |
|        |          |      |        |          |      |
| 99.70  | 2.90     | .30  |        |          |      |
| .72    | 2.94     | .28  | Examples: |       |      |
| .74    | 2.97     | .26  |        |          |      |
| .76    | 3.01     | .24  | 99.29% gives 2.47 |  |    |
| .78    | 3.06     | .22  | 0.37% gives −2.80 |  |    |

**P) PROBLEMS**

2a) What is the froot corresponding to 87%? The flog?

2b) What is the % corresponding to a froot of 1.00?

2c) What is the froot corresponding to a flog of −2.48? The % corresponding?

2d) What is the % corresponding to a flog of 4.00? The froot corresponding?

2e) What is the froot corresponding to 99.975%? The flog?

2f) For what % or %'s does/do the corresponding froot and flog differ by .60? By 3.80?

While a number of more complicated expressions have been--and may well continue to be--popular, we can usually do quite well by folding some very simple expression. By analogy with all our other inquiries into good ways to express data, we find it natural to begin with a square root and a logarithm. We can use the root--or the log--of any multiple of $f$. If we are to match folded percents near 50%, it turns out that we should fold $\sqrt{2f}$ and $\frac{1}{2}\log_e f = 1.15 \log_{10} f$. (Here $\log_e$ indicates the so-called natural or Napierian logarithm --the logarithm to the base $e = 2.71828\ldots$). [That this works will be clear when we come to exhibit 2.]

We need names for these expressions--both formal names and working names. Clearly, *folded roots* and *folded logs* are both formal and descriptive, while *froots* and *flogs* are convenient. Specifically, then, we plan to consider (notice that we are meeting logs of roots again):

$$\text{froot} = \sqrt{2f} - \sqrt{2(1-f)}$$
$$\text{flog} = \tfrac{1}{2}\log_e f - \tfrac{1}{2}\log_e (1-f)$$
$$= \log_e \sqrt{f} - \log_e \sqrt{1-f}$$
$$= 1.15 \log_{10} f - 1.15 \log_{10} (1-f)$$
$$= 1.15 \log_{10} (f/(1-f))$$

as alternatives to

$$\text{plurality} = f - (1-f)$$

or

$$\text{fcent} = (\%\text{ yes}) - (\%\text{ no}).$$

Exhibit **2** sets out two-decimal values of pluralities, froots, and flogs for every % from 10 to 90 and at closer intervals for smaller and larger %'s. Notice that--to the accuracy given, which is usually sufficient--the three expressions agree exactly for fractions between 38% and 62%--and to within 0.02 for fractions between 30% and 70%. (It was just to have this happen that we put the $\sqrt{2}$'s in our froots and the $\frac{1}{2}$'s in our flogs.) Unless we have to deal with some fractions below 20% or above 80% it is likely to make only trivial differences which of these expressions we use for our counted fractions (so long as we are consistent).

**froots and flogs were brought in to stretch the ends of the scale. If the ends of the scale are not needed--because there is no data there--it is most convenient and satisfactory that froots and flogs behave very, very much like pluralities or fcents.**

As always, it is nice to look at changes of expression graphically as well as in numbers. In this situation, where we think and write of "stretching the tails", it is unusually desirable to compare the different scales side by side and both watch and feel--to the extent that the feeling can be given--the tails stretch more and more as we change from pluralities--first to froots and then to flogs. Exhibit **3** is intended to do just this.

### review questions

How should we treat "those who are" and "those who aren't" when we re-express fractions? How can we do this easily? What is a trivial re-expression? In which of the two re-expressions are we most interested? Can we match re-expressions of fractions? How? What is '"folding"? How do our matched expressions behave in the middle of the scale (near 50%)? At the ends of the scale (near 0% or 100%)? What does tail-stretching mean? Can there be a picture of it? Looking like what? Do we express ss-fractions? How? What is a froot? And a flog?

### 15C. Quicker calculation

Most of the time we will want to use flogs and froots for ss-fractions, for fractions of the form

$$\frac{\text{count} + \frac{1}{6}}{\text{total} + \frac{1}{3}}.$$

exhibit **3** of chapter 15: comparative display

**The scales of pluralities (fcents), froots and flogs compared**

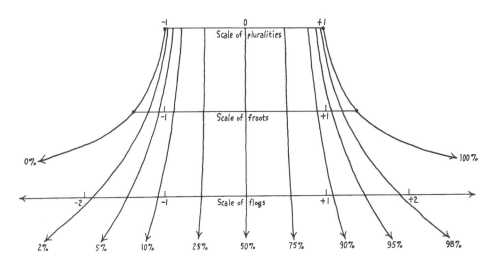

We may as well do as much as we can to make the overall calculation simple. For flogs we have

$$\text{flog} = \tfrac{1}{2}\log_e \frac{\text{count} + \tfrac{1}{6}}{\text{total} + \tfrac{1}{3}} - \tfrac{1}{2}\log_e \frac{\text{remainder} + \tfrac{1}{6}}{\text{total} + \tfrac{1}{3}}$$

$$= \tfrac{1}{2}\log_e (\text{count} + \tfrac{1}{6}) - \tfrac{1}{2}\log_e (\text{remainder} + \tfrac{1}{6})$$

$$= \log_e \sqrt{\text{count} + \tfrac{1}{6}} - \log_e \sqrt{\text{remainder} + \tfrac{1}{6}}$$

(where the last form arises because the log of a ratio is the difference of the logs, and the $\pm\tfrac{1}{2}\log_e (\text{total} + \tfrac{1}{3})$ terms cancel out). Exhibit **4** gives values of $\tfrac{1}{2}\log_e (\text{count} + \tfrac{1}{6}) = \log_e \sqrt{\text{count} + \tfrac{1}{6}}$ for each count from 0 to 200, and every 10th count from 200 to 3090. (More detail is usually worthless.)

exhibit **4** of chapter 15: reference table

**Values of $\log_e \sqrt{\text{count} + \tfrac{1}{6}}$ (a) for integer counts from 0 to 199, (b) for every 10th count from 200 to 3090, (c) for larger counts, (d) for half-integers ≤ 20**

A) TABLE for COUNTS up to 199--see panel D for half integers up to 20

|     | 0    | 1    | 2    | 3    | 4    | 5    | 6    | 7    | 8    | 9    |
|-----|------|------|------|------|------|------|------|------|------|------|
| 00  | -.90 | .08  | .39  | .58  | .71  | .82  | .91  | .98  | 1.05 | 1.11 |
| 10  | 1.16 | 1.21 | 1.25 | 1.29 | 1.33 | 1.36 | 1.39 | 1.42 | 1.45 | 1.48 |
| 20  | 1.50 | 1.53 | 1.55 | 1.57 | 1.59 | 1.61 | 1.63 | 1.65 | 1.67 | 1.69 |
| 30  | 1.70 | 1.72 | 1.74 | 1.75 | 1.77 | 1.78 | 1.79 | 1.81 | 1.82 | 1.83 |
| 40  | 1.85 | 1.86 | 1.87 | 1.88 | 1.89 | 1.91 | 1.92 | 1.93 | 1.94 | 1.95 |
| 50  | 1.96 | 1.97 | 1.98 | 1.99 | 2.00 | 2.01 | 2.01 | 2.02 | 2.03 | 2.04 |
| 60  | 2.05 | 2.06 | 2.06 | 2.07 | 2.08 | 2.09 | 2.10 | 2.10 | 2.11 | 2.12 |
| 70  | 2.13 | 2.13 | 2.14 | 2.15 | 2.15 | 2.16 | 2.17 | 2.17 | 2.18 | 2.19 |
| 80  | 2.19 | 2.20 | 2.20 | 2.21 | 2.22 | 2.22 | 2.23 | 2.23 | 2.24 | 2.25 |
| 90  | 2.25 | 2.26 | 2.26 | 2.27 | 2.27 | 2.28 | 2.28 | 2.29 | 2.29 | 2.30 |
| 100 | 2.30 | 2.31 | 2.31 | 2.32 | 2.32 | 2.33 | 2.33 | 2.34 | 2.34 | 2.35 |
| 110 | 2.35 | 2.36 | 2.36 | 2.36 | 2.37 | 2.37 | 2.38 | 2.38 | 2.39 | 2.39 |
| 120 | 2.39 | 2.40 | 2.40 | 2.41 | 2.41 | 2.41 | 2.42 | 2.42 | 2.43 | 2.43 |
| 130 | 2.43 | 2.44 | 2.44 | 2.45 | 2.45 | 2.45 | 2.46 | 2.46 | 2.46 | 2.47 |
| 140 | 2.47 | 2.47 | 2.48 | 2.48 | 2.49 | 2.49 | 2.49 | 2.50 | 2.50 | 2.50 |
| 150 | 2.51 | 2.51 | 2.51 | 2.52 | 2.52 | 2.52 | 2.53 | 2.53 | 2.53 | 2.53 |
| 160 | 2.54 | 2.54 | 2.54 | 2.55 | 2.55 | 2.55 | 2.56 | 2.56 | 2.56 | 2.57 |
| 170 | 2.57 | 2.57 | 2.57 | 2.58 | 2.58 | 2.58 | 2.59 | 2.59 | 2.59 | 2.59 |
| 180 | 2.60 | 2.60 | 2.60 | 2.61 | 2.61 | 2.61 | 2.61 | 2.62 | 2.61 | 2.62 |
| 190 | 2.62 | 2.63 | 2.63 | 2.63 | 2.63 | 2.64 | 2.64 | 2.64 | 2.64 | 2.65 |

Table continues →

exhibit 4 of chapter 15 (continued)

**B) TABLE for COUNTS from 200 to 3090 -- grab nearest; don't interpolate**

|      | 00    | 10    | 20    | 30    | 40    | 50    | 60    | 70    | 80    | 90    |
|------|-------|-------|-------|-------|-------|-------|-------|-------|-------|-------|
| 200  | 2.65  | 2.67  | 2.70  | 2.72  | 2.74  | 2.76  | 2.78  | 2.80  | 2.82  | 2.84  |
| 300  | 2.85  | 2.87  | 2.88  | 2.90  | 2.91  | 2.93  | 2.94  | 2.96  | 2.97  | 2.98  |
| 400  | 3.00  | 3.01  | 3.02  | 3.03  | 3.04  | 3.05  | 3.07  | 3.08  | 3.09  | 3.10  |
| 500  | 3.11  | 3.12  | 3.13  | 3.14  | 3.15  | 3.16  | 3.16  | 3.17  | 3.18  | 3.19  |
| 600  | 3.20  | 3.21  | 3.21  | 3.22  | 3.23  | 3.24  | 3.25  | 3.25  | 3.26  | 3.27  |
| 700  | 3.28  | 3.28  | 3.29  | 3.30  | 3.30  | 3.31  | 3.32  | 3.32  | 3.33  | 3.34  |
| 800  | 3.34  | 3.35  | 3.35  | 3.36  | 3.37  | 3.37  | 3.38  | 3.38  | 3.39  | 3.40  |
| 900  | 3.40  | 3.41  | 3.41  | 3.42  | 3.42  | 3.43  | 3.43  | 3.44  | 3.44  | 3.45  |
| 1000 | 3.45  | 3.46  | 3.46  | 3.47  | 3.47  | 3.48  | 3.48  | 3.49  | 3.49  | 3.50  |
| 1100 | 3.50  | 3.51  | 3.51  | 3.52  | 3.52  | 3.52  | 3.53  | 3.53  | 3.54  | 3.54  |
| 1200 | 3.545 | 3.549 | 3.553 | 3.557 | 3.562 | 3.566 | 3.569 | 3.573 | 3.577 | 3.581 |
| 1300 | 3.585 | 3.589 | 3.593 | 3.597 | 3.600 | 3.604 | 3.608 | 3.611 | 3.615 | 3.619 |
| 1400 | 3.622 | 3.626 | 3.629 | 3.633 | 3.636 | 3.640 | 3.643 | 3.647 | 3.650 | 3.653 |
| 1500 | 3.657 | 3.660 | 3.663 | 3.667 | 3.670 | 3.673 | 3.676 | 3.679 | 3.683 | 3.686 |
| 1600 | 3.689 | 3.692 | 3.695 | 3.698 | 3.701 | 3.704 | 3.707 | 3.710 | 3.713 | 3.716 |
| 1700 | 3.719 | 3.722 | 3.725 | 3.728 | 3.731 | 3.734 | 3.737 | 3.739 | 3.742 | 3.745 |
| 1800 | 3.748 | 3.751 | 3.753 | 3.756 | 3.759 | 3.762 | 3.764 | 3.767 | 3.770 | 3.772 |
| 1900 | 3.775 | 3.777 | 3.780 | 3.783 | 3.785 | 3.788 | 3.790 | 3.793 | 3.795 | 3.798 |
| 2000 | 3.800 | 3.803 | 3.805 | 3.808 | 3.810 | 3.813 | 3.815 | 3.818 | 3.820 | 3.822 |
| 2100 | 3.825 | 3.827 | 3.830 | 3.832 | 3.834 | 3.837 | 3.839 | 3.841 | 3.844 | 3.846 |
| 2200 | 3.848 | 3.850 | 3.853 | 3.855 | 3.857 | 3.859 | 3.862 | 3.864 | 3.866 | 3.868 |
| 2300 | 3.870 | 3.873 | 3.875 | 3.877 | 3.879 | 3.881 | 3.883 | 3.885 | 3.887 | 3.890 |
| 2400 | 3.892 | 3.894 | 3.896 | 3.898 | 3.900 | 3.902 | 3.904 | 3.906 | 3.908 | 3.910 |
| 2500 | 3.912 | 3.914 | 3.916 | 3.918 | 3.920 | 3.922 | 3.924 | 3.926 | 3.928 | 3.930 |
| 2600 | 3.932 | 3.934 | 3.935 | 3.937 | 3.939 | 3.941 | 3.943 | 3.945 | 3.947 | 3.949 |
| 2700 | 3.951 | 3.952 | 3.954 | 3.956 | 3.958 | 3.960 | 3.962 | 3.963 | 3.965 | 3.967 |
| 2800 | 3.969 | 3.970 | 3.972 | 3.974 | 3.976 | 3.978 | 3.979 | 3.981 | 3.983 | 3.985 |
| 2900 | 3.986 | 3.988 | 3.990 | 3.991 | 3.993 | 3.995 | 3.997 | 2.998 | 4.000 | 4.002 |
| 3000 | 4.003 | 4.005 | 4.007 | 4.008 | 4.010 | 4.011 | 4.013 | 4.015 | 4.016 | 4.018 |

**C) For COUNTS above 3000**

| First cut: | Divide by | Look in panel | Add (or) |
|------------|-----------|---------------|----------|
|            | 10        | B             | 1.151 (1.15) |
|            | 100       |               | 2.303 (2.30) |
|            | 1000      |               | 3.454 (3.45) |

Example: Given 4567, to 457 (rounded), find 3.07 at 460, add 1.15, to conclude with 4.22.

For froots, we have no equally simple device. We can often gain, however, from a table of $\sqrt{(\text{split count}) + \frac{1}{6}}$, given as exhibit 5, since we can write

$$\text{froot} = \sqrt{2f} - \sqrt{2(1-f)}$$

$$= \sqrt{2\left(\frac{\text{split count left} + \frac{1}{6}}{\text{total count} + \frac{1}{3}}\right)} - \sqrt{2\left(\frac{\text{split count right} + \frac{1}{6}}{\text{total count} + \frac{1}{3}}\right)}$$

$$= \frac{\sqrt{\text{split count left} + \frac{1}{6}} - \sqrt{\text{split count right} + \frac{1}{6}}}{\sqrt{\text{half total count} + \frac{1}{6}}}$$

exhibit **4** of chapter 15 (continued)

If more accuracy is required (quite unlikely), go to more-place log table. If table of natural logs is available, use $\frac{1}{2} \log_e$ count. If only table of common logs is at hand, use $1.1513 \log_{10}$ count.

**D) TABLE for HALF-INTEGERS < 20—for > 20, interpolate in panel A**

|    | 0h   | 1h   | 2h   | 3h   | 4h   | 5h   | 6h   | 7h   | 8h   | 9h   |
|----|------|------|------|------|------|------|------|------|------|------|
| 00 | −.20 | .26  | .49  | .65  | .77  | .87  | .95  | 1.02 | 1.08 | 1.13 |
| 10 | 1.18 | 1.23 | 1.27 | 1.31 | 1.34 | 1.38 | 1.41 | 1.44 | 1.46 | 1.49 |

**E) EXAMPLES—"remainder" is right count**

left count =    43 , remainder =   28 , hence flog = 1.88  − 1.67  =    .21
left count =   504 , remainder =   13 , hence flog = 3.11  − 1.29  =   1.82
left count =  1272 , remainder = 3057 , hence flog = 3.573 − 4.013 =  −.440
left count =    27 , remainder = 1304 , hence flog = 1.65  − 3.585 = −1.94
left count =   18h, remainder =   32h, hence flog = 1.46  − 1.74  =  −.28

Note that 504 was looked up as 500, 1272 as 1270, 3057 as 3060, and 1304 as 1300. Note further that, when one look-up has 3 decimals and one has 2, only 2 decimals are kept in the difference. Note still further that the fact that 1304 > 1300 was allowed to creep through when rounding 3.585 (to 3.59 instead of 3.58).

**P) PROBLEMS**

4a) What is $\log_e \sqrt{\text{count} + \frac{1}{6}}$ when count = 7? = 77? = 777?

4b) What is the flog of each of the ss-fractions for the following pairs of counts? (2, 27); (20, 270); (200, 2700). Comment.

4c) What is the flog of each of the following pairs of counts? (17, 9h); (26h, 203h); (42h, 197h).

4d) What count has a log $\sqrt{\text{count} + \frac{1}{6}}$ nearest to 2.68? To 3.566?

4e) What multiple of $\log(\text{count} + \frac{1}{6})$ is $\log \sqrt{\text{count} + \frac{1}{6}}$?

exhibit 5 of chapter 15: reference table

**Values of $\sqrt{\text{count} + \frac{1}{6}}$ (a) for integer counts from 0 to 199, (b) for every 10th count from 200 to 3090, (c) for half-integer counts < 20**

A) TABLE for COUNTS up to 199--see panel C for half-integers < 20

|     | 0 | 1 | 2 | 3 | 4 | 5 | 6 | 7 | 8 | 9 |
|-----|------|------|------|------|------|------|------|------|------|------|
| 00  | .41  | 1.08 | 1.47 | 1.78 | 2.04 | 2.27 | 2.48 | 2.68 | 2.86 | 3.03 |
| 10  | 3.19 | 3.34 | 3.49 | 3.63 | 3.76 | 3.89 | 4.02 | 4.14 | 4.26 | 4.38 |
| 20  | 4.49 | 4.60 | 4.71 | 4.81 | 4.92 | 5.02 | 5.12 | 5.21 | 5.31 | 5.40 |
| 30  | 5.49 | 5.58 | 5.67 | 5.76 | 5.85 | 5.93 | 6.01 | 6.10 | 6.18 | 6.26 |
| 40  | 6.34 | 6.42 | 6.49 | 6.57 | 6.65 | 6.72 | 6.79 | 6.87 | 6.94 | 7.01 |
| 50  | 7.08 | 7.15 | 7.22 | 7.29 | 7.36 | 7.43 | 7.49 | 7.56 | 7.63 | 7.69 |
| 60  | 7.76 | 7.82 | 7.88 | 7.95 | 8.01 | 8.07 | 8.13 | 8.20 | 8.26 | 8.32 |
| 70  | 8.38 | 8.44 | 8.50 | 8.55 | 8.61 | 8.67 | 8.73 | 8.78 | 8.84 | 8.90 |
| 80  | 8.95 | 9.01 | 9.06 | 9.12 | 9.17 | 9.23 | 9.28 | 9.34 | 9.39 | 9.44 |
| 90  | 9.50 | 9.55 | 9.60 | 9.65 | 9.70 | 9.76 | 9.81 | 9.86 | 9.91 | 9.96 |
| 100 | 10.01| 10.06| 10.11| 10.16| 10.21| 10.26| 10.30| 10.35| 10.40| 10.45|
| 110 | 10.50| 10.54| 10.59| 10.64| 10.68| 10.73| 10.78| 10.82| 10.87| 10.92|
| 120 | 10.96| 11.01| 11.05| 11.10| 11.14| 11.19| 11.23| 11.28| 11.32| 11.37|
| 130 | 11.41| 11.45| 11.50| 11.54| 11.58| 11.63| 11.67| 11.71| 11.75| 11.80|
| 140 | 11.84| 11.88| 11.92| 11.97| 12.01| 12.05| 12.09| 12.13| 12.17| 12.21|
| 150 | 12.25| 12.29| 12.34| 12.38| 12.42| 12.46| 12.50| 12.54| 12.58| 12.62|
| 160 | 12.66| 12.70| 12.73| 12.77| 12.81| 12.85| 12.89| 12.93| 12.97| 13.01|
| 170 | 13.04| 13.08| 13.12| 13.16| 13.20| 13.24| 13.27| 13.31| 13.35| 13.39|
| 180 | 13.42| 13.46| 13.50| 13.53| 13.57| 13.61| 13.64| 13.68| 13.72| 13.75|
| 190 | 13.79| 13.83| 13.86| 13.90| 13.93| 13.97| 14.01| 14.04| 14.08| 14.11|

B) TABLE for COUNTS from 200 to 3090--grab nearest; don't interpolate

|      | 00    | 10    | 20    | 30    | 40    | 50    | 60    | 70    | 80    | 90    |
|------|-------|-------|-------|-------|-------|-------|-------|-------|-------|-------|
| 200  | 14.15 | 14.50 | 14.84 | 15.17 | 15.50 | 15.82 | 16.13 | 16.44 | 16.74 | 17.03 |
| 300  | 17.33 | 17.61 | 17.89 | 18.17 | 18.44 | 18.71 | 18.98 | 19.24 | 19.50 | 19.75 |
| 400  | 20.00 | 20.25 | 20.50 | 20.74 | 20.98 | 21.22 | 21.45 | 21.68 | 21.91 | 22.14 |
| 500  | 22.36 | 22.59 | 22.81 | 23.03 | 23.24 | 23.46 | 23.67 | 23.88 | 24.09 | 24.29 |
| 600  | 24.50 | 24.70 | 24.90 | 25.10 | 25.30 | 25.50 | 25.69 | 25.89 | 26.08 | 26.27 |
| 700  | 26.46 | 26.65 | 26.84 | 27.02 | 27.21 | 27.39 | 27.57 | 27.75 | 27.93 | 28.11 |
| 800  | 28.29 | 28.46 | 28.64 | 28.81 | 28.99 | 29.16 | 29.33 | 29.50 | 29.67 | 29.84 |
| 900  | 30.00 | 30.17 | 30.33 | 30.50 | 30.66 | 30.82 | 30.99 | 31.15 | 31.31 | 31.47 |
| 1000 | 31.63 | 31.78 | 31.94 | 32.10 | 32.25 | 32.41 | 32.56 | 32.71 | 32.87 | 33.02 |
| 1100 | 33.17 | 33.32 | 33.47 | 33.62 | 33.77 | 33.91 | 34.06 | 34.21 | 34.35 | 34.50 |
| 1200 | 34.64 | 34.79 | 34.93 | 35.07 | 35.22 | 35.36 | 35.50 | 35.64 | 35.78 | 35.92 |
| 1300 | 36.06 | 36.20 | 36.33 | 36.47 | 36.61 | 36.74 | 36.88 | 37.02 | 37.15 | 37.28 |

Table continues →

exhibit **5** of chapter 15 (continued)

| | | | | | | | | | |
|---|---|---|---|---|---|---|---|---|---|
| 1400 | 37.42 | 37.55 | 37.69 | 37.82 | 37.95 | 38.08 | 38.21 | 38.34 | 38.47 | 38.60 |
| 1500 | 38.73 | 38.86 | 38.99 | 39.12 | 39.24 | 39.37 | 39.50 | 39.63 | 39.75 | 39.88 |
| 1600 | 40.00 | 40.13 | 40.25 | 40.38 | 40.50 | 40.62 | 40.75 | 40.87 | 40.99 | 41.11 |
| 1700 | 41.23 | 41.35 | 41.47 | 41.60 | 41.72 | 41.83 | 41.95 | 42.07 | 42.19 | 42.31 |
| 1800 | 42.43 | 42.55 | 42.66 | 42.78 | 42.90 | 43.01 | 43.13 | 43.25 | 43.36 | 43.48 |
| 1900 | 43.59 | 43.71 | 43.82 | 43.93 | 44.05 | 44.16 | 44.27 | 44.39 | 44.50 | 44.61 |
| 2000 | 44.72 | 44.83 | 44.95 | 45.06 | 45.17 | 45.28 | 45.39 | 45.50 | 45.61 | 45.72 |
| 2100 | 45.83 | 45.94 | 46.05 | 46.15 | 46.26 | 46.37 | 46.48 | 46.59 | 46.69 | 46.80 |
| 2200 | 46.91 | 47.01 | 47.12 | 47.22 | 47.33 | 47.44 | 47.54 | 47.65 | 47.75 | 47.86 |
| 2300 | 47.96 | 48.06 | 48.17 | 48.27 | 48.38 | 48.48 | 48.58 | 48.68 | 48.79 | 48.89 |
| 2400 | 48.99 | 49.09 | 49.20 | 49.30 | 49.40 | 49.50 | 49.60 | 49.70 | 49.80 | 49.90 |
| 2500 | 50.00 | 50.10 | 50.20 | 50.30 | 50.40 | 50.50 | 50.60 | 50.70 | 50.80 | 50.89 |
| 2600 | 50.99 | 51.09 | 51.19 | 51.29 | 51.38 | 51.48 | 51.58 | 51.67 | 51.77 | 51.87 |
| 2700 | 51.96 | 52.06 | 52.16 | 52.25 | 52.35 | 52.44 | 52.54 | 52.63 | 52.73 | 52.82 |
| 2800 | 52.92 | 53.01 | 53.11 | 53.20 | 53.29 | 53.39 | 53.48 | 53.57 | 53.67 | 53.76 |
| 2900 | 53.85 | 53.95 | 54.04 | 54.13 | 54.22 | 54.32 | 54.41 | 54.50 | 54.59 | 54.68 |
| 3000 | 54.77 | 54.86 | 54.96 | 55.05 | 55.14 | 55.23 | 55.32 | 55.41 | 55.50 | 55.59 |

C) TABLE for HALF-INTEGER COUNTS < 20 -- for > 20, interpolate in panel A

| | |0h| |1h| |2h| |3h| |4h| |5h| |6h| |7h| |8h| |9h| |
|---|---|---|---|---|---|---|---|---|---|
| 00 | .82 | 1.29 | 1.63 | 1.91 | 2.16 | 2.38 | 2.58 | 2.77 | 2.94 | 3.11 |
| 10 | 3.27 | 3.42 | 3.56 | 3.70 | 3.83 | 3.96 | 4.08 | 4.20 | 4.32 | 4.43 |

D) EXAMPLES--"remainder" is right count

left count = 43, remainder = 28, hence half total = 35h and
   froot = (6.57 − 5.3)/5.97 = .21

left count = 504, remainder = 13, hence half total = 258h and
   froot = (22.36 − 3.63)/16.13 = 1.16

left count = 18h, remainder = 32h, hence half total = 25h and
   froot = (4.32 − 5.72)/5.07 = −.28

Note that 504 was looked up as 500, and that 35h, 258h, 32h, and 25h were interpolated in panel A.

P) PROBLEMS

5a) What is $\sqrt{\text{count} + \frac{1}{6}}$ when count = 7? = 77? = 777?

5b) What is the froot of each of the ss-fractions for the following pairs of counts? (2, 27); (20, 270); (200, 2700)? Comment.

5c) What is the froot of each of the following pairs of counts? (17h, 9h); (26h, 203h); (42h, 197h)?

When the total count is constant, we may be able to avoid the division and use

$$\sqrt{\text{split count left} + \tfrac{1}{6}} - \sqrt{\text{split count right} + \tfrac{1}{6}},$$

which is then a constant multiple of the froot. (Especially when table 5 is not at hand, it can help to use a slide rule efficiently and easily, as explained in section 15G.)

### review questions

Can we simplify the arithmetic for re-expressing ss-fractions as froots? As flogs? Do we need to keep track of every digit in a count when we re-express it? Why/why not?

## 15D. Examples where careful expression clearly pays off

### a 2 × 2 opinion example

Being careful in how one expresses counts or counted fractions gives us better analyses in many, many instances. It is not always easy to be sure that this is happening to us; often the improvement is not obvious. If the data contains no two portions whose analysis we can usefully compare, it is hard to be sure that we have indeed gained by care in choosing the expression to be analyzed. Even if one portion can be checked against another, we are likely to learn little, so long as everything is very nearly the same. In such a situation, the effect of a choice of expression on the two portions will be almost the same; consequently the comparison of the two for any one expression is likely to be very similar to that for any other.

To learn about choice of expression from internal comparisons--from comparisons within one set of data--we need a situation where we can compare the behavior of one portion with that of another where:

◊our attention is directed toward differences, so that satisfactory agreement means similar differences.

◊the differences that are formed--and found to be nearly the same--are differences of smaller numbers in one portion, of larger numbers in the other.

The simplest such situation, of course, is a two-way table in which both factors show substantial effects.

Let us start with a 2 × 2 example. Under the headline "Protestants shift support to Kennedy," the *Washington Post* for February 2, 1962 reported the following Gallup poll results for November 1960 and January 1962:

|        | Protestants | | Catholics | |
|--------|-------------|------|-----------|------|
| Date   | 11/60 | 1/62 | 11/60 | 1/62 |
| Kennedy | 38% | 59% | 78% | 89% |
| Nixon   | 62% | 41% | 22% | 11% |

From an engineering (i.e., practical political) viewpoint, George Gallup was correct in saying that "one of the major--reasons for the President's popularity has been his success in allaying anti-Catholic sentiment while not losing the support of his fellow Catholics." An improvement of 21% among four-fifths of the population is many more votes than an improvement of 11% among the remaining one-fifth.

But we might want to try to understand (numerically) what is really going on. We can be very sure that this situation demands scale stretching near the ends. (It is surely easier to change 38% to 42% than to change 89% to 93%.) So let us go to flogs, and look at the same data again (with $f$ = fraction for Kennedy). The flogs are:

|       | Protestants | Catholics | Difference |
|-------|-------------|-----------|------------|
| 11/60 | −0.24 | +0.63 | +0.87 |
| 1/62  | +0.18 | +1.05 | +0.87 |
| Change | up 0.42 | up 0.42 | |

so that the Protestant and Catholic changes are very similar and mutually supporting.

Understanding of what happened to John F. Kennedy's popularity between November 1960 and January 1962--a scientific matter--is much better served by the flog analysis. We can say: "JFK's popularity improved by +0.42 on the flog scale; you would have got this same answer had you asked only Catholics or only Protestants."

**It is this constancy, present--in this example--in flogs but not in % or pluralities, that shows the advantages--in this example--of a choice of flogs.**

We cannot count on such nice things happening when we go to flogs. Sometimes going to flogs makes things wonderful; sometimes merely good, sometimes no better, sometimes worse. On balance, though, they seem to help often enough so that the writer usually tries flogs even before percents (plain or folded) or pluralities--flogs are the first-aid bandages for counted fractions.

## two-way tables of counts

We often have larger two-way tables, each cell of which contains a counted fraction. Opinion polls on several dates, reported separately for four different regions of the country, offer one example.

Two-way tables with froots or flogs as entries also arise in quite a different way. If we have a response that can reasonably be cut at two or more different places, then we can calculate a different flog or froot for each. Anticipating our next example a little, consider the counts in exhibit **6**. For each of the cuts that are feasible for the data given--Very large/Large, Large/Medium, Medium/Small, and Small/Very small--there is a flog--respectively $-4.20$, $-2.78$, $-2.22$, $-1.76$. If we had another such set of counts--as we soon will--we could write down a $2 \times 4$ table of flogs and analyze it. (If we had a third such set, we could have a $3 \times 4$ table, and so on.)

In 1954--in the course of a vigorous discussion with Gerschenkron--Landes published matched figures on the sizes of various kinds of business and industrial establishments in France and Germany as of 1906-07. Here "size of establishment" is given--by number of employees--in five broad groups. Instead of worrying how these sizes ought to be expressed in our analysis, we can begin by calling them very small, small, medium, large, and very large, and

**exhibit 6 of chapter 15: illustrative**

**Several flogs for a single set of counts (sizes of textile establishments in France, 1906)**

A) CALCULATION

| Size | Count | Cum down | Cum up | Exhibit 4* down | up | flog |
|---|---|---|---|---|---|---|
| Very large | 48 | | | | | |
| | | 48 | 213,579 | 1.94 | 6.136 | −4.20 |
| Large | 766 | | | | | |
| | | 814 | 212,813 | 3.35 | 6.134 | −2.28 |
| Medium | 1,676 | | | | | |
| | | 2,490 | 211,137 | 3.910 | 6.130 | −2.220 |
| Small | 3,636 | | | | | |
| | | 6,126 | 207,501 | 4.360 | 6.121 | −1.761 |
| Very small | 207,501 | | | | | |

* Exhibit 4 used thrice, table of logs used five times.

P) PROBLEMS

6a) Just how does the number 2,490 arise?

6b) Same for 212,813?

6c) And for 6.130?

6d) And for 3.35?

considering each and all of the four possible cuts. The data for all textile establishments in France in 1906 has already been flogged in exhibit 6; similar data for Germany in 1907 is given, with its flogs, in exhibit **7**. The difference between French and German flogs is of consistent sign and roughly constant size--say .4--indicating not only a consistent tendency for the German textile establishments to be somewhat larger, but a quite consistent difference, in flog terms.

The difference is consistent enough to encourage the belief that, if one more census could be found for either country,

⋄ it would show a relatively consistent difference over time.

⋄ an appropriate fraction of this time difference could be used either to adjust the French data to 1907 or to adjust the German data to 1906, so as to make a cleaner comparison of the two countries (one based on the same date).

exhibit **7** of chapter 15: textile establishments

### Sizes of textile establishments in France and Germany, 1906–1907

A) DATA (for France in '06, see exhibit 6) and CALCULATIONS

| Count of employees | | # estab | Germany, 1907 flog | France, 1906 flog flog* | diff | |
|---|---|---|---|---|---|---|
| | 1001 up | 71 | | | | |
| 1000.5 | | | 3.77 | 4.20 | .43 | |
| | 201–100 | 1,013 | | | | |
| 200.5 | | | 2.41 | 2.780 | .37 | |
| | 51–200 | 2,748 | | | | median |
| 50.5 | | | 1.757 | 2.220 | .463 | = .44 |
| | 11–50 | 4,984 | | | | |
| 10.5 | | | 1.321 | 1.761 | .440 | |
| | 1–10 | 123,768 | | | | |

* Note that, in this exhibit, we have calculated the flogs from below upwards--not from above downward--thus changing the sign in comparison with exhibit 6, whence come the French flogs.

P) PROBLEMS

7a) What expression of establishment size gives a simple behavior of the flogs for either country or both countries?

7b) Plot the two sets of flogs against this expression of establishments.

7c) Seek out data to allow comparison of France and Germany at the same date (see text and Landes).

7d) Make such a comparison.

S) SOURCE

D. S. Landes 1954. "Some attitudes, entrepreneurship, and economic development: A comment," **Explorations in Entrepreneurial History 6**: 245–272. Appendix table 2 on pages 268–269.

The pessimist will ask "Did we really gain by taking flogs--what if we had stuck to %?" For the largest cut--at 1000.5 employees--the %'s are 0.022 for France and 0.053 for Germany, a difference of 0.031%. For the smallest cut--at 10.5 employees--they are 2.9% for France and 6.6% for Germany--a difference of 3.7%. One is about 1/30 of a percent, the other nearly 4 percent. Clearly there was no agreement about

Germany MINUS France

in percent (or in fcents or in plurality), although we have just seen good agreement when we work in flogs.

A few further examples are set as problems in exhibit **8**.

### review questions

What was the first example of this section? How did we treat it? Why did we like/hate the re-expressions? How do you think re-expression as froots would have worked: compared to raw fractions? compared to re-expressed fractions? Can we treat larger sets of data similarly? For instance? Is this the

exhibit **8** of chapter 15: problems

**Problems involving double flogs**

P) PROBLEMS

8a) Brownlee gave the following counts for each of 6 grades of quality for the production on each of three shifts:

|         | Qual. a | Qual. b | Qual. c | Qual. d | Qual. e | Qual. f |
|---------|---------|---------|---------|---------|---------|---------|
| Shift A | 11      | 23      | 8       | 5       | 18      | 18      |
| Shift B | 17      | 29      | 10      | 17      | 7       | 15      |
| Shift C | 6       | 21      | 8       | 24      | 15      | 9       |

Analyze as in exhibit 7. Comment on the results.

8b) Clausen gave the following %'s of each level of definiteness of plan to return to previous employer for each of four durations of previous employment for soldiers leaving the U.S. Army in July 1945:

| Years in job | Definite plans | Tentative plans | Considering return | Not considering | (total) |
|---|---|---|---|---|---|
| <1  | 12% | 9%  | 20% | 59% | (100%)   |
| 1-2 | 22% | 11% | 18% | 49% | (100%)   |
| 2-5 | 31% | 12% | 17% | 40% | (100%)   |
| >5  | 49% | 16% | 14% | 42% | (121%)?? |

Analyze as in exhibit 7. Comment on the results. (Notice, here and below, that we are taking the row labels as circumstances.)

→

only way in which tables full of froots or flogs arise? Why/why not? What example did we look at next? To what class of data sets did it belong? What did we do with it? How well did this work? Why do you say so? How does this behavior affect using a third census to match dates more closely?

### 15E. Double folding--the 2 × 2 case

We dealt in Section 12E/F with a two-way table of counts, where the counts are really responses and both ways describe circumstances. In the last section, we looked at two-way tables of counts where one way is a level of cutting the response while the other is a circumstance--or is treated as such.

exhibit **8** of chapter 15 (continued)

8c) The same authors also gave the following figures for soldiers leaving the army in September 1945:

| Years in job | Definite plans | Tentative plans | Considering return | Not considering | (total) |
|---|---|---|---|---|---|
| <1  | 18% | 12% | 17% | 43% | (100%) |
| 1-2 | 35% | 24% | 19% | 22% | (100%) |
| 2-5 | 47% | 17% | 13% | 23% | (100%) |
| >5  | 69% | 16% | 13% | 7%  | (100%) |

Analyze as in exhibit 7. Comment on the results.

8d) (Double weight) Combine the July and September data and analyze all together as above.

8bc2) Compare the results of (8b) and (8c). Are they as you would have expected? (Consider the % not considering return.) Can you combine the effects of choice of cutting point for the two analyses? How would you analyze them?

8bcde2) Compare the analysis "all together" (8c) with the combination of the two analyses (8d). Discuss the differences/similarities.

**S) SOURCES**

K. A. Brownlee 1949. *Industrial Experimentation* (4th ed.) Chemical Rubber. Table on page 46.

J. A. Clausen 1950. "Studies of the postwar plans of soldiers: a problem in prediction." Chapter 15 (pages 568 to 708) of *Measurement and Prediction*, volume 4 of *The American Soldier* (ed. S. A. Stouffer), Princeton. Table on page 629.

We dealt with these by introducing a cut between each adjacent pair of levels and making a two-way table whose entries were froots or flogs, whose divisions were one in terms of circumstances and one in terms of cutting a response. It is time now to deal with two-way tables of counts, both of whose ways are levels of cutting a response--tables rightfully called "contingency tables with ordered margins".

The prototype is a double cutting of the total count into 2 × 2 = 4 counts, thus

If we want to seize the information in this 2-by-2 table that is not seized by the two margins--the two sets of combined counts, for rows: $a + b$ above and $c + d$ below; for columns: $a + c$ on the left and $b + d$ on the right, which we can schematize as

$$\underline{\frac{a+b}{c+d}} \qquad a+c \mid b+d$$

(or, equivalently, by the two marginal fractions and the total)--we do well to use logs and fold twice (once each way--once horizontally and once vertically), reaching some multiple of

$$\log \frac{a}{a+b+c+d} - \log \frac{b}{a+b+c+d} - \log \frac{c}{a+b+c+d} + \log \frac{d}{a+b+c+d},$$

which can be more simply written as

$$\log a - \log b - \log c + \log d = \log \frac{ad}{bc}$$

or, possibly more revealingly, as either

$$\left(\log \frac{a}{a+b} - \log \frac{b}{a+b}\right) - \left(\log \frac{c}{c+d} - \log \frac{d}{c+d}\right)$$

or, equally, as

$$\left(\log \frac{a}{a+c} - \log \frac{c}{a+c}\right) - \left(\log \frac{b}{b+d} - \log \frac{d}{b+d}\right)$$

In practice, we don't want to divide, so we prefer

$$\log a - \log b - \log c + \log d$$

as a basis for our arithmetic.

What is the simplest behavior? Suppose we have 100 football stars and 100 senior hospital nurses, where we know 102 are male and 98 are female. If we knew nothing about football or hospitals--and if we had to guess--we might say to ourselves "half football players and half nurses, eh? Maybe 51 male football players and 51 male nurses (and 49 female nurses and 49 female football players)." We all know too much about football and hospitals to pay any attention to simplicity in this example, but we would take it seriously if we knew nothing.

How do our doubly folded logs behave in relation to this example? The simple version would be

| Male | Female |
|------|--------|
| 51   | 49     |
| 51   | 49     |

with $\log a - \log b - \log c + \log d = 0$

and the likely version would be

| Male | Female |   |
|------|--------|---|
| 100  | 0      | football players |
| 2    | 98     | nurses |

for which we have trouble with log 0, and thus go to

$$\log(a + \tfrac{1}{6}) - \log(b + \tfrac{1}{6}) - \log(c + \tfrac{1}{6}) + \log(d + \tfrac{1}{6})$$
$$= 2.00 \; - \; (-.78) \; - \; (.34) \; + \; 1.99$$
$$= 4.43.$$

In simple language, the zero value of the doubly folded difference of logs for the "simple" table announced no apparent connection between sex and occupation, while the large value for the likely table announced a very strong relation between sex and occupation.

In general we can take a zero value of the doubly folded difference as indicating "no apparent relation".

When we dealt with counted fractions we had two counts--and we started both. Moreover, we threw in a factor of $\tfrac{1}{2}$ by taking logs (to base $e$) of the square roots of the (started) counts. It seems very natural to start all four counts here and to throw in the factor of $\tfrac{1}{2}$ by looking at square roots. As a result, we turn to

$$\log_e \sqrt{a + \tfrac{1}{6}} - \log_e \sqrt{b + \tfrac{1}{6}} - \log_e \sqrt{c + \tfrac{1}{6}} + \log_e \sqrt{d + \tfrac{1}{6}}$$

(This actually keeps our life simple, since we can look up each of the four terms in the table of exhibit 4. In the last example, this gives 5.10 instead of 4.43--just the factor of 1.15 again.)

The result we call

a double flog

which we often write as a

fflog.

Exhibit **9** rings some changes on data of Karl Pearson on intelligence and athletic power in school boys. However we calculate it--as a double difference of exhibit 4 contributions, as a difference of flogs one way, as a difference of flogs the other--we always get .27 for the fflog. The appearances are that "intelligence" and "athletic power" come together more frequently than the "simplest situation" of no relationship. The value of .27 shows about how much they appear to do this.

How big do we think .27 is? How are we to judge the size of flogs? One easy scale is offered by the quantitive relation between relatives (which works about as well for mental measurements as for physical). Some standard values, roughly correct, are:

great grandfather with great grandson: fflog = .15
cousin with cousin: fflog = .15
grandfather with grandson: fflog = .31
aunt with niece: fflog = .31
father with son: fflog = .67

Perhaps this scale will help you think about what a fflog of .27 means.

review questions

How many kinds of two-way tables of counts have we already looked at? Are there more? Why/why not? What kind of tables do we deal with in this section? What are they rightfully called? What is the simplest case? What are "$a$", "$b$", "$c$", and "$d$"? What is one way to seize upon the information missed by the margins? Is it natural to start the counts? To insert a factor? If yes, what factor? If not, why not? What is a double flog? A fflog? Is it easy to calculate a fflog from the four counts? Why/why not? What did we choose as an example? How did it come out? What is double folding? Why is it so called? Are there various ways to write a double flog?

15F. Double folding--larger cases

When we have more than two levels of either response--or of both responses--we have merely to work with all possible $2 \times 2$ tables made by using one of the available cuts for each response. For each paired cut, there will be four counts and a fflog. Since the paired cuts naturally fall into a two-way table, this means that we get a two-way table of fflogs to describe the apparent relationship.

exhibit **9** of chapter 15: school boys

**Intelligence and athletic power in school boys**

A) The DATA as COUNTS--and their fflogs

|  | Intelligent | Not intelligent | (sum) |
|---|---|---|---|
| **Athletic** | 581 | 567 | (1148) |
| **Not athletic** | 209 | 351 | (560) |
| (sum) | (790) | (918) | ((1708)) |

fflog = 3.18 − 3.17 − 2.67 + 2.93 = .27
(Found by entering exhibit 4 with 580, 570, 210, and 350.)

B) The DATA as % ATHLETIC

|  | Among intelligent | Among not intelligent | (All) (combined) |
|---|---|---|---|
| % | 73.5% | 61.8% | (67.2%) |
| flogs | .51 | .24 | (.36) |

Difference in flogs = .51 − .24 = .27

C) The DATA as % INTELLIGENT

|  | % | flog |
|---|---|---|
| **Among athletic** | 50.6% | .01 |
| **Among not athletic** | 37.3% | −.26 |
| (All combined) | (46.3%) | (−.07) |

Difference in flogs = .01 − (−.26) = .27

D) The DATA RE-EXPRESSED

Total count: 1708
**Intelligent:** 46.3%  or  flog = −.07
  **Athletic:** 67.2%  or  flog = .36
    **fflog:** .27

P) PROBLEMS

9a/b/c/d) The data of exhibit 10, below, can be condensed to read, in part

(for 9a) $\dfrac{821 \;|\; 232}{231 \;|\; 1958}$   (for 9b) $\dfrac{1018 \;|\; 35}{1744 \;|\; 445}$

(for 9c) $\dfrac{1009 \;|\; 179}{43 \;|\; 471}$   (for 9d) $\dfrac{2579 \;|\; 149}{183 \;|\; 333}$

Carry through the analog of the analysis above, giving full detail.

S) SOURCE

K. Pearson 1906. "On the relationship of intelligence to size and shape of head, and to other physical and mental characters." **Biometrika** 5: 105–146. Condensed from table xxxiii on page 144.

exhibit **10** of chapter 15: two eyes

**Calculation of double flogs for unaided distance vision in the two eyes of 3242 men aged 32–39, employed in U.K. Royal Ordnance Factories 1943–1946**

**A) The DATA**

| Right eye | Vision of left eye | | | |
|---|---|---|---|---|
| | Highest | Second | Third | Lowest |
| Highest | 821 | 112 | 85 | 35 |
| Second | 116 | 494 | 145 | 27 |
| Third | 72 | 151 | 583 | 87 |
| Lowest | 43 | 34 | 100 | 331 |

**B) Same numbers, with CORNER TOTALS**

| Highest | | Second | | Third | | Lowest | |
|---|---|---|---|---|---|---|---|
| 3242 | 1052 | 2190 | 1843 | 1399 | 2762 | 480 | 3242 |
| | 821 | | 112 | | 85 | | 35 |
| 1053 | 821 | 232 | 933 | 120 | 1018 | 35 | 1053 |
| 2189 | 231 | 1958 | 910 | 1279 | 1744 | 445 | 2189 |
| | 116 | | 494 | | 145 | | 27 |
| 1835 | 937 | 898 | 1543 | 292 | 1773 | 62 | 1835 |
| 1407 | 115 | 1292 | 300 | 1107 | 989 | 418 | 1407 |
| | 72 | | 151 | | 583 | | 87 |
| 2728 | 1009 | 1719 | 1766 | 962 | 2579 | 149 | 2728 |
| 514 | 43 | 471 | 77 | 437 | 183 | 331 | 514 |
| | 43 | | 34 | | 106 | | 331 |
| 3242 | 1052 | 2190 | 1843 | 1399 | 2762 | 480 | 3242 |

**C) The FFLOGS in 0.01, FLOGS in 0.01 (at \*'s), and TOTAL COUNT**

| Right eye | Vision of left eye | | | | | | | |
|---|---|---|---|---|---|---|---|---|
| | (Highest) | $\alpha$ | (second) | $\beta$ | (Third) | $\gamma$ | (Lowest) | $\rightarrow$ * |
| (Highest) $\alpha$ | | 170 | | 120 | | 100 | | −37 |
| (Second) $\beta$ | | 123 | | 149 | | 125 | | 14 |
| (Third) $\gamma$ | | 93 | | 117 | | 172 | | 83 |
| (Lowest) $\stackrel{\rightarrow}{*}$ | | −37 | | 14 | | 87 | | 3242 |

**P) PROBLEMS**

**10a)** Compare the single flogs for rows (right eye) and columns (left eye). What might the differences mean?

**10b)** Describe the 3 × 3 table of double flogs in words.

**10c)** What kind of numbers for the double flogs would have made a 3 × 3 table simplest to describe? Next simplest?

➔

Exhibit **10** goes through this program for data on vision of the left eye and vision of the right eye, each cut at three places. Panel A shows the data. Panel B shows the result of adding up things in each of several ways-- each internal block of four values giving the corner totals for a paired cut whose crossing point is at that location. Thus, for example, the following paired cut

$$
\begin{array}{c}
\phantom{\alpha}\quad\quad\quad\quad\beta\\
\phantom{\alpha}\quad 821\quad 112\quad 85\quad 35\\
\alpha\\
\phantom{\alpha}\quad 116\quad 494\quad 145\quad 27\\
\phantom{\alpha}\quad\; 72\quad 151\quad 583\quad 87\\
\phantom{\alpha}\quad\; 43\quad\; 34\quad 106\quad 331
\end{array}
$$

leads to corner totals of

$$
\begin{array}{c}
\phantom{\alpha}\quad\quad\beta\\
\phantom{\alpha}\quad 933\quad 120\\
\alpha\\
\phantom{\alpha}\quad 910\quad 1279
\end{array}
$$

and hence (panel C) to a double flog of 1.20.

exhibit **10** of chapter 15 (continued)

10d1) Kendall and Stuart (Table 33.5 at page 586 of volume 2) also give counts according to unaided distance vision of 7427 women employed by the U.K. Royal Ordnance factories in 1943–46. For decreasing left-eye strengths, the counts are: right eye highest (1520, 266, 124, 66, total 1926); for right eye second (234, 1512, 432, 78, total 2256); for right eye third (117, 362, 1712, 205, total 2456); for right eye lowest (36, 82, 179, 492, total 789), with row totals (1907, 2220, 2507, 841). Parallel panel A.

10d2) Repeat the analysis of panels B and C on this data.

10d3) Compare the results for men with those for women, separately for: (i) size, (ii) single flogs, (iii) double flogs. Discuss your results.

10e) Return to problem 8b, take "years in job" as a response, as well as definitions of plans, and repeat the analysis of panels B and C.

10f) The same for the data of problem 8c.

10g) Compare the analysis of 10e and 10f. Discuss your comparison.

10h) (Double weight). The same for the data of problem 8d.

**S) SOURCE**

M. G. Kendall and A. Stuart 1961. *The Advanced Theory of Statistics* (3 volume edition). Table 33.3 at page 564 of volume 2.

Letting the same adding-up process extend outside the table leads to pairs of counts, as when

|  | | β | | |
|---|---|---|---|---|
| 821 | 112 | 85 | 35 |
| 116 | 494 | 145 | 27 |
| 72 | 151 | 583 | 87 |
| 43 | 34 | 106 | 331 |

gives

|  | β |
|---|---|
| 1843 | 399 |

corresponding to a single flog of 0.14. Thus our two-way table of double flogs is naturally bordered by a row and a column of single flogs.

If we fill in the total count where the additional row crosses the additional column, we will, it turns out, have completed giving enough information to allow reconstructing the given counts. Thus panel C is honestly an ANALYSIS of the given data, one in which information about size of collection, distribution of the separate responses, and mutual relationship has been separated.

### review questions

What do we do when more than one cut is available for either response or both responses? What else does adding-up arithmetic, if used freely, give us? All put together, is this enough to reconstruct the data?

### 15G. Easy froots and flogs with a slide rule (optional)

Tables of froots and flogs are not always available. In any event, they require conversion of counts into fractions or percent before use. Accordingly, there can be advantages to more direct calculation, especially when a slide rule is available. This is particularly so when the froots or flogs to be found come in sets with the same total, something that will, in particular, occur whenever we are dealing with several alternative cutting points for a response.

Suppose, then, that, for some cutting point, our counts, each increased by $\frac{1}{6}$, are $a$ and $b$ so that

$$f = \frac{a}{a+b}$$

$$1 - f = \frac{b}{a+b}$$

$$\text{froot} = \sqrt{2f} - \sqrt{2(1-f)}$$

$$= \sqrt{\frac{a}{(a+b)/2}} - \sqrt{\frac{b}{(a+b)/2}}$$

or

$$= \sqrt{2a/(a+b)} - \sqrt{2b/(a+b)}$$

Thus, if $a = 43s$, $b = 471s$, $a + b = 514ss$, $(a + b)/2 = 257s$ (where, as above, s stands for $\frac{1}{6}$), the corresponding froot is

$$\sqrt{\frac{43s}{257s}} - \sqrt{\frac{471s}{257s}}.$$

One setting of the slide rule (with many settings of the cursor) suffice to read out all the numbers needed for all calculations with the same total. (One resetting may be needed to overlap the other end of the scale.)

After we have set 257s on B against 1 on DF (or, if we prefer, 514ss on B against 2 on DF), against 43s on B we read 0.409 on DF and against 471s on B we read 1.353 on DF. (On a rule without a DF scale, we use the D scale BOTH TIMES instead.) Since

$$0.409 = \sqrt{\frac{43s}{257s}}$$

we have

$$\text{froot} = 0.409 - 1.353 = -0.944$$

With three or more froots to compute with the same total, this approach should be faster than using a table to convert %'s into froots, the %'s having been found with the slide rule.

For flogs on the slide rule we write

$$\text{flog} = 1.15\left(\log_{10}\frac{a}{a+b} - \log_{10}\frac{b}{a+b}\right)$$
$$= 1.15(\log_{10} a - \log_{10} b)$$

and calculate according to the last line. For our example we have

$$\log 43 = 1.634$$
$$\log 471 = 2.673$$
$$\text{flog} = 1.15(1.634 - 2.673) = 1.15(-1.039) = -1.196$$

(The multiplications by 1.151 may either all be postponed to the end of the calculations--and then be done together--or (if D and L scales are both available on the body of rule) the rule can be left set for multiplication by 1.151.)

This procedure, too, is likely to be easier than using a log table but not as easy as using exhibit 4 (when available).

A good hand-held calculator is even easier.

### review questions

In what form do we write the formula for froots in preparation for slide-ruling it? Which scales do we use? What do we set against what? What do we set the cursor to? Might we have to use a different scale? Which one? How would it be used? And what about flogs?

### 15H. How far have we come?

This chapter has focussed on effective expressions of counted fractions and the extension to two-way tables according to ordered variables.

We are now ready to:

◊ recognize and use ss- (split and started) counts and fractions.

◊ use any or all of three folded expressions of fractions that are matched near 50%: pluralities (fcents), froots, flogs.

◊ take advantage of tables and shortcuts (and, in optional section 15G, a slide rule) in calculating these folded expressions.

We are also ready to:

◊ find, and make use of, several flogs (or several values of another folded expression) for a single set of counts.

◊ recognize, and use, fflogs (double flogs) as expressions of disproportionality in the 2 × 2 table formed by intersecting cuts (paired cuts).

◊ do this for a single paired cut, several paired cuts (in a streak), or a two-way array of paired cuts.

In the process, we have come to understand more clearly the importance of:

◊ seeking constancy of difference in re-expression (here, of counted fractions).

# Better smoothing  16

chapter index on next page

In chapter 7 we learned to make smooths quite smooth. We left to the present chapter the question of how good were the roughs that went with our so-smooth smooths. We could do this because there we were looking to our smooths for, basically, only rather qualitative results: Did the curve seem to go up, go down, oscillate, curve one way or another? We did try to:

◇ look hard at the roughs that went with the smooths (though we used them to size our blurs and identify outside points).

◇ ask whether the smooths were as good smooths as we could easily make.

We concentrated on getting the details out of the way, on unhooking our eyes from irregularities and corners--so that we might see some sort of "big picture".

In this present chapter, we are going to find out how to do a "better" job of smoothing--how to stick much closer to the data without making the smooth much rougher. (We will need this in the next chapter.)

We did NOT postpone doing a still better job of smoothing to this chapter because this is unimportant. We postponed it because there was already enough detail in chapter 7 and we thought we could get away with postponement. Accordingly, we regard this chapter as something that belongs in the mainstream of our techniques.

Clearly, it is important to understand chapter 7 quite well before tackling the present chapter.

### review questions

What did we do in chapter 7? What did we postpone? What did we not do in chapter 7 that let us postpone? What are we going to do in this chapter? Is making better smooths part of our mainstream? Why/why not? What is needed before tackling this chapter?

### 16A. Reroughing

We are now ready to look hard at roughs to see what they tell us about better smooths. We start with simple illustrative cases and go on from there. While hanning does do a fair amount to make smooths smoother, it has its disadvantages. Consider the sequence

..., 36, 51, 64, 75, 84, 91, 96, 99, 100, 99, 96, 91, 84, 75, 64, 51, 36, ...

# index for chapter 16

review questions 523

**16A. Reroughing** 523
reroughing 526
twicing 526
review questions 526

**16B. Some examples** 526
comparison of smooths 527
review questions 527

**16C. If we want things still smoother** 531
review questions 533

**16D. Further possibilities** 534
new components 534
heavier smoothers 534
problems 535
review questions 541

**16H. How far have we come?** 542

| EXHIBIT | PAGE |
|---|---|
| 16A | |
| 1★ | 525 |
| 16B | |
| 2★ | 528 |
| 16C | |
| 3 | 532 |
| 4 | 532 |
| 5 | 533 |
| 6 | 534 |
| 16D | |
| 7★ | 535 |
| 8★ | 541 |
| 16H | |

(the law of whose formation is easily found by subtracting each entry from 100). If we hann this, we find, first for the skip means and then for the completed hanning:

..., 63, 74, 83, 90, 95, 98, 99, 98, 95, 90, 83, 74, 63, ...
..., 63h, 74h, 83h, 90h, 95h, 98h, 99h, 98h, 95h, 90h, 83h, 74h, 63h, ...

The corresponding rough is

..., h, h, h, h, h, h, h, h, h, h, h, h, h, h, h, h, h, h, ...

A solid sequence of one-halves (h's) is not a very satisfactory rough. It could easily be much more nearly centered at zero.

What we need to do is to smooth the rough, and add it back to the smooth we started with. The scheme is as shown in exhibit 1, which also sets out the verbal algebra.

exhibit 1 of chapter 16: technique schematic

**The scheme and algebra of reroughing**

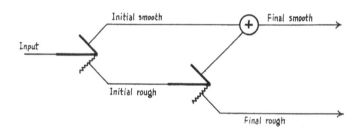

Since
$$\text{start} \equiv \text{smooth PLUS rough},$$
and
$$\text{rough} \equiv (\text{smooth of rough}) \text{ PLUS } (\text{rough of rough}),$$
we must have, first substituting and then redefining:
$$\text{start} = \text{smooth PLUS} \begin{pmatrix} \text{smooth} \\ \text{of rough} \end{pmatrix} \text{PLUS} \begin{pmatrix} \text{rough} \\ \text{of rough} \end{pmatrix},$$
$$\text{start} = \text{final smooth PLUS final rough},$$
where
$$\text{final smooth} \equiv \text{smooth PLUS} \begin{pmatrix} \text{smooth} \\ \text{of rough} \end{pmatrix}$$
and
$$\text{final rough} \equiv \text{rough of rough}.$$

→

If we do add together:

◇ the smooth of the start, and

◇ the smooth of the (first) rough,

then we get a final smooth whose rough is precisely the rough of the rough of the original sequence--hence the name

reroughing.

If the two smoothers are exactly the same, we will have applied the same smoother "twice", so that it is rather natural to think and write of

twicing

"Twicing" or use of some specified smoother twice, as in "3R,twice" is just a special case of reroughing. Specifically, "3R,twice" is just the same as "3R,reroughed with 3R". Symbolically, we represent either repetition with a comma (following the tag of the smoother used first, and preceding either "reroughed by ---" or the word "twice" as may be appropriate). Notice that stringing the names of two smoothers together without a comma means using the second to smooth still further the smooth of the first (to do resmoothing).

review questions

When does hanning slip a little? How is this revealed? What can we do about it? What is reroughing? Why is it so called? What is twicing? Why so called? How are these two represented in a tag?

**16B. Some examples**

Exhibit **2** shows some of the calculation for turning a 3RSS smooth of the bituminous-coal production example of chapter 7 into two alternative smooths:

exhibit **1** of chapter 16 (continued)

**P) PROBLEMS**

1a) Draw the corresponding diagram for resmoothing, as when 3R is followed by SSH3.

1b) Draw the corresponding diagram for resmoothing twice, as when 3R is followed by SS is followed by H3.

1c) Draw the corresponding diagram for reroughing twice, as when the result of reroughing one smoother with a second is, regarded as a unit, reroughed with a third smoother.

◊a smooth that should be tagged either 3RSSH twice or 3RSSH reroughed by 3RSSH

◊a smooth obtained by applying (H,twice) to 3RSS--note that the combination of these two is further smoothing, not reroughing (which already took place within ("H,twice").

### comparison of smooths

Panel C shows, semischematically, four smooths (and the original data). This type of display does not try to imitate a graph, but rather tries to emphasize whether the sequence is going up or going down, the right-and-left displacement corresponding to this sort of pattern:

```
                high
         down
         down
                low
                    up
                    up
                    up
                high
                low
                    up
                    up
                high
```

When we scan panel C, we notice the great irregularity of the starting sequence and the noticeable irregularity of 3RSS(H,twice). The latter does not seem to be a smooth we should bother with.

This leaves us comparing the other three smooths. Panel D shows their values at tops and bottoms. We can see there what we ought to expect:

◊as we go from 3RSS to 3RSSH, there is a tendency to pull bottoms up, and peaks down.

◊this tendency is reversed, somewhat more vigorously, as we go from 3RSSH to "3RSSH,twice".

After all, we know that hanning does tend to pull tops down and bottoms up. And we twiced things to get away from this effect, both in H and in the 3RSS to which H was applied. It looks as if, in this example, twicing--a special case of reroughing--was doing what we would like it to do.

### review questions

What example did we try? What reroughings did we try? How did we compare them? What did we conclude?

exhibit 2 of chapter 16: coal production

## Some examples of reroughing based on the 3RSS smooth of bituminous coal production (see exhibits 8 and 17 of chapter 7)

### A) FINDING "3RSSH,twice"

| Start | Its 3RSS | Hanning > | H | 1st rough | Its 3R | Its 3RSS | Hanning > | H | Sum of Smooths | Final rough | (>)* |
|---|---|---|---|---|---|---|---|---|---|---|---|
| 569 | 422 |  | 422 | 147 | ? | −6 |  | −6 | 416 | 153 | (416) |
| 416 | 422 | 422 | 422 | −6 | −6 | −6 | −6 | −6 | 416 | 0 | (424) |
| 422 | 422 | 453 | 438 | −16 | −6 | −6 | −6 | −6 | 432 | −10 | (445) |
| 565 | 484 | 471 | 478 | 87 | −6 | −6 | −3 | −4 | 474 | 91 | (470) |
| 484 | 520 | 502 | 511 | −27 | 0 | 0 | −3 | −2 | 509 | −25 | (498) |
| 520 | 520 | 520 | 520 | 0 | 0 | 0 | 2 | 0 | 521 | −1 | (516) |
| 573 | 520 | 519 | 520 | 53 | 3 | 3 | 3 | 1 | 522 | 51 | (520) |
| 518 | 518 | 512 | 515 | 3 | 3 | 3 | 3 | 3 | 518 | 0 | (516) |
| 501 | 505 | 510 | 508 | −7 | 3 | 3 | 3 | 3 | 511 | −10 | (508) |
| 505 | 501 | 486 | 494 | 11 | 11 | 3 | 3 | 3 | 497 | 8 | (484) |
| 468 | 468 | 442 | 455 | 13 | 11 | 3 | −2 | 0 | 455 | 13 | (442) |
| 382 | 382 | 401 | 392 | −10 | −10 | −6 | −2 | −4 | 388 | −6 | (398) |
| 310 | 334 | 358 | 346 | −36 | −10 | −6 | −6 | −6 | 340 | −30 | (362) |
| 334 | 334 | 346 | 340 | −6 | −6 | −6 | −4 | −5 | 335 | −1 | (347) |
| 359 | 359 | 353 | 356 | 3 | −2 | −2 | −2 | −2 | 354 | 5 | (356) |
| 372 | 372 | 377 | 374 | −2 | 3 | 3 | 0 | 2 | 376 | −4 | (374) |
| 439 | 395 | 384 | 390 | 49 | 49 | 3 | 3 | 3 | 393 | 46 | (388) |
| 446 | 395 | 396 | 396 | 50 | 49 | 3 | 3 | 3 | 399 | 47 | (398) |
| 349 | 398 | 417 | 408 | −59 | −39 | −7 | −2 | −4 | 404 | −55 | (416) |
| 395 | 439 | 430 | 434 | −39 | −39 | −7 | −2 | −4 | 430 | −35 | (433) |
| 461 | 461 | 475 | 468 | −7 | −7 | −7 | −6 | −6 | 462 | −1 | (472) |
| 511 | 511 | 522 | 516 | −5 | −5 | −5 | 0 | −2 | 514 | −3 | (515) |
| 583 | 588 | 547 | 565 | 18 | 6 | 6 | 0 | 3 | 568 | 15 | (552) |
| 590 | 583 | 586 | 584 | 8 | 6 | 6 | 6 | 6 | 590 | 0 | (581) |
| 620 | 590 | 586 | 588 | 32 | 6 | 6 | 6 | 6 | 594 | 26 | (593) |
| 578 | 590 | 590 | 590 | −12 | −12 | 6 | 6 | 6 | 596 | −18 | (592) |
| 534 | 590 | 584 | 587 | −53 | −12 | 6 | 3 | 4 | 591 | −57 | (582) |
| 631 | 578 | 553 | 566 | 65 | 65 | 0 | 3 | 2 | 568 | 63 | (562) |
| 600 | 516 | 547 | 532 | 68 | 65 | 0 | 0 | 0 | 532 | 68 | (542) |
| 438 | 516 | 516 | 516 | −78 | 0 | 0 | 0 | 0 | 516 | −78 | (524) |
| 516 | 516 | 516 | 516 | 0 | 0 | 0 | 0 | 0 | 516 | 0 | (510) |
| 534 | 516 | 492 | 504 | 30 | 0 | 0 | −2 | −1 | 503 | 31 | (495) |
| 467 | 467 | 486 | 476 | −9 | −3 | −3 | −2 | −2 | 474 | −7 | (480) |
| 457 | 457 | 462 | 460 | −3 | −3 | −3 | −3 | −3 | 457 | 0 | (464) |
| 392 | 457 | 457 | 457 | −65 | −3 | −3 | −3 | −3 | 454 | −62 | (450) |
| 461 | 457 | 434 | 446 | 21 | 21 | −3 | −2 | −2 | 444 | 23 | (438) |
| 500 | 412 | 434 | 423 | 77 | 77 | 0 | −2 | −1 | 422 | 78 | (428) |
| 493 | 412 | 412 | 412 | 81 | 77 | 0 | 0 | 0 | 412 | 81 | (417) |
| 410 | 412 | 412 | 412 | −2 | 0 | 0 | 0 | 0 | 412 | −2 | (412) |
| 412 | 412 | 412 | 412 | 0 | 0 | 0 | 0 | 0 | 412 | 0 | (412) |
| 416 | 412 | 414 | 413 | 3 | 0 | 0 | −4 | −2 | 411 | 5 | (411) |
| 403 | 416 | 417 | 416 | −13 | −8 | −8 | −4 | −6 | 410 | −7 | (416) |
| 422 | 422 | 438 | 430 | −8 | −8 | −8 | −8 | −8 | 422 | 0 | (428) |
| 459 | 459 | 444 | 452 | 7 | −8 | −8 | −2 | −5 | 447 | 12 | (449) |
| 467 | 467 | 486 | 476 | −9 | 3 | 3 | −2 | 0 | 476 | −9 | (478) |
| 512 | 512 | 500 | 506 | 6 | 3 | 3 | 3 | 3 | 509 | 3 | (505) |
| 534 | 534 | 528 | 531 | 3 | 3 | 3 | 3 | 3 | 534 | 0 | (527) |
| 552 | 545 | 540 | 542 | 10 | 3 | 3 | 3 | 3 | 545 | 7 | (541) |
| 545 | 545 |  | 545 | 0 | ? | 3 |  | 3 | 548 | −3 | (548) |

Examples (1st line): 569 given, 422 given, 422 = 422, 147 = 569 − 422, −6 from end-value rule, −6 = −6, −6 = −6, 416 = (422) + (−6), 153 = 569 − 416. Note: In "Its 3R" column, → points to values changed by the second running median of 3.

* Values in parentheses, in last column, will be discussed in a later section.

exhibit 2 of chapter 16 (continued)

B) Finding "3RSS(H,twice)"--so sad

| The 3RS | Hanning > | Hanning → | Local rough | Hanning > | Hanning → | Sum of smooths | Final rough |
|---|---|---|---|---|---|---|---|
| 422 |  | 422 | 0 |  | 0 | 422 | 147 |
| 422 | 422 | 422 | 0 | -8 | -4 | 418 | -2 |
| 422 | 453 | 438 | -16 | 3 | -6 | 432 | -10 |
| 484 | 471 | 478 | 6 | -4 | 1 | 479 | 86 |
| 520 | 502 | 511 | 9 | 3 | 6 | 517 | -33 |
| 520 | 520 | 520 | 0 | 4 | 2 | 522 | -2 |
| 520 | 519 | 520 | 0 | 2 | 0 | 521 | 52 |
| 518 | 512 | 515 | 3 | -2 | 1 | 515 | 3 |
| 505 | 510 | 508 | -3 | 5 | 3 | 509 | -8 |
| 501 | 486 | 494 | 7 | 5 | 6 | 500 | 5 |
| 468 | 442 | 455 | 13 | -2 | 5 | 461 | 7 |
| 382 | 401 | 392 | -10 | 0 | -5 | 387 | -5 |
| 334 | 358 | 346 | -12 | -8 | -10 | 336 | -26 |
| 334 | 346 | 340 | -6 | -4 | -5 | 335 | -1 |
| 359 | 353 | 356 | 3 | -4 | 0 | 356 | 3 |
| 372 | 377 | 374 | -2 | 4 | 1 | 375 | -3 |
| 395 | 384 | 390 | 5 | -2 | 2 | 392 | 47 |
| 395 | 396 | 396 | -1 | -2 | -2 | 394 | 52 |
| 398 | 417 | 408 | -10 | 2 | -4 | 404 | -55 |
| 439 | 430 | 434 | 5 | -8 | -2 | 432 | -37 |
| 461 | 475 | 468 | -7 | 0 | -4 | 464 | -3 |
| 511 | 522 | 516 | -5 | 6 | 0 | 516 | -5 |
| 583 | 547 | 565 | 18 | 3 | 8 | 573 | 10 |
| 583 | 586 | 584 | -1 | 10 | 5 | 588 | 2 |
| 590 | 586 | 588 | 2 | 0 | 0 | 589 | 31 |
| 590 | 590 | 590 | 0 | 2 | 0 | 591 | -13 |
| 590 | 584 | 587 | 3 | 6 | 4 | 591 | -57 |
| 578 | 553 | 566 | 12 | -6 | 4 | 564 | 67 |
| 516 | 547 | 532 | -16 | 6 | -4 | 527 | 73 |
| 516 | 516 | 516 | 0 | -8 | -4 | 512 | -74 |
| 516 | 516 | 516 | 0 | 6 | 3 | 519 | -3 |
| 516 | 492 | 504 | 12 | -4 | 4 | 508 | 26 |
| 467 | 486 | 476 | -9 | 4 | 0 | 474 | -7 |
| 457 | 462 | 460 | -3 | -4 | 0 | 456 | 1 |
| 457 | 457 | 457 | 0 | 4 | 4 | 459 | -67 |
| 457 | 434 | 446 | 11 | -6 | 4 | 448 | 19 |
| 412 | 434 | 423 | -11 | 6 | -2 | 419 | 79 |
| 412 | 412 | 412 | 0 | -6 | -2 | 409 | 84 |
| 412 | 412 | 412 | 0 | 0 | 0 | 412 | -2 |
| 412 | 412 | 412 | 0 | 0 | 0 | 412 | 0 |
| 412 | 414 | 413 | -1 | 0 | 0 | 413 | 3 |
| 416 | 417 | 416 | 0 | -4 | -2 | 414 | -11 |
| 422 | 438 | 430 | -8 | 4 | -3 | 424 | -2 |
| 459 | 444 | 452 | 7 | -8 | 0 | 452 | 7 |
| 467 | 486 | 476 | -9 | 6 | 2 | 474 | -7 |
| 512 | 500 | 506 | 6 | -3 | 2 | 508 | 4 |
| 534 | 528 | 531 | 3 | 4 | 2 | 535 | -1 |
| 545 | 540 | 542 | 3 | 2 | 2 | 544 | 8 |
| 545 |  | 545 | 0 |  | 0 | 545 | 0 |

Examples (3rd line): 422 given, 453 = ½(422 + 484), 438 = ½(422 + 453), −16 = 422 − 438, 3 = ½(0 + 6), −6 = ½(−6 + 3), 432 = (438) + (−6), −10 = 422 (given in A) − 432.

exhibit 2 of chapter 16 (continued)

## C) SCHEMATIC COMPARISON

| (*) | 3RSS | 3RSSH | 3RSS(H,twice) | 3RSSH,twice | (start) |
|---|---|---|---|---|---|
|   | 422 | 422 | 422 | 416 | 569 |
| * | 422 | 422 | 418 | 416 | 416 |
|   | 422 | 438 | 432 | 432 | 422 |
|   | 484 | 478 | 479 | 472 | 565 |
|   | 520 | 511 | 517 | 509 | 484 |
|   | 520 | 520 | 522 | 521 | 520 |
| * | 520 | 520 | 521 | 522 | 573 |
|   | 518 | 515 | 515 | 518 | 518 |
|   | 505 | 508 | 509 | 511 | 501 |
|   | 501 | 494 | 500 | 497 | 505 |
|   | 468 | 455 | 460 | 455 | 468 |
|   | 382 | 392 | 387 | 388 | 382 |
|   | 334 | 346 | 336 | 340 | 310 |
| * | 334 | 340 | 335 | 335 | 334 |
|   | 359 | 356 | 356 | 354 | 359 |
|   | 372 | 374 | 375 | 376 | 372 |
|   | 395 | 390 | 392 | 393 | 439 |
|   | 395 | 396 | 394 | 399 | 446 |
|   | 398 | 408 | 404 | 404 | 349 |
|   | 439 | 434 | 432 | 430 | 395 |
|   | 461 | 468 | 464 | 462 | 461 |
|   | 511 | 516 | 516 | 514 | 511 |
|   | 583 | 566 | 573 | 568 | 583 |
|   | 583 | 584 | 588 | 590 | 590 |
|   | 590 | 588 | 589 | 594 | 620 |
| * | 590 | 590 | 591 | 596 | 578 |
|   | 590 | 587 | 591 | 591 | 534 |
|   | 578 | 566 | 564 | 568 | 631 |
|   | 516 | 532 | 527 | 532 | 600 |
|   | 516 | 516 | 512 | 516 | 438 |
|   | 516 | 516 | 519 | 516 | 516 |
|   | 516 | 504 | 508 | 503 | 534 |
|   | 467 | 476 | 474 | 474 | 467 |
|   | 457 | 460 | 456 | 457 | 457 |
|   | 457 | 457 | 459 | 454 | 392 |
|   | 457 | 446 | 448 | 444 | 467 |
|   | 412 | 423 | 419 | 422 | 500 |
|   | 412 | 412 | 409 | 412 | 493 |
| * | 412 | 412 | 412 | 412 | 410 |
|   | 412 | 412 | 412 | 412 | 412 |
|   | 412 | 413 | 413 | 411 | 416 |
|   | 416 | 416 | 414 | 410 | 403 |
|   | 422 | 430 | 424 | 422 | 422 |
|   | 459 | 452 | 452 | 447 | 459 |
|   | 467 | 476 | 487 | 476 | 467 |
|   | 512 | 506 | 508 | 513 | 512 |
|   | 534 | 531 | 535 | 534 | 534 |
|   | 545 | 542 | 544 | 545 | 552 |
| * | 545 | 545 | 545 | 548 | 545 |

→

## 16C. If we want things still smoother

What if "3RSSH,twice" is not smooth enough for us? As exhibit 3 shows, this might well be the case. The general appearance of this 3RSSH,twice smooth is quite smooth--if we allowed ourselves even a slightly free hand, we could make this very smooth very easily--but the details are not yet very smooth. One natural thing to do would be to hann again. Exhibit 4 shows the result--better but not excellent. Since hanning got us about half-way there, perhaps we should just skip-mean our 3RSSH,twice smooth, leaving out the line-mean. Exhibit 5 shows us that this does help, at least in this example. (The numbers are in the last column, headed (>), of panel A of exhibit 4.)

exhibit **2** of chapter 16 (continued)

**D) TOPS and BOTTOMS**--on lines marked "*" in panel C

| 3RSS | 3RSSH | 3RSSH,twice |
|------|-------|-------------|
| 422  | 422   | 416         |
| 520  | 520   | 523         |
| 334  | 340   | 335         |
| 590  | 590   | 596         |
| 412  | 412   | 412         |
| 545  | 545   | 548         |

**P) PROBLEMS**

2a) Find a sequence at least 20 long of interest to you, and smooth it with 3RSSH,twice.

2b) Do the same with a sequence at least 50 long.

2c) Do the same with a sequence suggested by your instructor.

2d) Starting with the (>) column of panel A, find the corresponding rough, smooth it by 3RSSH, add the result to (>), and plot the sum.

2e) Try 3RSS>-twice applied to the start above (leave out the → columns, and the columns of line means. How do you like the results? Does it do well or poorly in avoiding flat tops?

2e2) Draw a hand-smoothed curve through the points found in (2e). Compare with exhibit 7. Comment on the difference.

2f) Starting with the "Final rough" column of panel A, do another 3RSSH smooth and add the result back to the "sum of smooths" column. Comment.

2f2) Collect the results of the smooth of (2f) according to the values of the last H column of panel A. Comment further.

2g) Take the 3RSSH smooth (left column of panel A) on to a 3RSSH3RSSH smooth, and find its rough. Comment.

2g2) Take the result from 2g and do a 3RSSH smooth on it. What now?

exhibit **3** of chapter 16: coal production

**The 3RSSH,twice smooth of bituminous coal production**

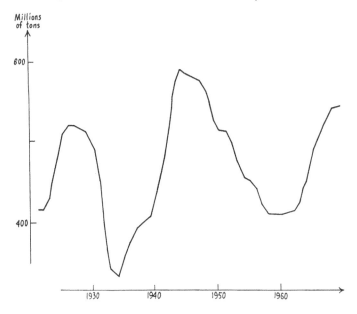

exhibit **4** of chapter 16: coal production

**The result of hanning the smooth of exhibit 3**

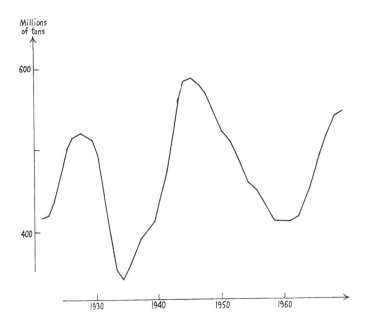

By now we have the smooth so smooth as to make us wonder how close to the data we are sticking. Whenever this worries us, it is time to rerough. Exhibit **6** shows the result of finding the rough that goes with exhibit 5, smoothing it with 3RSSH, and adding the smooth back. Now we have a highly polished smooth which we are sure has not wandered very far from the data.

Notice the comment on exhibit 6 about drawing a smooth curve through the points, rather than using a broken line. We could have made exhibit 5 look just as smooth as--maybe even a shade smoother than--exhibit 6, had we drawn a curve through its points. The true difference between exhibit 5 and exhibit 6 is that we are surer that exhibit 6 goes as high on the hills and as low in the valleys as we would like--not that the points are smoother.

### review questions

How can we make our smooth smoother? Should we worry how close we then stick to the data? What can we do?

exhibit **5** of chapter 16: coal production

**The result of skip-meaning the 3RSSH,twice smooth of exhibit 3**

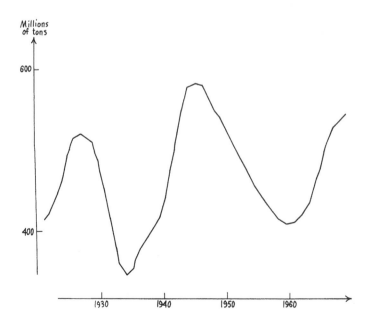

## 16D. Further possibilities

**new components**

Clearly the pieces from which we have made our smoothers have close relatives. We could use running medians of 5,7,9, or what have you. We could use » for a double-skip-mean--one that looks exactly two steps back and two steps forward--and then use I for double-skip mean followed by line mean, and so on. We shall not need any of these here, though some readers may occasionally find them helpful. Note that, as we lengthen the span of the pieces of our smoothers, the need for reroughing will increase.

**heavier smoothers**

We learned, even in chapter 7, that one way to make a smooth still smoother was to smooth it some more. We can perfectly well begin with 3RSSH3RSSH3, if we are energetic enough, find its rough, and smooth the rough with 3RSSH3RSSH3. If arithmetic is no effort--perhaps because we have a friendly computer all programmed--we can be serious about

exhibit **6** of chapter 16: coal production

**The result of reroughing the smooth of exhibit 5 with 3RSSH. (This time we have drawn a smooth curve through the points instead of a broken line.)**

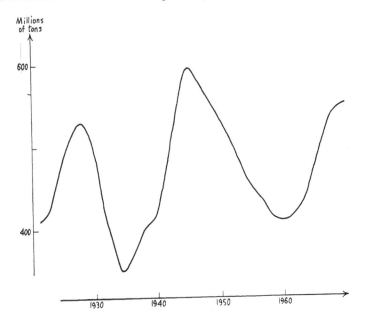

trying "3RSSH3RSSH3,twice" or even "(3RSSH3RSSH3,3RSSH3RSSH3), 3RSSH3RSSH3" which we naturally call "3RSSH3RSSH,thrice".

### problems

Exhibit **7** presents some problems. Exhibit **8** presents some optional problems.

exhibit **7** of chapter 16: data and problems

**Further data and problems on reroughing**

**A) DATA before 1776**

| Year | Z123 (Klb) | Z225 (Klb) | Z254 (Hlb) | Z350 (s/g) | Z357 (*) |
|---|---|---|---|---|---|
| 1775 | 1122 | 834 | 80 | — | 166.0 |
| 4 | 747 | 1191 | 302 | 2.17 | 169.7 |
| 3 | 721 | 964 | 2063 | 2.20 | 165.8 |
| 2 | 747 | 684 | 1512 | 2.19 | 161.2 |
| 1 | 434 | 1136 | 2829 | 2.19 | 165.6 |
| 0 | 550 | 190 | 859 | 2.19 | 154.0 |
| 1769 | 403 | 203 | 860 | 2.16 | 158.3 |
| 8 | 498 | 88 | 2919 | 2.23 | 166.4 |
| 7 | 530 | 44 | 1524 | 2.08 | 166.2 |
| 6 | 492 | 114 | 1190 | 2.23 | 165.4 |
| 5 | 336 | 704 | 1754 | 2.04 | 171.6 |
| 4 | 529 | 765 | 1432 | 2.05 | 172.4 |
| 3 | 439 | 647 | 375 | 2.59 | 173.1 |
| 2 | 255 | 2226 | 516 | 2.79 | 175.8 |
| 1 | 384 | 796 | 70 | 3.04 | 174.1 |
| 0 | 508 | 989 | — | 3.54 | 160.3 |
| 1759 | 696 | 120 | — | 3.94 | 154.7 |
| 8 | 563 | 273 | | 3.12 | 159.2 |
| 7 | 876 | 369 | | 2.74 | 166.0 |
| 6 | 223 | 289 | | 2.35 | 172.5 |
| 5 | 304 | 241 | | 2.27 | 168.9 |
| 4 | 130 | 830 | | 2.44 | 168.2 |
| 3 | 29 | 451 | | 2.47 | 168.0 |
| 2 | 4 | 83 | | 2.39 | 166.7 |
| 1 | 20 | 162 | | 2.51 | 170.6 |
| 0 | 63 | 12 | | 2.53 | 171.1 |
| 1749 | 138 | 321 | | 2.72 | 172.4 |
| 8 | 62 | 393 | | 3.60 | 174.3 |
| 7 | 138 | 287 | | 3.62 | 184.6 |
| 6 | — | 81 | | 2.69 | 179.2 |
| 5 | | | | 2.65 | 175.7 |
| 4 | | 35 | | 2.51 | 167.4 |
| 3 | | 515 | | 2.36 | 160.3 |
| 2 | | 558 | | 2.84 | 159.7 |
| 1 | | 70 | | 2.46 | 145.2 |
| 1740 | | 49 | | 1.81 | 160.1 |

exhibit 7 of chapter 16 (continued)

B) FIRST SECTIONS (to 1895 or 1920) of TEN SEQUENCES

| Year | J16 (H#) | M259 (#) | P196 (M) | X56 (M$) | X286 (M$) | Year | D606 ($/y) | M117 (M$) | M251 (Kt) | X274 (M$) |
|---|---|---|---|---|---|---|---|---|---|---|
| 1860 | — | — | — | — | — | 1884 | — | 4.91 | — | — |
| 1 | — | — | — | — | — | 5 | — | 5.1 | — | — |
| 2 | — | — | — | — | — | 6 | — | 6.8 | — | — |
| 3 | 8.2 | — | — | 10 | — | 7 | — | 7.6 | — | — |
| 4 | 8.4 | — | — | 147 | — | 8 | — | 8.5 | — | — |
| 5 | 8.9 | — | — | 614 | — | 9 | — | 10.3 | — | — |
| 6 | 15.4 | — | — | 695 | 10.5 | 1890 | 560 | 11.5 | — | — |
| 7 | 17.0 | — | — | 685 | 18.7 | 1 | 554 | 10.4 | — | — |
| 8 | 23.7 | — | — | 745 | 17.6 | 2 | 563 | 12.0 | — | 14 |
| 9 | 25.6 | — | — | 716 | 30.0 | 3 | 563 | 9.5 | — | 14 |
| 1870 | 34.0 | 211 | 16 | 706 | 32.1 | 4 | 546 | 9.2 | — | 14 |
| 1 | 39.8 | 210 | 20 | 791 | 17.8 | 1895 | 546 | 13.3 | — | 13 |
| 2 | 38.7 | 223 | 24 | 805 | 26.4 | 6 | 544 | 11.8 | 1 | 15 |
| 3 | 31.6 | 263 | 28 | 836 | 34.3 | 7 | 543 | 13.3 | 1 | 16 |
| 4 | 29.1 | 260 | 35 | 828 | 18.0 | 8 | 542 | 16.0 | 1 | 53 |
| 5 | 20.7 | 260 | 59 | 897 | 17.5 | 9 | 543 | 19.7 | 2 | 76 |
| 6 | 25.1 | 254 | 113 | 812 | 24.2 | 1900 | 548 | 20.5 | 3 | 99 |
| 7 | 18.1 | 244 | 147 | 818 | 32.3 | 1 | 549 | 21.8 | 3 | 99 |
| 8 | 35.6 | 260 | 210 | 814 | 24.9 | 2 | 562 | 25.4 | 3 | 124 |
| 9 | 41.0 | 389 | 371 | 1090 | 15.3 | 3 | 593 | 25.3 | 3 | 147 |
| 1880 | 47.3 | 280 | 533 | 1085 | 8.0 | 4 | 600 | 23.7 | 4 | 110 |
| 1 | 37.0 | 410 | 595 | 1364 | 5.8 | 1905 | 589 | 32.2 | 5 | 75 |
| 2 | 45.3 | 502 | 599 | 1365 | 5.0 | 6 | 607 | 36.4 | 7 | 89 |
| 3 | 56.6 | 593 | 844 | 1337 | 59.8 | 7 | 661 | 40.8 | 8 | 180 |
| 4 | 55.0 | 762 | 920 | 1233 | 71.1 | 8 | 667 | 16.0 | 5 | 130 |
| 5 | 60.9 | 574 | 1080 | 1420 | 126.7 | 9 | 644 | 39.3 | 15 | 70 |
| 6 | 61.6 | 530 | 1607 | 1459 | 76.0 | 1910 | 677 | 41.7 | 18 | 54 |
| 7 | 52.0 | 535 | 1865 | 1650 | 91.2 | 1 | 705 | 35.6 | 19 | 48 |
| 8 | 46.2 | 728 | 2212 | 1716 | 121.1 | 2 | 721 | 44.0 | 21 | 58 |
| 9 | 42.2 | 668 | 2413 | 1920 | 117.1 | 3 | 760 | 46.3 | 24 | 49 |
| 1890 | 40.2 | 733 | 2505 | 1979 | 130.8 | 4 | 795 | 34.6 | 29 | 66 |
| 1 | 37.6 | 956 | 3137 | 1974 | 120.1 | 1915 | 815 | 41.6 | 45 | 48 |
| 2 | 55.1 | 991 | 3282 | 2327 | 141.1 | 6 | 817 | 54.5 | 58 | 140 |
| 3 | 48.4 | 958 | 3661 | 1939 | 92.6 | 7 | 989 | 55.6 | 65 | 1102 |
| 4 | 56.6 | 958 | 3621 | 2228 | 66.3 | 8 | 1424 | 56.5 | 62 | 1631 |
| 5 | 37.3 | 1142 | 4238 | 2279 | 48.4 | 9 | 1509 | 44.2 | 64 | 1016 |
| 6 | | | 4967 | 2141 | 42.2 | 1920 | 1817 | 51.3 | 69 | 329 |

exhibit **7** of chapter 16 (continued)

C) SECOND SECTIONS (from 1921 on) of the same TEN SEQUENCES

| Year | J16 (H#) | M259 (#) | P196 (M) | X56 (M$) | X280 (M$) | D606 ($/y) | M117 (M$) | M251 (Kt) | X274 (M$) |
|---|---|---|---|---|---|---|---|---|---|
| 1921 | 43.8 | 1995 | 52,770 | 15142 | 200.6 | 1032 | 25.3 | 27 | 461 |
| 2 | 29.3 | 1984 | 56,413 | 16323 | 173.3 | 1591 | 37.1 | 37 | 203 |
| 3 | 18.9 | 2462 | 67,239 | 16899 | 386.5 | 1585 | 57.0 | 69 | 361 |
| 4 | 13.9 | 2402 | 73,256 | 18349 | 801.4 | 1570 | 44.3 | 75 | 232 |
| 5 | 11.0 | 2234 | 82,712 | 19912 | 1004.8 | 1597 | 51.3 | 70 | 205 |
| 6 | 10.4 | 2518 | 92,523 | 20644 | 1057.4 | 1613 | 56.9 | 74 | 239 |
| 7 | 10.5 | 2231 | 100,260 | 21778 | 1007.1 | — | 51.1 | 82 | 256 |
| 8 | 10.4 | 2176 | 109,131 | 22645 | 1019.1 | — | 52.8 | 105 | 295 |
| 9 | 11.6 | 2187 | 122,822 | 21586 | 935.0 | — | 59.9 | 114 | 417 |
| 1930 | 12.7 | 2063 | 124,193 | 23235 | 995.0 | — | 48.0 | 115 | 345 |
| 1 | 12.6 | 1463 | 117,402 | 22164 | 996.5 | — | 33.5 | 89 | 486 |
| 2 | 10.6 | 1207 | 106,915 | 17428 | 715.7 | — | 21.8 | 52 | 422 |
| 3 | 7.53 | 1064 | 115,087 | 16742 | 265.5 | — | 27.6 | 43 | 887 |
| 4 | 7.51 | 1226 | 130,287 | 19896 | 149.7 | — | 31.8 | 37 | 1797 |
| 5 | 3.30 | 1242 | 190,147 | 22477 | 117.2 | — | 35.1 | 60 | 913 |
| 6 | 1.21 | 1342 | 159,076 | 26153 | 100.8 | — | 46.3 | 112 | 1832 |
| 7 | .56 | 1413 | 170,171 | 26716 | 88.1 | — | 52.4 | 146 | 759 |
| 8 | .45 | 1105 | 171,842 | 26763 | 78.5 | — | 32.5 | 143 | 1459 |
| 9 | .38 | 1078 | 180,828 | 29416 | 71.9 | — | 44.3 | 164 | 1736 |
| 1940 | .35 | 1388 | 189,508 | 33014 | 66.8 | — | 57.0 | 206 | 1062 |
| 1 | .40 | 1266 | 218,083 | 37273 | 62.9 | — | 65.2 | 309 | 1733 |
| 2 | .28 | 1471 | 257,657 | 40533 | 59.4 | — | 70.6 | 521 | 2127 |
| 3 | .21 | 1451 | 296,305 | 54590 | 56.9 | — | 71.7 | 920 | 8503 |
| 4 | .16 | 1298 | 323,734 | 65585 | 54.0 | — | 74.0 | 776 | 20156 |
| 5 | .18 | 1068 | 332,345 | 76534 | 52.1 | — | 67.3 | 495 | 24980 |
| 6 | .14 | 968 | 350,132 | 80212 | 50.2 | — | 58.5 | 410 | 14249 |
| 7 | .47 | 1158 | 369,763 | 77146 | 47.8 | — | 73.4 | 572 | 2123 |
| 8 | .64 | 999 | 386,916 | 78753 | 45.2 | — | 74.9 | 623 | 4108 |
| 9 | .68 | 585 | 385,046 | 7849 | 42.7 | — | 63.6 | 603 | 2742 |
| 1950 | .52 | 643 | 392,025 | 82430 | 40.8 | — | 72.7 | 719 | 4751 |
| 1 | .36 | 785 | 418,872 | 86589 | 39.1 | — | 79.3 | 837 | 6649 |
| 2 | .46 | 548 | 435,616 | 92719 | 32.9 | — | 68.3 | 937 | 6454 |
| 3 | .48 | 461 | 423,129 | 94475 | 36.6 | — | 78.8 | 1252 | 4074 |
| 4 | .47 | 396 | 401,849 | 99358 | 35.5 | — | 59.7 | 1461 | 6770 |
| 5 | .48 | 420 | 412,309 | 98631 | 34.5 | — | 75.3 | 1566 | 5796 |
| 6 | .45 | 448 | 424,247 | 100820 | 33.5 | — | 75.5 | 2679 | 6059 |
| 7 | .66 | — | 442,328 | 100972 | 32.5 | — | — | — | 4098 |

→

exhibit **7** of chapter 16 (continued)

**D) TWELVE SEQUENCES at 5-year INTERVALS**

| | (100 log sum of 5 years*) | | | | | | | | (sum of 5 years*) | | | |
|---|---|---|---|---|---|---|---|---|---|---|---|---|
| Year | C97 (H log) | C99 (H log) | C102 (H log) | C123† (H log) | C124† (H log) | E101 ($/b) | E109 (R/Hp) | E111 ($/g) | U6 (M$) | U9 (M$) | U182 (M$) | W74 (#) |
| 1975 | — | — | — | — | — | — | — | — | — | — | −16 | — |
| 1800 | — | — | — | — | — | 1.82 | 10.7 | 2.50 | — | — | −3 | — |
| 5 | — | — | — | — | — | 1.95 | 10.5 | 3.61 | — | — | 19 | — |
| 10 | — | — | — | — | — | 1.80 | 9.5 | 3.94 | — | — | −10 | — |
| 5 | — | — | — | — | — | 1.56 | 12.5 | 4.48 | — | — | 5 | — |
| 20 | — | — | — | (1.29) | (1.31) | .93 | 9.8 | 2.37 | — | — | −108 | — |
| 5 | 1.61 | 2.40 | — | 3.74 | 3.35 | .92 | 7.3 | .40 | — | 12 | 6 | — |
| 30 | 1.53 | 2.20 | — | 3.94 | 3.89 | 1.07 | 5.5 | .29 | 5 | −17 | −3 | — |
| 5 | 2.37 | 3.28 | — | 4.49 | 4.15 | 1.22 | 6.0 | .55 | 11 | −30 | −103 | — |
| 40 | 1.71 | 2.55 | — | 9.76 | 4.59 | 1.06 | 5.5 | .27 | 11 | −13 | −130 | — |
| 5 | 2.35 | 2.83 | — | 4.80 | 4.77 | 1.04 | 4.8 | .40 | −9 | 0 | 7 | 44 |
| 1850 | 2.52 | 2.63 | — | 5.27 | 5.32 | 1.28 | 3.7 | .33 | −11 | 0 | −14 | 297 |
| 5 | 1.32 | 3.50 | — | 5.47 | 5.59 | 2.44 | 4.1 | .43 | 169 | 2 | −213 | 709 |
| 60 | 2.64 | 3.54 | — | 5.07 | 5.19 | 1.50 | 3.1 | .42 | 256 | −10 | −66 | 1321 |
| 5 | 2.80 | 3.54 | — | 4.83 | 5.25 | 2.16 | 7.1 | 1.52 | 205 | 4 | −380 | 2194 |
| 70 | 3.27 | 3.46 | — | 5.18 | 5.54 | 1.37 | 4.4 | .43 | 193 | 70 | −590 | 2307 |
| 5 | 4.19 | 4.43 | 1.15 | 5.17 | 5.57 | 1.40 | 3.4 | .35 | 205 | 111 | −688 | 4335 |
| 80 | 4.38 | 4.46 | 1.72 | 5.04 | 5.02 | 1.06 | 3.7 | .38 | −61 | 47 | 656 | 3196 |
| 5 | 4.79 | 5.10 | .70 | 5.36 | 5.83 | .86 | 2.3 | .35 | −105 | 52 | 23 | 4359 |
| 90 | 5.18 | 5.30 | 3.35 | 5.14 | 5.81 | .89 | 3.0 | .41 | 17 | 66 | −776 | 3983 |
| 5 | 5.39 | 5.46 | 4.12 | 5.20 | 5.76 | .60 | 2.1 | .29 | 230 | 100 | −258 | 4585 |
| 1900 | 5.41 | 5.56 | 4.33 | 5.07 | 5.65 | .70 | 2.6 | .48 | −118 | 134 | 1973 | 8757 |
| 5 | 5.82 | 5.98 | 4.48 | 5.64 | 6.09 | 1.01 | 1.9 | .63 | 7 | 113 | 1613 | 3943 |
| 10 | 5.97 | 6.04 | 4.67 | 6.07 | 6.03 | 1.10 | 1.9 | .68 | −74 | 70 | 1091 | 3279 |
| 5 | 5.95 | 5.97 | 4.86 | 6.00 | 5.90 | 1.29 | 1.7 | .46 | −13 | 114 | 2709 | 7271 |
| 20 | 4.45 | 5.23 | 3.85 | 4.86 | 5.36 | 2.46 | 4.2 | 1.73 | −535 | 426 | 17426 | 8458 |
| 5 | 4.88 | 5.57 | 4.27 | 5.05 | 5.64 | 1.67 | 2.8 | 1.01 | −1323 | 49 | 6399 | 12290 |
| 30 | 4.15 | 4.92 | 2.58 | 4.99 | 5.27 | .90 | 2.2 | .47 | −167 | 93 | 5456 | 13781 |
| 5 | 3.55 | 4.54 | 2.42 | 3.65 | 3.97 | 1.04 | 2.6 | .50 | −2399 | −471 | 2010 | 15071 |
| 40 | 3.59 | 4.53 | 1.82 | 3.30 | 3.99 | .87 | 2.6 | .37 | −12955 | −600 | 4488 | 25455 |
| 5 | 3.11 | 2.97 | 2.05 | 2.91 | 3.58 | 1.66 | 2.8 | .79 | −416 | 91 | 38354 | 18880 |
| 1950 | 3.48 | 4.75 | 2.03 | 3.81 | 4.32 | 2.22 | 6.3 | .53 | −4171 | −269 | 34813 | 18016 |
| 5 | 3.43 | 4.86 | 2.06 | 4.30 | 4.68 | 2.26 | 8.2 | .64 | −249 | −382 | 24243 | 12642 |

\* 100 times the log to the base 10 of sum for 5 years.
† Values for every 5th year.

exhibit 7 of chapter 16 (continued)

E) TWELVE SEQUENCES FOR 1920–1957

| Year | C123 (#) | C124 (#) | D47 (%) | J194 (in) | L147 (Mlb) | L172 (Mlb) | M215 (t) | U6 (M$) | U9 (M$) | W77 (H#) | V99 (%) | Y268 (M$) |
|---|---|---|---|---|---|---|---|---|---|---|---|---|
| 1920 | 15257 | 83496 | 4.0 | 22.3 | 119 | — | 103 | −95 | 26 | 103 | 9.9 | 104 |
| 1 | 32400 | 162859 | 11.9 | 27.5 | 59 | — | — | −667 | −12 | 116 | 6.5 | 154 |
| 2 | 10529 | 33797 | 7.6 | 37.7 | 93 | — | — | −238 | −8 | 129 | 7.4 | 139 |
| 3 | 25905 | 86617 | 3.2 | 46.4 | 159 | — | 114 | −294 | −2 | 148 | 6.7 | 127 |
| 4 | 27492 | 112344 | 4.4 | 21.8 | 243 | — | 269 | −258 | 36 | 157 | 7.4 | 103 |
| 25 | 16022 | 36610 | 4.0 | 31.2 | 315 | — | 566 | 134 | 35 | 138 | 8.2 | 169 |
| 6 | 17390 | 45199 | 1.9 | 31.6 | 287 | 20 | 658 | −98 | 23 | 150 | 8.0 | 119 |
| 7 | 23698 | 55969 | 4.1 | 20.6 | 342 | 30 | 554 | −6 | 21 | 146 | 7.9 | 101 |
| 8 | 24161 | 37904 | 4.4 | 36.8 | 420 | 30 | 575 | 392 | 19 | 141 | 8.2 | 60 |
| 9 | 19849 | 21873 | 3.2 | 38.4 | 652 | 36 | 395 | −175 | 19 | 145 | 7.8 | 62 |
| 1930 | 13736 | 18080 | 8.7 | 26.9 | 449 | 48 | 334 | −280 | 11 | 132 | 4.0 | 65 |
| 1 | 3422 | 4806 | 15.9 | 37.8 | 300 | 29 | 668 | −145 | −2 | 114 | −1.5 | 48 |
| 2 | 254 | 1157 | 23.8 | 42.7 | 312 | 29 | 188 | 446 | −6 | 96 | −5.0 | 47 |
| 3 | 134 | 887 | 24.9 | 29.7 | 510 | 35 | 426 | 173 | −41 | 91 | −9.6 | 34 |
| 4 | 233 | 1154 | 21.7 | 321 | 1146 | 47 | 975 | −1134 | −86 | 114 | −5.2 | 113 |
| 35 | 408 | 1355 | 20.1 | 33.2 | 1168 | 60 | 1140 | −1739 | −336 | 109 | 5.1 | 212 |
| 6 | 324 | 1195 | 16.9 | 34.9 | 1503 | 64 | 1243 | −1117 | −171 | 107 | 10.0 | 379 |
| 7 | 378 | 1904 | 14.3 | 23.3 | 1140 | 76 | 1666 | −1586 | −80 | 112 | 7.1 | 306 |
| 8 | 609 | 2411 | 19.0 | 21.1 | 1110 | 66 | 1448 | −1974 | −223 | 102 | 6.1 | 417 |
| 9 | 415 | 2070 | 17.2 | 16.7 | 1241 | 87 | 2040 | −3574 | −71 | 105 | 7.4 | 361 |
| 1940 | 252 | 2120 | 14.6 | 33.0 | 914 | 100 | 2531 | −4744 | −55 | 98 | 7.0 | 360 |
| 1 | 129 | 732 | 9.9 | 47.5 | 1328 | 62 | 3125 | −982 | −41 | 85 | 7.5 | 407 |
| 2 | 92 | 493 | 4.7 | 40.0 | 975 | 52 | 4441 | −316 | −39 | 68 | 8.6 | 433 |
| 3 | 164 | 681 | 1.9 | 33.6 | 997 | 56 | 5684 | −69 | 3 | 56 | 9.1 | 447 |
| 4 | 203 | 1030 | 1.2 | 27.4 | 1147 | 60 | 4893 | 845 | 104 | 60 | 10.0 | 511 |
| 45 | 225 | 886 | 1.9 | 25.7 | 850 | 89 | 2633 | 106 | 64 | 75 | 11.0 | 643 |
| 6 | 189 | 1473 | 3.9 | 37.1 | 531 | 94 | 2471 | −311 | −21 | 81 | 10.1 | 677 |
| 7 | 442 | 381 | 3.6 | 19.3 | 272 | 117 | 1472 | −1866 | −37 | 90 | 8.6 | 779 |
| 8 | 946 | 4826 | 3.4 | 19.9 | 373 | 140 | 1919 | −1680 | −58 | 115 | 7.6 | 899 |
| 9 | 933 | 6192 | 5.5 | 35.5 | 634 | 141 | 1316 | −686 | −50 | 160 | 8.2 | 797 |
| 1950 | 3976 | 5693 | 5.0 | 13.9 | 715 | 176 | 2294 | 372 | −103 | 168 | 8.7 | 706 |
| 1 | 4972 | 5481 | 30 | 25.5 | 329 | 159 | 2986 | 550 | −94 | 174 | 7.8 | 730 |
| 2 | 6289 | 8969 | 2.7 | 32.2 | 14 | 180 | 3622 | −684 | −62 | 161 | 8.2 | 833 |
| 3 | 1538 | 5369 | 2.5 | 19.3 | 9 | 189 | 4564 | −2 | −86 | 166 | 7.9 | 891 |
| 4 | 1622 | 10061 | 5.0 | 15.4 | 137 | 207 | 6515 | −16 | −75 | 159 | 9.6 | 935 |
| 5 | 5486 | 17518 | 4.0 | 19.4 | 146 | 196 | 7810 | −97 | −65 | 182 | 8.1 | 936 |
| 6 | 9050 | 27807 | 3.8 | 19.3 | 70 | 233 | 7014 | −106 | −122 | 208 | 7.9 | 1171 |
| 57 | 4585 | 21826 | 4.3 | 40.1 | — | — | — | −104 | −147 | 175 | 8.3 | 1378 |

➡

exhibit 7 of chapter 16 (continued)

**I) IDENTIFICATION of SEQUENCES for PANEL A**

- **Z123:** Indigo exported from South Carolina in thousands of pounds.
- **Z225:** Tobacco imported by England from the Carolinas in thousands of pounds.
- **Z254:** Tea imported from England to the American colonies in hundreds of pounds.
- **Z350:** Average price of New England rum in Philadelphia in shillings per gallon.
- **Z357:** Average amount of Pennsylvania currency for 100 pounds sterling.

**J) IDENTIFICATION of SEQUENCES for PANELS B and C**

- **J16:** U.S. homestead entries, except on ceded Indian lands, in hundreds.
- **M259:** Number of fatal injuries in U.S. coal mining.
- **P196:** Millions of cigarettes produced in U.S.
- **X56:** Total liabilities of national banks in millions of dollars.
- **X286:** Gold certificates in circulation in millions of dollars.
- **D606:** Average annual earnings of wage earners in steam railroading in dollars per years.
- **M117:** Coke production in millions of short tons.
- **M251:** Primary production of aluminum (all ores) in thousands of short tons.
- **X274:** U.S. Government deposits in all banks in millions of dollars.

**K) IDENTIFICATION of SEQUENCES in PANEL D**

- **C97:** Number of immigrants from USSR and Baltic states.
- **C99:** Number of immigrants from Italy.
- **C102:** Number of immigrants from Turkey in Asia.
- **C123:** Number of farm laborers and foremen (for years before 1899, **C129** "Farmers"). (As immigrants.)
- **C124:** Laborers, excluding farm and mine. (For years before 1899, **C131** "Laborers".) (As immigrants.)
- **E101:** Wholesale price of wheat in dollars per bushel.
- **E109:** Wholesale price of nails in dollars per hundred pounds.
- **E111:** Wholesale price of turpentine in dollars per gallon.
- **W74:** Number of design patents issued.

**L) IDENTIFICATION of SEQUENCES in PANEL E**

- **C123:** Number of immigrants: Farm laborers and foremen.
- **C124:** Number of immigrants: Laborers, except farm and mine.
- **D47:** Unemployment as a percent of the civilian labor force.
- **J194:** Annual precipitation in inches at Beeville Experiment Station, Texas.
- **L147:** Landed catch of sardines, Pacific Coast states, in millions of pounds.
- **L172:** Production of canned tuna and tuna-like fish, in millions of pounds of net weight.
- **M215:** Output of tungsten concentrates, in short tons.
- **U6:** Excess of exports (+) or imports (−) of gold in millions of dollars.
- **U9:** Excess of exports (+) or imports (−) of silver, in millions of dollars.
- **W77:** Number of trade marks registered, in hundreds.
- **X199:** Net profits of national banks as percent of total capital accounts.
- **Y268:** Estate and gift taxes, in millions of dollars.

### review questions

Are these close relatives of the pieces we are now accustomed to put together to make a smoother? How could we try to make our smooths still smoother? Need we allow it to get away from the data? Why/why not?

exhibit **7** of chapter 16 (continued)

**P) PROBLEMS**

7a/b/c/d/e) Analyze sequences **Z123/Z225/Z254/Z350/Z357** from panel A.

7f/g/h/i/j/k/l/m/n) Analyze sequences **J16/M259/P196/X56/X286/D606/M117/M251/X274** from panels B and C.

7o/p/q/r/s/t/u/v/w/x/y/z) Analyze sequences **C97/C99/C102/C123/C124/E101/E109/E111/U6/U9/U182/W74** from panel D.

7aa/ab/ac/ad/ae/af/ag/ah/ai/aj/ak/al) Analyze sequences **C123/C124/D47/J194/L147/L172/M215/U6/U9/W77/X199/Y268** from panel E.

7am) Analyze the difference of sequences **C97** and **C99**.

7an) Analyze the ratio of sequences **C123** to **C124**.

More: Other sequences in the same source that could be considered for analysis include **C101, C107, E103, E105, M225, M249, P190, Q185, Q186, X215, X274**. In the earlier (1945) version of the source (numbering unrelated), the following, among others, could be considered: **E114, F193, G102, J169, N21, P66, P67**, the ratio **P66/P67, P111**, App. 30 (Aug).

**S) SOURCE**

*Historical Statistics of the United States: Colonial Times to 1957*. U.S. Dept. of Commerce, 1960.

exhibit **8** of chapter 16: data and problems

## Some optional problems for the energetic (and the computerized)

8a/b/c/d/e/f/g/h/i/j/k/l) Turn to the source mentioned in section S of exhibit 7 and analyze the full sequence for
**C97/C99/C102/C123/C124/E101/E109/E111/U6/U9/U182/W74**.

8m/n/o/p/q) Turn to the source mentioned in section S of exhibit 7 and analyze the full sequence for **J16/M259/P196/X56/X286**.

8r) Find a long sequence of interest to yourself, and analyze it.

### 16H. How far have we come?

The basic idea in the more careful smoothing is that better roughs require more care in taking what might be smooth away from the rough (and putting it in the smooth). Both idea and process are somewhat complementary to what we have already done to make smooths better, namely, to search harder in the smooth for what might be rough (and then move it to the smooth).

We are now ready to:

◇clean up our roughs by smoothing them again, adding what we get out to the original smooth ("twicing", "reroughing").

We now understand more clearly that:

◇whenever we are concerned about a smooth sticking closely enough to the input, we should smooth the corresponding rough, ending up by re-roughing the smooth.

◇more powerful smoothing components are available if and when we need them.

# COUNTS in BIN after BIN  17

chapter index on next page

We are now starting a group of chapters concerned with how values are distributed. This chapter approaches one of the most general cases, one where we are endeavoring to learn how the values are distributed for a sequence of counts, one for each of a system of cells or bins (such as, for example, the counts of family units with 0, 1, 2, 3,...children, or the counts of NBA basketball players with heights falling in each inch of height).

We want to begin by leaning only on the given data. To this end, we find that things we have already learned something about (going at least to square roots of counts, using the improved smoothing techniques of the last chapter) help us considerably.

Having a good smooth, and having had a hard look at a good rough, we will then be ready to try to make simple fits to our smooth. In this chapter-- and throughout the book--we shall confine ourselves, with a few exceptions, to fits symmetrical around a peak. We find that these do well for us here.

### review questions

What new area are we beginning on? How general a case do we consider in this chapter? What is it? What will we try to do first? Will earlier chapters help us? What will we try next? Under what limitations?

### 17A. Root smooth and root rough--bins of equal width

We are now ready to begin dealing with strings of counts. How do such strings arise? Usually because the possible values for some observation--the height of a basketball player, the occurrence of a flash of light, the number of books in a home--are divided up into intervals--often called cells or bins-- and a record is kept of how many events fall into each bin. (We use "bin" instead of "cell" because it is one letter shorter and has fewer alternative meanings.)

Often the bins are thought of as equal in width. We would surely do this, for example, when we binned basketball players by the inch of height. Equal widths is the easy case to begin with, and will be the subject of this section.

The general principle still applies:

**Since we are dealing with COUNTS, we always START by at least taking (square) ROOTS.**

We begin with counts based on measurements.

# index for chapter 17

review questions 543

### 17A. Root smooth and root rough--bins of equal width 543
lengths of forearms 545
symmetry 546
concentrations of gold 547
review questions 550

### 17B. Counts of basic counts 550
corn borers in field corn 550
octaves for logs 552
breeding pairs of birds 553
more problems 554
review questions 554

### 17C. Fitting to smoothed roots 555
finding the peak 555
choosing an expression 555
making the fit 556
review questions 557

### 17D. Corn borers, wheat prices, and Student's simulations 561
Student's simulations 563
wheat price changes 566
review questions 569

### 17E. Bins of unequal width 570
Student's $t$ 571
fitting the smooth 573
review questions 576

### 17F. Double roots 576
table 576
polonium scintillations 577
review questions 582

### 17G. Cautionary examples 582
forearm lengths again 583
peak polishing (rest of section is optional) 584
polonium scintillations again 584
comment 584
review questions 586

### 17H. How far have we come? 587

| EXHIBIT | PAGE |
|---|---|
| 17A | |
| 1★ | 545 |
| 2 | 547 |
| 3 | 547 |
| 4★ | 548 |
| 5 | 549 |
| 6 | 549 |
| 17B | |
| 7★ | 550 |
| 8 | 551 |
| 9★ | 552 |
| 10 | 553 |
| 11★ | 554 |
| 17C | |
| 12 | 556 |
| 13 | 557 |
| 14 | 559 |
| 15 | 560 |
| 17D | |
| 16★ | 562 |
| 17 | 563 |
| 18★ | 564 |
| 19 | 566 |
| 20★ | 567 |
| 21 | 569 |
| 17E | |
| 22★ | 570 |
| 23 | 572 |
| 24★ | 573 |
| 25 | 576 |
| 17F | |
| 26★ | 578 |
| 27★ | 579 |
| 28 | 581 |
| 17G | |
| 29★ | 582 |
| 30 | 585 |
| 31 | 585 |
| 32 | 586 |
| 17H | |

## lengths of forearms

Exhibit 1 shows the arithmetic for data of Pearson and Lee for the lengths of forearms of 1050 fathers. The smoothing calculations are by now standard. (See Chapters 7 and 16.) We have added two lines of zeros above and below the lines with nonzero counts; we could do this since shorter or longer forearms than observed are surely possible.

**exhibit 1 of chapter 17: forearm lengths**

**Smoothed root and its rough for Pearson and Lee's data on forearm lengths of 1050 fathers**

**A) DATA, ROOTS, SMOOTHS and ROUGHS**

| Bin # | Lengths (inches) of forearm | Count of fathers Raw | Root | 3RSS | 3RSSH | 1st rough | 3RSSH | Back add* | "Final rough" |
|---|---|---|---|---|---|---|---|---|---|
|   |         | 0     | 0    |      | 0    | 0    | 0    | 0    | 0    |
|   |         | 0     | 0    |      | .2   | -.2  | 0    | .2   | -.2  |
| 1 | 15–15.5 | 1     | 1.0  |      | 1.1  | -.1  | -.1  | 1.0  | 0    |
|   | 15.5–16 | 6.5   | 2.5  |      | 2.6  | -.1  | -.1  | 2.5  | 0    |
| 3 | 16–16.5 | 17    | 4.1  |      | 4.4  | -.3  | -.1  | 4.3  | -.1  |
|   | 16.5–17 | 49    | 7.0  |      | 7.3  | -.3  | -.1  | 7.1  | -.2  |
| 5 | 17–17.5 | 125.5 | 11.2 |      | 10.9 | .3   | .2   | 11.1 | .1   |
|   | 17.5–18 | 200   | 14.1 | 14.1 | 13.4 | .7   | .5   | 13.9 | .2   |
| 7 | 18–18.5 | 235.5 | 15.3 | 14.1 | 14.0 | 1.3  | .6   | 14.6 | .7   |
|   | 18.5–19 | 183.5 | 13.5 |      | 13.1 | .4   | .4   | 13.5 | 0    |
| 9 | 19–19.5 | 125.5 | 11.3 |      | 11.0 | .3   | .3   | 11.3 | 0    |
|   | 19.5–20 | 57.5  | 7.6  |      | 8.0  | -.4  | .2   | 8.2  | -.6  |
| 11| 20–20.5 | 31.5  | 5.6  |      | 5.4  | .2   | 0    | 5.4  | .2   |
|   | 20.5–21 | 8     | 2.8  |      | 3.3  | -.5  | -.1  | 3.2  | -.4  |
| 13| 21–21.5 | 3.5   | 1.9  |      | 2.0  | -.1  | -.1  | 1.9  | 0    |
|   | 21.5–22 | 2     | 1.4  | 1.6  | 1.6  | -.2  | -.1  | 1.5  | -.1  |
| 15| 22–22.5 | 2.5   | 1.6  | 1.4  | 1.1  | .5   | 0    | 1.1  | .5   |
|   |         | 0     | 0    |      | .4   | -.4  | 0    | .4   | -.4  |
|   |         | 0     | 0    |      | 0    | 0    | 0    | 0    | 0    |
|   | (Sum)   | (1050)|      |      |      |      |      |      |      |

\* "Back add" is final smooth = sum of both 3RSSH columns. "Final rough" = Root MINUS Back add.

*Note:* Half counts arise from dividing borderline fathers, putting one-half in each bin.

**B) LETTER VALUES for ROUGH--"zero" line at each end included**

#17

|     |    |    |    |   |     |       |      |
|-----|----|----|----|---|-----|-------|------|
| M8  |    | 0  |    |   |     | .4h   |      |
| H4h | .1 |    | -.2| .3| f   | .5h   | -.6h |
| E2h | .2 |    | -.4|   | one |       | xxx  |
|     |    |    |    |   | F   | 1.0   |      |

→

The results are pictured in exhibits **2** (the smooth) and **3** (the rough). Notice the smoothness and near symmetry of exhibit 2. Note the way in which fences, eighths, hinges, and median are shown in exhibit 3. The horizontal marks nearest the rough are from panel B of exhibit 1; those further away are from a set of reference values for any root rough, namely:

$$M = 0$$
$$H = \pm .34$$
$$E = \pm .58$$
$$f = \pm 1.36$$

Experience suggests that the actual behavior of such a rough will be somewhat more compressed than these standard values--but not as much more compressed as in this example, which we can thus suspect of being selected to give an unusually good fit.

**symmetry**

We clearly make it easier to describe a pattern of counts if that pattern is symmetrical. If a simple re-expression will do this for us, it is almost certain to be worthwhile to use it. In practice, a rather substantial fraction of all sets of measurements will need re-expression.

Chemical concentrations that are neither too large nor too restricted in range are a good example, whether the concentrations are in rocks, water, blood, grain, or tissue. It should become almost automatic to try logs on any such concentration.

exhibit **1** of chapter 17 (continued)

**P) PROBLEMS**

1a) Make a stem-and-leaf for the final rough and check the letter-value display given in panel B.

1b) Rerough with 3RSSH one more time. Are the residuals improved? Why/why not?

1b2) Do a letter-value display for the rough of 1b and compare with panel 3. Comment

1c) Check the first 3RSSH of panel A.

1d) Check the second 3RSSH of panel A.

1e) Do a modified (section 7E) letter-value display. Comment.

**S) SOURCE**

K. Pearson and A. Lee 1903. On the laws of inheritance in man. I. Inheritance of physical characters. *BMTA* 2: 357–462. Text table at page 367

#### concentrations of gold

Exhibit **4** deals with 1536 measurements of gold concentrations--expressed as pennyweight (dwt) values (per ton)--in assays of samples from the City Deep mine, Central Witwatersrand, South Africa, as reported by Koch and Link.

Exhibits **5** and **6** show the smooth and the rough, respectively. We can see essentially nothing about symmetry, since the smooth we have does not even

exhibit 2 of chapter 17: forearm lengths

**The smooth (smoothed roots of fathers per half-inch for the Pearson and Lee data; based on exhibit 1)**

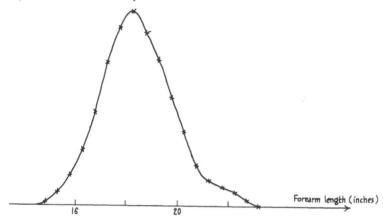

exhibit 3 of chapter 17: forearm lengths

**The rough--in a standard format (for the smooth of exhibit 2)--(includes ONE zero line at each end)**

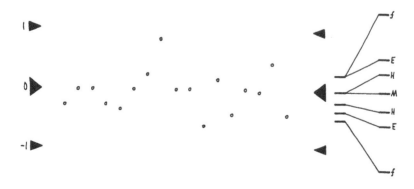

exhibit 4 of chapter 17: gold assays

**Smoothed roots and their rough for 1536 gold values (in dwt) for the City Deep mine**

A) DATA, ROOTS, SMOOTHS and ROUGHS

| Bin # | Interval | Count of samples | Root of count | 3RSSH | First rough | 3RSSH | Add back | Final rough |
|---|---|---|---|---|---|---|---|---|
|   | (0 to 5) | (910) |   |   |   |   |   |   |
| 1 | 5 to 10 | 208 | 14.4 | 14.4 | 0 | 0 | 14.4 | 0 |
| 2 | 10 to 20 | 198 | 14.1 | 13.4 | .7 | .2 | 13.6 | .5 |
| 3 | 20 to 40 | 124 | 11.1 | 10.8 | .3 | .3 | 11.1 | 0 |
| 4 | 40 to 80 | 52 | 7.2 | 7.8 | −.6 | .2 | 8.0 | −.8 |
| 5 | 80 to 160 | 34 | 5.8 | 5.4 | .4 | 0 | 5.4 | .4 |
| 6 | 160 to 320 | 8 | 2.8 | 3.2 | −.4 | 0 | 3.2 | −.4 |
| 7 | 320 to 6400 | 2 | 1.4 | 1.4 | 0 | 0 | 1.4 | 0 |
| 8 |   |   | 0 | .4 | −.4 | 0 | .4 | −.4 |
| (Sum) |   | (1536) | 0 | 0 | 0 | 0 | 0 | 0 |

Notes

1. We cannot add even one line at the low side, analogous to the two added at the high side, because we do not know how many of the 910 samples assaying between 0 and 5 actually assayed between 2.5 and 5.0. (The data was given in intervals of length 5 for raw assays.)
2. Using intervals corresponding to doubling of the raw values, conveniently called octaves, is often convenient when raw measurements are given and log measurements are to be analyzed.

B) LETTER VALUES for FINAL ROUGH

```
#8                             .8
M4h        0
H2h  .2          −.4  .6   f  1.1        −1.3
E1h  .4h         −.6 1.0h     xxx         xxx
```

P) PROBLEMS

4a) S. B. Flacker and H. A. Brischle, 1944. "Measuring the local distribution of *Ribes*," *Ecology* **28**: 288–303; (Figure 3 on page 29) give counts of 2.5 acre blocks (in Clearwater National Forest), showing different numbers of *Ribes* plants as follows:

| Plants: | 0 | 1 | 2 or 3 | 4 to 7 | 8 to 15 |
|---|---|---|---|---|---|
| # Blocks: | 4 | 3 | 7 | 4 | 8 |

| Plants: | 16 to 31 | 32 to 63 | 64 to 127 | 128 to 255 | 256 up |
|---|---|---|---|---|---|
| # Blocks: | 8 | 16 | 6 | 4 | 4 |

Analyze as above.

S) SOURCE

George S. Koch, Jr. and Richard F. Link, 1970. *Statistical Analysis of Geological Data*, John Wiley and Sons, New York. Table 6.5 at page 216.

reach the peak. Nevertheless the rough (i) looks just reasonably irregular and (ii) has its letter values spread out about as we might expect--somewhat more pulled in than the standard letter values, but not nearly as much less as in exhibit 3. So far so good.

exhibit 5 of chapter 17: gold assays

**The smooth (of root counts)**

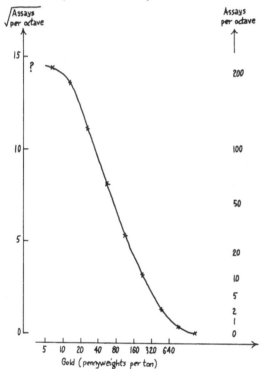

exhibit 6 of chapter 17: gold assays

**The rough (for the smooth of exhibit 5)**

### review questions

How do strings of counts arise? Why do we say "bin"? What is the general principle for dealing with counts? What example did we start with? What did we do with it? What did we see? What is our standard format for roughs associated with smoothed roots? How should we use our reference values for fEHMHEf? Is symmetry useful in a smooth? Why/why not? What do we expect to do with chemical concentrations? What example was next? Did we treat it differently? Why/why not? What did we see?

## 17B. Counts of basic counts

One case where the bins seem made for us arises when the observations are themselves counts--number of books per home, number of children per family, number of flashes of light in an eighth of a minute. Since we now have two kinds of counts, it pays us to be careful, calling the observations "basic counts" so that we are dealing with "counts of basic counts".

Sometimes we can go right ahead taking each basic count as a bin, and treating each bin as of equal width. (In a later section of this chapter we will learn an alternative approach.)

#### corn borers in field corn

Exhibit 7 shows the calculation for the numbers of corn borers (pupae and 5th instar larvae) in 3205 plants of field corn, as counted by McGuire, Brindley, and Bancroft. We have to be a little careful at the low end, because there CANNOT be LESS than zero borers in a plant. There MIGHT have

exhibit 7 of chapter 17: field corn

**Smoothed roots and their rough for 3205 counts of borers--pupae and 5th instar larvae only--in plants of field corn**

### A) DATA, ROOTS, SMOOTHS and ROUGHS

| Counts | Root of counts | | | Final rough | (letters values) | |
|---|---|---|---|---|---|---|
|  | raw | smoothed | twiced |  |  |  |
| 588 | 24.2 | 24.2 | 25.5 | −1.3 |  |  |
| 807 | 28.4 | 26.4 | 27.7 | .7 |  |  |
| 741 | 27.2 | 25.9 | 27.0 | .2 | (−1.7h | f) |
| 479 | 21.9 | 22.3 | 22.9 | −1.0 | (−1.0 | E) |
| 328 | 18.1 | 17.6 | 17.6 | .5 | ( −.4h | H) |
| 159 | 12.6 | 12.9 | 12.6 | 0 | ( .0h | M) |
| 67 | 8.2 | 8.4 | 8.1 | .1 | ( .4 | H) |
| 22 | 4.7 | 5.0 | 4.7 | 0 | ( .7 | E) |
| 5 | 2.2 | 3.0 | 2.8 | −.6 | ( .7 | f) |
| 7 | 2.6 | 2.1 | 2.2 | .4 | (out: xxx) |  |
| 2 | 1.4 | 1.2 | 1.4 | .0 |  |  |
| 0 | 0 | .4 | .4 | −.4 |  |  |
| (3205) |  |  |  | (0) |  |  |

been fathers with forearm lengths between 14.5 and 15 inches, so we could and should adjoin a 0 below our first count in exhibit 1. We cannot and should not adjoin 0 at −1 in exhibit 7.

Plotting the rough is left for the reader to do. This time the rough is about the size it ought to be--perhaps even a shade larger. (If the low residuals at basic counts of 0 and 3 were strays, we could sense a trend, but we have no reason to believe they are strays.) Exhibit **8** shows the smooth. This time we have lost not all, but most, of one side of a possibly symmetric smooth.

exhibit **7** of chapter 17 (continued)

**P) PROBLEMS**

7a) Try applying 3RSSH to the final rough of panel A, and adding back again.

7a2) What are the letter values now? How do they compare with (i) the reference values, (ii) the letter values of panel A?

7a3) Plot the corresponding rough.

7b) McGuire *et al.* also give the results of counting total borers per plant. The bin counts, beginning with bin zero, were: 355, 600, 781, 567, 411, 2105, 135, 42, 17, 11, 11. Copy the analysis for this data.

**S) SOURCE**

J. U. McGuire, T. A. Brindley, and T. A. Bancroft (1957). The distribution of European corn borer larvae *Pyrausta nubilalis* (Hbn.), in field corn. *Biometrics* **13**: 65–78. Distribution 2 on page 75.

exhibit **8** of chapter 17: field corn

**The smooth for exhibit 7**

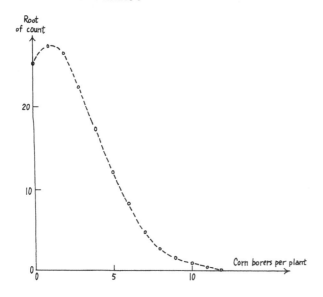

Sometimes, when we are dealing with counts of basic counts, it is not satisfactory to use equal widths for the bins. This usually happens because the pattern, if we did this, would be very long-tailed and hard to use. To avoid this, we often need to go over to roots of basic counts or logs of basic counts. When we go over to roots, we usually need to use unequal widths (in roots) for our bins, something we have not yet come to. (We will give an example shortly.)

### octaves for logs

When we go to started logs of basic counts, however, we can have equal intervals by working in octaves. Labelling the bins 1, 2 to 3, 4 to 7, 8 to 15, 16 to 31, and so on, corresponds to making cuts at 0.5, 1.5, 3.5, 7.5, 15.5, 31.5, and so on. For "counts plus 0.5" the cuts are now at 1, 2, 4, 8, 16, 32, and so on, so the bins for "log of (count plus 0.5)" are of equal widths. Doing this is convenient, and often works well.

exhibit 9 of chapter 17: valley birds

**Smoothed roots and their rough for numbers of breeding pairs of birds by octave, Quaker Run Valley, New York**

A) DATA, ROOTS, SMOOTHS and ROUGHS

| Bin # | Number of breeding pairs | Count of species | Root of count raw | 3RSSH | twice | Final rough | Letter values | |
|---|---|---|---|---|---|---|---|---|
| 1 | 0.5 to 1.5 | 2 | 1.4 | 1.4 | 1.4 | 0 | | |
| 2 | 1.5 to 3.5 | 5 | 2.2 | 2.2 | 2.3 | −.1 | (out: xxx) | |
| 3 | 3.5 to 7.5 | 9 | 3.0 | 2.8 | 3.0 | 0 | (−1.0 | f) |
| 4 | 7.5 to 15.5 | 10 | 3.2 | 3.0 | 3.2 | 0 | (−1.0 | f) |
| 5 | 15.5 to 31.5 | 8 | 2.8 | 3.0 | 3.2 | −.4 | (−.4 | E) |
| 6 | 31.5 to 63.5 | 13 | 3.6 | 3.0 | 3.1 | .5 | (−.3 | H) |
| 7 | 63.5 to 127.5 | 6 | 2.4 | 3.0 | 3.0 | −.6 | (0 | M) |
| 8 | 127.5 to 255.5 | 9 | 3.0 | 3.0 | 3.0 | 0 | (.1h | H) |
| 9 | 255.5 to 511.5 | 11 | 3.3 | 2.9 | 2.8 | .5 | (.5 | E) |
| 10 | 511.5 to 1023.5 | 4 | 2.0 | 2.5 | 2.2 | −.2 | (.8h | f) |
| 11 | 1023.5 to 2047.5 | 3 | 1.7 | 1.5 | 1.4 | .3 | (out: xxx) | |
| 12 | 2047.5 to 4095.5 | 0 | 0 | .4 | .4 | −.4 | | |
| | (sum) | (80) | 0 | 0 | 0 | 0 | | |

P) PROBLEMS

9a) Try 3RSSH'ing the final rough. What do you conclude?

S) SOURCE

F. W. Preston (1948). "The commonness, and rarity, of species," *Ecology* **29**: 254–283. (Table 1–A on page 258.)

A. A. Saunders (1936). Ecology of the birds of Quaker Run Valley, Alleghany State Park, New York. *New York State Museum Handbook* **16**.

**breeding pairs of birds**

Exhibit **9** shows the arithmetic, where the basic counts are the numbers of breeding pairs of each species of bird in Quaker Run Valley, New York, as reported by Saunders via Preston. We have not tried to add zero lines for fractional numbers of species.

With these small numbers, exhibit **10** easily accommodates both the smooth and the rough. Again, we have a reasonably symmetric smooth--and a well-behaved rough.

Consider carefully the mess we would have had if we had stayed with raw basic counts. We see that 2 species had 1 pair each, and a total of 16 species had up to 7 pairs each, almost 2 species per basic count. Between 512 and 1023 pairs there were 4 species, and between 1024 and 2047 there were 3. No fixed interval for raw numbers of pairs makes any sense. Either it is too short to help us above 511, or so long as to pack almost all the pairs into the first interval. Without octaves, our situation would have been unhappy.

exhibit **10** of chapter 17: valley birds

**The smooth and the rough (of roots of counts of species by numbers of breeding pairs, from exhibit 9)**

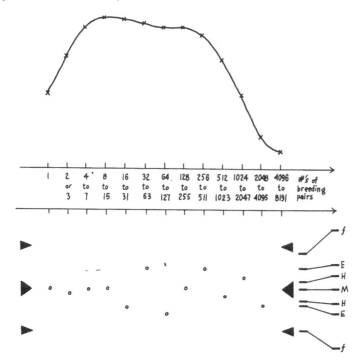

## more problems

Exhibit 11 contains data and wording for a variety of problems.

## review questions

Can we treat binned counts just like binned measurements? Always? What example was next? What did we see? And after that?

### exhibit 11 of chapter 17: problems

**Data and further problems on smoothed roots and their roughs**

**A) COUNTS of YEAST CELLS in a HAEMOCYTOMETER**

| Basic Count | 0 | 1 | 2 | 3 | 4 | 5 | 6 | 7 |
|---|---|---|---|---|---|---|---|---|
| # counted | 103 | 202 | 268 | 143 | 32 | 17 | 10 | 0 |

**B) COUNTS of PLANTS of *Salicornia europea***

in 625cm² squares
```
 0* | 23334455556789
 1  | 1122344778
 2  | 002336888999
 3  | 01146799
 4* | 00235789
 5  | 000011222345
 6  | 011234578889
 7  | 00334567889
 8* | 228
 9  | 116
10  | 04
11  | 07
12* | 7
13  | 23
```

in 2500cm² squares
```
 0* | 3445555567889
 1  | 12334566789
 2  | 11122356
 3  | 00355689
 4* | 7
 5  | 8
 6  | 169
 7  |
 8* | 0
 9  |
1** | 01, 02, 04, 29, 35, 39
  · | 52, 71, 73
2** | 25, 32, 35
```

### P) PROBLEMS

11a) Can we use octaves for the data of panel A? Why/why not? Do the best you can.

11b/c) Same questions and instructions for the left/right data set of panel B.

### S) SOURCES

For A: Student, 1907. "On the error of counting with a haemocytometer," *Biometrika* 5: 351–360. Example (ii) on page 356.

For B: E. Ashby, 1935, "The quantitative analysis of vegetation," *Annals of Botany* 47: 779–803. Table III (page 190) and Table IV.

### 17C. Fitting to smoothed roots

Our smooths make nice pretty pictures, so much so that we want to go further and try to fit some simple shape to them--or to a re-expression of them. The simplest formula that gives a smooth symmetrical peak will be one that, in the Pearson and Lee data, for example, has a simple straight-line dependence of

(some expression of root count)

ON

(forearm length MINUS peak forearm length)$^2$

(The square accounts for both the peaking and the symmetry, as we shall see.)

We shall write "shift" for the difference between where we are and the peak. Thus, for the present,

$$B = \text{peak} \quad \text{PLUS} \quad \text{shift},$$

$$\text{shift} = B \quad \text{MINUS} \quad \text{peak},$$

$$\text{fit} = \text{constant} \quad \text{PLUS} \quad (\text{tilt})(\text{shift})^2$$

In any given set of data, then, there are three things for us to do:

◇ find the peak.

◇ choose the re-expression,

◇ fit the line.

#### finding the peak

Using the values near the peak to find where the peak lies is often very difficult. In particular, small irregularities can throw us far off. How does our eye judge where the peak is, GIVEN that the smooth looks symmetrical? Mainly by moving down from the peak, choosing two points at the same height, and splitting the difference. For simplicity (especially when we come to arithmetic), we always start from a plotted point. Exhibit **12** shows how this can be done graphically for the Pearson and Lee data (exhibits 1 to 3) on forearm lengths. Notice a slight tendency for the peak count to be lower when we work at a higher level. Exhibit **13** shows how the same thing can be done arithmetically for the same data.

The left side of panel B shows the resulting picture of smoothed roots against squared shift. Note that we have used one character, ×, on one side of the peak, and another, o, on the other. This helps us to see whether one side is systematically behaving differently from the other (a little of this shows here)--or whether the behavior on one side, or on both, is only irregular. We see a reasonably well-defined curve in this exhibit, but nothing like a straight line. Clearly, a different expression of the root smooth is needed.

#### choosing an expression

We know how to set about choosing an expression (see chapter 6)--beginning with a few chosen points. Looking at exhibit 13, we first select

(0, 15), (10, 7.5), (20, 3.9). Looked at hard, the three immediately suggest taking logs of the root smooth--since 0, 10, 20 are equally spaced, while 15, 7.5, 3.9 change by nearly a constant factor. (The numerical values of the logs are at the right of exhibit 13.) The righthand side of panel B shows a plot of log smoothed roots against squared shift. This picture clearly calls for a straight line. We have a reasonable re-expression and are ready to press on. (But notice, toward the right of the picture, that all the o's are higher and all the ×'s are lower. This speaks of a tendency to lack of symmetry about our chosen peak.)

### making the fit

Exhibit 14 shows the same picture, with a straight line drawn in. Since this line goes through (0, 1.17), (50, −.24), its equation is

$$\log(\text{smoothed roots}) = 1.17 - 0.0282 \text{ (squared difference)}.$$

The result of using this is set out in exhibit 15 with the residuals shown in panel B. While this plot shows moderate irregularity, it feels a little trendy--and its

exhibit **12** of chapter 17: forearm length

**Locating the peak through midpoints (shown "∧") of secants**

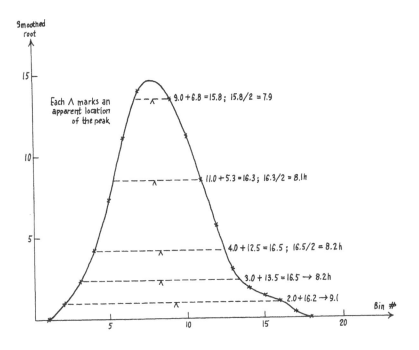

letter values are broader than the reference. Thus we wonder whether the fit is good enough. Another way to ask this question is to smooth the residuals--arithmetic in panel A and display in panel C. The smoothed residuals are large enough, as the last picture shows--going out to the hinges of the unsmoothed residuals--to make us wonder about a better fit.

### review questions

What do we want our fits to look like? What will we call "shift"? What dependence on shift will be simplest, among those that might look right? What will depend simply on shift$^2$? How do we find a peak? Would this work for an unsymmetrical distribution. With what example did we start? With what results? What did we use three selected points for?

exhibit **13** of chapter 17: forearm lengths

**Finding the peak for the Pearson and Lee data of exhibit 1**

A) ROOT SMOOTH, its MATCHING ACROSS, and SQUARED DIFFERENCE

| Bin # | Smoothed roots | Matched step | Sum of steps | Step minus 8.25 | Square of same | log smoothed roots |
|---|---|---|---|---|---|---|
| 1 | .2 | 17.0 | 18.0 | −7.25 | 52.6 | −.70 |
|  | 1.0 | 16.2 | 18.2 | −6.25 | 39.1 | .00 |
| 3 | 2.5 | 13.5 | 16.5 | −5.25 | 27.6 | .40 |
|  | 4.3 | 12.5 | 16.5 | −4.25 | 18.1 | .63 |
| 5 | 7.3 | 11.3 | 16.3 | −3.25 | 10.6 | .86 |
|  | 11.2 | 10.1 | 16.1 | −2.25 | 5.1 | 1.05 |
| 7 | 13.8 | -- |  | −1.25 | 1.6 | 1.14 |
|  | 14.4 | -- |  | −.25 | .1 | 1.16 |
| 9 | 13.5 | 6.9 | 15.9 | .75 | .6 | 1.13 |
|  | 11.3 | 6.1 | 16.1 | 1.75 | 3.1 | 1.05 |
| 11 | 8.2 | 5.2 | 16.2 | 2.75 | 7.6 | .91 |
|  | 5.4 | 4.4 | 16.4 | 3.75 | 14.1 | .73 |
| 13 | 3.2 | 3.4 | 16.4 | 4.75 | 22.6 | .50 |
|  | 1.9 | 2.6 | 16.6 | 5.75 | 33.1 | .28 |
| 15 | 1.5 | 2.4 | 17.4 | 6.75 | 45.6 | .18 |
|  | 1.1 | 2.1 | 18.1 | 7.75 | 60.1 | .04 |
| 17 | .4 | 1.5 | 18.5 | 8.75 | 76.6 | −.40 |
|  | 0 | median | 16.5 |  |  |  |
|  |  | peak is | 8.25 |  |  |  |

Example: 7.3 comes at 5, and again between 11 and 12, where interpolation suggests it comes at 11.4. Next 5 PLUS 11.4 = 16.4, given in the "sum of steps" column. Note. The increase of the "sum-of-steps" value toward the tail indicates a tendency to lack of symmetry about 8.25, the positive tail appearing longer.

➡

exhibit **13** of chapter 17 (continued)

**B) PLOTS AGAINST SQUARED SHIFT**

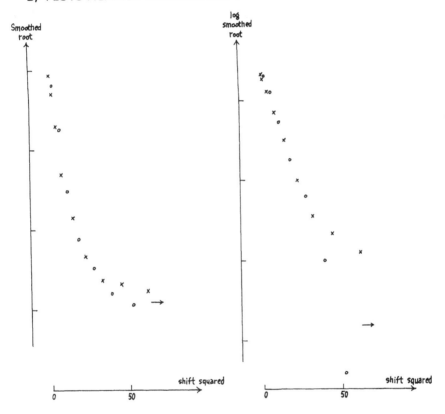

exhibit **14** of chapter 17: forearm lengths

**The final plot, with a straight line inserted**

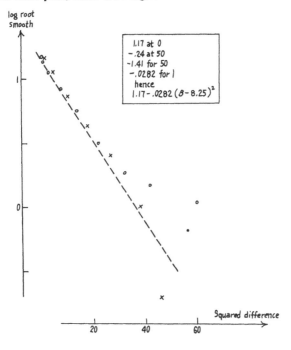

1.17 at 0
−.24 at 50
−1.41 for 50
−.0282 for 1
hence
$1.17 - .0282(\delta - 8.25)^2$

exhibit **15** of chapter **17**: forearm lengths

**Calculation of the fit corresponding to the straight line of exhibit 14 (first-fit)**

A) ARITHMETIC--peak at 8.25, fit as $1.17 - .0282\text{(difference)}^2$

| Interval | Bin # | Fitted values log root | Fitted values root | observed root | The rough* | Res- idual | 3RSSH | (letter values**) |
|---|---|---|---|---|---|---|---|---|
| 15–15.5 | 1 | −.31 | .5 | 0 | 0 .2 | −.5 | −.2 | |
|  |  | .07 | 1.2 | 1.0 | 0 | −.2 | −.2 | |
|  | 3 | .39 | 2.5 | 2.5 | 0 | 0 | −.2 | (letter values**) |
|  |  | .66 | 4.6 | 4.1 | −.2 | −.5 | 0 | |
|  | 5 | .87 | 7.4 | 7.0 | −.3 | −.4 | .4 | (out: none) |
|  |  | 1.03 | 10.7 | 11.2 | 0 | .5 | .5 | (−1.5    f) |
| 18–18.5 | 7 | 1.12h | 13.3 | 14.1 | .3 | .8 | .5 | ( −.6    E) |
|  |  | 1.17 | 14.8 | 15.3 | .9 | .5 | .2 | ( −.3    H) |
|  | 9 | 1.15h | 14.3 | 13.5 | 0 | −.8 | −.5 | (   0    M) |
|  |  | 1.08h | 12.2 | 11.3 | 0 | −.9 | −.8 | (  .5    H) |
|  | 11 | .95h | 9.0 | 7.6 | −.6 | −.7 | −.8 | (  .5    E) |
|  |  | .77h | 6.0 | 5.6 | .2 | −.4 | −.6 | ( 1.7    f) |
|  | 13 | .53h | 3.4 | 2.8 | −.4 | −.6 | −.3 | (out: none) |
|  |  | .24 | 1.7 | 1.9 | 0 | .2 | .2 | |
| 22–22.5 | 15 | −.11 | .8 | 1.4 | −.1 | .6 | .5 | |
|  |  | −.52h | .3 | 1.6 | .5 | 1.3 | .6 | |
|  | 17 | −.99 | .1 | 0 | −.4 | −.1 | .6 | |

Examples: $1.12h = 1.17 - (.0282)(1.25)^2$

\* from exhibit 1
\*\* for unsmoothed residuals

B) The RESIDUALS PICTURED--as is

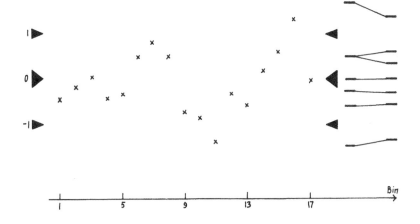

## 17D. Corn borers, wheat prices, and Student's simulations

Let us turn back next to exhibit 7, with its counts of corn borers per plant. The root smooth comes to a maximum and turns over, going down for just one point. Zero borers gives 24.2, which interpolates to about 2.7 borers on the other side of the smooth. So we may as well start with a trial peak at 1.3 borers. Exhibit **16** has the arithmetic. The left side of panel B shows log smoothed roots against shift-squared, where we see that our line is at best a first approximation.

Going back to panel A, we calculate the residuals from the line, and then plot them again--at the right side of panel B. We now see that we can fit a reasonably good line to the first six points--marked "o"--if we forget the other five--marked "x". Doing this makes some sense, because the first six involve the larger counts and the more stable logs.

We have now fitted, writing $B$ for borers per plant:

$$1.45 - .031(\text{shift})^2 = 1.45 - .031(B - 1.3)^2$$

$$\text{PLUS}$$

$$-.005 + .002(B - 1.3)^2$$

to our log smoothed roots, a total of

$$1.435 - .029(B - 1.3)^2.$$

With the new fit, the residuals are not beautiful, as exhibit **17** pictures. Two things are notable:

⋄ the residuals for the two points closest to the peak are high--one is outside.

⋄ the residuals as a whole are too positive.

exhibit **15** of chapter 17 (continued)

### C) The SMOOTHED RESIDUALS PICTURED

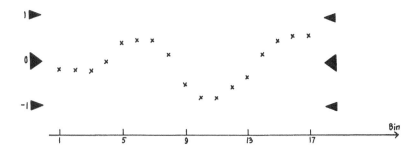

Both of these are natural consequences of the peak being too narrow for our smoothing technology. Fitting to "log raw roots" may, in this case, be better than fitting to "log smoothed roots".

If we had wished, we could have tried an additional fit linear in $B$, rather than linear in $(B - 1.3)^2$. Adding such a fit would shift the fitted peak.

### exhibit 16 of chapter 17: field corn

**A trial fit to the corn borer data (smooth from exhibit 15, peak at 1.3 from text)**

#### A) FITS and RESIDUALS

| Borers per plant | Smoothed root | log smoothed root | (shift)² | 1.45 −.031 × (shift)² | Resid. log root | .005 −.002 (shift)² | Improved fit log root | root | Residual | (Letter values) | |
|---|---|---|---|---|---|---|---|---|---|---|---|
| 0 | 25.5 | 1.41 | 1.7 | 1.40 | .01 | −.00 | 1.40 | 25.1 | .9 | | |
|   | 27.7 | 1.44 | .1  | 1.44 | .00 | −.01 | 1.45 | 28.2 | .2 | | |
| 2 | 27.0 | 1.43 | .5  | 1.43 | .00 | −.00 | 1.43 | 26.9 | .3 | (out: one) | |
|   | 22.9 | 1.36 | 2.9 | 1.37 | −.01 | −.00 | 1.37 | 23.4 | −1.5 | (1.7 | f) |
| 4 | 17.6 | 1.25 | 7.3 | 1.27 | −.02 | .01 | 1.26 | 18.2 | −.1 | ( .7 | E) |
|   | 12.6 | 1.10 | 13.7 | 1.12 | .02 | −.02 | 1.10 | 12.6 | 0 | ( .5h | H) |
| 6 | 8.1 | .91 | 22.1 | .92 | .01 | −.04 | .88 | 7.8 | .6 | ( .2 | M) |
|   | 4.7 | .67 | 32.5 | .68 | −.01 | −.06 | .62 | 4.2 | .5 | (−.2 | H) |
| 8 | 2.8 | .45 | 44.9 | .39 | .06 | −.08 | .31 | 2.0 | .2 | (−.9 | E) |
|   | 2.2 | .34 | 59.3 | .05 | .29 | −.11 | .06 | .9 | 1.3 | (1.3h | f) |
| 10 | 1.4 | .15 | 75.7 | −.33 | .48 | −.15 | −.18 | .7 | .7 | (out: xxx) | |
|   | .4 | −.40 | 94.1 | −.76 | .36 | −.18 | −.94 | .1 | .3 | | |
| 12 |  (0) | (L) | | | | | | | | | |

#### B) PLOTS

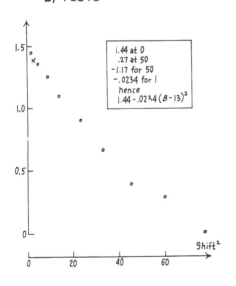

1.44 at 0
.27 at 50
−1.17 for 50
−.0234 for 1
hence
$1.44 - .0234(B - 13)^2$

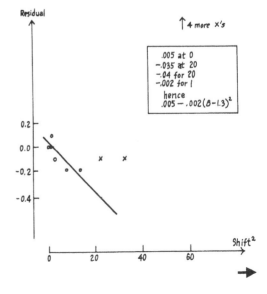

.005 at 0
−.035 at 20
−.04 for 20
−.002 for 1
hence
$.005 - .002(B - 1.3)^2$

## Student's simulations

In 1908 Student--a chemist, later brewmaster for Guinness--published two papers which did much to found modern statistics. In both he simulated the behavior of certain quantities one might like to calculate from batches of numbers. He did this by preparing 750 batches of four numbers (or 750 batches of 8) and calculating the values of the interesting quantity for each batch. He used the heights and the left middle finger measurements of 3000 criminals, writing these on 3000 pieces of cardboard, shuffling well and drawing them out "at random". (For the batches of eight, each criminal appeared in two batches.) Some of his results are shown in exhibit **18**. Since two of his collections of counts are for bins of unequal widths, their analysis will have to wait till the next section. The third can now be treated in just the same way we have treated other sets of counts, bin after bin.

exhibit **16** of chapter 17 (continued)

### P) PROBLEMS

16a) Carry one extra decimal through panel A, beginning with the log smoothed root. Is the increased accuracy worth the effort?

16b) (Class exercise). Try a variety of different fits of the same form, calculating and plotting residuals for each.

16c) Plot the final residuals against B, rather than $(B - 1.3)^2$, fit a line, add to fit. Discuss the effect.

16d) Plot the unsmoothed log roots against $(B - 1.3)^2$. Complete the fit. Discuss the results.

exhibit **17** of chapter 17: field corn

**Residuals from the new fit**

## exhibit 18 of chapter 17: simulations

**Results of two of Student's simulations**

A) The DATA

| Bin | Values of correlation coefficient | Counts of these |
|---|---|---|
| 1 | (to −.98) | (1) |
| 2 | −.97 to −.93 | 0 |
| 3 | −.92 to −.88 | 0 |
| 4 | −.87 to −.83 | 1 |
| 5 | −.82 to −.78 | 7 |
| 6 | −.77 to −.73 | 10 |
| 7 | −.72 to −.68 | 10 |
| 8 | −.67 to −.63 | 15 |
| 9 | −.62 to −.58 | 14 |
| 10 | −.57 to −.53 | 15 |
| 11 | −.52 to −.48 | 18 |
| 12 | −.47 to −.43 | 24 |
| 13 | −.42 to −.38 | 18 |
| 14 | −.37 to −.33 | 27 |
| 15 | −.32 to −.28 | 36 |
| 16 | −.27 to −.23 | 33 |
| 17 | −.22 to −.18 | 43 |
| 18 | −.17 to −.13 | 45 |
| 19 | −.12 to −.08 | 26h |
| 20 | −.07 to −.03 | 26h |
| 21 | −.02 to +.02 | 34 |
| 22 | +.03 to +.07 | 42h |
| 23 | +.08 to +.12 | 27h |
| 24 | +.13 to +.17 | 34 |
| 25 | +.18 to +.22 | 23h |
| 26 | +.23 to +.27 | 36h |
| 27 | +.28 to +.32 | 33h |
| 28 | +.33 to +.37 | 28h |
| 29 | +.38 to +.42 | 19 |
| 30 | +.43 to +.47 | 24 |
| 31 | +.48 to +.52 | 22 |
| 32 | +.53 to +.57 | 15 |
| 33 | +.58 to +.62 | 13h |
| 34 | +.63 to +.67 | 9 |
| 35 | +.68 to +.72 | 2h |
| 36 | +.73 to +.77 | 7 |
| 37 | +.78 to +.82 | 5 |
| 38 | +.83 to +.87 | 3 |
| 39 | +.88 to +.92 | — |
| 40 | +.93 to +.97 | — |
| 41 | (.98 up) | — |

| Values of "Student's t" | Counts of t for heights | for fingers |
|---|---|---|
| to −3.05 | 9 | 4 |
| −3.05 to −2.05 | 14h | 15h |
| −2.05 to −1.55 | 11h | 18 |
| −1.55 to −1.05 | 33 | 33h |
| −1.05 to −.75 | 43h | 44 |
| −.75 to −.45 | 70h | 75 |
| −.45 to −.15 | 119h | 122 |
| −.15 to .15 | 151h | 138 |
| .15 to .45 | 122 | 120h |
| .45 to .75 | 67h | 71 |
| .75 to 1.05 | 49 | 46h |
| 1.05 to 1.55 | 26h | 36 |
| 1.55 to 2.05 | 16 | 11 |
| 2.05 to 3.05 | 10 | 9 |
| 3.05 on | 6 | 6 |

Note. 750 batches of eight for the correlation coefficient; two sets of 750 batches of four for what is now called "Student's t".

➡

The result of taking the peak at 0.0 is shown in exhibit **19**. Clearly, we do NOT need to take logs in this case. We find

$$6.3 - .0158(\text{shift in bins})^2$$

for our fit, and we find the residuals shown in exhibit 19. These look quite reasonable, being of the right size and suitably irregular. (When they are smoothed they show rather neat, small wiggles, suggesting no trace of remaining structure.)

exhibit **18** of chapter 17 (continued)

**B) PLOT--for the correlation coefficient**

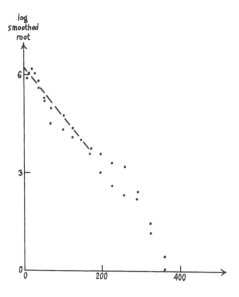

**P) PROBLEMS**

18a) Find roots, smoothed roots and their roughs for the correlation coefficient counts, using bins given. Compare results with exhibit 24.

18b) Do the same with half as many bins, each twice as wide.

**S) SOURCES**

Counts of "*t*": Student, 1908. "The probable error of a mean." *Biometrika* **6**: 1–25.
Counts of correlation coefficients: Student, 1908. "Probable error of a correlation coefficient," *Biometrika* **6**: 302–310.
See alternatively: pp. 1–10 and pp. 11–34 of *"Student's" Collected papers* edited by E. S. Pearson and John Wishart, n.d. Biometrika Office, Cambridge University press.

### wheat price changes

In 1953, Maurice Kendall published a study of price series from which we can take counts of weekly changes in wheat prices from 1883 to 1934. Exhibit **20** has the arithmetic. The first rough has values of 3.4 and 2.9 for the two center bins--in words, the peak is too sharp for our smoother. This is easily dealt with by splitting the smooth between these two. Taking the peak at zero, we see that log smoothed roots do not give a straight line against $B^2$, but that reciprocal smoothed roots may. Fitting a line by eye gives fitted roots which leave reasonable residuals. Exhibit **21** shows both root rough and residuals. In each case, both plain and modified letter values are shown--modified next to plot, and plain between these and the reference values. The fit is not wonderful, but not unsatisfactory.

exhibit **19** of chapter 17: simulations

**The residuals, alone above and with their smooth below (numbers based on exhibit 18).**

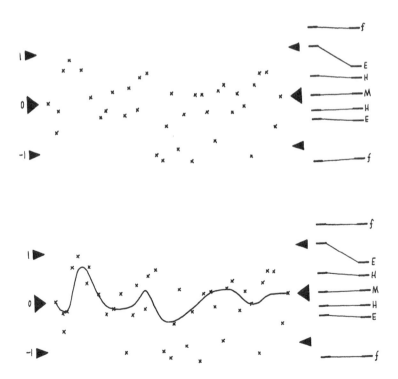

exhibit **20** of chapter 17: price changes

**Changes in wheat prices from one week to the next, 1883–1934. Bins are 2 cents a bushel wide. Bin numbers are all odd and are centers of changes. (Thus −13 covers −14 to −12, etc.) Smooth 3RSSH-twice. Peak assumed at zero.**

**A) ARITHMETIC**

| Bin #* | Count in bin | Root of count | Smoothed root | Their rough | Split smoothed roots | Their rough |
|---|---|---|---|---|---|---|
| L   | (2)  |      | .0   | .0   | .0   | .0   |
| −17 | h    | 0.7  | .8   | −.1  | .8   | −.1  |
| −15 | 5    | 2.2  | 2.0  | .2   | 2.0  | .2   |
| −13 | 8h   | 2.9  | 2.6  | .3   | 2.6  | .3   |
| −11 | 7h   | 2.7  | 3.3  | −.6  | 3.3  | −.6  |
| −9  | 24   | 4.9  | 4.5  | .4   | 4.5  | .4   |
| −7  | 47h  | 6.9  | 6.9  | .0   | 6.9  | .0   |
| −5  | 114h | 10.7 | 10.7 | .0   | 10.7 | .0   |
| −3  | 263  | 16.2 | 17.0 | −.8  | 17.2 | −1.0 |
| −1  | 722  | 26.9 | 23.5 | 3.4  | 26.4 | .5   |
| 1   | 708h | 26.6 | 23.7 | 2.9  | 25.9 | .7   |
| 3   | 284h | 16.9 | 17.1 | −.2  | 16.2 | .7   |
| 5   | 100  | 10.0 | 10.5 | −.5  | 9.3  | .7   |
| 7   | 50   | 7.1  | 6.5  | .6   | 6.5  | .6   |
| 9   | 17   | 4.1  | 4.1  | .0   | 4.1  | .0   |
| 11  | 10   | 3.2  | 3.0  | .2   | 3.0  | .2   |
| 13  | 5    | 2.2  | 2.2  | .0   | 2.2  | .0   |
| 15  | 4h   | 2.1  | 2.1  | .0   | 2.1  | .0   |
| 17  | 4h   | 2.1  | 1.9  | .2   | 1.9  | .2   |
| 19  | 1h   | 1.2  | 1.5  | −.3  | 1.5  | −.3  |
| 21  | 3h   | 1.9  | 0.5  | 1.4  | 0.5  | 1.4  |
| H   | (2)  |      | .2   | −.2  | .2   | −.2  |

\* Midvalue as a change in cents per bushel--changes on a division value assigned half to each bin.

**B) LETTER VALUES for ROUGH--plain and modified**

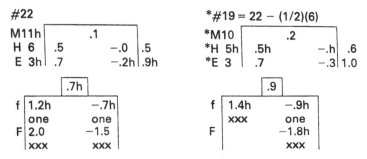

outside both = −1.0, outside one = 1.4

exhibit **20** of chapter 17 (continued)

### C) ARITHMETIC CONTINUED

| squared shift | log smoothed roots | 1000/ smoothed roots | 37 + 2.256 × squared shift | Fitted root | Root residual |
|---|---|---|---|---|---|
| 361 |  |  | 851.5 | 1.2 | −1.2 |
| 289 | −.10 | 1250 | 689.0 | 1.5 | −.8 |
| 225 | .30 | 500 | 544.6 | 1.8 | .4 |
| 169 | .41 | 385 | 418.3 | 2.4 | .5 |
| 121 | .51 | 303 | 310.0 | 3.2 | −.5 |
| 81 | .65 | 222 | 219.8 | 4.5 | .4 |
| 49 | .84 | 145 | 147.6 | 6.8 | .1 |
| 25 | 1.03 | 93 | 93.4 | 10.7 | .0 |
| 9 | 1.24 | 58 | 57.3 | 17.5 | −1.3 |
| 1 | 1.42 | 38 | 39.3 | 25.5 | 1.4 |
| 1 | 1.41 | 39 | 39.3 | 25.5 | 1.1 |
| 9 | 1.21 | 62 | 57.3 | 17.5 | −.6 |
| 25 | .97 | 108 | 93.4 | 10.7 | −.7 |
| 49 | .81 | 154 | 147.6 | 6.8 | .3 |
| 81 | .61 | 244 | 219.8 | 4.5 | −.4 |
| 121 | .48 | 333 | 310.0 | 3.2 | .0 |
| 169 | .34 | 455 | 418.3 | 2.4 | −.2 |
| 225 | .32 | 476 | 544.6 | 1.8 | .3 |
| 289 | .28 | 526 | 689.0 | 1.5 | .6 |
| 361 | .18 | 667 | 851.5 | 1.2 | .0 |
| 441 | −.30 | 2000 | 1032.0 | 1.0 | .9 |
| 529 | −.70 | 5000 | 1230.4 | 0.8 | −.8 |

### D) LETTER VALUES for RESIDUALS

#22

| M 11h | 0 |  |  |
| --- | --- | --- | --- |
| H 6 | .4 | −.6h | 1.0h |
| E 3h | .7h | −.8 | 1.5h |

|  | 1.6 |  |
| --- | --- | --- |
| f | 1.9 | −2.1 |
|  | xxx | xxx |

*#18h = 22 − (1/2)(3)

| *M 9h | 0 |  |  |
| --- | --- | --- | --- |
| *H 5 | .4h | −.6h | 1.2 |
| *E 3 | .9 | −.8 | 1.7 |

|  | 1.6h |  |
| --- | --- | --- |
| f | 2.1 | −2.3 |
|  | xxx | xxx |

### P) PROBLEMS:

**20a)** Kendall also gave data on changes of investment share prices as follows:
(21): 13, 15, 25, 46, 66, 81; 91; 44, 32, 14, 15, 8; (14) where these counts apply to intervals centered as follows:
(Low); −11.5 to −1.5; 0.5; 2.5 to 10.5; (High).
Make an analysis parallel to the three panels above.

### S) SOURCE

M. G. Kendall, 1953."The analysis of economic time-series---Part I: Prices." *J. Roy. Stat. Soc.* **A116** 11-25.
(Reprinted in P. Cootner, ed. 1964, 1969. *The Random Character of Stock Market Prices*, MIT Press.)

**review questions**

What example came first? How did we guess a peak? Why did we have to guess? What did we do next? With what result? Whose data did we turn to next? What were these counts based upon? What result did we reach? What example came next? With what result? How did rough and residuals compare? What would you suspect from this?

exhibit **21** of chapter 17: price changes

**Root rough and root residuals for exhibit 20**

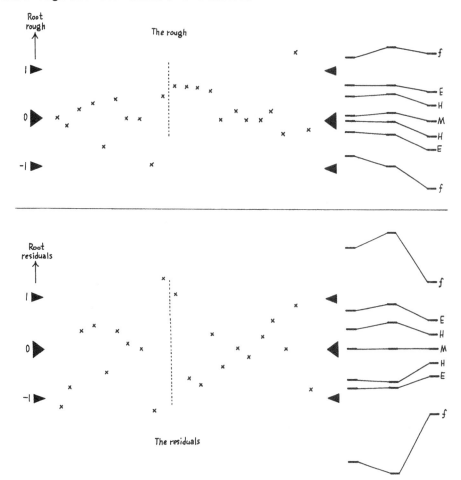

### 17E. Bins of unequal width

When the widths of our bins differ, we must work with

$$\sqrt{\frac{\text{counts}}{\text{bin width}}},$$

instead of just

$$\sqrt{\text{counts}}.$$

To see that this is so, make up an equal-width bin example where the counts vary between 90 and 110, and experiment by combining two or three small bins into one large one. You will find that dividing counts by bin width is essential.

Since the bins are not of equal width, their centers are not equally spaced.

**exhibit 22** of chapter 17: simulation

**Root smooth and root rough for Student's simulation of *t* using heights**

#### A) BIN ARITHMETIC

| Bin # | Bin dimensions | | | | 3RSSH | Getting back |
|---|---|---|---|---|---|---|
| | Start | End | Width | Center | | |
| 1 | Low | −3.05 | --- | --- | | |
|   | −3.05 | −2.05 | 1.0 | −2.55 | −2.55 | hold |
| 3 | −2.05 | −1.55 | .5 | −1.80 | −1.86 | go 6/54 toward #4 |
|   | −1.55 | −1.05 | .5 | −1.30 | 1.32 | go 2/40 toward #5 |
| 5 | −1.05 | −.75 | .3 | −.90 | −.92 | go 2/32 toward #6 |
|   | −.75 | −.45 | .3 | −.60 | −.60 | hold |
| 7 | −.45 | −.15 | .3 | −.30 | −.30 | hold |
|   | −.15 | .15 | .3 | .00 | .00 | hold |
| 9 | .15 | .45 | .3 | .30 | .30 | hold |
|   | .45 | .75 | .3 | .60 | .60 | hold |
| 11 | .75 | 1.05 | .3 | .90 | .92 | go 2/32 toward #10 |
|   | 1.05 | 1.55 | .5 | 1.30 | 1.32 | go 2/40 toward #11 |
| 13 | 1.55 | 2.05 | .5 | 1.80 | 1.86 | go 6/54 toward #12 |
|   | 2.05 | 3.05 | 1.0 | 2.55 | 2.55 | hold |
| 15 | 3.05 | High | --- | --- | | |

Notes.

1. Bins with one open side (#1, #15) have no useful definitions of width or center.

2. After hanning, bin #3 ends up at −1.86, #4 at −1.32. To bring #3 after hanning, back to −1.80, where it started, requires moving .06, while going all the way to #4 would move 0.54. Hence moving .06/.54 = 1/9 of the way to #4 should do the trick. And so on.

→

This leaves us concerned about the use of hanning. We could:

◇ drop hanning and use only 3RSS twice.
◇ whenever we hann, hann the sequence of bin centers as well as the sequence of partly smoothed roots, and interpolate back.

We will leave the first of these almost entirely to problems.

### Student's t

Let us try the second approach for another of Student's simulations, as already shown in exhibit 18. Exhibit **22** has the arithmetic. Note that since we smooth

$$\sqrt{\frac{\text{count}}{\text{bin width}}},$$

we must multiply the final rough by the square root of the bin width in order to find a rough in terms of root counts. Exhibit **23** shows the rough, which looks suitably irregular and of about the right spread.

exhibit **22** of chapter 17 (continued)

**B) SECOND SMOOTHING**--first smoothing in panel C of exhibit 24.

| Bin # | First rough | 3RSSH | Moved back* | Final smooth | Final rough | Square root width | Rough** (in roots) |
|---|---|---|---|---|---|---|---|
| 2 | 0 | −.3 | (−.3) | 3.5 | .3 | 1.00 | .3 |
| 3 | −.9 | −.3 | −.3 | 5.4 | −.6 | .71 | −.4 |
| 4 | −.3 | −.3 | −.3 | 8.1 | 0 | .71 | 0 |
| 5 | 0 | −.3 | −.3 | 11.7 | .3 | .55 | .2 |
| 6 | −.3 | 0 | (0) | 15.6 | −.5 | .55 | −.3 |
| 7 | 1.1 | .8 | (.8) | 19.7 | .3 | .55 | .2 |
| 8 | 2.3 | 1.2 | (1.2) | 21.4 | 1.1 | .55 | .6 |
| 9 | 1.3 | 1.1 | (1.1) | 20.0 | .2 | .55 | .1 |
| 10 | −.8 | .6 | (.6) | 16.4 | −1.4 | .55 | −.8 |
| 11 | .5 | .1 | .1 | 12.4 | .4 | .55 | .2 |
| 12 | −1.2 | 0 | 0 | 8.5 | 1.2 | .71 | −.8 |
| 13 | 0 | 0 | 0 | 5.7 | 0 | .71 | 0 |
| 14 | 0 | 0 | 0 | 3.2 | 0 | 1.00 | 0 |

\* No move as large as ±0.05 is called for.
\*\* Rough (in roots) = (final rough) TIMES (root of bin width)

Note that the final smooth and final rough above are in root of (counts/width), NOT yet in root of counts.

→

exhibit 22 of chapter 17 (continued)

C) LETTER VALUES for ROOT ROUGH

#13         (three zeros)

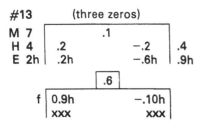

| M 7 | | .1 | | |
| H 4 | .2 | | −.2 | .4 |
| E 2h | .2h | | −.6h | .9h |

|   | .6 |   |
|---|----|---|
| f | 0.9h | −.10h |
|   | xxx  | xxx   |

P) PROBLEMS

22a) Repeat for Student's data based on finger measurements (righthand column of exhibit 18).

22b) H. Chavez, M. Contreras G., and T. P. E. Hernandez D. (1968, "On the coast of Tamaulipas." *International Turtle and Tortoise Society Journal*, vol. 2, no. 5, pp. 16 to 19 and 27 to 34.) gave the following counts for the duration (in days) of incubation for 1664 eggs of the ridley turtle.

| days:   | 50 | 51  | 52 | 53  | 54  | 55  | 56  | 57 | 58 | 59 | 60 | 61 | // | 64 | 65 |
|---------|----|-----|----|-----|-----|-----|-----|----|----|----|----|----|----|----|----|
| # eggs: | 77 | 122 | 10 | 321 | 725 | 180 | 162 | 21 | 14 | 6  | 9  | 14 |    | 1  | 1  |

Carry out a careful analysis using √days. Comment.

22c) Carry out a careful analysis using logs. Comment.

22d) Do the same using $\sqrt{-49 + \text{time in days}}$. Comment.

exhibit 23 of chapter 17: simulations

**The rough of the fit of exhibit 22?**

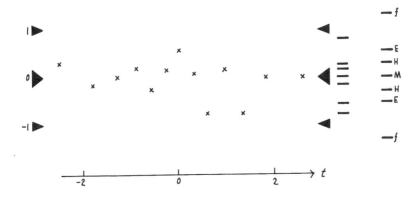

### fitting the smooth

Exhibit **24** carries on this analysis, doing the usual peak-finding procedure. The left side of panel D shows the plot of smoothed raw counts against shift-squared--very curved. The center of the same panel shows log smoothed roots against shift-squared--plenty curved. So we go on to reciprocals, as in the right side of the same panel. At last our picture is straight, so we proceed to calculate first a fitted reciprocal smooth, then a fitted root smooth, then residuals on a "root of count over width" basis, then residuals on a root-of-count basis. Exhibit **25** plots the latter residuals, which are reasonably irregular and of about the right width. (They tend to be slightly positive, and to tilt a little, suggesting that a modified fit might be a shade better.)

**exhibit 24** of chapter 17: simulation

**Fitting the smooth of exhibit 22 (peak found to be at 8.00)**

**A) The ARITHMETIC**

| Bin # | Final smooth | Match. bin # | Haif sum | Bin center | (shift)² | 100/ smooth root | Fit = 4.8 + 4.32 × (B − 8)² | Fit to smooth | Root of Resid. | Root of bin width | Resid. (in roots) |
|---|---|---|---|---|---|---|---|---|---|---|---|
|   | 3.5 | 13.8 | 7.9 | −2.55 | 6.50 | 28.6 | 32.9 | 3.0 | .8 | 1.0 | .8 |
| 3 | 5.4 | 13.1 | 8.1 | −1.8 | 3.24 | 18.5 | 18.8 | 5.3 | −.5 | .71 | −.4 |
|   | 8.1 | 12.1 | 8.0 | −1.3 | 1.69 | 12.3 | 12.1 | 8.3 | −.2 | .71 | −.1 |
| 5 | 11.7 | 11.1 | 8.0 | −.9 | .81 | 8.5 | 8.3 | 12.0 | 0 | .55 | 0 |
|   | 15.8 | 10.1 | 8.0 | −.6 | .36 | 6.3 | 6.4 | 15.6 | −.3 | .55 | −.2 |
| 7 | 19.7 | 9.1 | 8.0 | −.3 | .09 | 5.1 | 5.2 | 19.2 | .8 | .55 | .4 |
|   | 21.4 |   |   | 0 | 0 | 4.7 | 4.8 | 20.8 | 1.7 | .55 | .9 |
| 9 | 20.0 | 7.2 | 8.1 | .3 | .09 | 5.0 | 5.2 | 19.2 | 1.0 | .55 | .6 |
|   | 16.4 | 6.2 | 8.0 | .6 | .36 | 6.1 | 6.4 | 15.6 | −.6 | .55 | −.3 |
| 11 | 12.1 | 5.1 | 8.0 | .9 | .81 | 8.3 | 8.3 | 12.0 | .8 | .55 | .4 |
|   | 8.5 | 4.1 | 8.0 | 1.3 | 1.69 | 11.8 | 12.1 | 8.3 | −1.0 | .71 | −.7 |
| 13 | 5.7 | 3.1 | 8.0 | 1.8 | 3.24 | 17.5 | 18.8 | 5.3 | .4 | .71 | .3 |
|   | 3.2 |   |   | 2.55 | 6.50 | 31.2 | 32.9 | 3.0 | .2 | 1.00 | .2 |

**B) LETTER VALUES for the RESIDUALS**

#13    one zero

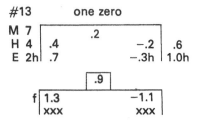

exhibit **24** of chapter 17 (continued)

**C) ROOTING and FIRST SMOOTHING**

| Bin # | Bin width | Bin count | Ratio | Root | 3RSSH | Moved back* | First rough |
|---|---|---|---|---|---|---|---|
|   | 1.0 | 14h | 14h | 3.8 | 3.8 | (3.8) | 0 |
| 3 | .5 | 11h | 23 | 4.8 | 5.4 | 5.7 | −.9 |
|   | .5 | 33 | 66 | 8.1 | 8.2 | 8.4 | −.3 |
| 5 | .3 | 43h | 145 | 12.0 | 11.8 | 12.0 | 0 |
|   | .3 | 70h | 235 | 15.3 | 15.6 | (15.6) | −.3 |
| 7 | .3 | 119h | 398 | 22.0 | 18.9 | (18.9) | 1.1 |
|   | .3 | 151h | 505 | 22.5 | 20.2 | (20.2) | 2.3 |
| 9 | .3 | 122 | 407 | 20.2 | 18.9 | (18.9) | 1.3 |
|   | .3 | 67h | 225 | 15.8 | 15.8 | (15.8) | −.8 |
| 11 | .3 | 49 | 163 | 12.8 | 12.0 | 12.3 | .5 |
|   | .5 | 26h | 53 | 7.3 | 8.2 | 8.5 | −1.2 |
| 13 | .5 | 16 | 32 | 5.7 | 5.4 | 5.7 | 0 |
|   | 1.0 | 10 | 10 | 3.2 | 3.2 | (3.2) | 0 |

* Unchanged values in ( ).

5.6 = 5.3 + (1/9)(8.2 − 5.3),

8.4 = 8.2 + (1/15)(11.8 − 8.2), etc

**D) three plots**

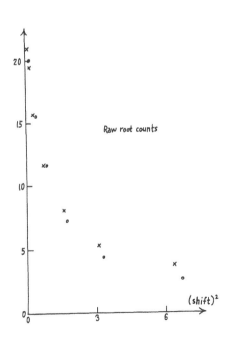

Raw root counts

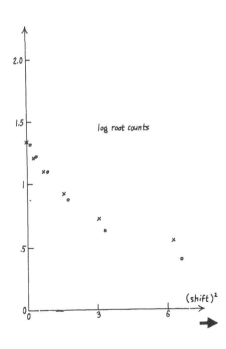

log root counts

exhibit **24** of chapter 17 (continued)

(panel D continued)

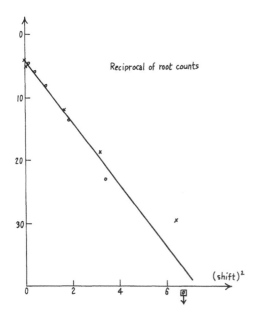

**P) PROBLEMS**

24a) Try .03 as the peak location, and carry through to a plot of residuals. Compare with exhibit 44.

24b) Do the same for −.03 as the peak location.

24c) Fit a straight line to log smoothed roots instead, and plot residuals. Comment.

24d) Do panel A for Student's data on fingers (from panel A of exhibit 18).

exhibit **25** of chapter 17: simulations

The root residuals (plotted against bin number)--(numbers in exhibit 24)

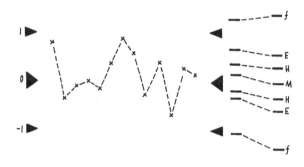

The form of the fit that we have, which is much better than we could have found using either raw roots or log roots, is

$$\frac{1}{\sqrt{\text{count/width}}} = 4.8 + 4.32\,(\text{shift})^2,$$

a form different from most of those we have so far seen.

### review questions

What do we work with when the widths of our bins differ? Why? What concerns us about our smoothing? What are our choices? What could we do? What did we do? To what example did we turn? What changes in our calculations are needed? Can we use the final rough as it stands? Why/why not? How did our rough look? What about fitting? Can we use the residuals as they stand? What form did we fit? Had we met it before?

### 17F. Double roots

Many sets of counts of basic counts deserve expression by some sort of square root. Since bin widths are important, we need to use such an expression not only for the integer counts that we observe, but also for the half-integer counts that mark the bin ends. Convenience and ease of understanding is particularly worthwhile when provided for the half-integer bin ends.

It is fortunate, accordingly, that the most frequently useful expression is the so-called double-root

$$\sqrt{2 + 4\,(\text{count})},$$

for which the endpoints will reduce to

$$\sqrt{2 + 4\,(\text{integer} + 0.5)} = \sqrt{4 + 4\,(\text{integer})} = 2\sqrt{\text{integer} + 1}.$$

Thus, the bin for basic count = 0, a basic count which would be expressed as $\sqrt{2}$, is thought of as stretching from $\sqrt{0} = 0$ to $\sqrt{4} = 2$. (If we had not started our roots, the left end of the bin for basic count = 0 would have been at the square root of a negative number--something hard to plot.)

### table

Exhibit **26** gives all the most useful numbers for double roots up to a basic count of 25. For each basic count, there are given, in order:

◇the value of the double root.
◇the value of the hanned double root.
◇the fraction of move-back required.
◇the bin width (one endpoint minus the other).
◇the root of the bin width (convenient if root count is given).
◇the start of each bin (its end is given as the start of the next).
◇the center of each bin.

The third item is quite helpful. Double-roots force us to unequal bins and thus to move back. Having the move-back fractions already calculated saves us much arithmetic.

When we come to peaking and fitting our double roots, it is optional-- and seems usually desirable--to treat the bins as if they were placed at the hanned double roots instead of at bin centers. (This shifts the "zero" bin by .41, the "one" bin by .04, the "two" bin by .02, the next four by .01, and the rest by .00. Except for the "zero" bin, all changes are either quite small or very small, and the "zero" bin change often helps.

### polonium scintillations

Exhibit **27** gives the arithmetic for counts by Rutherford and Geiger of the number of scintillations (flashes of light) in intervals of time of fixed length (1/8 minute). Each scintillation indicated the arrival of one $\alpha$-particle from the radioactive decay of a single atom of polonium in their sample. This body of data, and its interpretation by Rutherford, Geiger, and (in an appendix) Bateman, are classics in the study of radioactive decay and the establishment of its random character. The upper panel of exhibit **28** shows the rough, which (i) is moderately irregular, (ii) is not excessively wide, (iii) shows the effects of the failure of our smooth to follow as closely as we might wish the rather narrow peak (in root ratios) presented by the data.

exhibit 26 of chapter 17: reference table

**Useful numbers for double roots up to a basic count of 25**

| Basic count | Double root | | Move back | Width | | Bin start | Bin center |
|---|---|---|---|---|---|---|---|
| | raw | hanned | | raw | root | | |
| 0 | 1.41 | 1.41 | hold | 2.000 | 1.414 | 0.00 | 1.00 |
| 1 | 2.45 | 2.37 | up 1/9.4 | .828 | .910 | 2.00 | 2.41 |
| 2 | 3.16 | 3.13 | up 1/17.8 | .636 | .798 | 2.83 | 3.15 |
| 3 | 3.74 | 3.72 | up 1/26 | .536 | .732 | 3.46 | 3.73 |
| 4 | 4.24 | 4.23 | up 1/34 | .472 | .687 | 4.00 | 4.24 |
| 5 | 4.69 | 4.68 | up 1/42 | .427 | .653 | 4.47 | 4.69 |
| 6 | 5.10 | 5.09 | up 1/50 | .393 | .627 | 4.90 | 5.10 |
| 7 | 5.48 | 5.47 | up 1/58 | .365 | .604 | 5.29 | 5.47 |
| 8 | 5.83 | 5.83 | up 1/66 | .343 | .586 | 5.66 | 5.83 |
| 9 | 6.16 | 6.16 | up 1/74 | .325 | .570 | 6.00 | 6.16 |
| 10 | 6.48 | 6.48 | up 1/82 | .309 | .556 | 6.32 | 6.48 |
| 11 | 6.78 | 6.78 | up 1/90 | .295 | .543 | 6.63 | 6.78 |
| 12 | 7.07 | 7.07 | up 1/98 | .283 | .532 | 6.93 | 7.07 |
| 13 | 7.35 | 7.35 | up 1/106 | .272 | .522 | 7.21 | 7.35 |
| 14 | 7.62 | 7.62 | up 1/114 | .263 | .513 | 7.48 | 7.61 |
| 15 | 7.87 | 7.87 | up 1/122 | .254 | .504 | 7.75 | 7.87 |
| 16 | 8.12 | 8.12 | up 1/130 | .246 | .496 | 8.00 | 8.12 |
| 17 | 8.37 | 8.36 | up 1/138 | .239 | .489 | 8.25 | 8.37 |
| 18 | 8.60 | 8.60 | up 1/146 | .233 | .482 | 8.49 | 8.60 |
| 19 | 8.83 | 8.83 | up 1/154 | .226 | .475 | 8.72 | 8.83 |
| 20 | 9.06 | 9.06 | up 1/162 | .221 | .470 | 8.94 | 9.05 |
| 21 | 9.27 | 9.27 | up 1/170 | .216 | .464 | 9.16 | 9.27 |
| 22 | 9.49 | 9.49 | up 1/178 | .211 | .459 | 9.38 | 9.49 |
| 23 | 9.70 | 9.70 | up 1/186 | .206 | .454 | 9.59 | 9.69 |
| 24 | 9.90 | 9.90 | up 1/194 | .202 | .449 | 9.80 | 9.90 |
| 25 | 10.10 | 10.10 | up 1/202 | .198 | .445 | 10.00 | 10.10 |

*Note.* For basic counts of 3 or more, "move back" = $1/(2 + 8 \text{ (basic count)})$.

**P) PROBLEMS**

26a) show the calculations for basic count = 1, keeping enough decimals to check the values given.

26b) Do the same for basic count = 4.

26c) And for basic count = 17.

exhibit **27** of chapter 17: polonium scintillations

**Rutherford and Geiger's data on the radioactive decay of polonium (events are scintillations caused by alpha-particles)**

A) COUNTS, ROOTS, ROOT RATIOS, and FIRST SMOOTH

| Events per 1/8 minute | Counts for bin | Root count | Root ratio* | 3RSSH | Move back | First rough |
|---|---|---|---|---|---|---|
| 0 | 57 | 7.6 | 5.3 | 5.3 | | 0 |
|   | 203 | 14.2 | 15.6 | 15.3 | 16.2 | .6 |
| 2 | 383 | 19.6 | 24.6 | 24.0 | 24.3 | .3 |
|   | 525 | 22.9 | 31.3 | 29.6 | 29.7 | 1.6 |
| 4 | 532 | 23.1 | 33.6 | 31.2 | 31.2 | 2.4 |
|   | 408 | 20.2 | 30.9 | 29.8 | 29.7 | 1.2 |
| 6 | 273 | 16.5 | 26.4 | 25.8 | 25.7 | .7 |
|   | 139 | 11.8 | 19.5 | 19.2 | 19.1 | .4 |
| 8 | 45 | 6.7 | 11.5 | 12.9 | 12.8 | −1.3 |
|   | 27 | 5.2 | 9.1 | 8.8 | | .3 |
| 10 | 10 | 3.2 | 5.7 | 6.0 | | −.3 |
|   | 4 | 2.0 | 3.7 | 3.8 | | −.1 |
| 12 | 0 | 0 | 0 | 2.4 | | 2.4 |
|   | 1 | 1.0 | 1.9 | 1.9 | | 0 |
| 14 | 1 | 1.0 | 2.0 | 1.4 | | .6 |
|   | 0 | 0 | 0 | .5 | | −.5 |
| 16 | 0 | 0 | 0 | 0 | | |
| (total) | (2607) | | | | | |

* Root ratio $= \sqrt{\dfrac{\text{count}}{\text{width}}} = \dfrac{\text{Root count}}{\sqrt{\text{width}}}$ where "$\sqrt{\text{width}}$" comes from exhibit 26.

B) SECOND SMOOTH—starting with 3RSSH of first rough

| Bin # | 3RSSH | Moved back | Added back | Rough | Rough (in roots) | Adjusted smooth* |
|---|---|---|---|---|---|---|
| 0 | 0 | T | 5.3 | 0 | 0 | 5.3 |
| 1 | .1 | h | 16.3 | −.7 | −.7 | 16.3 |
| 2 | .4 | e | 24.7 | −.1 | −.1 | 24.7 |
| 3 | .8 |   | 30.5 | .8 | .6 | 31.1 |
| 4 | 1.1 | S | 32.3 | 1.3 | .9 | 33.5 |
| 5 | 1.1 | a | 30.8 | .1 | .1 | 31.4 |
| 6 | .6 | m | 26.5 | −.1 | −.1 | 26.5 |
| 7 | .4 | e | 19.5 | 0 | 0 | 19.5 |
| 8 | .2 |   | 13.0 | −1.5 | −.9 | 3.0 |
| 9 | 0 |   | 8.8 | .3 | .2 | 8.8 |
| 10 | −.1 |   | 5.9 | −.2 | −.1 | 5.9 |
| 11 | −.1 |   | 3.7 | 0 | 0 | 3.7 |
| 12 | 0 |   | 2.4 | −2.4 | 1.3 | 2.4 |
| 13 | 0 |   | 1.9 | 0 | 0 | 1.9 |
| 14 | 0 |   | 1.4 | .6 | .3 | 1.4 |
| 15 | 0 |   | 0.5 | −.5 | −.3 | .5 |
| 16 | 0 |   | 0 | 0 | 0 | 0.0 |

* Adjusted by judgment (a) at the peak, (b) in the far right tail—principally here as a basis for fitting.

→

Before going on to fitting, done in panel C of exhibit 27, we have adjusted the smooth in two places, as noted at the foot of panel B.

The lower panel of exhibit **28** shows the residuals, which are neither good nor horrible. In order to fit the smoothed (and adjusted) root ratios moderately far away from the peak, we had to move the peak of the fit about one bin number lower than the peak of the data, thus creating the large hill in the plot of residuals. Clearly, the residuals are wider than we would like, and two points are outside the reference fences (though not the self fences).

exhibit **27** of chapter 17 (continued)

C) FIT--using $1.52 - .12(\text{hanned} - 4.0)^2$ for log (root) ratio

| Bin # | Hanned dbl. rt. | Shift | Shift² | log* | ↓ | Fitted log | Fitted ratio | Residual ratio | Residual root |
|---|---|---|---|---|---|---|---|---|---|
| 0  | 1.41 | −2.59 | 6.7  | .72  | P | .72  | 5.2  | .1   | .1   |
| 1  | 2.37 | −1.63 | 2.7  | 1.21 | L | 1.20 | 15.0 | −.2  | −.2  |
| 2  | 3.13 | −.87  | .8   | 1.39 | O | 1.42 | 26.3 | −1.7 | −1.4 |
| 3  | 3.72 | −.28  | .1   | 1.49 |   | 1.51 | 32.4 | −1.1 | −.8  |
| 4  | 4.23 | .23   | .1   | 1.53 | T | 1.51 | 32.4 | 1.2  | .8   |
| 5  | 4.68 | .68   | .5   | 1.50 |   | 1.46 | 28.8 | 2.1  | 1.4  |
| 6  | 5.09 | 1.09  | 1.2  | 1.42 | U | 1.38 | 24.0 | 2.4  | 1.5  |
| 7  | 5.47 | 1.47  | 2.2  | 1.29 |   | 1.26 | 18.2 | 1.3  | .8   |
| 8  | 5.83 | 1.83  | 3.3  | 1.11 | S | 1.12 | 13.2 | −1.7 | −1.0 |
| 9  | 6.16 | 2.16  | 4.7  | .94  | E | .96  | 9.1  | 0    | 0    |
| 10 | 6.48 | 2.48  | 6.2  | .77  | D | .78  | 6.0  | −.3  | −.2  |
| 11 | 6.78 | 2.78  | 7.7  | .51  |   | .60  | 4.0  | −.3  | −.2  |
| 12 | 7.07 | 3.07  | 9.4  | .38  | H | .39  | 2.4  | −2.4 | −1.3 |
| 13 | 7.35 | 3.35  | 11.2 | .28  | E | .18  | 1.5  | .4   | −.2  |
| 14 | 7.62 | 3.62  | 13.1 | .15  |   | −.05 | .9   | 1.1  | .6   |
| 15 | 7.87 | 3.87  | 15.0 | .30  | R | −.28 | .5   | −.5  | −.2  |
| 16 | 8.12 | 4.12  | 17.0 | L    | E | −.52 | .3   | −.3  | −.1  |

*log of adjusted fit from panel B

P) PROBLEMS

27a) Make the plot used in panel C and check--or improve on--the choice of fit.

27b) Find a similar set of data and repeat all these calculations.

27c) Replot exhibit 28 against hanned location, rather than against bin number.

S) SOURCE

E. Rutherford and H. Geiger, 1910. "The probability variations in the distribution of α-particles." (*The London, Edinburgh and Dublin*) *Philosophical Magazine and Journal of Science* 20: 698–704. Table on page 701.)

The double-root

$$\text{location} = \sqrt{2 + 4 \cdot \text{count}},$$
$$\text{width} = \sqrt{4 + 4 \cdot \text{count}} - \sqrt{4 \cdot \text{count}},$$

has allowed us to do (a) a reasonably good job of smoothing the Rutherford and Geiger data and (b) a job of fitting it that may be bearably good. (There seem to be further systematic things going on in the residuals, but it is hard to put words to them.)

exhibit **28** of chapter 17: polonium scintillations

**The rough and the residuals from exhibit 27**

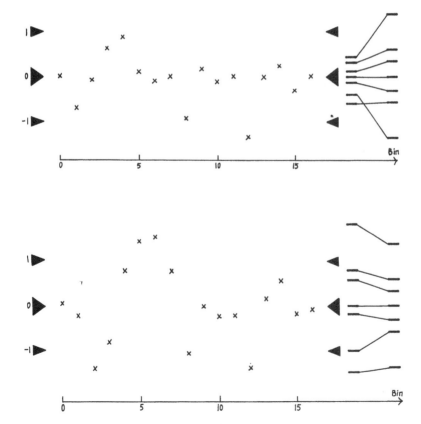

### review questions

What is a double-root? Why might we want one? What are the bin boundaries for basic counts 1, 2, and 3? Of what quantities did we give a table? How are they used? Where do we usually treat double-root values as located? Is this unusual? To what example did we then turn? What were the results? What did we do before fitting? How well did we do with our fit? What do we think of double-roots?

### 17G. Cautionary examples

It is human, it would seem, to jump from "thus-and-so is a satisfactory fit" to "thus-and-so is the (only) proper fit". The best way to inhibit such natural--but dangerous--leaping at conclusions seems to be to look at some examples where it cannot be right to jump. The simplest such are examples where TWO DIFFERENT FITS are reasonably SATISFACTORY. (It cannot be right to jump to the conclusion that BOTH are the ONLY proper fit!) Let us do a pair of examples, choosing cases where we were marginally satisfied with previous fits. At the same time, we can show how we can polish up our fits by finding and using "desired peaks".

**exhibit 29 of chapter 17: forearm lengths**

Calculations for root-reciprocals of actual root counts

A) FAST FIT taking PEAK at 7 -- shift$^2$ = $(B - 7)^2$

| B = Bin # | Recip-root of root count* | shift$^2$ | | .255 + .0127 × shift$^2$ | Fitted root count | Root resid |
|---|---|---|---|---|---|---|
| 1 | 1.000 | 36 | M | .712 | 2.0 | −1.0 |
|   | .632  | 25 | A | .572 | 3.1 | −.6  |
| 3 | .494  | 16 |   | .458— | 4.8 | −.7 |
|   | .378  | 9  | K | .369 | 7.3  | −.3 |
| 5 | .299  | 4  | E | .306 | 10.7 | .5  |
|   | .266  | 1  |   | .268 | 13.9 | .2  |
| 7 | .256  | 0  |   | .255 | 15.4 | −.1 |
|   | .272  | 1  | A | .268 | 13.9 | −.4 |
| 9 | .297  | 4  |   | .306 | 10.7 | .6  |
|   | .363  | 9  |   | .369 | 7.3  | .3  |
| 11| .423  | 16 | P | .458 | 4.8  | .8  |
|   | .598  | 25 | L | .572 | 3.1  | −.3 |
| 13| .725  | 36 |   | .712 | 2.0  | −.1 |
|   | .845  | 49 | O | .877 | 1.3  | +.1 |
| 15| .791  | 64 | T | 1.068| 0.9  | +.7 |

* This is 1/√root count later called just "recip-root".

Note that .255 + .0127(shift)$^2$ comes from plot called for after the "shift$^2$ column".

→

### forearm lengths again

We fitted the counts of fathers' forearm lengths moderately well with

$$\log \sqrt{\text{count}} = 1.17 - .0282(B - 8.25)^2.$$

Let us try a gently differing expression, replacing "log" by "1/root". Exhibit **29** has the computations, starting by assuming "peak" = 7 as a reasonable eye-catching and eye-chosen start. The plotting for a fit is left to the reader. The residuals from

$$\text{recip-root of } \sqrt{\text{count}} = .255 + .0127(B - 7)^2$$

then seem--see the end of panel A of exhibit 29--to be quite satisfactory. Indeed, they compare favorably with those of exhibit 15, where we fitted to log root instead of recip-root. But we might do still better!

exhibit **29** of chapter 17 (continued)

B) DESIRED SHIFT and PEAK, then REFIT with PEAK = 7.15--shift² = $(B = 7.15)^2$.

| Desired | | ↓ | new shift² | .255 + .0127 shift² | Fitted root count | Root resid |
|---|---|---|---|---|---|---|
| shift* | peak | ↓ | | | | |
| 7.66 | 8.66 | M | 37.8 | .735 | 1.8 | −.8 |
| 5.45 | 7.45 | A | 26.5 | .592 | 2.9 | −.4 |
| 4.34 | 7.34 |   | 17.2 | .473 | 4.5 | −.4 |
| 3.11 | 7.11 | K | 9.9 | .381 | 6.9 | .1 |
| 1.86 | 6.86 | E | 4.6 | .313 | 10.2 | 1.0 |
| .93 | 6.93 |   | 1.3 | .272 | 13.5 | .6 |
| .28 | (6.72) |   | .0 | .255 | 15.4 | −.1 |
| 1.16 | 6.84 | A | .7 | .264 | 14.3 | −1.8 |
| 1.82 | 7.18 |   | 3.4 | .298 | 11.3 | 0 |
| 2.92 | 7.08 |   | 8.1 | .358 | 7.8 | −.2 |
| 3.64 | 7.36 | P | 14.8 | .443 | 5.1 | .5 |
| 5.20 | 6.80 | L | 23.5 | .554 | 3.3 | −.5 |
| 6.08 | 6.92 | O | 34.2 | .689 | 2.1 | −.2 |
| 6.82 | 7.18 |   | 46.9 | .851 | 1.4 | 0 |
| 6.50 | 8.50 | T | 61.6 | 1.037 | .9 | .7 |

*recip-root = .255 + .0127 $(B - \text{desired shift})^2$, so that desired shift = $\sqrt{\text{recip-root} - .255/.0127}$.

Note that 7.15 comes from plot called for after the "desired peak" column.

P) PROBLEMS

29a) Calculate and plot the desired peaks for the fit of exhibit 23 to this same data.

29a2) How do you suggest continuing? Try what you suggest.

### peak polishing (rest of section is optional)

To see if we can do better, we try a convenient polishing technique--asking, for each bin, which value of "peak" in

$$.255 + .0127(B - peak)^2$$

will reproduce the observed count. As the equations at the foot of panel B (of exhibit 29) make clear, it is relatively easy to calculate the "desired shift" and the corresponding "desired peak". This is done in panel B, plotting of the results being left to the reader. When we plot--or when we look hard at panel B--we see that all but the first and last points are in relatively good agreement with one another.

The median of the twelve well-behaved values of desired peak is 7.15, so the remainder of panel B calculates the fit for "peak = 7.15", with results slightly better than for "peak = 7".

**The real merit of this desired-peak analysis is not the change from 7.00 to 7.15, but rather the fact that a constant peak does very well, even when we try to look very hard, leaving the peak free to vary.**

The upper panel of exhibit **30** shows the final residuals, which are quite well behaved--one low point being an exception. These are to be compared with panel B of exhibit 15, where we were left wondering about a better fit. Perhaps we now have one.

### polonium scintillations again

We can now apply the desired peak technique to the polonium-scintillations example, where (exhibit 27) we earlier found

$$\log \sqrt{\frac{count}{width}} = 1.52 - .12(L - peak)^2$$

to be a reasonable fit.

Let us see what a careful look can dredge up. Starting from the original log root ratios, and these two fitted constants, we find the arithmetic at the left of exhibit **31**. Exhibit **32** shows the desired peaks--and a fit by a broken straight line. The rest of exhibit 31 carries this fit through to root residuals, which are plotted in the lower part of exhibit 30. The two-straight-line fit, which corresponds to using one fit on the lower and a different one on the upper end, is clearly quite good. When we compare with the lower panel of exhibit 28, where we called the residuals "neither good or horrible," we see that we have again improved the fit considerably.

### comment

Clearly, calculating desired peaks offers us a useful way to look more closely at our fits, sometimes going along with exactly the sort of simple fit we

exhibit 30 of chapter 17: forearm lengths and polonium scintillations

**Residuals from two alternative fits (see exhibits 29 and 31 for numbers)**

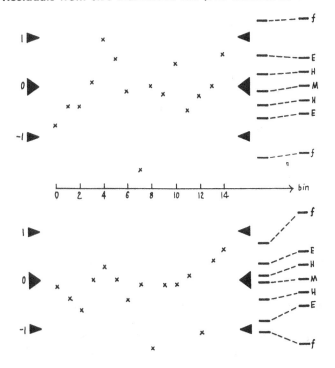

exhibit 31 of chapter 17: polonium scintillations

**The desired peak calculation for another example**

A) The ARITHMETIC -- using $1.52 - .125$ (hanned $- 4.0)^2$

| B=<br>Bin # | L=<br>Bin loc | log root ratio | Desired shift | peak | | Fitted peak | | $1.52 -$<br>$.125x$<br>$(L -$<br>$fit)^2$ | Fitted root ratio | Root ratio resid | Root resid |
|---|---|---|---|---|---|---|---|---|---|---|---|
|   |   |   |   |   |   | 3.77<br>+<br>.11L | 4.63<br>−<br>.092L |   |   |   |   |
| 0 | 1.41 | .727 | 2.52 | 3.93 | R | 3.92 |   | .732 | 5.4 | −.1 | −.1 |
|   | 2.45 | 1.195 | 1.61 | 4.06 | N | 4.04 |   | 1.204 | 16.0 | −.4 | −.4 |
|   | 3.16 | 1.390 | 1.02 | 4.18 |   | 4.12 |   | 1.405 | 25.4 | −.8 | −.6 |
|   | 3.74 | 1.496 | .44 | 4.18 | T | 4.18 |   | 1.496 | 31.3 | 0 | 0 |
| 4 | 4.24 | 1.526 | imag | 4.24 | O | 4.24 | 4.24 | 1.520 | 33.1 | .5 | .3 |
|   | 4.69 | 1.490 | .49 | 4.20 | E | 4.29 | 4.20 | 1.490 | 30.9 | 0 | 0 |
| 6 | 5.10 | 1.421 | .89 | 4.21 | X |   | 4.16 | 1.410 | 25.7 | −.7 | −.4 |
|   | 5.48 | 1.290 | 1.36 | 4.12 | H |   | 4.13 | 1.292 | 19.6 | −.1 | −.1 |
| 8 | 5.83 | 1.059 | 1.92 | 3.91 | I |   | 4.09 | 1.142 | 13.9 | −2.4 | −1.4 |
|   | 6.16 | .960 | 2.12 | 4.04 | B |   | 4.06 | .969 | 9.3 | −.2 | −.1 |
| 10 | 6.48 | .755 | 2.47 | 4.01 | I |   | 4.03 | .770 | 5.9 | −.2 | −.1 |
|   | 6.78 | .566 | 2.76 | 4.02 | T |   | 4.01 | .561 | 3.6 | .1 | .1 |
| 12 | 7.07 | -- |   |   |   |   | 3.98 | .326 | 2.1 | −2.1 | −1.1 |
|   | 7.35 | -- |   |   | 3 |   | 3.95 | .075 | 1.2 | .7 | .4 |
| 14 | 7.62 | -- |   |   | 2 |   | 3.93 | −.182 | .7 | 1.3 | −.7 |

started with, sometimes calling for a more complex fit. We can be happy to have such a tool. The main issue of this section, however, is broader than a specific technique.

We have used desired peaks to show that finding a reasonable fit does not mean that we are unlikely to do better with some other fit. The changes we have considered are quite appreciable: A log expression of root count was replaced by a recip-root expression; a single log root ratio fit was replaced by a pair of log root ratio fits involving quite different constants. The moral is clear: **If such large changes need not greatly distort the fit, smaller changes need not distort the fit appreciably. Even when we see a very good fit--something we know has to be a very useful summary of the data--we dare not assume that what is fitted is of exactly the right form; we dare not believe we have found a natural law.**

### review questions

What two purposes does this section have? How did we try to meet them? What example did we start with? What did we fit? How well did it do? Compared to what? What is a "desired shift"? A "desired peak"? Can we use them to improve a fit? How? What happened in the example? Where did we turn next? What did we do? What happened? Can we use part of two fits? Why/why not? How good was our result? By itself? Compared to an earlier fit? What is the moral toward which this section points? How does it point?

exhibit **32** of chapter 17: polonium scintillations

**Analysis of the desired peaks from exhibit 31 and display of the resulting residuals**

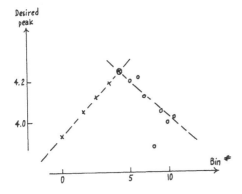

### 17H. How far have we come?

This chapter has focused on the analysis of sequences of counts in successive bins and a technique of fitting designed to work for peaked sequences.

The fitting technique first notes that symmetrically peaked curves have to fall away from their peak by an amount determined by the SQUARE of the shift away from the location of the peak, and then seeks a re-expression of the curve's height so that this dependence on the squared shift is linear.

As the many examples have shown, these two techniques together work surprisingly well, especially as we have seen by using a display of the roughs that is particularly fitted for examining roughs arising from counts.

We are now ready:

⋄ to start in on a sequence of bin counts by taking (square) roots and smoothing the results carefully.

⋄ to plot the rough and comparing its fEHMHEf with some standard values.

⋄ to locate the peak by splitting the difference between points of equal height.

⋄ to plot (varied expressions of) the smooth against the squared shift (the square of location MINUS peak) and find (hopefully) an expression for the smooth for which a straight line is an adequate fit.

⋄ to deal with bins of unequal width by replacing $\sqrt{count}$ by $\sqrt{count/width}$, whether the inequality of widths was inevitable or the result of our own choice.

⋄ to deal with basic counts in terms of $\sqrt{2 + 4 \text{ (basic count)}}$.

⋄ to use an analysis in terms of "desired peaks" to improve a first fit.

We now understand that:

⋄ bin boundaries can be described in terms of amounts, balances, or basic counts.

⋄ bin widths are sometimes fixed (e.g., numbers of children) and sometimes at our choice (e.g., octaves for larger counts).

⋄ that "thus and so is a satisfactory fit" DOES NOT mean that "thus and so" is "the proper fit", or "the best fit", or "a natural law".

# 588 index for chapter 18

review questions 589

**18A. Sizes and counts 589**
"Zipf's law" 590
c'rank 590
a possible approach 594
review questions 594

**18B. Product-ratio analysis 594**
half-octave calculation 595
display 595
review questions 595

**18C. Forcing the unusual to be noticed 598**
review questions 602

**18D. Comparisons between collections 602**
review questions 603

**18E. Looking at the smallest basic count 604**
review questions 604

**18F. When zeros are counted 605**
review questions 608

**18G. Under the microscope 608**
review questions 610

**18H. How far have we come? 612**

| EXHIBIT | PAGE |
|---|---|
| **18A** | |
| 1★ | 591 |
| 2 | 593 |
| **18B** | |
| 3★ | 596 |
| 4 | 597 |
| 5★ | 597 |
| **18C** | |
| 6 | 599 |
| 7 | 600 |
| 8 | 601 |
| 9★ | 601 |
| **18D** | |
| 10 | 603 |
| 11★ | 603 |
| **18E** | |
| 12 | 604 |
| **18F** | |
| 13 | 605 |
| 14 | 606 |
| 15 | 607 |
| **18G** | |
| 16 | 608 |
| 17 | 609 |
| 18 | 609 |
| 19★ | 610 |
| **18H** | |

# Product-ratio plots -- going beyond bins

**18**

One problem that arises in a variety of fields is dealing with very stretched-tailed patterns of basic counts--or amounts. The sizes of different cities and towns in a region--the occurrences of different words in a play, novel, or text--even the numbers of species in each genus of some kind of plant or animal--these are but three examples. Two goals are appropriate:

◇ we want to be able to picture such distributions so that we can look at and compare them.

◇ we would like to find some way to describe most of their behavior in some very simple way.

(Of course, once we have done this, we will want to look at the "rest" of the behavior, also.) By now, these kinds of goals should be quite familiar.

Techniques we have already learned do moderately well in producing usable pictures. Octave bins for basic counts tame many long-tailed distributions, but they miss detail that is often meaningful. This is not surprising. We should be able to learn rather more from a count of 1647 when we preserve it in detail than when we say only that it is between 1024 and 2047.

In this chapter, then, we look at procedures that focus on individual counts, rather than bins--at least for the largest counts--striving first for pictures we can compare and then for simple descriptions and the consequent residuals.

These techniques can also be used for other very stretch-tailed distributions, like those of the areas of lakes and ponds (in a country or a continent), that do not involve counts. We shall leave such examples to the reader.

### review questions

What problem arises? What goals are reasonable? Are they familiar? How well will what we have already learned do? Will we use bins in this chapter?

### 18A. Sizes and counts

We mentioned sizes of lakes and ponds--clearly not basic counts--as a possible example. Even if such sizes are not basic counts, they often share, because they are given to limited precision, a difficulty that is very common with basic counts: tied basic counts or tied sizes. The classical approach to this kind of data, put forward by George Kingsley Zipf, spoke of "rank" and "size" and proceeded to their product, which

### "Zipf's law"

said was constant. The question of "which rank" was never clearly addressed, perhaps because of emphasis on large sizes of basic counts. When many basic counts of "1" occur, as is so often the case, their "rank" requires careful definition. (In the Pushkin example to which we will shortly turn, words of basic count 1 occupy ranks 2400 to 4783. What rank should we use for "size 1"?)

We will use the "completed rank", which we will often abbreviate as

### c'rank,

the largest rank assigned to that basic count (or size), the number of individuals attaining at least that basic count, or size.

Our next exhibits will look at three examples where the basic counts are:

◇ the number of those papers, abstracted in *Science Abstracts*, A, for 1961 and classified as on Electrical Properties of Solids, that appeared in a given journal.

◇ the number of those papers, abstracted in *Science Abstracts*, A, for 1961 and classified as on Atomic and Molecular Physics, that appeared in a given journal.

◇ the number of appearances of a given Russian word (grammatically defined) in Pushkin's *A Captain's Daughter*.

The long-tailedness of all three is shown by the following information about extreme occurrences:

|  | "Most" | "1's" | "Total" | "#" |
|---|---|---|---|---|
| Elec. Prop. of Solids: | 255 | 46 | 1342 | 118 |
| Atomic and Molec. Phys: | 372 | 39 | 1339 | 107 |
| *The Captain's Daughter:* | 1160 | 2384 | 29345 | 4783 |

Thus, for *The Captain's Daughter*, one word occurred 1160 times, while 2384 words appeared only once and there were a total of 29,345 occurrences of 4,783 different words.

Exhibit **1** has the numbers; exhibit **2** shows the three root-residual plots for the fits indicated in panel B of exhibit 1, which give root-count-per-octave --or its logarithm--as linear in bin number. We see nothing distinctive, nothing alerting. Yet at least one of these three examples has a peculiarity well worth noticing.

The five highest basic counts in each example are:

| | | | | | |
|---|---|---|---|---|---|
| Elec. Prop. of Solids: | 255 | 130 | 128 | 75 | 69 |
| Atomic and Molec. Phys: | 372 | 79 | 61 | 53 | 50 |
| *The Captain's Daughter:* | 1160 | 777 | 724 | 582 | 479 |

exhibit 1 of chapter 18: three sets of occurrences

**Three long-tailed counts by octave bins**

A) DATA by OCTAVE

| Bin # B | Basic counts | Count | Root smooth | Root fit | Root resid |
|---|---|---|---|---|---|
|   | 1 | 46 | 6.8 | 5.6 | 1.2 |
| 1 | 2 to 3 | 23 | 4.8 | 4.9 | −.1 |
|   | 4 to 7 | 15 | 3.9 | 4.3 | −.4 |
| 3 | 8 to 15 | 16 | 3.6 | 3.6 | .4 |
|   | 16 to 31 | 11 | 3.0 | 3.0 | .3 |
| 5 | 32 to 63 | 2 | 1.9 | 2.3 | −.9 |
|   | 64 to 127 | 2 | 1.4 | 1.7 | −.3 |
| 7 | 128 to 255 | 3 | 1.0 | 1.0 | .7 |
|   | 256 to 511 | 0 | .4 | .4 | −.4 |
| 9 | 512 to 1023 |   |   |   |   |
|   | 1024 to 2047 |   |   |   |   |

| Bin # B | Basic counts | Count | Root smooth | Root fit | Root resid |
|---|---|---|---|---|---|
|   | 1 | 39 | 6.2 | 5.6 | .6 |
| 1 | 2 to 3 | 20 | 4.5 | 4.9 | −.4 |
|   | 4 to 7 | 13 | 3.6 | 4.2 | −.6 |
| 3 | 8 to 15 | 13 | 3.6 | 3.5 | .1 |
|   | 16 to 31 | 13 | 3.4 | 2.8 | .8 |
| 5 | 32 to 63 | 7 | 2.4 | 2.1 | .5 |
|   | 64 to 127 | 1 | 1.4 | 1.4 | −.4 |
| 7 | 128 to 255 | 0 | .5 | .7 | −.7 |
|   | 256 to 511 | 1 | 0 | 0 | 1.0 |
| 9 | 512 to 1023 |   |   |   |   |
|   | 1024 to 2047 |   |   |   |   |

| Bin # B | Basic counts | Count | Root smooth | Root fit | Root resid |
|---|---|---|---|---|---|
|   | 1 | 2384 | 48.8 | 50.1 | −1.3 |
| 1 | 2 to 3 | 1280 | 35.8 | 34.7 | 1.1 |
|   | 4 to 7 | 580 | 24.1 | 24.0 | .1 |
| 3 | 8 to 15 | 280 | 16.7 | 16.6 | .1 |
|   | 16 to 31 | 133 | 11.5 | 11.5 | 0 |
| 5 | 32 to 63 | 70 | 8.4 | 7.9 | .5 |
|   | 64 to 127 | 31 | 5.6 | 5.5 | .1 |
| 7 | 128 to 255 | 11 | 3.3 | 3.8 | −.5 |
|   | 256 to 511 | 10 | 3.2 | 2.6 | .6 |
| 9 | 512 to 1023 | 3 | 1.7 | 1.8 | −.1 |
|   | 1024 to 2047 | 1 | 1.3 | 1.3 | −.3 |
|   |   | (0) | (0) | .9 | (−.9) |

→

exhibit 1 of chapter 18 (continued)

**B) The FITS**

Bins are (started) octaves; #0 contains basic count 1, #1 contains basic counts 2 and 3.

Electrical Prop. of Solids: $\sqrt{\text{(counts/bin)}} = 5.6 - .65B$

Atomic and Molecular Physics: $\sqrt{\text{(counts/bin)}} = 5.6 - .7B$

The Captain's Daughter: $\log_{10} \sqrt{\text{(counts/bin)}} = 1.70 - .16B$

**C) The LETTER VALUES**

(Elec. Prop. of Solids)

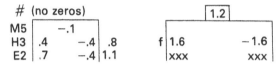

(Atomic and Molec. Phys)

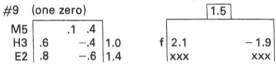

(The Captain's Daughter)

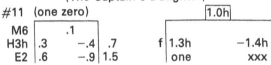

**P) PROBLEMS**

1a) In view of the steady--and slight--tendency to positive residuals for *The Captain's Daughter* (how shown?) try $\log_{10} \sqrt{\text{counts/bin}} = .01 + 1.70 - 16B$, finding residuals and letter values. Is it an improvement?

**S) SOURCES**

(for Elec. Prop. of Solids, and Atomic and Molec. Phys.): Tables 20 and 22, respectively, of Stella Keenan and Pauline Atherton, 1964, *The Journal Literature of Physics* (AIP/DRP PA1(1964)), 156 pp., American Institute of Physics, New York.

(for "The Captain's Daughter"): G. Herdan, 1966. *The Advanced Theory of Langauge as Choice and Chance* (Kommunikation und Kybernetik in Einzeldarstellungen. Band 4), Springer, New York. Table 32 at page 97.

His source: Harry H. Josselson, 1953. *The Russian Word Count: (and Frequency Analysis of Grammatical Categories of Standard Literary Russian)*, Wayne University Press, Detroit.

Clearly, the highest basic count for Atomic and Molecular Physics is much more detached from the other high basic counts than is the case for either of the other two examples. We ought to have some way to trap such indications. How?

We can get a better idea if we look at the roots of the five high basic counts, together with their differences, where we find, as roots:

| | | | | | |
|---|---|---|---|---|---|
| Elec. Prop. of Solids (root counts): | 16.0 | 11.4 | 11.3 | 8.7 | 8.3 |
| Atomic and Molec. Phys. (root counts): | 19.3 | 8.9 | 7.8 | 7.3 | 7.1 |
| *The Captain's Daughter* (root counts): | 34.1 | 27.9 | 26.9 | 24.1 | 21.9 |

exhibit **2** of chapter 18: three sets of occurrences

**The residuals from fits to octave bins for the three examples of exhibit 1**

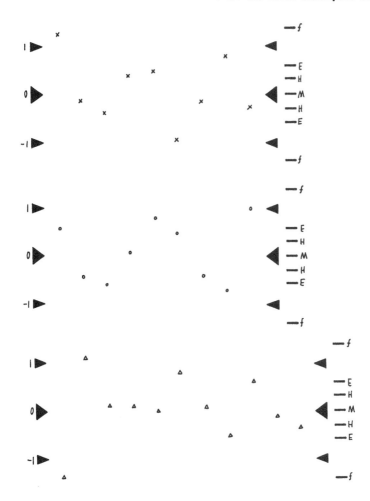

and, as differences of these roots:

| | | | | |
|---|---|---|---|---|
| Elec. Prop. of Solids (diffs): | 4.6 | 0.1 | 2.6 | 0.4 |
| Atomic and Molec. Phys. (diffs): | 10.4 | 1.1 | 0.5 | 0.2 |
| *The Captain's Daughter* (diffs): | 6.2 | 1.0 | 2.8 | 2.2 |

In each case the highest and the second highest are quite well separated--so well separated that asking about "the highest" is very close to asking about a specific name. If we were asking about a name, we would expect to begin by taking the square root of its count.

The total number of species--here journal names or Russian words--involved is another count. Again we would naturally take a square root.

### a possible approach

At one extreme many individuals (many journals, many words) appear only once; at the other extreme, one individual (one journal, one word) appears very many times. There is a vague symmetry between two counts:

⋄ the number of appearances = the basic count.

⋄ the number of individuals appearing at least that many times = the completed rank (from above).

Perhaps we should find an analysis that

⋄ combines these pairs of counts.

⋄ treats them symmetrically.

⋄ involves the square root of the largest number of appearances (= the root of the largest basic count) at one extreme and the square root of the total number of individuals (= the root of the largest rank) at the other.

### review questions

Why do we need another way to look at long-tailed distributions of counts? What examples did we start with? What did the roughs show? What did the largest individual values show? How could we let them show this effectively? What two counts available for each basic count have a vague symmetry? What is a c'rank?

## 18B. Product-ratio analysis

What are the easy ways to combine $x$ and $y$, treating them symmetrically? One is to form $x + y$ and $x - y$, which do not sound promising here. Another, when $x$ and $y$ are both never zero, is to form $xy$ and $x/y$--or, better than $x/y$ itself, if both $x$ and $y$ are always positive (as here), $xy$ and $\log(x/y)$. This is a little more promising start. When $x = 1$, we want to involve $\sqrt{y}$. When $y = 1$, we want to involve $\sqrt{x}$. Clearly, $\sqrt{xy}$ will do just this for us.

So we ought to look at what we get when we look at

$$\sqrt{\text{(basic count) (completed rank above)}}$$

in terms of

$$\log(\text{(basic count)}/\text{(completed rank)})$$

--at root of PRODUCT against log of RATIO. This seems even more plausible when we recall that Zipf stressed the PRODUCT. (We are not as optimistic as he was. We hope that the product will be manageable, but hardly that it will prove to be constant.)

Exhibit **3** illustrates the operations of such a product-ratio analysis on a slightly smaller example. This exhibit combines a detailed computation using all detail--in the left four columns--and computation based on two sets of half-octaves--to be described below. Note that the "root of prod" column, whose values are the square roots of the products of

◇ each basic count (basic), and

◇ the total number of cases (here authors) with at least that basic count (c'rank),

BEGINS with $\sqrt{46}$, where 46 is the largest basic count, and ENDS with $\sqrt{721}$, where 721 is the total number of authors (as we planned).

### half-octave calculation

It is enough to take (basic, c'rank) pairs at or near the half-octave values --e.g., 1, (1h), 2, 3, 4, 6, 8, 11, 16, 22, ..., of both basic and c'rank, stopping each once basic = c'rank has been crossed. In practice, we would make the calculations in the lines of exhibit 3 only for the two sets of half-octaves, thus saving eight of the lines started in exhibit 3--and saving many more lines in larger examples.

### display

Exhibit **4** plots "root of PRODUCT" against log of RATIO. The straightness of the plot is notable. Giving the two ends corresponding to 721 authors with at most 46 contributions must tell most of the story. (Here the root of PRODUCT has been very manageable, and yet far from constant.)

Some related problems are set in exhibit **5**.

### review questions

How did we combine "basic" and "c'rank"? Why choose these combinations? What is a product-ratio analysis? Do we have to look at every basic count? How can we use half-octave steps? Are half-octaves sacred? What example did we try? How did the product-ratio plot look?

exhibit 3 of chapter 18: econometric activity

**Numbers of authors contributing various basic counts of papers, 1933–1952--counts combine meetings of the Econometric Society and the pages of its journal *Econometrica*.**

A) The NUMBERS--detailed and half-octave calculations

| | Detailed calculation | | | | values for half-octaves only* | | | |
|---|---|---|---|---|---|---|---|---|
| Contr | Count | | √ | log | √PROD | | log of RATIO | |
| = basic | of | Ind. | of† | of‡ | | | | |
| count | authors | c'rank | PROD | RATIO | basic | c'rank | basic | c'rank |
| 46 | (1) | 1 | 6.8 | 1.66 | | 6.8 | | 1.66 |
| 37 | (1) | 2 | 8.6 | 1.27 | | 8.6 | | 1.27 |
| 30 | (1) | 3 | 9.5 | 1.00 | | 9.5 | | 1.00 |
| 28 | (2) | 5 | 11.8 | .75 | | | | |
| 24 | (1) | 6 | 12.0 | .60 | | 12.0 | | .60 |
| 23 | (1) | 7 | 12.7 | .52 | | | | |
| 18 | (1) | 8 | 12.0 | .35 | | 12.0 | | .35 |
| 17 | (2) | 10 | 13.0 | .23 | | | | |
| 16 | (1) | 11 | 13.3 | .16 | 13.3 | | .16 | |
| 14 | (2) | 13 | 13.5 | .03 | | | | |
| 13 | (3) | 16 | 14.4 | −.09 | | 14.4 | | −.09 |
| 12 | (2) | 18 | 14.7 | −.18 | | | | |
| 11 | (4) | 22 | 15.6 | −.30 | 15.6 | | −.30 | |
| 9 | (1) | 23 | 14.4 | −.41 | | | | |
| 8 | (11) | 34 | 16.5 | −.63 | 16.5 | | −.63 | |
| 7 | (6) | 40 | 16.7 | −.76 | | | | |
| 6 | (23) | 63 | 19.4 | −1.02 | 19.4 | | −1.02 | |
| 5 | (14) | 77 | 19.6 | −1.19 | | | | |
| 4 | (40) | 117 | 21.6 | −1.47 | 21.6 | | −1.47 | |
| 3 | (61) | 178 | 23.1 | −1.77 | 23.1 | | −1.77 | |
| 2 | (107) | 285 | 23.9 | −2.15 | 23.9 | | −2.15 | |
| 1 | (436) | 721 | 26.9 | −2.86 | 26.9 | | −2.86 | |

* Half-octave sequences of PROD and RATIO combined.
† Examples: $6.8 = \sqrt{(48)\,1}$, $8.6 = \sqrt{(37)(2)}$, and so on to $26.9 = \sqrt{(1)(721)}$.
‡ Examples: $1.66 = \log 46/1$, $1.27 = \log 37/2$, and so on to $-2.86 = \log 1/721$.

P) PROBLEMS

3a) Fit a straight line to (1.66, 6.8) and (−2.86, 26.9) and use to give fitted roots. Find and plot residuals. Comment.

3b) Interpolate values for basic = 1h and c'rank = 1h, and smooth the resulting basic and c'rank sequences. Compare with raw, and comment.

S) SOURCE

D. H. Leavens, 1953. (Communication) *Econometrica* 21: 630–632.

exhibit **4** of chapter 18: econometric activity

**Plot of root of PRODUCT against log of RATIO (numbers in exhibit 3)**

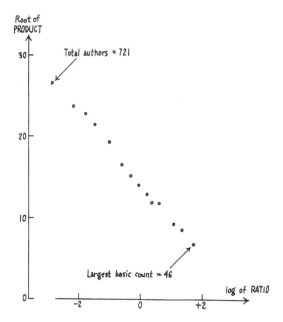

exhibit **5** of chapter 18: three participations.

**Half-octave summaries of three other sets of participation data**

A) HIGH-BASIC-COUNT ENDS

| c'rank | # of contributions per individual = basic | | |
|---|---|---|---|
| | Math† | Physics† | Chemistry† |
| 1  | 70     | --  | 346    |
| 2  | 42     | 48  | 114    |
| 3  | 39     | 37  | 109    |
| 4  | 35     | 34  | 107    |
| 6  | 27     | 27  | 84     |
| 8  | 21     | 25  | 78     |
| 11 | 20(10)*| 24  | 68     |
| 16 | 13(15) | 18  | 57     |
| 22 |        | 16  | 54(21) |
| 32 |        |     | 47     |
| 44 |        |     | 41     |

➔

## 18C. Forcing the unusual to be noticed

Exhibit **6** gives the data, in half-octave form, for a total of nine different subfields of physics, including the two that appeared in exhibit 1. Exhibit **7** shows the plot of "root of PRODUCT" vs. "log of RATIO" for the example "Atomic and Molecular Physics" that attracted our attention earlier. We see a reasonably peaky plot, with one point--that for the most popular journal--far away from a continuation of the others. We seem to have had a "whopper" thrust upon us, without having to look for it specifically. (This is of course what we want our displays to do.)

exhibit **5** of chapter 18 (continued)

**B) LOW-BASIC-COUNT ENDS**

| Basic count | # with as many or more contributions = c'rank | | |
|---|---|---|---|
| | Math | Physics | Chemistry |
| 1 | 278 | 1325 | 6891 |
| 2 | 145 | 541 | 2900 |
| 3 | 102 | 337 | 1841 |
| 4 | 78 | 210 | 1348 |
| 6 | 55 | 127 | 877 |
| 8 | 36 | 80 | 633 |
| 11 | 23 | 48 | 419 |
| 16 | 12 | 22 | 250 |
| 22 | | 14 | 149 |
| 32 | | | 68 |
| 44 | | | 41 |

\* Entries in parenthesis (after columns) are the actual c'ranks used--see Note (1).

†For mathematics: Members of the Chicago Section of the American Mathematical Society, 1897-1992.

†For physics: Up to 1900 from Auerbach's *Geschichtstafeln der Physik*.

†For chemistry: Authors with last names beginning with A or B, in *Chemical Abstracts*, 1907-1916.

Notes
1. Since our definitions are in terms of basic counts, half-octave values of ranks may fall in the middle of a basic count, and thus be unusable. In such cases, we have used the nearest usable value LESS than the half-octave value.

2. Note that columns in panel B do NOT correspond directly to those in panel A.

**P) PROBLEMS**

5a/b/c) Make a plot of the raw root-of-product and log-of-ratio values for Math/Physics/Chemistry.

5d/e/f) (helped by (5a/b/c)). Make a plot of the smoothed root-of-product and log-of-ratio values for Math/Physics/Chemistry.

exhibit 6 of chapter 18: physics papers

**Half-octave summaries for nine subfields of physics--all papers in each abstracted in *Science Abstracts*, A, for 1961**

A) HIGH-BASIC-COUNT END--c'rank and corresponding basic counts; data for ranks 5, 7, 9, and 10 included for curiosity (see previous exhibit for use of ( )'s).

| c'rank | Astro-phys. | Fluids + Gases | Elec. + Mag. | Nuclear Phys. | Elem. Part. | Atomic + Molec. | Solid State | ElProp Solids | MagProp Solids |
|---|---|---|---|---|---|---|---|---|---|
| 1 | 99 | 116 | 152 | 391 | 174 | 372 | 219 | 255 | 168 |
| 2 | 77 | 83 | 135 | 368 | 160 | 79 | 208 | 130 | 165 |
| 3 | 71 | 60 | 132 | 231 | 157 | 61 | 196 | 128 | 122 |
| 4 | 55 | 57 | 119 | 184 | 123 | 53 | 160 | 95 | 57 |
| 6 | 43 | 36 | 2 | 127 | 73 | 41 | 152 | 36 | 53 |
| 8 | 34 | 31(7) | 75 | 89 | 29 | 38 | 110 | 27 | 45 |
| 11 | 24 | 25 | 64(9) | 54 | 18 | 27 | 84 | 24 | 38 |
| 16 | 16 | 17(15) | 43(14) | 40(15) | 12 | 22(15) | 49 | 19(15) | 18 |
| 22 | 8(21) | 15(21) | 31(21) | 30(21) |  | 16 | 38 | 14(20) | 12 |
| 32 |  |  | 22 | 17 |  |  | 20(31) |  |  |
| (Total) | (906) | (1206) | (2342) | (2902) | (1238) | (1339) | (3059) | (1342) | (1245) |
| 5 | 49 | 52 | 83 | 155 | 110 | 50 | 155 | 69 | 53 |
| 7 | 40 | 31 | 81 | 102 | 31 | 40 | 117 | 33 | 45 |
| 9 | 31 | 30 | 74 | 75 | 27 | 35 | 100 | 26 | 40 |
| 10 | 27 | 29 | 64 | 59 | 21 | 31 | 94 | 24 | 35 |

B) HIGH-C'RANK END--basic count and corresponding c'ranks.

| | | | | | | | | | |
|---|---|---|---|---|---|---|---|---|---|
| 1 | 115 | 154 | 182 | 141 | 97 | 107 | 184 | 118 | 87 |
| 2 | 73 | 104 | 124 | 81 | 60 | 68 | 125 | 72 | 62 |
| 3 | 51 | 80 | 102 | 78 | 48 | 57 | 102 | 58 | 49 |
| 4 | 39 | 68 | 81 | 67 | 42 | 46 | 84 | 49 | 38 |
| 6 | 27 | 50 | 69 | 66 | 31 | 36 | 68 | 40 | 31 |
| 8 | 21 | 41 | 58 | 52 | 24 | 33 | 55 | 34 | 28 |
| 11 | 17 | 31 | 48 | 45 | 17 | 27 | 42 | 29 | 21 |
| 16 | 16 | 17 | 35 | 34 | 12 | 22 | 37 | 18 | 16 |
| 22 | 13 | 11 | 32 | 30 |  | 15 | 30 | 12 | 15 |
| 32 |  |  | 20 | 20 |  |  | 26 |  |  |

S) SOURCE

Stella Keenan and Pauline Atherton 1964. *The Journal Literature of Physics* (AIP/DRP PAI (1964)). American Institute of Physics, New York.

Specifically: Table 8 on pages 12-13 for Astrophysics, Table 12 on pages 18-20 for Physics of Fluids and Gases, Table 16 on pages 24-26 for Electricity and Magnetism (also Plasma), Table 17 on pages 27-29 for Nuclear Physics, Table 18 on pages 29-30 for Elementary Particles, Table 20 on pages 32-33 for Atomic and Molecular Physics, Table 21 on pages 34-36 for Solid State Physics, Table 22 on pages 37-38 for Electrical Properties of Solids, Table 24 on pages 40-41 for Magnetic Properties of Solids.

The reasonable conclusion is that this journal--the *Journal of Chemical Physics*--plays a special role for this subfield. One natural approach is to set this journal--with its 372 papers on Atomic and Molecular Physics--aside, and analyze the others. Doing this will reduce every rank by one, changing both the "root of the PRODUCT" and "log of the RATIO" values for each cut. (It will also move the high-basic-count ends away from being half-octaves, but we agreed to do such things freely.) Thus the low-basic end will involve $(1, 106)$, $(2, 67), \ldots$, $(22, 14)$, while the high-basic end will involve $(79, 1)$, $(61, 2)$, $(53, 3)$, $(41, 5)$, $(38, 7)$, $(27, 10)$, $(22, 14)$ and $(16, 21)$. Exhibit **8** shows the corresponding plot, which is now quite symmetrical, with a peak close to "log of ratio" = 0.0. Clearly, once *J. Chem. Phys.* is removed, the other journals show coherent, though two-segment, behavior. So our best description to date of the distribution of those papers on Atomic and Molecular Physics abstracted in 1961 thus consists of two parts:

⋄ one special journal, *J. Chem. Phys.* with 372 papers.

⋄ a mass of other journals--106 of which had at least one of the 967 remaining papers--whose "$\sqrt{\text{product}}$ − log ratio" plot is well described by the two straight lines joining $(-\log 106, \sqrt{106})$ to $(0, \sqrt{335})$ to $(\log 79, \sqrt{79})$.

exhibit **7** of chapter 18: physics papers

**Product–ratio plot for papers in atomic and molecular physics**

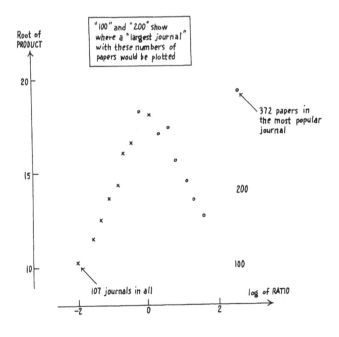

Interestingly enough, when we examine in similar detail the data for Electrical Properties of Solids, which we early took as a reference standard of nice behavior, we find not one very special journal but five somewhat special journals. (See exhibit **9** for this, and other problems.)

exhibit **8** of chapter 18: physics papers

**The plot for Atomic and Molecular Physics after removal of the *Journal of Chemical Physics***

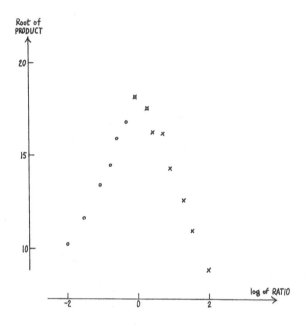

exhibit **9** of chapter 18: problems

**Some problems involving product–ratio plots**

9a/b/c/d) Make plots like exhibit 7 for "Astrophys"/"Fluids and Gases"/"Elec. + Mag."/"Nuclear Physics", and identify any unusual values.

9a2/b2/c2/d2) Omit the unusual journals, if any, found in (9a/b/c/d), and make a plot for the others. Comment.

9e/f/g/h) As in (9a/b/c/d), for "Elem. Part. "/"Solid State"/"Elect. Solids"/"Mag. Prop. Solids".

9e2/f2/g2/h2) As in (9a2/b2/c2/d2), for the corresponding category.

**review questions**

What example attracted attention earlier? How did its product–ratio plot behave? What was the diagnosis? What did we do then? With what outcome? How would you describe the position we reached?

### 18D. Comparisons between collections

Turning back to exhibit 5, we would clearly like to compare these three cases to see to what extent they behave similarly and to what extent differently. Direct comparison of the kind of plots we are so far making will not help us much. With 6891 contributions in chemistry and only 278 in mathematics, we can certainly not expect either the largest basic count--or the number of occurrences with basic count one--to be anywhere near the same.

Clearly, we need to make an allowance for the size of the body of data we are looking at. This we can, and will, do in different ways. All will be such that, for example, doubling the count for every basic count, thus doubling each c'rank, will not affect the plot. We like this property, because it means that "getting twice as much data, exactly similar" has no effect on our plot.

The easy way to do this is to choose a number $a$ and a basic count $b$ and to divide every c'rank by $a$ times the c'rank for $b$ (for that set of data).

By replacing c'rank by c'rank/$ac_b$, we replace PRODUCT, the root of the product of "basic" and "c'rank", by

$$\frac{\text{PRODUCT}}{\sqrt{ac_b}}$$

and RATIO, the log of the ratio of basic to c'rank, by

$$\text{RATIO} + \log(ac_b).$$

Thus we have shrunk the vertical coordinate of our plot by a constant factor and shifted its horizontal coordinate by an additive constant. The shape of our plot is thus left unchanged.

If we take $a = b$, we will get a point at (log $a^2$, 1.000) for any set of data. If we do the same for several sets of data, their (raw) plots will have to cross (or touch) at this point.

If we take $a = b = 1$, we will divide every c'rank by the c'rank for basic count 1, which is the total number of cases, thus using

$$\text{completed fraction} = \frac{\text{completed rank}}{\text{total \# of cases}}$$

in place of the c'rank. This is sometimes nice, but doesn't always do what we might like, as we shall soon see.

Exhibit **10** shows the result of taking $a = b = 6$ for the three data sets of exhibit 4. Clearly we can do a very good job of making comparisons here. For basic counts above 6, the three plots are rather similar, with Physics falling fastest and Chemistry showing appreciable curvature. For basic counts below 6, all three follow straight lines moderately well, though the slopes of the lines are (a) quite different from one another and (b) appreciably, moderately, or very different from the slopes for the same data for basic counts greater than 6.

Exhibit **11** sets some problems.

### review questions

Do we need to allow for differences in size of data collections? Why/why not? What were $a$ and $b$? How were they used? What happened when $a = 6$? What examples did we try? What did they show?

exhibit **10** of chapter 18: three participations

**Three collections matched at basic count 6 ($a = b = 6$)**

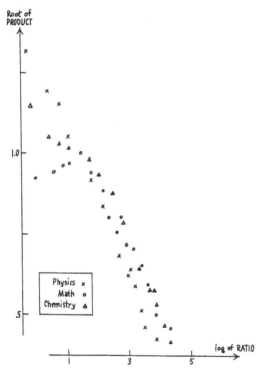

exhibit **11** of chapter 18: problems

**Some problems on matched plots, suitable for teamwork**

11a) (can be divided among 4; 3 calculations and 1 plot). Take $a = b = 10$ and make the three plots like those in exhibit 10. Discuss and comment.

11b) (can be divided among 4; 3 calculations and 1 plot). Do the same for $a = b = 3$.

### 18E. Looking at the smallest basic count

Now that we are used to looking at our plots with more and more care, we notice that the point for the smallest basic count, so far basic count one, is not always on the line through the neighboring points. It is easy to experiment with the results of changing the total count, which is, of course, the c'rank for the lowest basic count, and seeing where the point goes. This amounts to deleting (or adding) some "species" with the smallest basic count.

Exhibit 12 shows a part of exhibit 10 with such alternatives plotted. For chemistry, we might find 6251, corresponding to 3351 (=6251 − 2900) authors of one contribution, a better fit to the line than the actual 6891, corresponding to 3991 (=6891 − 2900) such authors. Since $\sqrt{3351} = 58$ and $\sqrt{3991} = 63$, we are likely to find the difference worth appreciable comment (5 is a considerable difference for two square roots of actual counts). For mathematics, on the other hand, 270 giving 125, instead of 278 giving 133, corresponds to $\sqrt{270} = 16.4$ in place of $\sqrt{278} = 16.7$, which does not deserve comment.

#### review questions

What did we start to do in this section? Was it easy? Was it effective?

exhibit **12** of chapter 18: three participations

**Part of exhibit 10 with points for alternative c'ranks for basic count 1**

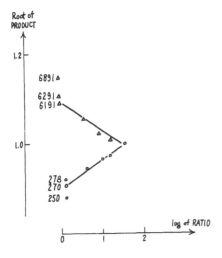

## 18F. When zeros are counted

There is another situation in which product–ratio plots are helpful, at least for comparative purposes. If we have a binning of counts where zeros can occur and tails are quite stretched out, as for instance in these instances for which data is given in exhibit **13**:

- numbers of fleas on each of 209 live rats,
- numbers of isopods found under each of 122 boards,
- numbers of bacterial clumps seen in each of 400 microscope fields of milk, or
- numbers of lice on the heads of each of 1083 Hindu male prisoners,

we can start our basic count, so that the log of the ratio gives no trouble. This seems most likely to be helpful when there are more "zeros" than any other single basic count in the original data. (Thus the milk-film example may be marginal.) Exhibit **14** shows the standard (but unsmoothed) product–ratio plots for the four examples. Clearly, lice are behaving differently, while fleas, isopods, and clumps show related patterns.

exhibit **13** of chapter 18: ecological counts

**Some data where started counts give helpful product–ratio plots**

| Rats, with ≥ fleas | | Boards, with ≥ isopods | | Fields, with ≥ clumps | | Heads, with ≥ headlice | |
|---|---|---|---|---|---|---|---|
| Basic | c'rank | Basic | c'rank | Basic | c'rank | Basic | c'rank |
| 0 | 209 | 0 | 122 | 0 | 400 | 0 | 1083 |
| 1 | 146 | 1 | 94 | 1 | 344 | 1 | 461 |
| 2 | 114 | 2 | 66 | 2 | 240 | 2 | 355 |
| 3 | 84 | 3 | 52 | 3 | 160 | 3 | 305 |
| 5 | 58 | 5 | 33 | 5 | 56 | 5 | 243 |
| 7 | 39 | 7 | 20 | 7 | 20 | 7 | 209 |
| 10 | 26 | 10 | 11 | 8 | 11 | 10 | 168 |
| 15 | 14 | 11 | 8 | 9 | 6 | 15 | 130 |
| 20 | 8 | 13 | 6 | 10 | 3 | 21 | 96 |
| 26 | 6 | 14 | 5 | | | 14 | 5 |
| 46 | 4 | 15 | 3 | | | 40 | 44 |
| 48 | 3 | 17 | 2 | | | 47 | 33 |
| 61 | 2 | 10 | 7 | | | 58 | 22 |
| 83 | 1 | | | | | 74 | 16 |
| | | | | | | 11 | 11 |
| | | | | | | 129 | 8 |
| | | | | | | 149 | 6 |
| | | | | | | 239 | 4 |
| | | | | | | 170 | 3 |
| | | | | | | 303 | 2 |
| | | | | | | 385 | 1 |

→

exhibit **13** of chapter 18 (continued)

S) SOURCES

For rat fleas on 209 live rats in Mobile, Alabama: L. C. Cole, 1946, "A theory for analyzing contagiously distributed populations." *Ecology* 27: 329–341. Table VI on page 338.

For *Trachelipus Rathkei* (an isopod) under 122 boards: L. C. Cole, 1946. "A theory for analyzing contagiously distributed populations." *Ecology* 27: 329–341. Table V at page 337. His source: L. C. Cole, Ecological Monographs 16.

For bacterial clumps in each of 400 microscope fields in a milk farm: M. E. Morgan, P. MacLeod, E. O. Anderson, and C. I. Bliss, 1951. "A sequential procedure for grading milk by microscopic counts." *Storrs Agricultural Experiment Station Bulletin* 276. Used by: C. I. Bliss, 1953. "Fitting the negative binomial distribution to biological data." *Biometrics* 9: 176–196. Table 2 on page 186.

For lice of all stages on the heads of 1083 Hindu male prisoners in Cannamore, South India in 1937–39: C. I. Bliss, 1953. "Fitting the negative binomial distribution to biological data." *Biometrics* 9: 176–196. Table 8 on page 194. His source: F. J. Anscombe, 1950."Sampling theory of the negative binomial and logarithmic series of distributions." (Full data at page 11 of the source for exhibit 19, below).

Original source: P. A. Buxton 1940. "Studies on populations of headlice. III." Material from South India *Parasitology* 32: 296.

exhibit **14** of chapter 18: ecological counts

**Fleas on rats with five different starts added to all basic counts (matched with $a = b = 3$)**

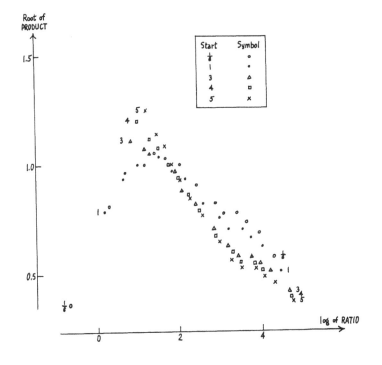

We can use any start that seems helpful, 1/6, 1/4, 1/2, 1, 2, 3, 4, 5, or what have you. Exhibit 14 shows what five possibilities do with the rat data from exhibit 13. (For later convenience we used $a = b = 3$.) All of them give reasonably manageable (= reasonably easily describable) results, but a start of 4 seems to give the simplest result.

Exhibit **15** shows the plots for start 4 (and $a = b = 3$) for the four sets of data in exhibit 13. We see that:

◇ most of each plot is crudely straight.
◇ the points for basic count = 1 tend to look discrepant.
◇ the tilt of the "heads" plot is appreciably less than the others there.

The effectiveness, in exhibit 14, of a substantial start in straightening out our plot makes us wonder whether some of our earlier plots, where a start was NOT forced by counts of zero, might also be straightened out. Inquiry into this is left to the reader.

exhibit **15** of chapter 18: ecological counts

**All four sets of counts, started by 4 and matched at basic count 3**

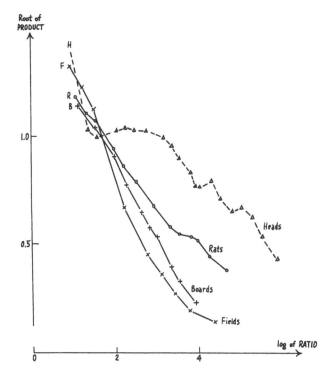

**review questions**

What is a started basic count? Why might we want to use it? To what examples did we apply it? With what results? Where else might we have tried it?

### 18G. Under the microscope

Three of the plots in exhibit 15 are somewhat similar in slope. A rough fit gives

$$\text{PROD} = 1.000 - 0.36 (\text{RATIO} - 1.69) = 1.608 - 0.36L,$$

where PROD is root of product and RATIO is log of ratio using a start of 4 and $a = b = 4$. Given a fit, it is by now almost automatic for us to take residuals, with the result shown in exhibit **16**. We can now distinguish the behavior of the three data sets quite sharply.

The smallest basic count, here zero, is acting up again, so exhibit **17** sets some problems.

**exhibit 16 of chapter 18: ecological counts**

**Residuals (of root of product with start = 4, $a = b = 3$) from a general straight-line fit (formula in text)**

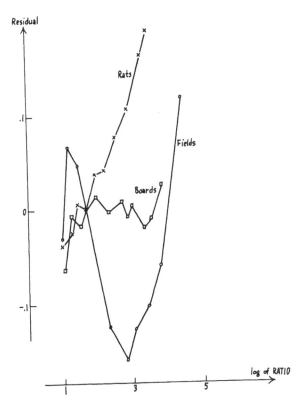

While something could be gained by fitting a line to the "heads" data (see problem (17d)), we do not need a microscope to see how far the zero-basic-count point is out of line for this data. As exhibit 18 shows, a reduction in the zero-basic-count count of from 300 to 400 would be required to bring this point in line. These correspond to differences in roots of

$$\sqrt{622} - \sqrt{322} = 7.0,$$

$$\sqrt{622} - \sqrt{222} = 10.0,$$

exhibit 17 of chapter 18: ecological counts

**Problems on "using the microscope" (started counts)**

17a/b/c) Find what c'rank--and hence what count for basic count = 0--would bring the basic-count-zero point into line on exhibit 16 for the rats/boards/fields data. Note the difference in roots, $\sqrt{\text{assigned}}$ # of zeros − $\sqrt{\text{given}}$ # of zeros for your case, and comment.

17d) Fit a line to the corresponding product–ratio plot for "heads", and plot the residuals (as in exhibit 16).

exhibit 18 of chapter 18: ecological counts

**Headlice per head plot (started by 4 and matched at 3) with two possible straight lines and some other c'ranks for basic count one**

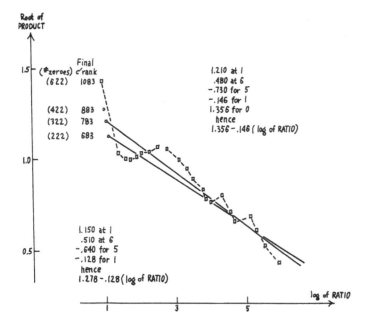

and can only be considered large. A reasonable interpretation would be that the criminals involved were divided into two groups: one, providing about a third of the heads, being nearly free from lice, while the remainder had louse infestations that behaved simply in our product–ratio plot.

Exhibit **19** poses some closing problems.

### review questions

What standard process lets us "use the microscope"? What did we try it on? How did it work?

### exhibit 19 of chapter 18: data and problems

**Some problems involving data on species per genus assembled by Williams**

#### A) INSECTS

| MacGillivray | | Beare + D | | Stainton | | South + E | | Edwards | | Butler + C | |
|---|---|---|---|---|---|---|---|---|---|---|---|
| c'rank | basic | c'rank | basic | c'rank | basic | c'rank | basic | c'rank | basic | c'rank | basic |
| 1 | 200 | 1 | 155 | 1 | 40 | 1 | 43 | 1 | 56 | 1 | 41 |
| 2 | 70 | 2 | 77 | 2 | 26 | 2 | 18 | 2 | 26 | 2 | 33 |
| 3 | 60 | 3 | 64 | 3 | 23 | 4 | 17 | 3 | 18 | 4 | 21 |
| 4 | 31 | 4 | 53 | 4 | 18 | 6 | 13 | 4 | 15 | 5 | 20 |
| 8 | 30 | 6 | 40 | 6 | 15 | 7 | 11 | 6 | 12 | 7 | 15 |
| 12 | 27 | 9 | 37 | 7 | 13 | 10 | 10 | 7 | 11 | 8 | 13 |
| 16 | 21 | 11 | 33 | 10 | 11 | 15 | 8 | 11 | 8 | 11 | 9 |
| 20 | 17 | 16 | 25 | 16 | 8 | 22 | 6 | 15 | 6 | 13 | 8 |
| 43 | 11 | 22 | 21 | 27 | 6 | 42 | 4 | 17 | 4 | 25 | 8 |
| 59 | 8 | 35 | 16 | 52 | 4 | 67 | 3 | 19 | 3 | 25 | 4 |
| 70 | 6 | 70 | 11 | 77 | 3 | 118 | 2 | 28 | 2 | 30 | 3 |
| 107 | 4 | 96 | 8 | 117 | 2 | 357 | 1 | 47 | 1 | 40 | 2 |
| 129 | 3 | 134 | 6 | 249 | 1 | | | | | 71 | 1 |
| 171 | 2 | 211 | 4 | | | | | | | | |
| 352 | 1 | 286 | 3 | | | | | | | | |
| | | 412 | 2 | | | | | | | | |
| | | 803 | 1 | | | | | | | | |

MacGillivray: *Coccidae* (scale insects) of the world (p. 22 of source)
Beare + D: *Coleoptera* of Britain, 1904 (p. 24 of source)
Stainton: *Macrolepidoptera* (excl. Butterflies) of Britain, 1857 (p. 24 of source)
South + E: *Macrolepidoptera* (excl. Butterflies) of Britain, 1939 (p. 24 of source)
Edwards: *Cicadina* of Britain, 1894 (p. 28 of source)
Butler + C: *Cicadina* of Britain, 1942 (p. 28 of source)

exhibit **19** of chapter 18 (continued)

**B) BIRDS and PLANTS**

| Witherby ++ | | Bentham + H | | Babington | | Willis | |
|---|---|---|---|---|---|---|---|
| c'rank | basic | c'rank | basic | c'rank | basic | c'rank | basic |
| 1 | 15 | 1 | 47 | 1 | 97 | 1 | 525 |
| 5 | 13 | 2 | 21 | 2 | 74 | 2 | 500 |
| 6 | 12 | 3 | 17 | 3 | 48 | 3 | 325 |
| 11 | 11 | 4 | 16 | 5 | 31 | 4 | 320 |
| 17 | 8 | 7 | 15 | 6 | 29 | 6 | 250 |
| 22 | 6 | 8 | 14 | 8 | 21 | 8 | 190 |
| 43 | 4 | 12 | 12 | 10 | 15 | 12 | 180 |
| 60 | 3 | 16 | 11 | 16 | 13 | 16 | 150 |
| 92 | 2 | 22 | 10 | 26 | 11 | 21 | 125 |
| 199 | 1 | 32 | 8 | 38 | 8 | 31 | 105 |
| | | 45 | 6 | 60 | 6 | 42 | 65 |
| | | 88 | 4 | 117 | 4 | 63 | 60 |
| | | 139 | 3 | 162 | 3 | 88 | 45 |
| | | 223 | 2 | 255 | 2 | 123 | 30 |
| | | 479 | 1 | 510 | 1 | 151 | 22 |
| | | | | | | 195 | 16 |
| | | | | | | 264 | 11 |
| | | | | | | 350 | 8 |
| | | | | | | 400 | 6 |
| | | | | | | 488 | 4 |
| | | | | | | 565 | 3 |
| | | | | | | 691 | 2 |
| | | | | | | 1000 | 1 |

Witherby ++: Birds, including subspecies, of Britain, 1941 (p. 29 of source)
Bentham + H: Flowering plants of Britain, 1906 (p. 30 of source)
Babington: Flowering plants of Britain, 1922 (p. 30 of source)
Willis: Plants and ferns of the world (1000 genera, p. 30 of source)

**P) PROBLEMS**

19a/b/c/d/e/f) Make an unstarted product–ratio plot with $a = b = 1$ for the MacGillivray/Beare + D/Stainton/South + E/Edwards/Butler + C data.

19g/h/i/j) Do the same for the Witherby ++/Bentham + H/Babington/Willis data.

19a2/b2/c2/d2/e2/f2) Choose a start between 2 and 6, and superpose the corresponding plot on that made in problem (19a/b/c/d/e/f). Compare; conjecture what start would give a reasonably straight line, if any.

19g2/h2/i2/j2) Do the same on the result of (19g/h/i/j).

19a3/b3/c3/d3/e3/f3) Use the start conjectured in problem (19a2/b2/c2/d2/e2/f2), making the plot, and commenting on its straightness.

19g3/h3/i3/j3) Do the same for 19g2/h2/i2/j2.

19cd4) Compare the results of (19c3) and (19d3). Comment. Make any additional plot needed.

### 18H. How far have we come?

This chapter has treated long-tailed distributions, mainly of counts, with attention to the individual largest values. It has centered on the relation of root-of-PRODUCT to log-of-RATIO, of

$$\sqrt{(\text{basic count or size})(\text{completed rank})}$$

to

$$\log(\text{basic count or size})/(\text{completed rank}),$$

which often proves to be simply describable and almost always plots usefully.

We are now ready:

⋄ to make product–ratio plots, using either all (basic, c'rank) pairs or only those corresponding (perhaps roughly) to half-octave sequences of "basic" and "c'rank".

⋄ to use $a = b = $ (a convenient integer) to bring several such plots through a common point (by shrinking or magnifying vertically and sliding horizontally).

⋄ to start our basic counts, either when necessary (zeros counted) or convenient (simplifying our plot's appearance).

⋄ to form simple residuals and thus put the "shape" of stretched-tail distributions under a microscope.

#### commentary

There may be readers who will react rather negatively to the plots of this chapter, on the ground that they "do not know what the axes mean", an argument with some--but only very little--face plausibility. It is, the writer

exhibit **19** of chapter 18 (continued)

19ef4) Same for (19c3) and (19f3).

19hi4) Same for (19h) and (19i).

19k) (teaser) Look up the source and plot some of the data there on number of generations in families.

#### S) SOURCE

C. B. Williams 1944 "Some applications of the logarithmic series and the index of diversity to ecological problems." *J. Ecology* 32: 1–44. (See this source for references to original sources.)

believes, fair to argue that very meaningful analyses have taken place in this chapter--meaningful in that:

◇ the bulk of the behavior of several stretched-tail distributions has been concentrated into a few numbers.

◇ the rest has been exhibited in the form of simple residuals.

In our analysis of fleas on rats, for instance, we could summarize in terms of

◇ a start of 4,

◇ a largest basic count of 84 out of 209 rats,

and present the rest of the behavior in terms of residuals (compare exhibit 16) from a line through the points corresponding to the (basic, c'rank) pairs (209, 1) and (1, 84).

**This has taken us a long way forward.** We can compare two such distributions quite effectively, and can detect many of their idiosyncrasies. **We can do this without requiring "a feeling" or "an intuitive understanding" of what the coordinates in our plot mean.**

# index for chapter 19

review questions 615

**19A. Looking at shapes of distribution 616**
Chevrolet prices 616
shapes 616
binned data 620
review questions 620

**19B. The Gaussian reference 623**
review questions 626

**19C. Using letter values to look at shapes of distribution 626**
*pseudospreads* 630
review questions 637

**19D. Pushback technique (optional section) 637**
a half-octave version 639
review questions 643

**19H. How far have we come? 644**

| EXHIBIT | PAGE |
|---|---|
| **19A** | |
| 1★ | 617 |
| 2 | 618 |
| 3 | 618 |
| 4 | 619 |
| 5★ | 621 |
| 6 | 622 |
| **19B** | |
| 7 | 624 |
| 8★ | 626 |
| **19C** | |
| 9★ | 627 |
| 10★ | 629 |
| 11 | 631 |
| 12 | 632 |
| 13★ | 633 |
| 14★ | 635 |
| 15 | 636 |
| **19D** | |
| 16★ | 638 |
| 17 | 639 |
| 18 | 640 |
| 19★ | 643 |
| 20 | 644 |
| **19H** | |

# Shapes of distribution 19

Suppose we have a batch of values, and that we want to talk about how they are "distributed." Chapter 17 took a rather flexible approach involving a fair amount of mechanism: bins, counts in bins, bin densities = counts/width, square roots of bin densities, smoothing, and then, in the special case of near symmetry, simple fitting to our usual re-expressions. Chapter 18 took a rather different approach to a special case--to the case where there are many instances of very small values, but a small number of very large ones. In the present chapter, we will take a third approach, based on comparing the observed distribution with a reference standard, so that the residuals can tell us how what is before us deviates from that standard. Again we deal with only a special case, but this time one which is much more common, one involving batches of data whose distribution is not too different, perhaps when suitably re-expressed, from the chosen reference standard.

To speak of a distribution is to speak of where the values are. We are now familiar (see chapter 15) with:

◇ re-expressing fractions.

◇ looking at several different cuts.

If we know the fraction of values "to the left"--or "to the right"--of each and every possible cut, we know where all the values are. So we can tackle the distribution problem by working with the parallel fractions "above" or "below" each cut. Working with these parallel fractions will help us look at shapes of distribution and show us how to exhibit actual behavior--actual shape of distribution--by comparing observed counted fractions with a standard reference shape, thus generating pictures of the deviation of the actual from the standard.

**As we ought to expect, exact matching of the standard almost never happens.**

We will also, in this chapter, go back to chapter 2, one of our very early chapters, and build upon the use of letter values as one way to conveniently summarize how the values of our batch are spread out. As well as halving outwards in terms of letter values, we will, alternatively, also work inwards by half-octaves.

**review questions**

What approaches have which previous chapters taken toward looking at shapes of distribution? What approach will we take in the present chapter?

How will what we are to do be related to chapter 15's counted fractions? What will we need as a basis of comparison? What will we compare with it? How often do we expect very close agreement with our reference standard? What tool will we take from which very early chapter? What related approach will we also follow?

### 19A. Looking at shapes of distribution

We want simple expressions related to distributions--expressions that come near enough for rough descriptions of the residuals. To ask for raw fractions or ss-fractions to behave simply in terms of cut position would be too much. What we need is a tailing off of the values at either end, a situation where extreme fractions change much more slowly than middle fractions. What we need is tail-stretching.

Perhaps the re-expressions of fractions we have been using will help us. Either froots or flogs--presumably based on ss-fractions--might behave simply enough for us to be able to use at least frequently:

◇ a simple fit of froots or flogs to cut position,

◇ AND a description of the residuals

as an adequate description of distributions of values.

#### Chevrolet prices

Exhibit **1** tries this out, for the batch of seventeen Chevrolet prices of exhibit 3 of chapter 1. The first results are plotted in exhibit **2**, where we see a fairly steady rise. Showing the results for all possible cuts, however, gives us both isolated points, wherever one or more values fall at that cut, and horizontal line segments (which together cover all other values). From the point of view of a first understanding of what we are doing, showing both points and horizontal lines seems good. Once we get the general picture and want to look harder, however, we gain (both in visibility of result and in ease of plotting) by showing only the points--either isolated or, as in exhibit **3**, lightly connected.

Taking the eye-fitted line of exhibit 3 back to panel A of exhibit 1, we find the first residuals, shown in exhibit **4**. The additional fitting that is now strongly indicated is left to a problem, as is the repetition of the whole process using froots instead of flogs. Those who do these problems will find a much closer fit by a straight line when froots are used. How are we to interpret this?

#### shapes

We have an intuitive idea of what we mean by the shape of a distribution of values--more precisely, perhaps an intuitive idea of what does not change a shape. To take a toy example with only five values, the sets 1, 4, 9, 16, 25, and 10, 40, 90, 160, 250, and 101, 104, 109, 116, 125 all have the same shape, one

exhibit **1** of chapter 19: Chevrolet prices

**Calculations for displaying re-expressed fractions against cut values, including first residuals**

### A) INITIAL CALCULATIONS

| Cut | Less than PLUS 1/2 equal | Same +1/6* | ss-fraction | flog | | first fit† | first residual | |
|---|---|---|---|---|---|---|---|---|
| 149 | 0 | s | .0096 | −2.32 | T | | | T |
| 150 | h | hs | .038 | −1.61 | U | −1.48 | −.13 | U |
| 151 | 1 | 1s | .067 | −1.31 | R | | | R |
| 249 | 1 | 1s | .067 | −1.31 | N | | | N |
| 250 | 1h | 1hs | .096 | −1.12 | T | −1.32 | +.20 | T |
| 251–687 | 2 | 2s | .125 | −.97 | O | | | O |
| 688 | 2h | 2hs | .154 | −.85 | | −.60 | −.25 | |
| 689–694 | 3 | 3s | .183 | −.75 | E | | | E |
| 695 | 3h | 3hs | .211 | −.66 | X | −.58 | −.08 | X |
| 696–794 | 4 | 4s | .240 | −.58 | H | | | H |
| 795 | 5 | 5s | .298 | −.43 | I | −.42 | −.01 | I |
| 796–894 | 6 | 6s | .356 | −.30 | B | | | B |
| 895 | 7h | 7hs | .442 | −.12 | I | −.25 | +.13 | I |
| 896–1098 | 9 | 9s | .529 | .06 | T | | | T |
| 1099 | 9h | 9hs | .558 | .12 | S | .08 | +.04 | 4 |
| 1100–1165 | 10 | 10s | .587 | .17 | 2 | | | |
| 1166 | 10h | 10hs | .615 | .23 | A | .19 | +.04 | |
| 1167–1332 | 11 | 11s | .644 | .30 | N | | | |
| 1333 | 11h | 11hs | .673 | .36 | D | .47 | −.11 | |
| 1334–1498 | 12 | 12s | .702 | .43 | | | | |
| 1499 | 12h | 12hs | .731 | .50 | 3 | .74 | −.24 | |
| 1500–1692 | 13 | 13s | .760 | .58 | | | | |
| 1693 | 13h | 13hs | .789 | .66 | | 1.06 | −.40 | |
| 1694–1698 | 14 | 14s | .817 | .75 | | | | |
| 1699 | 14h | 14hs | .846 | .85 | | 1.07 | −.22 | |
| 1700–1774 | 15 | 15s | .875 | .98 | | | | |
| 1775 | 15h | 15hs | .904 | 1.13 | | 1.20 | −.08 | |
| 1776–1894 | 16 | 16s | .933 | 1.31 | | | | |
| 1895 | 16h | 16hs | .982 | 1.61 | | 1.40 | +.21 | |
| 1896– | 17 | 17s | .9904 | 2.32 | | | | |

*Here "s" = "and 1/6" and the divisor is 17ss = 17 1/3.
† See exhibit 3.

### P) PROBLEMS

1a) Use the additional fit found in exhibit 14 to find second residuals. Plot and comment.

1b) Repeat the calculations using froots instead of flogs. Plot, compare, comment.

1c) Find a batch of 15 to 25 values of interest to you and analyze it as above.

**exhibit 2** of chapter 19: Chevrolet prices

**Complete plot (all possible cuts) of flog against cut values (from exhibit 11)**

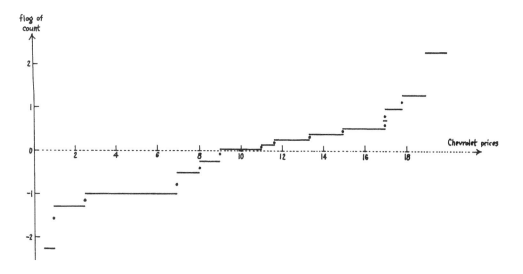

**exhibit 3** of chapter 19: Chevrolet prices

**Working plot (points only, lightly connected) and eye-fitted line of flog against cut value for 17 Chevrolet prices (from exhibit 11)**

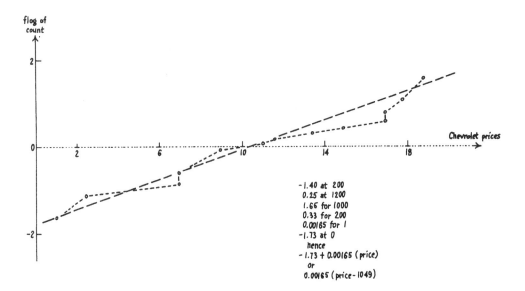

that is tighter on the left than on the right. We insist that the sorts of changes we have just made:

◇ multiplying each value by the same constant (in our example, by 10), or
◇ adding the same constant (in our example, 100) to each value,

do not change shape. If this is so, then no combination of such changes will affect shape either. (Why?) **Shape** is essentially that which is NOT affected by TRIVIAL re-expression.

If a particular re-expression of "fraction above" is a "linear function of cut position"--is (constant) PLUS ((another constant) times (cut position))--then the shape of our distribution is the same as when the constants are zero and one, as when

the same re-expression of fraction above = cut position,

which we can take as defining the standard distribution corresponding to that particular re-expression.

Recall now the result of problem (1b).

exhibit **4** of chapter 19: Chevrolet prices

**Residuals from the eye-fitted line of exhibit 13 and a supplementary line**

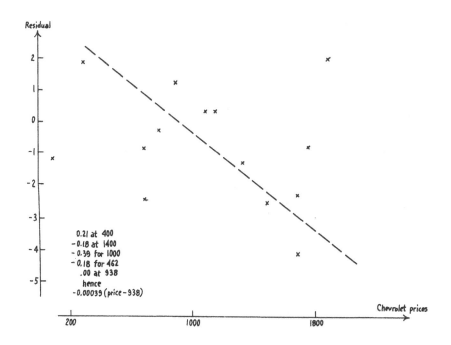

We can now say: The shape of the distribution of the seventeen Chevrolet prices is closer to the shape corresponding to froots than it is to the shape corresponding to flogs. In general, we can compare the shape of any batch with either or both of these standard shapes.

### binned data

We have carried through the analysis of the 17 Chevrolet prices in the fullest possible detail. We already noticed that it was surely enough to work with only the values that actually occurred. In most cases we do not want to go into even that much detail, and we frequently could not, even if we wished to, since many sorts of data are binned before we can get our hands on them.

One sort of example where unbinned data is not available--and would be unmanageable if it were--is data on individual income tax returns. Exhibit 5 deals with the 59,085,182 individual income-tax returns for 1958 reported by the Internal Revenue Service, specifically at their binning by adjusted gross income. As exhibit 6 shows:

⋄ for incomes up to $10,000, the flogs are roughly linear in the square root of the cut.

⋄ for incomes above $15,000, the flog is roughly linear in the log of the cut.

⋄ both "roughly linear" behaviors involve a detectable curvature (clear enough if residuals are found and plotted).

The words we have just used give a qualitative description of the distribution of adjusted gross income as reported by those filing income-tax returns. This can be made quantitative by specifying fitted lines--and could be made still more precise by fitting quadratics to the segments (in place of lines).

The purpose of this example is NOT intended to show a new and wonderful representation for the shape of the adjusted gross income distribution--this we heartily disclaim. (Accordingly, we do not expect this particular representation to work particularly well for other data.) Rather it is to show:

⋄ **that binned data can be dealt with by the same approach as individual data.**

⋄ **that even fairly complex shapes of distribution can be seized upon with reasonable approximation when we use the simplest of methods.**

These are its true purposes.

### review questions

What does "how distributed" mean? Have we expressed an answer before? Where? How? Can we usefully describe patterns of distribution with (counted) fractions? Why/why not? If we can, are the fractions likely to need

exhibit **5** of chapter 19: adjusted gross income

**Analysis of a binning of U.S. individual income tax returns for 1958 according to their adjusted gross income**

A) DATA and CALCULATIONS

| Upper edge of bin | Count | | | Information for cuts | | | | log |
|---|---|---|---|---|---|---|---|---|
| | | Cum ↓ (*) | Cum ↑ (*) | Cum ↓ | Cum ↑ | flog | √cut | cut |
| $600 | 3,950,030 | 3,950,030 | | 6.73% | | −1.32 | .78 | |
| $1,000 | 3,060,247 | 7,010,277 | | 11.9% | | −1.00 | 1.00 | |
| $1,500 | 4,120,276 | 11,130,553 | | 19.0% | | −.72 | 1.22 | |
| $2,000 | 3,570,536 | 14,701,089 | | 25.0% | | −.55 | 1.41 | |
| $2,500 | 3,689,218 | 18,390,307 | | 31.3% | | −.39 | 1.58 | |
| $3,000 | 3,723,909 | 22,114,216 | | 37.7% | | −.23 | 1.73 | |
| $3,500 | 3,742,848 | 25,857,064 | | 44.0% | | −.12 | 1.87 | |
| $4,000 | 3,729,578 | 29,586,642 | 29,114,282 | 50.4% | 49.6% | +.01 | 2.00 | 3.60 |
| $4,500 | 3,745,242 | | 25,369,040 | | 43.2% | .14 | 2.12 | 3.65 |
| $5,000 | 3,639,977 | | 21,729,063 | | 37.0% | .27 | 2.24 | 3.70 |
| $6,000 | 6,375,555 | | 15,353,508 | | 26.2% | .82 | 2.45 | 3.78 |
| $7,000 | 4,676,947 | | 10,676,561 | | 18.2% | .75 | 2.65 | 3.84 |
| $8,000 | 3,226,844 | | 7,449,717 | | 12.7% | .96 | 2.83 | 3.90 |
| $9,000 | 2,171,701 | | 5,278,016 | | 8.99% | 1.16 | 3.00 | 3.95 |
| $10,000 | 1,452,594 | | 3,825,422 | | 6.52% | 1.33 | 3.16 | 4.00 |
| $15,000 | 2,488,095 | | 1,337,327 | | 2.28% | 1.88 | 3.87 | 4.18 |
| $20,000 | 588,262 | | 749,065 | | 1.28% | 2.17 | | 4.30 |
| $25,000 | 264,732 | | 484,333 | | .825% | 2.39 | | 4.40 |
| $50,000 | 369,939 | | 114,394 | | .195% | 3.12 | | 4.70 |
| $100,000 | 91,715 | | 22,679 | | .0386% | 3.93 | | 5.00 |
| $150,000 | 14,080 | | 8,599 | | .0146% | 4.42† | | |
| $200,000 | 3,863 | | 4,736 | | .0081% | 4.72 | | |
| $500,000 | 3,956 | | 780 | | .00133% | 5.62 | | |
| $1,000,000 | 536 | | 244 | | .00042% | 6.20 | | |
| | 244 | | | | | | | |

(Total = 59,085,182 which includes 384,258 with no gross income that have been excluded. Thus working total = 58,700,924.)

* Bothering with the added 1/6 would obviously be foolish.
† The last four values in this column have to be found from flog = 1.1513 log₁₀ (f/(1 − f)).

P) PROBLEMS

5a) Find a set of 8 to 12 binned counts of interest to you and analyze it along the lines just illustrated.

5b) Do the same for a set of 13 to 17 binned counts.

5c) Do the same for a set of 18 or more binned counts.

S) SOURCE

U.S. Internal Revenue Service, 1958.

re-expression? Why/why not? If we can't, what should we do instead? What did we take as an example? Did it need re-expression? Why/why not? And when we displayed the results? Using how many styles of display? Why/why not? Did there seem to be a reasonable fit? What did we do next? Why? What was the overall result? What do we want to mean by "shape of distribution"? What ought not change such a shape? What is a standard distribution corresponding to a re-expression? How can we compare the shape of distribution of a batch with such a standard? What do we now learn from the previous data? Do we need individual values or can we use counts in bins? What example did we try next? Could we have worked with individual values? How did things come out? Could we fit the left tail reasonably? How? Could we fit the right tail reasonably? How? Need the re-expression that works well in one tail also work well in the other? Why/why not? Did we think the representation we found for our data ought to be used for other similar data? Why/why not?

exhibit **6** of chapter 19: adjusted gross income

**The two tails plotted--and eye fitted--separately (59,085,182 individual in-come-tax returns in 1958, from exhibit 15)**

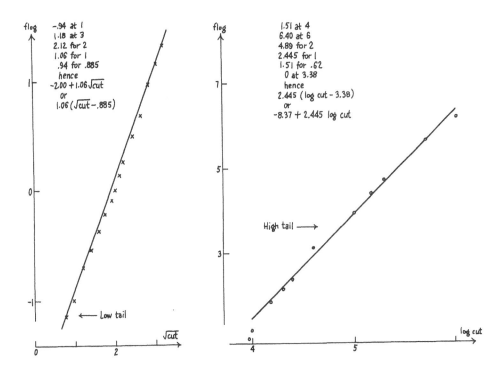

## 19B. The Gaussian reference

Although the shape of distribution associated with the froot re-expression of fractions matched the shape of our seventeen Chevrolet prices quite well, this is not a very common occurrence. Accordingly, this shape would not be a good choice as a single standard shape.

The shape of distribution associated with flog re-expression of fractions is much more often a close match to data. Indeed, so far as data goes, this shape of distribution is one that might quite reasonably be taken as the single standard. Both for specific reasons--mainly more closely associated with mathematical theory than with data--and, perhaps, because of accidents, this was not done.

The reference standard for shapes of distribution has long been the shape associated with the name of Gauss, who combined mathematical genius with great experience with the highest-quality data of his day--that of surveying and astronomy. Later writers have made the mistake of thinking that the Gaussian (sometimes misleadingly called normal) distribution was a physical law to which data must adhere--rather than a reference standard against which its discrepancies are to be made plain. It has taken a long time to recover from such unwarranted simplification, but much progress has been made.

Today we can use the Gaussian shape of distribution in a variety of ways to our profit. We can:

⋄ use it freely as a reference standard, as a standard against which to assess the actual behavior of real data--doing this by finding and looking at deviations.

⋄ use it, cautiously but frequently, as a crude approximation to the actual behavior, both of data itself and of quantities derived from data.

In using the Gaussian shape as such an approximation, we owe it to ourselves to keep in mind that real data will differ from the Gaussian shape in a variety of ways, so that treating the Gaussian case is only the beginning.

Mathematically, it would be much neater if both:

⋄ we knew EXACTLY how values are distributed.

⋄ this exact knowledge could be expressed SIMPLY.

In the real world we can never expect either of these to hold. **We are indeed lucky if:**

⋄ we know APPROXIMATELY how values are distributed.

⋄ this approximation is itself NOT TOO FAR from being simple.

To accept these facts is never an easy task, never something rapidly learned. **It is, however, an inevitable necessity.**

Exhibit 7 sets the Gaussian shape out as a re-expression of fractions. We have chosen, in order to tie ourselves to something more easily assessed, to scale this re-expression so that 25% and 75% correspond to ±1 (rather than

exhibit 7/19: Shapes of distribution

exhibit 7 of chapter 19: reference table

**The Gaussian shape as a re-expression of fractions with 25% and 75% re-expressed as ±1 (take sign of answer from head of column giving %)**

| + | .0% | .2% | .4% | .6% | .8% | 1.0% — |
|---|---|---|---|---|---|---|
| 50% | .000 | .007 | .015 | .022 | .030 | .037 **49%** |
| 51 | .037 | .045 | .052 | .059 | .067 | .074 **48** |
| 52 | .074 | .082 | .089 | .097 | .104 | .112 **47** |
| 53 | .112 | .119 | .126 | .134 | .141 | .149 **46** |
| 54 | .149 | .156 | .164 | .171 | .179 | .186 **45** |
| 55% | .186 | .203 | .201 | .209 | .216 | .224 **44%** |
| 56 | .224 | .231 | .239 | .246 | .254 | .262 **43** |
| 57 | .262 | .269 | .277 | .284 | .292 | .299 **42** |
| 58 | .299 | .307 | .314 | .322 | .330 | .337 **41** |
| 59 | .337 | .345 | .351 | .360 | .368 | .376 **40** |
| 60% | .376 | .383 | .391 | .397 | .406 | .414 **39%** |
| 61 | .414 | .422 | .430 | .437 | .445 | .453 **38** |
| 62 | .453 | .461 | .469 | .476 | .484 | .492 **37** |
| 63 | .492 | .500 | .508 | .514 | .523 | .532 **36** |
| 64 | .532 | .539 | .547 | .555 | .563 | .571 **35** |
| 65% | .571 | .579 | .587 | .595 | .603 | .612 **34%** |
| 66 | .612 | .620 | .628 | .636 | .644 | .652 **33** |
| 67 | .652 | .660 | .669 | .677 | .685 | .693 **32** |
| 68 | .693 | .702 | .710 | .718 | .727 | .735 **31** |
| 69 | .735 | .744 | .752 | .760 | .769 | .777 **30** |
| 70% | .777 | .786 | .795 | .803 | .812 | .820 **29%** |
| 71 | .820 | .829 | .838 | .846 | .855 | .864 **28** |
| 72 | .864 | .872 | .882 | .890 | .900 | .909 **27** |
| 73 | .909 | .918 | .927 | .936 | .945 | .954 **26** |
| 74 | .954 | .962 | .972 | .981 | .991 | 1.000 **25** |

| + | .0% .5% | .1% .6% | .2% .7% | .3% .8% | .4% .9% | .5% 1.0% — |
|---|---|---|---|---|---|---|
| 94 | 2.305 | 2.318 | 2.330 | 2.343 | 2.356 | 2.369 **5.5** |
| 94.5 | 2.311 | 2.382 | 2.396 | 2.410 | 2.424 | 2.438 **5** |
| 95% | 2.438 | 2.493 | 2.468 | 2.483 | 2.498 | 2.514 **4.5%** |
| 95.5 | 2.514 | 2.529 | 2.545 | 2.562 | 2.579 | 2.596 **4** |
| 96 | 2.595 | 2.613 | 2.630 | 2.649 | 2.667 | 2.686 **3.5** |
| 96.5 | 2.686 | 2.705 | 2.726 | 2.746 | 2.767 | 2.782 **3** |
| 97 | 2.782 | 2.766 | 2.833 | 2.857 | 2.881 | 2.906 **2.5** |
| 97.5% | 2.906 | 2.932 | 2.958 | 2.986 | 3.015 | 3.045 **2** |
| 98 | 3.045 | 3.076 | 3.109 | 3.143 | 3.179 | 3.217 **1.5** |
| 98.5 | 3.217 | 3.258 | 3.301 | 3.346 | 3.396 | 3.449 **1** |
| 99 | 3.449 | 3.507 | 3.571 | 3.643 | 3.724 | 3.819 **.5** |
| 99.5 | 3.819 | 3.932 | 4.074 | 4.276 | 4.582 | X **0** |
|  | 1.0% .5% | .9% .4% | .8% .3% | .7% .2% | .6% .1% | .5% .0% |

| + | .00% .05% | .02% .07% | .04% .06% | .06% .04% | .08% .03% | .10% — |
|---|---|---|---|---|---|---|
| 98.0% | 3.045 | 3.051 | 3.057 | 3.063 | 3.070 | 3.076 **1.9%** |
| 98.1 | 3.076 | 3.082 | 3.089 | 3.096 | 3.102 | 3.109 **.8** |
| 98.2 | 3.109 | 3.116 | 3.123 | 3.129 | 3.136 | 3.143 **.7** |
| 98.3 | 3.143 | 3.150 | 3.157 | 3.165 | 3.172 | 3.179 **.6** |
| 98.4 | 3.179 | 3.187 | 3.194 | 3.202 | 3.210 | 3.217 **.5** |
| 98.5% | 3.217 | 3.225 | 3.233 | 3.241 | 3.249 | 3.258 **1.4%** |
| 98.6 | 3.258 | 3.266 | 3.274 | 3.283 | 3.292 | 3.301 **.3** |
| 98.7 | 3.301 | 3.309 | 3.318 | 3.328 | 3.337 | 3.346 **.2** |

| | | | | | | | | |
|---|---|---|---|---|---|---|---|---|
| 98.8 | 3.346 | 3.356 | 3.366 | 3.376 | 3.386 | 3.396 | 3.396 | .1 |
| 98.9 | 3.396 | 3.406 | 3.417 | 3.427 | 3.438 | 3.449 | 3.449 | 1.0 |
| 99.0% | 3.449 | 3.460 | 3.472 | 3.483 | 3.495 | 3.507 | 3.507 | .9% |
| 99.1 | 3.507 | 3.520 | 3.532 | 3.545 | 3.558 | 3.571 | 3.571 | .8 |
| 99.2 | 3.571 | 3.585 | 3.599 | 3.613 | 3.628 | 3.643 | 3.643 | .7 |
| 99.3 | 3.643 | 3.659 | 3.674 | 3.690 | 3.707 | 3.724 | 3.724 | .6 |
| 99.4 | 3.724 | 3.742 | 3.760 | 3.779 | 3.799 | 3.819 | 3.819 | .5 |
| 99.5% | 3.819 | 3.840 | 3.861 | 3.884 | 3.908 | 3.932 | 3.932 | .4% |
| 99.6 | 3.932 | 3.958 | 3.984 | 4.013 | 4.042 | 4.074 | 4.074 | .3 |
| 99.7 | 4.074 | 4.107 | 4.143 | 4.181 | 4.222 | 4.276 | 4.276 | .2 |
| 99.8 | 4.276 | 4.316 | 4.370 | 4.431 | 4.501 | 4.582 | 4.582 | .1 |
| 99.9 | 4.582 | 4.679 | 4.802 | 4.971 | 5.249 | X | 5.249 | 0.0% |
| | .10% | .08% | .06% | .04% | .02% | .00% | | |
| + | .000% | .001% | .002% | .003% | .004% | .005% | | — |
| | .005% | .006% | .007% | .008% | .009% | .010% | | |
| 99.50% | 3.819 | 3.829 | 3.840 | 3.851 | 3.861 | 3.872 | 3.872 | .45% |
| 99.55 | 3.872 | 3.884 | 3.895 | 3.908 | 3.919 | 3.932 | 3.932 | .4 |
| 99.60 | 3.932 | 3.945 | 3.958 | 3.971 | 3.984 | 3.998 | 3.998 | .35 |
| 99.65 | 3.998 | 4.013 | 4.027 | 4.042 | 4.058 | 4.074 | 4.074 | .3 |
| 99.70 | 4.074 | 4.090 | 4.107 | 4.125 | 4.143 | 4.162 | 4.162 | .25 |
| 99.75 | 4.162 | 4.181 | 4.201 | 4.222 | 4.244 | 4.267 | 4.267 | .2% |
| 99.80 | 4.276 | 4.291 | 4.316 | 4.341 | 4.370 | 4.400 | 4.400 | .15 |
| 99.85 | 4.400 | 4.431 | 4.465 | 4.501 | 4.539 | 4.582 | 4.582 | .1 |
| 99.90 | 4.582 | 4.628 | 4.679 | 4.736 | 4.802 | 4.878 | 4.878 | .05 |
| 99.95 | 4.878 | 4.971 | 5.065 | 5.249 | 5.513 | X | 5.513 | 0.0% |
| | .010% | .009% | .008% | .007% | .006% | .005% | | |
| | .005% | .004% | .003% | .002% | .001% | .000% | | |

| | | | | | | | | |
|---|---|---|---|---|---|---|---|---|
| 75% | 1.000 | 1.009 | 1.019 | 1.028 | 1.036 | 1.047 | 1.047 | 24% |
| 76 | 1.047 | 1.056 | 1.066 | 1.076 | 1.085 | 1.095 | 1.095 | 23 |
| 77 | 1.095 | 1.105 | 1.115 | 1.125 | 1.135 | 1.145 | 1.145 | 22 |
| 78 | 1.145 | 1.154 | 1.165 | 1.175 | 1.185 | 1.195 | 1.195 | 21 |
| 79 | 1.195 | 1.209 | 1.216 | 1.226 | 1.237 | 1.248 | 1.248 | 20 |
| 80% | 1.248 | 1.258 | 1.269 | 1.279 | 1.291 | 1.302 | 1.302 | 19% |
| 81 | 1.302 | 1.313 | 1.324 | 1.335 | 1.346 | 1.351 | 1.351 | 18 |
| 82 | 1.357 | 1.368 | 1.380 | 1.391 | 1.402 | 1.414 | 1.414 | 17 |
| 83 | 1.414 | 1.426 | 1.438 | 1.450 | 1.462 | 1.474 | 1.474 | 16 |
| 84 | 1.474 | 1.487 | 1.499 | 1.511 | 1.524 | 1.537 | 1.537 | 15 |
| 85% | 1.537 | 1.549 | 1.562 | 1.575 | 1.588 | 1.602 | 1.602 | 14% |
| 86 | 1.602 | 1.615 | 1.629 | 1.642 | 1.656 | 1.670 | 1.670 | 13 |
| 87 | 1.670 | 1.684 | 1.698 | 1.713 | 1.727 | 1.742 | 1.742 | 12 |
| 88 | 1.742 | 1.757 | 1.772 | 1.787 | 1.804 | 1.818 | 1.818 | 11 |
| 89 | 1.818 | 1.834 | 1.850 | 1.867 | 1.883 | 1.900 | 1.900 | 10 |
| 90% | 1.900 | 1.917 | 1.934 | 1.951 | 1.970 | 1.988 | 1.988 | 9% |
| 91 | 1.988 | 2.006 | 2.025 | 2.044 | 2.063 | 2.083 | 2.083 | 8 |
| 92 | 2.083 | 2.103 | 2.124 | 2.144 | 2.166 | 2.184 | 2.184 | 7 |
| 93 | 2.188 | 2.210 | 2.233 | 2.257 | 2.280 | 2.305 | 2.305 | 6 |
| 94 | 2.305 | 2.330 | 2.356 | 2.382 | 2.410 | 2.438 | 2.438 | 5 |
| | 1.0% | .8% | .6% | .4% | .2% | .0% | | |
| + | .0% | .1% | .2% | .3% | .4% | .5% | | — |
| | .5% | .6% | .7% | .8% | .9% | 1.0% | | |
| 90% | 1.000 | 1.909 | 1.917 | 1.926 | 1.943 | 1.943 | 1.943 | 9.5% |
| 90.5 | 1.943 | 1.951 | 1.960 | 1.970 | 1.978 | 1.988 | 1.988 | 9 |
| 91 | 1.988 | 1.997 | 2.006 | 2.016 | 2.025 | 2.034 | 2.034 | 8.5 |
| 91.5 | 2.034 | 2.044 | 2.054 | 2.063 | 2.073 | 2.083 | 2.083 | 8 |
| 92 | 2.083 | 2.093 | 2.103 | 2.113 | 2.124 | 2.134 | 2.134 | 7.5 |
| 92.5% | 2.134 | 2.144 | 2.155 | 2.166 | 2.171 | 2.188 | 2.188 | 7% |
| 93 | 2.188 | 2.199 | 2.210 | 2.222 | 2.233 | 2.244 | 2.244 | 6.5 |
| 93.5 | 2.244 | 2.257 | 2.269 | 2.280 | 2.293 | 2.305 | 2.305 | 6 |
| | 1.0% | .9% | .8% | .7% | .6% | .5% | | |
| | .5% | .4% | .3% | .2% | .1% | .0% | | |

matching $f - (1 - f)$ near 50%, for example). Exhibit **8** sets out some problems to be treated using this re-expression. (Chapter 20, below, will relate this shape to certain of the fits of chapter 14.)

### review questions

What is a Gaussian distribution? What is it useful for? What is it dangerous for? Have there been any superstitions about it? What would be mathematically neat? Is the real world neat? What task does all this set us? How do we make a Gaussian distribution available for use?

### 19C. Using letter values to look at shapes of distribution

We have become accustomed to summarizing batches in terms of letter values --of a median, two hinges, two eighths, and so on. For most purposes "a hinge is a hinge" and "an eighth is an eighth". For our present purposes we find it convenient to look somewhat closer.

exhibit **8** of chapter 19: data and problems

**Some problems using the Gaussian re-expression of fractions and the Gaussian standard with which distribution shapes are often compared**

**A) AGES at MARRIAGE of 235,252 spinsters**

| Age range | Isserlis data |
|---|---|
| 15–19 | 17546 |
| 20–24 | 118542 |
| 25–29 | 70411 |
| 30–34 | 20241 |
| 35–39 | 5873 |
| 40–44 | 1706 |
| 45–49 | 636 |
| 50–54 | 171 |
| 55–59 | 64 |
| 60–64 | 28 |
| 65–69 | 23 |
| 70 over | 11 |

**P) PROBLEMS**

8a1) Convert the Isserlis data first to cumulative fractions, and reexpress these fractions on the Gaussian scale.

8a2) Plot these reexpressed fractions in as many ways as are necessary to (i) make a good fit and (ii) reveal the characteristics of the residuals.

**S) SOURCE**

L. Isserlis 1917. "On the representation of statistical data." *Biometrika* 11: 418–425. Table VI at page 423.

Were we to work with batches of 7, 8, 9, 10, 11, or 12 observations, the ss-fractions corresponding to median (M), hinges (H), eighths (E) and so on (D, C, B, ...) are as indicated in exhibit **9**. The ss-fraction for M is always 0.500--right on the nose. The ss-fractions for H's, which we might like to have = .250, run

$$.295, .260, .286, .258, .279, .257.$$

exhibit **9** of chapter 19: example

**The ss-fractions corresponding to MHE..., for batches of 7 to 12 (using both $h = 1/2$ and $s = 1/6$, so that $ss = 1/3$)**

A) BATCHES of 7 or 8

| depth | ss-count | ss-fraction |
|---|---|---|
| # 7 | | |
| M 4 | 3hs | .500 |
| H 2h | 2s | .295 (vs .250) |
| E 1h | 1s | .159 (vs .125) |
| D 1 | hs | .091 (vs .062) |
| | (7ss) | |

| depth | ss-count | ss-fraction |
|---|---|---|
| # 8 | | |
| M 4h | 4s | .500 |
| H 2h | 2s | .260 (vs .250) |
| E 1h | 1s | .140 (vs .125) |
| D 1 | hs | .080 (vs .062) |
| | (8ss) | |

B) BATCHES of 9 or 10

| depth | ss-count | ss-fraction |
|---|---|---|
| # 9 | | |
| M 5 | 4hs | .500 |
| H 3 | 2hs | .286 (vs .250) |
| E 2 | 1hs | .179 (vs .125) |
| D 1h | 1s | .125 (vs .062) |
| C 1 | hs | .072 (vs .031) |
| | (9ss) | |

| depth | ss-count | ss-fraction |
|---|---|---|
| # 10 | | |
| M 5h | 5s | .500 |
| H 3 | 2hs | .258 (vs .250) |
| E 2 | 1hs | .161 (vs .125) |
| D 1h | 1s | .133 (vs .062) |
| C 1 | hs | .065 (vs .031) |
| | (10ss) | |

C) BATCHES of 11 or 12

| depth | ss-count | ss-fraction |
|---|---|---|
| # 11 | | |
| M 6 | 5hs | .500 |
| H 3h | 3s | .279 (vs .250) |
| E 2 | 1hs | .147 (vs .125) |
| D 1h | 1s | .103 (vs .062) |
| C 1 | hs | .059 (vs .031) |
| | (11ss) | |

| depth | ss-count | ss-fraction |
|---|---|---|
| # 12 | | |
| M 6h | 6s | .500 |
| H 3h | 3s | .257 (vs .250) |
| E 2 | 1hs | .135 (vs .125) |
| D 1h | 1s | .095 (vs .062) |
| C 1 | hs | .054 (vs .031) |
| | (12ss) | |

Those for E's (which we might like to be .125) run

.159, .140, .179, .161, .147, .135.

The pattern is clear. As the batch size creeps up, these ss-fractions run down toward their nominal values, and then jump away again (every 2 in count for H, every 4 for E, every 8 for D, etc.). On the whole, though, they get steadily closer to the target values.

When we want to be extra (perhaps sometimes unduly) careful, then, we have a choice:

⋄ work with H, E, ... as they are, making allowance for what ss-fraction each corresponds to.

⋄ go back and replace our H-, E-, etc., values by values at other depths (for E for a batch of 7, for example, we want the ss-count to be $(.125)(7.33) = 0.91$, so that we need to work at a depth about .34/.68 nearer the upper extreme, namely 1.24 (instead of at 1h = 1.50).

Taking the latter choice is a lot of repeated work, since we have to go 24% of the way from depth 1 to depth 2 every time we deal with a batch of 7. Going the former route is a lot of work, but it is work that can be over and done with, since we can, for each batch size, first ask what ss-fraction corresponds to the H, E, D, or whatever (for that batch size) and then re-express the resulting ss-fraction once and for all.

Exhibit 12 will, when we reach it, show the result for batches of sizes 1 to 100, while exhibit 13 will give approximate formulas to be used for batches of more than 100. Before we turn to these exhibits, however, we shall look at how the results are to be used, taking the $\sqrt{\text{height}}$ of 219 volcanoes--given in stem-and-leaf form in exhibit 11 of chapter 3--as our example. Exhibit **10** has the calculations, first for H's, E's, D's, and C's separately (panels A to D), then for all as part of a letter-value-display (panel E), and finally all this with mid's included (panel F). The final step, shown in panel G, is to list smoothed values

exhibit **9** of chapter 19 (continued)

### P) PROBLEMS

9a) Provide similar information for batches of 13 or 14.

9b) Provide similar information for batches of 15 or 16.

9c) Provide similar information for batches of 17 or 18.

9d) Provide similar information for batches of 19 or 20.

9abcd2) Compare results for batches of 13 to 20 with what you expected from 7 to 12. Discuss.

of mids and "25% pseudospreads". Each 25% pseudospread would be 2.00 for a distribution (of an infinite number of values) with exactly Gaussian shape, and 25% points at ±1.00. Equally, each would be the 25% spread of the distribution if the batch followed a Gaussian distribution exactly (to do which it would have to include infinitely many values). We know that it almost never

exhibit **10** of chapter 19: volcano heights

**Using M, H, E, ... to look at shape of distribution for 219 volcano √height's**

**A)** The HINGES--at depth 55h
```
hinges:     61    and    98
H-spread:         37
   divisor:      0.995      (1.00 would do)
     ratio:      37.2
```

**B)** The EIGHTHS--at depth 28
```
eighths:    49    and    110
E-spread:         61
   divisor:      1.705      (1.71 would do)
     ratio:      35.8
```

**C)** The D's--at depth 14h
```
    D's:    40    and    118
D-spread:         78
   divisor:      2.254      (2.27 would nearly do)
     ratio:      34.6
```

**D)** The C's--at depth 7h
```
    C's:    30    and    132
C-spread:         102
   divisor:      2.738      (2.76 would probably do)
     ratio:      37.3
```

**E)** the EXTENDED LETTER-VALUE DISPLAY

|   | # 219 |     |     | spr |         | pseudo-spread |
|---|-------|-----|-----|-----|---------|---------------|
| M | 110   |  80 |     |     |         |               |
| H | 55h   | 61  | 98  | 37  | (0.995) | 37.2          |
| E | 28    | 49  | 110 | 61  | (1.705) | 35.8          |
| D | 14h   | 40  | 118 | 78  | (2.254) | 34.7          |
| C | 7h    | 30  | 132 | 102 | (2.738) | 37.4          |
| B | 4     | 25  | 140 | 115 | (3.165) | 36.3          |
| A | 2h    | 23h | 140 | 116h| (3.482) | 33.4          |
| Z | 1h    | 18  | 140 | 122 | (3.806) | 32.1          |
| Y | 1     | 14  | 140 | 126 | (3.95)  | 31.9          |

➔

will follow a Gaussian distribution exactly. (In particular, no human ever worked with a batch with infinitely many values.) So we call such values

### pseudospreads

to remind us that they will tell us about a particular spread--here the 25% spreads--when a very rare thing happens. They are useful to us for all that,

exhibit 10 of chapter 19 (continued).

**F) The Same with MIDS INCLUDED**

| #219 | | | Mid | | Spr | | | 25% Pseudo-Spread |
|---|---|---|---|---|---|---|---|---|
| M | 110 | | 80 | | | | | |
| H | 55h | 61 | (79h) | 98 | 37 | | ( .995) | 37.2 |
| E | 28 | 49 | (79) | 110 | 61 | | (1.705) | 35.8 |
| D | 14h | 40 | (79) | 118 | 78 | | (2.244) | 37.6 |
| C | 7h | 30 | (81) | 132 | 102 | | (2.738) | 37.3 |
| B | 4 | 25 | (82h) | 140 | 115 | | (3.165) | 36.3 |
| A | 2h | 23h | (82) | 140 | 116h | | (3.482) | 33.4 |
| Z | 1h | 18 | (79) | 140 | 122 | | (3.806) | 32.1 |
| Y | 1 | 14 | (77) | 140 | 126 | | (3.95) | 31.9 |

**G) The MIDS and 25%-PSEUDOSPREADS after SMOOTHING--by 3RSS**

| letter | mid | 25% | div² |
|---|---|---|---|
| M | 79h | (35.8) | 0 |
| H | 79h | 35.8 | .99 |
| E | 79h | 35.8 | 2.91 |
| D | 79h | 35.8 | 5.08 |
| C | 81 | 35.8 | 7.5 |
| B | 82 | 35.8 | 10.0 |
| A | 82 | 33.4 | 12.1 |
| Z | 79 | 32.1 | 14.5 |
| Y | 77 | 31.9 | 15.6 |

**P) PROBLEMS**

10a) Do the same for the heights (not $\sqrt{\text{heights}}$) of the 219 volcanoes. Discuss the result.

10b) Do the same for the log heights (not $\sqrt{\text{heights}}$) of the 219 volcanoes. Discuss the result.

10c) Do the same for the 82 areas of the counties of Mississippi (see exhibit 4 of chapter 1). Discuss the result.

10d) Find a batch of 80 to 150 values that interest you and do the same.

10e) Find a batch of 150 to 300 values that interest you and do the same.

**S) SOURCE**

Exhibit 11 of chapter 3.

because by their CHANGES they tell us how the shape of distribution of the batch compares with the Gaussian shape.

Exhibit 11 plots the (smoothed) mids and 25% pseudospreads, using the square of the divisors as the horizontal coordinate. We see the mids waving first one way and then the other, leaving us with no particular impression of skewness. The 25% pseudospreads start at about 37 and eventually decrease to 32, suggesting a possible (slight) tendency toward squeezed-in tails.

Exhibit 12, as advertised, contains the divisors for batches of up to 100; exhibit 13 contains approximate formulas for use with larger batches.

Exhibit 14 deals with a smaller example--one calling for exhibit 12 rather than for exhibit 13. The resulting mids and 25% pseudospreads are plotted in exhibit 15. Both show a trend. The trend in mids corresponds to skewness with high values stretched farther out than low values, the fitted line being least satisfactory at the median. The 25% pseudospreads have a trend corresponding to stretched tails; the fitted line fits fairly well (as might other shapes of curve).

exhibit 11 of chapter 19: volcano heights

**Plots of 25% pseudospread and mid against (Gaussian) ratio for √heights of 219 volcanoes (from exhibit 10)**

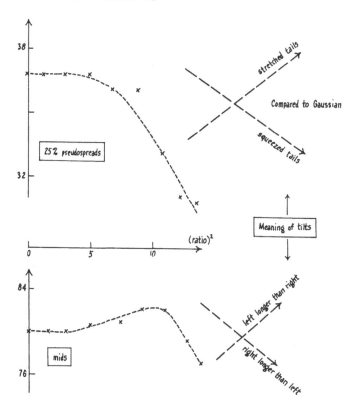

exhibit 12 of chapter 19: reference table

**Factors to get pseudospreads from HED... spreads for batches of ≤100**

| Batch size | H | E | D | C | B | A | Batch size | H | E | D | C | B | A | Z |
|---|---|---|---|---|---|---|---|---|---|---|---|---|---|---|
| 3 | .57 | 1.25 | X | X | X | X | 45 | .97 | 1.63 | 2.19 | 2.65 | 2.89 | 3.23 | |
| 4 | .91 | 1.51 | X | X | X | X | 46 | .99 | 1.65 | 2.21 | 2.67 | 2.90 | 3.24 | |
| 5 | .72 | 1.15 | 1.70 | X | X | X | 47 | .97 | 1.67 | 2.22 | 2.68 | 2.91 | 3.25 | |
| 6 | .94 | 1.33 | 1.86 | X | X | X | 48 | .99 | 1.69 | 2.24 | 2.69 | 2.93 | 3.26 | |
| 7 | .80 | 1.48 | 1.98 | X | X | X | 49 | .97 | 1.63 | 2.14 | 2.53 | 2.94 | 3.28 | |
| 8 | .95 | 1.60 | 2.08 | X | X | X | 50 | .99 | 1.65 | 2.16 | 2.54 | 2.95 | 3.29 | |
| 9 | .84 | 1.36 | 1.70 | 2.17 | X | X | 51 | .97 | 1.67 | 2.17 | 2.56 | 2.96 | 3.30 | X |
| 10 | .96 | 1.47 | 1.79 | 2.25 | X | X | 52 | .99 | 1.69 | 2.19 | 2.51 | 2.98 | 3.31 | X |
| 11 | .87 | 1.55 | 1.87 | 2.32 | X | X | 53 | .97 | 1.64 | 2.20 | 2.58 | 2.99 | 3.32 | X |
| 12 | .97 | 1.63 | 1.95 | 2.38 | X | X | 54 | .99 | 1.66 | 2.21 | 2.60 | 3.00 | 3.33 | X |
| 13 | .89 | 1.46 | 2.01 | 2.44 | X | X | 55 | .97 | 1.67 | 2.23 | 2.61 | 3.01 | 3.34 | X |
| 14 | .97 | 1.53 | 2.07 | 2.49 | X | X | 56 | .99 | 1.69 | 2.29 | 2.62 | 3.02 | 3.35 | X |
| 15 | .90 | 1.59 | 2.12 | 2.54 | X | X | 57 | .97 | 1.64 | 2.16 | 2.63 | 3.03 | 3.36 | X |
| 16 | .98 | 1.65 | 2.17 | 2.58 | X | X | 58 | .99 | 1.66 | 2.17 | 2.67 | 3.04 | 3.32 | X |
| 17 | .91 | 1.51 | 1.93 | 2.22 | 2.62 | X | 59 | .97 | 1.67 | 2.18 | 2.60 | 3.05 | 3.38 | X |
| 18 | .98 | 1.56 | 1.98 | 2.26 | 2.66 | X | 60 | .99 | 1.69 | 2.20 | 2.67 | 3.06 | 3.39 | X |
| 19 | .92 | 1.61 | 2.02 | 2.30 | 2.69 | X | 61 | .97 | 1.65 | 2.21 | 2.68 | 3.07 | 3.40 | X |
| 20 | .98 | 1.66 | 2.06 | 2.34 | 2.73 | X | 62 | .99 | 1.66 | 2.22 | 2.69 | 3.08 | 3.41 | X |
| 21 | .93 | 1.55 | 2.10 | 2.37 | 2.76 | X | 63 | .98 | 1.68 | 2.23 | 2.70 | 3.09 | 3.42 | X |
| 22 | .98 | 1.59 | 2.14 | 2.41 | 2.79 | X | 64 | .99 | 1.69 | 2.25 | 2.71 | 3.10 | 3.43 | X |
| 23 | .93 | 1.63 | 2.17 | 2.44 | 2.82 | X | 65 | .98 | 1.65 | 2.17 | 2.58 | 2.89 | 3.11 | 3.44 |
| 24 | .99 | 1.67 | 2.20 | 2.47 | 2.85 | X | 66 | .99 | 1.66 | 2.19 | 2.59 | 2.90 | 3.12 | 3.45 |
| 25 | .94 | 1.57 | 2.03 | 2.50 | 2.87 | X | 67 | .98 | 1.68 | 2.20 | 2.60 | 2.91 | 3.13 | 3.45 |
| 26 | .99 | 1.61 | 2.06 | 2.52 | 2.90 | X | 68 | .99 | 1.69 | 2.21 | 2.61 | 2.92 | 3.14 | 3.46 |
| 27 | .94 | 1.64 | 2.09 | 2.55 | 2.92 | X | 69 | .98 | 1.65 | 2.22 | 2.62 | 2.93 | 3.15 | 3.47 |
| 28 | .99 | 1.67 | 2.12 | 2.57 | 2.94 | X | 70 | .99 | 1.67 | 2.23 | 2.63 | 2.94 | 3.16 | 3.48 |
| 29 | .95 | 1.59 | 2.15 | 2.60 | 2.96 | X | 71 | .98 | 1.68 | 2.24 | 2.64 | 2.95 | 3.16 | 3.48 |
| 30 | .99 | 1.62 | 2.17 | 2.62 | 2.99 | X | 72 | .99 | 1.69 | 2.25 | 2.65 | 2.96 | 3.18 | 3.49 |
| 31 | .95 | 1.65 | 2.20 | 2.64 | 3.01 | X | 73 | .98 | 1.66 | 2.19 | 2.66 | 2.97 | 3.18 | 3.50 |
| 32 | .99 | 1.68 | 2.22 | 2.66 | 3.02 | X | 74 | .99 | 1.67 | 2.19 | 2.67 | 2.97 | 3.19 | 3.51 |
| 33 | .95 | 1.60 | 2.08 | 2.44 | 2.68 | 3.04 | 75 | .98 | 1.68 | 2.20 | 2.65 | 2.98 | 3.20 | 3.52 |
| 34 | .99 | 1.63 | 2.11 | 2.46 | 2.79 | 3.05 | 76 | .99 | 1.69 | 2.21 | 2.69 | 2.99 | 3.21 | 3.53 |
| 35 | .96 | 1.65 | 2.13 | 2.48 | 2.72 | 3.08 | 77 | .98 | 1.66 | 2.22 | 2.70 | 3.00 | 3.21 | 3.53 |
| 36 | .99 | 1.68 | 2.15 | 2.50 | 2.74 | 3.10 | 78 | 1.00 | 1.67 | 2.23 | 2.70 | 3.01 | 3.22 | 3.54 |
| 37 | .96 | 1.61 | 2.17 | 2.52 | 2.75 | 3.11 | 79 | .98 | 1.68 | 2.24 | 2.71 | 3.01 | 3.23 | 3.54 |
| 38 | .99 | 1.64 | 2.19 | 2.54 | 2.78 | 3.13 | 80 | 1.00 | 1.69 | 2.25 | 2.72 | 3.02 | 3.24 | 3.55 |
| 39 | .96 | 1.66 | 2.21 | 2.55 | 2.79 | 3.14 | 81 | .96 | 1.66 | 2.19 | 2.61 | 3.03 | 3.24 | 3.56 |
| 40 | .99 | 1.68 | 2.23 | 2.57 | 2.81 | 3.16 | 82 | 1.00 | 1.67 | 2.20 | 2.62 | 3.04 | 3.25 | 3.56 |
| 41 | .96 | 1.62 | 2.12 | 2.59 | 2.83 | 3.17 | 83 | .98 | 1.68 | 2.21 | 2.63 | 3.05 | 3.26 | 3.57 |
| 42 | .99 | 1.64 | 2.14 | 2.61 | 2.84 | 3.17 | 84 | 1.00 | 1.69 | 2.22 | 2.64 | 3.06 | 3.27 | 3.58 |
| 43 | .95 | 1.66 | 2.15 | 2.62 | 2.56 | 3.20 | 85 | .98 | 1.66 | 2.23 | 2.65 | 3.06 | 3.27 | 3.59 |
| 44 | .99 | 1.68 | 2.17 | 2.64 | 2.87 | 3.21 | 86 | 1.00 | 1.67 | 2.24 | 2.65 | 3.07 | 3.28 | 3.60 |

→

exhibit **12** of chapter 19 (continued)

| Batch size | H | E | D | C | B | A | Z | Batch size | H | E | D | C | B | A | Z |
|---|---|---|---|---|---|---|---|---|---|---|---|---|---|---|---|
| 87 | .98 | 1.68 | 2.25 | 2.66 | 3.07 | 3.28 | 3.60 | 93 | .98 | 1.67 | 2.23 | 2.71 | 3.11 | 3.32 | 3.63 |
| 88 | 1.00 | 1.69 | 2.25 | 2.67 | 3.08 | 3.29 | 3.60 | 94 | 1.00 | 1.68 | 2.24 | 2.71 | 3.12 | 3.33 | 3.64 |
| 89 | .98 | 1.66 | 2.20 | 2.67 | 3.09 | 3.30 | 3.61 | 95 | .98 | 1.69 | 2.25 | 2.72 | 3.13 | 3.33 | 3.64 |
| 90 | 1.00 | 1.67 | 2.21 | 2.68 | 3.09 | 3.30 | 3.61 | 96 | 1.00 | 1.70 | 2.25 | 2.73 | 3.13 | 3.34 | 3.65 |
| 91 | .98 | 1.68 | 2.22 | 2.69 | 3.10 | 3.31 | 3.62 | 97 | .98 | 1.67 | 2.21 | 2.64 | 2.98 | 3.35 | 3.66 |
| 92 | 1.00 | 1.69 | 2.22 | 2.70 | 3.11 | 3.32 | 3.63 | 98 | 1.00 | 1.68 | 2.21 | 2.64 | 2.98 | 3.35 | 3.66 |
|  |  |  |  |  |  |  |  | 99 | .98 | 1.69 | 2.72 | 2.65 | 2.99 | 3.36 | 3.67 |
|  |  |  |  |  |  |  |  | 100 | 1.00 | 1.70 | 2.23 | 2.66 | 3.00 | 3.37 | 3.67 |

exhibit **13** of chapter 19: reference table

**Factors to get 25% pseudospreads from HED... spreads for batches of size ≥100**

A) The APPROXIMATE FACTORS--divide into indicated spread to get pseudo-H-spread

for H-spreads: $1.00 - \frac{2}{n}\left(1 - .86\left(\frac{n}{2} \text{ Mod 1}\right)\right)$

for E-spreads: $1.71 - \frac{3}{n}\left(1 - .80\left(\frac{n}{4} \text{ Mod 1}\right)\right)$

for D-spreads: $2.27 - \frac{5}{n}\left(1 - .77\left(\frac{n}{8} \text{ Mod 1}\right)\right)$

for C-spreads: $2.76 - \frac{10}{n}\left(1 - .76\left(\frac{n}{16} \text{ Mod 1}\right)\right)$

for B-spreads: $3.19 - \frac{15}{n}\left(1 - .76\left(\frac{n}{32} \text{ Mod 1}\right)\right)$

for A-spreads: $3.58 - \frac{30}{n}\left(1 - .75\left(\frac{n}{64} \text{ Mod 1}\right)\right)$

for Z-spreads: $3.94 - \frac{55}{n}\left(1 - .75\left(\frac{n}{128} \text{ Mod 1}\right)\right)$

for Y-spreads: $4.28 - \frac{110}{n}\left(1 - .75\left(\frac{n}{256} \text{ Mod 1}\right)\right)$

for X-spreads: $4.59 - \frac{200}{n}\left(1 - .75\left(\frac{n}{512} \text{ Mod 1}\right)\right)$

for W-spreads: $4.89 - \frac{380}{n}\left(1 - .75\left(\frac{n}{1024} \text{ Mod 1}\right)\right)$

→

exhibit **13** of chapter 19 (continued)

**B) EXAMPLE of USE--for n = 137**

| | Halvings | Fractions* | 1 − (const)(fraction) | Resulting factor |
|---|---|---|---|---|
| H | n/2 = 68.5 | .5 | (1 −    ) = .57 | $1.00 - \frac{2}{137}(.57) = 0.99$ |
| E | n/4 = 34.25 | .25 | (1 −    ) = .80 | $1.71 - \frac{3}{137}(.80) = 1.69$ |
| D | n/8 = 17.125 | .125 | (1 −    ) = .90 | $2.27 - \frac{5}{137}(.90) = 2.24$ |
| C | n/16 = 8.5625 | .56 | (1 −    ) = .57 | $2.76 - \frac{10}{137}(.57) = 2.72$ |
| B | n/32 = 4.28125 | .28 | (1 −    ) = .79 | $3.19 - \frac{15}{137}(.79) = 3.10$ |
| A | n/64 = 2.1406 | .14 | (1 −    ) = .90 | $3.58 - \frac{30}{137}(.90) = 3.38$ |
| Z | n/128 = 1.0703 | .07 | (1 −    ) = .95 | $3.94 - \frac{55}{137}(.95) = 3.56$ |
| Y | n/256 = .5352 | .54 | (1 −    ) = .60 | $4.28 - \frac{110}{137}(.60) = 3.80$ |

* These are values of "n/divisor Mod 1" for "n/divisor" shown in first column.

**C) IMPORTANT NOTE**

The capital M in "x Mod 1" MEANS "x MINUS the next LOWER integer"; thus:

| Explicit form | Fraction |
|---|---|
| 4.001 Mod 1 = .001 | .001 |
| 4.000 Mod 1 = 1.000 | 1.000 |
| 3.999 Mod 1 = .999 | .999 |

and so on.

**D) ARITHMETIC SAVER--making successive division easy**

```
137 = 68 · 2  + 1        219 = 109 · 2  + 1
    = 34 · 4  + 1            =  54 · 4  + 3
    = 17 · 8  + 1            =  27 · 8  + 3
    =  8 · 16 + 9            =  13 · 16 + 11
    =  4 · 32 + 9            =   6 · 32 + 27
    =  2 · 64 + 9            =   3 · 64 + 27
    =  1 · 128 + 9           =   1 · 128 + 91
    =  0 · 128 + 137         =   0 · 256 + 219
```

exhibit **13** of chapter 19 (continued)

Examples of mental calculation:

$27 \cdot 8 = 26 \cdot 8 + 1 \cdot 8$
$\phantom{27 \cdot 8} = 13 \cdot 16 + 8$
$3 + 8 = 11$

$13 \cdot 16 = 12 \cdot 16 + 1 \cdot 16$
$\phantom{13 \cdot 16} = 6 \cdot 32 + 16$
$16 + 11 = 27$

**P) PROBLEMS**

13a) Find the factors for H, E, D, C, B, and A values, for $n = 137$. (Show your work.)

13b) Same for $n = 257$.

13c) Same for $n = 173$.

13d) Same for $n = 1079$.

exhibit **14** of chapter 19: county areas

**Mids and 25% pseudospreads for the logs of the areas of the 83 counties of Michigan**

**A) The FULL LETTER-VALUE DISPLAY**

|      |       | mid     |       | spr  |        | 25% pseudo-spread | div² |       |
|------|-------|---------|-------|------|--------|-------------------|------|-------|
| M42  |       | 1.76    |       |      |        |                   | 0.0  |       |
| H21h | 1.73h | (1.81h) | 1.89h | .16  | ( .98) | .16               | 1.0  |       |
| E11  | 1.69  | (1.83h) | 1.98  | .29  | (1.68) | .17               | 2.8  |       |
| D6   | 1.66  | (1.86h) | 2.07  | .41  | (2.21) | .19               | 4.9  |       |
| C3h  | 1.59  | (1.84)  | 2.10  | .51  | (2.63) | .19               | 6.9  |       |
| B2   | 1.54  | (1.87)  | 2.20  | .66  | (3.05) | .22               | 9.3  |       |
| A1h  | 1.52h | (1.88)  | 2.23  | .70h | (3.26) | .22               | 10.6 | 1     |
| Z1   | 1.51  | (1.88h) | 2.26  | .75  | (3.57) | .21               | 12.7 | 5     |

**P) PROBLEMS**

14a) Imitate panel A for the areas (not log areas) of the 83 counties.

14a2) Plot the results of (14a1).

exhibit **14** of chapter 19 (continued)

14b) Same for the root areas.

14b2) Plot the results of (14b1).

14ab3) Compare the three analyses of the 83-county batch (panel A, (14a2), (14b2)). Discuss the comparison. What re-expression would be most symmetrical?

14c) Find a batch of 80 to 125 values that interest you and imitate panel A.

14c2) Plot the results of (14c).

14d) Find a batch of 150 to 250 values that interest you and imitate panel A.

14d2) Plot the results of (14d).

**S) SOURCE**
Exhibit 5 of chapter 3


exhibit **15** of chapter 19: Michigan counties

**Plots of mids and 25% pseudospreads for the logs of the land areas of the 83 counties of Michigan (from exhibit 14)**

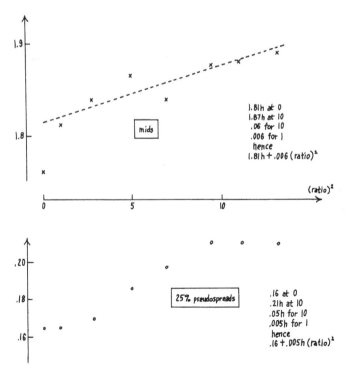

The relations

$$\text{mid} = 1.81h + .006\,(\text{div})^2$$

and

$$25\%\text{ pseudospread} = 0.16 + 0.005h\,(\text{div})^2$$

thus describe the shape, spread, and location of these 83 points with moderate precision.

### review questions

Can we assign ss-fractions to letter values? What happens if we try? In how many ways could we be extra careful? What are they? Which do we choose? Why? How hard is it to package the results? What example did we try the result on first? What did the results look like? What is a mid? What is a pseudospread? What do we plot mids against? Pseudospreads? How do these plots behave in our examples? What do we put in a table for use in finding 25% pseudospreads? What do we do for larger batches? What example did we do next? What happened? What formulas summarized our plot? Did we do a further example? What happened?

### 19D. Pushback technique (optional section)

Now that we have seen how hard we can look at shapes using mids (found directly, without a standard) and 25% spreads (found using a Gaussian standard), we would like to look at the individual observations--or individual letter values--as effectively as possible. A variety of simple pushback techniques let us do this easily in a way where we find ourselves doing other useful things.

We shall start from the letter-values, and the (Gaussian) 25% pseudospreads found from them. We let $Q$ be

one-half the median of the 25% pseudospreads.

For the example of exhibit 14--the log areas of the 83 counties of Michigan-- for instance, the median is 0.19, so that $Q = 0.10$. We can now proceed as in exhibit **16**, multiplying $Q$ by the appropriate value of "ratio", and subtracting the result from each of the letter values. Exhibit **17** shows the smoothed differences plotted against ratios. Two sections are reasonably well fitted by (segments of) straight lines, which, when we add back the amounts

$$Q\,(\text{ratio}) = .1\,(\text{ratio})$$

that we had subtracted before plotting exhibit 17, become

(for ratios below $-1.5$)    $1.86 + .1\,(\text{ratio})$

(for ratios above $+1.0$)    $1.741 + 0.048\,(\text{ratio})$.

(We could fit the middle--rather more roughly--with a third straight segment if we wished.)

If we want to describe the shape of distribution of these 83 logs of areas of counties of Michigan, we can now say, AMONG OTHER THINGS:

◦ each tail behaves rather like the Gaussian standard in shape.

◦ the righthand (positive) tail is longer than the lefthand (negative) one.

By dealing with the two tails separately, we are able, for this example, to say simpler things than we said in the last section, where we worked with location

**exhibit 16 of chapter 19: Michigan counties**

**Pushback analysis (letter-value version) of the logs of the areas of the 83 counties of Michigan** $(Q = .10)$

**A) CALCULATIONS**

| Depth | Letter | Obs'd | Ratio | Q× Ratio | Diff* | 3R | | S | | | Fits | Resid |
|---|---|---|---|---|---|---|---|---|---|---|---|---|
| 1 | Z | 1.51 | −3.57 | −.36 | 1.87 | 1.86 | | | | N O W | 1.503 | +.007 |
| 1h | A | 1.52h | −3.26 | −.33 | 1.85h | 1.85h | | | | | 1.534 | −.009 |
| 2 | B | 1.54 | −3.05 | −.31 | 1.85 | 1.85 | ⌉1.85 | | | L O K | 1.555 | −.015 |
| 3h | C | 1.59 | −2.63 | −.26 | 1.85 | 1.85 | ⌋1.86 | 1.85h | | | 1.597 | −.007 |
| 6 | D | 1.66 | −2.21 | −.22 | 1.88 | 1.86 | ⌉1.85 | 1.85h | | | 1.639 | +.021 |
| 11 | E | 1.69 | −1.68 | −.17 | 1.86 | 1.86 | ⌋1.86 | 1.85 | | A T | 1.692 | −.002 |
| 21h | H | 1.73h | −.98 | −.10 | 1.83h | 1.83h | | | | | (1.762) | (−.027) |
| 42 | M | 1.76 | .00 | .00 | 1.76 | 1.79h | ⌉1.79h | | | E X | (1.741) | (+.019) |
| 21h | H | 1.89h | .98 | .10 | 1.79h | 1.79h | ⌋1.79h | | | H B | 1.886 | +.009 |
| 11 | E | 1.98 | 1.68 | .17 | 1.81 | 1.81 | | | | I T | 1.990 | −.010 |
| 6 | D | 2.07 | 2.21 | .22 | 1.85 | 1.84 | | | | | 2.068 | +.002 |
| 3h | C | 2.10 | 2.63 | .26 | 1.84 | 1.85 | | | | | 2.130 | −.030 |
| 2 | B | 2.20 | 3.05 | .31 | 1.89 | 1.89 | | | | 17 | 2.192 | +.008 |
| 1h | A | 2.23 | 3.26 | .33 | 1.90 | 1.90 | | | | | 2.223 | +.007 |
| 1 | Z | 2.26 | 3.57 | .36 | 1.90 | 1.90 | | | | | 2.269 | −.009 |

Notes

1. "Ratio" contains signed values of "factors" from exhibit 12.
2. Area enclosed in ⌞⌟ could be omitted and the next column filled in by symmetry. (It is convenient to fill it in for use in plotting.)

* Diff = "obsd" MINUS "Q × Ratio"

**P) PROBLEMS**

16a) Make a similar calculation and plot for the root volcano height data of exhibit 10. How many straight lines seem needed? Discuss and explain.

**S) SOURCE**

Exhibit 14 above.

and spread--with mids and 25% pseudospreads. In other examples, the opposite can and will happen--the location and spread pictures will be simpler. We need to have both approaches.

**a half-octave version**

As a way of describing the general shape of a batch, using its letter values (as we have just done) is quite reasonable and effective. If we want to focus our attention on individual values, we can usually do somewhat better.

In exhibit 16, the depths involved were 1, 1h, 2, 3h, 6, 11, ... Clearly, 1, 2, 6, and 11 tell us about individual values, while 1h wastefully tells us about a mixture of depths 1 and 2. Moreover, and perhaps confusingly, 3h tells us about a mixture of depths 3 and 4. With the same amount of arithmetic we could deal with 1, 2, 3, 4, 6, 11, ..., instead, finding only results each of which tells us about an individual point. If our concern is with individual values, this is a better thing to do.

Exhibit **18** offers values of "ratio" for such a "half-octave" choice of depths and batch sizes up to 100. Exhibit **19** applies these to the logs of areas of counties of Michigan. Exhibit **20** illuminates this application.

exhibit **17** of chapter 19: Michigan counties

**Plot of adjusted letter values against (Gaussian) ratios (pushback analysis, from exhibit 16)**

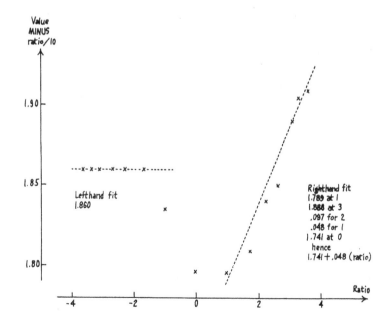

exhibit 18 of chapter 19: reference table

**Values of standard ratios for points at half-octave depths (standard is the Gaussian shape)**

| Batch count | \multicolumn{8}{l}{Ratio for point at depth} | | | | | | | |
|---|---|---|---|---|---|---|---|---|
| | 1 | 2 | 3 | 4 | 6 | 8 | 11 | 16 | 22 |
| 1 | .00 | X | X | X | X | X | X | X | X |
| 2 | .84 | X | X | X | X | X | X | X | X |
| 3 | 1.25 | .00 | X | X | X | X | X | X | X |
| 4 | 1.51 | .44 | X | X | X | X | X | X | X |
| 5 | 1.71 | .72 | .00 | X | X | X | X | X | X |
| 6 | 1.86 | .94 | .30 | X | X | X | X | X | X |
| 7 | 1.98 | 1.11 | .52 | .00 | X | X | X | X | X |
| 8 | 2.08 | 1.25 | .69 | .22 | X | X | X | X | X |
| 9 | 2.17 | 1.37 | .84 | .40 | X | X | X | X | X |
| 10 | 2.25 | 1.47 | .96 | .55 | X | X | X | X | X |
| 11 | 2.32 | 1.56 | 1.07 | .68 | .00 | X | X | X | X |
| 12 | 2.38 | 1.63 | 1.16 | .79 | .15 | X | X | X | X |
| 13 | 2.44 | 1.71 | 1.25 | .89 | .28 | X | X | X | X |
| 14 | 2.49 | 1.77 | 1.33 | .98 | .39 | X | X | X | X |
| 15 | 2.54 | 1.83 | 1.39 | 1.05 | .49 | .00 | X | X | X |
| 16 | 2.58 | 1.88 | 1.45 | 1.12 | .58 | .11 | X | X | X |
| 17 | 2.62 | 1.93 | 1.51 | 1.19 | .66 | .22 | X | X | X |
| 18 | 2.66 | 1.98 | 1.57 | 1.25 | .73 | .31 | X | X | X |
| 19 | 2.70 | 2.02 | 1.62 | 1.30 | .81 | .39 | X | X | X |
| 20 | 2.73 | 2.06 | 1.66 | 1.36 | .87 | .46 | X | X | X |
| 21 | 2.76 | 2.10 | 1.71 | 1.40 | .93 | .53 | .00 | X | X |
| 22 | 2.79 | 2.19 | 1.75 | 1.45 | .98 | .60 | .08 | X | X |
| 23 | 2.82 | 2.17 | 1.79 | 1.49 | 1.03 | .66 | .16 | X | X |
| 24 | 2.85 | 2.20 | 1.82 | 1.53 | 1.08 | .71 | .23 | X | X |
| 25 | 2.87 | 2.24 | 1.86 | 1.57 | 1.13 | .77 | .30 | X | X |
| 26 | 2.90 | 2.27 | 1,89 | 1.61 | 1.17 | .82 | .36 | X | X |
| 27 | 2.92 | 2.29 | 1.92 | 1.64 | 1.21 | .86 | .41 | X | X |
| 28 | 2.94 | 2.32 | 1.95 | 1.67 | 1.25 | .91 | .47 | X | X |
| 29 | 2.97 | 2.35 | 1.98 | 1.71 | 1.28 | .94 | .52 | X | X |
| 30 | 2.99 | 2.37 | 2.01 | 1.74 | 1.32 | .99 | .56 | X | X |
| 31 | 3.01 | 2.39 | 2.03 | 1.76 | 1.35 | 1.02 | .61 | .00 | X |
| 32 | 3.03 | 2.42 | 2.06 | 1.79 | 1.38 | 1.06 | .65 | .06 | X |
| 33 | 3.05 | 2.44 | 2.08 | 1.82 | 1.41 | 1.10 | .69 | .11 | X |
| 34 | 3.06 | 2.46 | 2.11 | 1.84 | 1.44 | 1.13 | .73 | .16 | X |
| 35 | 3.08 | 2.48 | 2.13 | 1.87 | 1.47 | 1.16 | .77 | .21 | X |
| 36 | 3.10 | 2.50 | 2.15 | 1.89 | 1.50 | 1.19 | .80 | .26 | X |
| 37 | 3.11 | 2.52 | 2.17 | 1.92 | 1.53 | 1.22 | .84 | .30 | X |
| 38 | 3.13 | 2.54 | 2.19 | 1.94 | 1.55 | 1.25 | .87 | .34 | X |
| 39 | 3.15 | 2.56 | 2.21 | 1.96 | 1.57 | 1.27 | .90 | .39 | X |
| 40 | 3.16 | 2.51 | 2.23 | 1.98 | 1.60 | 1.30 | .93 | .42 | X |
| 41 | 3.17 | 2.58 | 2.25 | 2.00 | 1.62 | 1.33 | .96 | .46 | X |
| 42 | 3.19 | 2.61 | 2.27 | 2.02 | 1.64 | 1.35 | .99 | .49 | X |

exhibit **18** of chapter 19(continued)

| Batch count | Ratio for point at depth | | | | | | | | | |
|---|---|---|---|---|---|---|---|---|---|---|
| | 1 | 2 | 3 | 4 | 6 | 8 | 11 | 16 | 22 | 32 | 44 |
| 43 | 3.20 | 2.62 | 2.29 | 2.04 | 1.66 | 1.37 | 1.02 | .53 | .00 | | |
| 44 | 3.22 | 2.64 | 2.30 | 2.06 | 1.69 | 1.40 | 1.04 | .56 | .04 | | |
| 45 | 3.23 | 2.65 | 2.32 | 2.07 | 1.71 | 1.42 | 1.07 | .59 | .08 | | |
| 46 | 3.24 | 2.67 | 2.34 | 2.09 | 1.73 | 1.49 | 1.09 | .62 | .12 | | |
| 47 | 3.25 | 2.68 | 2.35 | 2.11 | 1.74 | 1.46 | 1.12 | .65 | .16 | | |
| 48 | 3.27 | 2.70 | 2.37 | 2.13 | 1.76 | 1.48 | 1.14 | .68 | .19 | | |
| 49 | 3.28 | 2.71 | 2.38 | 2.14 | 1.78 | 1.40 | 1.16 | .70 | .23 | | |
| 50 | 3.29 | 2.72 | 2.40 | 2.16 | 1.79 | 1.52 | 1.19 | .73 | .26 | | |
| 51 | 3.30 | 2.74 | 2.41 | 2.17 | 1.82 | 1.54 | 1.21 | .73 | .29 | X | X |
| 52 | 3.31 | 2.75 | 2.43 | 2.19 | 1.83 | 1.56 | 1.23 | .76 | .32 | X | X |
| 53 | 3.32 | 2.76 | 2.44 | 2.20 | 1.85 | 1.58 | 1.25 | .80 | .35 | X | X |
| 54 | 3.33 | 2.77 | 2.45 | 2.22 | 1.86 | 1.59 | 1.27 | .83 | .38 | X | X |
| 55 | 3.34 | 2.79 | 2.47 | 2.23 | 1.88 | 1.61 | 1.29 | .85 | .41 | X | X |
| 56 | 3.35 | 2.80 | 2.48 | 2.24 | 1.90 | 1.63 | 1.31 | .87 | .43 | X | X |
| 57 | 3.36 | 2.81 | 2.49 | 2.26 | 1.91 | 1.64 | 1.32 | .89 | .46 | X | X |
| 58 | 3.37 | 2.82 | 2.50 | 2.27 | 1.92 | 1.66 | 1.34 | .91 | .49 | X | X |
| 59 | 3.38 | 2.83 | 2.51 | 2.28 | 1.94 | 1.68 | 1.36 | .94 | .51 | X | X |
| 60 | 3.39 | 2.84 | 2.53 | 2.30 | 1.95 | 1.69 | 1.38 | .96 | .53 | X | X |
| 61 | 3.40 | 2.85 | 2.54 | 2.31 | 1.97 | 1.71 | 1.39 | .97 | .56 | X | X |
| 62 | 3.41 | 2.86 | 2.55 | 2.32 | 1.98 | 1.72 | 1.41 | .99 | .58 | X | X |
| 63 | 3.42 | 2.87 | 2.56 | 2.33 | 1.99 | 1.73 | 1.42 | 1.01 | .60 | .00 | X |
| 64 | 3.43 | 2.88 | 2.57 | 2.34 | 2.01 | 1.75 | 1.44 | 1.03 | .62 | .03 | X |
| 65 | 3.44 | 2.89 | 2.58 | 2.35 | 2.02 | 1.76 | 1.45 | 1.05 | .65 | .06 | X |
| 66 | 3.45 | 2.90 | 2.59 | 2.37 | 2.03 | 1.78 | 1.47 | 1.07 | .67 | .08 | X |
| 67 | 3.45 | 2.91 | 2.60 | 2.38 | 2.04 | 1.79 | 1.48 | 1.08 | .69 | .11 | X |
| 68 | 3.46 | 2.92 | 2.61 | 2.39 | 2.05 | 1.80 | 1.50 | 1.10 | .71 | .14 | X |
| 69 | 3.47 | 2.93 | 2.62 | 2.40 | 2.07 | 1.81 | 1.51 | 1.12 | .72 | .16 | X |
| 70 | 3.48 | 2.94 | 2.63 | 2.41 | 2.08 | 1.83 | 1.53 | 1.13 | .74 | .19 | X |
| 71 | 3.49 | 2.95 | 2.64 | 2.42 | 2.09 | 1.84 | 1.54 | 1.15 | .76 | .21 | X |
| 72 | 3.49 | 2.96 | 2.65 | 2.43 | 2.10 | 1.85 | 1.55 | 1.16 | .78 | .23 | X |
| 73 | 3.50 | 2.97 | 2.66 | 2.44 | 2.11 | 1.86 | 1.57 | 1.18 | .80 | .25 | X |
| 74 | 3.51 | 2.97 | 2.67 | 2.45 | 2.12 | 1.87 | 1.58 | 1.19 | .91 | .28 | X |
| 75 | 3.52 | 2.98 | 2.68 | 2.46 | 2.13 | 1.89 | 1.59 | 1.21 | .83 | .30 | X |
| 76 | 3.52 | 2.99 | 2.69 | 2.47 | 2.14 | 1.90 | 1.60 | 1.22 | .85 | .32 | X |
| 77 | 3.53 | 3.00 | 2.70 | 2.48 | 2.15 | 1.91 | 1.62 | 1.23 | .86 | .34 | X |
| 78 | 3.54 | 3.01 | 2.71 | 2.49 | 2.16 | 1.92 | 1.63 | 1.25 | .88 | .36 | X |
| 79 | 3.54 | 3.01 | 2.71 | 2.50 | 2.17 | 1.93 | 1.64 | 1.26 | .89 | .38 | X |
| 80 | 3.55 | 3.02 | 2.72 | 2.50 | 2.18 | 1.94 | 1.65 | 1.27 | .91 | .40 | X |
| 81 | 3.56 | 3.03 | 2.73 | 2.51 | 2.19 | 1.95 | 1.66 | 1.29 | .92 | .42 | X |
| 82 | 3.57 | 3.04 | 2.74 | 2.52 | 2.20 | 1.96 | 1.67 | 1.30 | .94 | .43 | X |
| 83 | 3.57 | 3.05 | 2.75 | 2.53 | 2.21 | 1.97 | 1.68 | 1.31 | .95 | .45 | X |

exhibit **18** of chapter 19 (continued)

| Batch count | Ratio for point at depth |  |  |  |  |  |  |  |  |  |
|---|---|---|---|---|---|---|---|---|---|---|
| | 1 | 2 | 3 | 4 | 6 | 8 | 11 | 16 | 22 | 32 | 44 |
| 84 | 3.58 | 3.05 | 2.75 | 2.54 | 2.22 | 1.98 | 1.69 | 1.32 | .97 | .47 | X |
| 85 | 3.58 | 3.06 | 2.76 | 2.55 | 2.23 | 1.99 | 1.71 | 1.34 | .98 | .49 | X |
| 86 | 3.59 | 3.07 | 2.77 | 2.55 | 2.24 | 2.00 | 1.72 | 1.35 | 1.00 | .50 | X |
| 87 | 3.60 | 3.07 | 2.78 | 2.56 | 2.25 | 2.01 | 1.73 | 1.36 | 1.01 | .52 | .00 |
| 88 | 3.60 | 3.08 | 2.78 | 2.57 | 2.25 | 2.02 | 1.74 | 1.37 | 1.02 | .54 | .02 |
| 89 | 3.61 | 3.09 | 2.79 | 2.58 | 2.26 | 2.03 | 1.74 | 1.38 | 1.04 | .55 | .04 |
| 90 | 3.62 | 3.09 | 2.80 | 2.59 | 2.27 | 2.04 | 1.76 | 1.39 | 1.05 | .57 | .06 |
| 91 | 3.62 | 3.10 | 2.81 | 2.59 | 2.28 | 2.04 | 1.77 | 1.41 | 1.06 | .58 | .08 |
| 92 | 3.63 | 3.11 | 2.81 | 2.60 | 2.29 | 2.05 | 1.78 | 1.42 | 1.07 | .60 | .10 |
| 93 | 3.63 | 3.11 | 2.82 | 2.61 | 2.30 | 2.06 | 1.79 | 1.43 | 1.08 | .61 | .12 |
| 94 | 3.64 | 3.12 | 2.83 | 2.62 | 2.30 | 2.07 | 1.79 | 1.44 | 1.10 | .63 | .14 |
| 95 | 3.64 | 3.13 | 2.83 | 2.62 | 2.31 | 2.08 | 1.80 | 1.45 | 1.11 | .64 | .16 |
| 96 | 3.65 | 3.13 | 2.84 | 2.63 | 2.32 | 2.09 | 1.81 | 1.46 | 1.12 | .66 | .17 |
| 97 | 3.65 | 3.14 | 2.85 | 2.64 | 2.33 | 2.10 | 1.82 | 1.47 | 1.13 | .67 | .19 |
| 98 | 3.66 | 3.15 | 2.85 | 2.64 | 2.34 | 2.10 | 1.83 | 1.48 | 1.14 | .68 | .21 |
| 99 | 3.67 | 3.15 | 2.86 | 2.65 | 2.34 | 2.11 | 1.84 | 1.49 | 1.15 | .70 | .23 |
| 100 | 3.67 | 3.16 | 2.87 | 2.66 | 2.35 | 2.12 | 1.85 | 1.50 | 1.16 | .71 | .24 |

B) For BATCHES of MORE than 100 and DEPTH 1

Take "standard ratio" = $2.97\sqrt{-0.45 + \log_{10}(\text{batch size})}$

C) For BATCHES of MORE than 100 and DEPTHS >1

Use--in (A) or (B), as appropriate--the standard ratio for DEPTH = 1 and the following altered batch sizes:

| For depth | Batch size used for depth 1 |
|---|---|
| 2 | $\frac{1}{5}(2(\text{batch size}) - 1)$ |
| 3 | $\frac{1}{4}(\text{batch size} - 1)$ |
| 4 | $\frac{1}{11}(2(\text{batch size}) - 3)$ |
| 6 | $\frac{1}{17}(2(\text{batch size}) - 5)$ |
| 8 | $\frac{1}{23}(2(\text{batch size}) - 7)$ |
| 11 | $\frac{1}{16}(\text{batch size} - 5)$ |
| 16 | $\frac{1}{47}(2(\text{batch size}) - 15)$ |
| 22 | $\frac{1}{65}(2(\text{batch size}) - 21)$ |

Example and check: For batch size = 100, (B) above gives 3.70 for depth 1 (using (A) directly gives 3.67). For depth 2, we use batch size = $\frac{1}{5}(199) = 39.8$ whence (A) gives 3.16 (using (A) directly gives 3.16). For depth 3, we use batch size = $\frac{1}{4}(99) = 24.75$, whence (A) gives 2.87 (using (A) directly gives 2.87). For depth 4, we use batch size = $\frac{1}{11}(197) = 17.91$ which gives 2.66 (using (A) directly gives 2.66). For depth 6, we use $\frac{1}{17}(195) = 11.47$, whence (A) gives 2.35 (using (A) directly gives 2.35). For depth 8, we use $\frac{1}{23}(193) = 8.39$, whence (A) gives 2.12 (using (A) directly gives 2.12). For depth 11, $\frac{1}{16}(95) = 5.938$ gives 1.85 (direct: 1.85). For depth 16, $\frac{1}{47}(185) = 3.94$ gives 1.49 (direct: 1.50). For depth 22, $\frac{1}{65}(179) = 2.75$ gives 1.15 (direct: 1.16).

### review questions

What is our aim in this section? Where do we start from? What do we calculate? How do we proceed? How do the descriptions found here compare with those of the last section? Which is always better? Why/why not? If we want to concentrate more on individual values in the batch, how can we do this? Will it take much more effort? Why/why not? What did we put in a table to make calculation easy? What did we do next? Could we do pushbacks for every individual value in the batch? Would it be worthwhile?

exhibit **19** of chapter 19: county areas

**Pushback analysis (123468... version) of the logs of the areas of the counties of Michigan ($Q = 10$)**

### A) CALCULATIONS

| Depth or Pt | Obs'd | Ratio | Q × Ratio | Diff | Add'l fit | 2nd resid | 3R | S | Rough |
|---|---|---|---|---|---|---|---|---|---|
| 1 | 1.51 | −3.57 | −.36 | 1.87 | | | | 1.85 | +.02 |
| 2 | 1.54 | −3.05 | −.30 | 1.84 | | | 1.85 | | −.01 |
| 3 | 1.57 | −2.75 | −.28 | 1.85 | | | 1.85 | | .00 |
| 4 | 1.61 | −2.53 | −.25 | 1.86 | | | 1.86 | | .00 |
| 6 | 1.66 | −2.21 | −.22 | 1.88 | | | 1.88 | 1.88 | .00 |
| 8 | 1.68 | −1.97 | −.19 | 1.88 | | | 1.88 | 1.88 | .00 |
| 11 | 1.69 | −1.68 | −.17 | 1.86 | | | 1.86 | | .00 |
| H | 1.73h | −.98 | −.10 | 1.83h | | | | 1.84 | .00 |
| M | 1.76 | 0 | 0 | 1.76 | .00 | 1.76 | 1.76 | | .00 |
| H | 1.89h | .98 | .10 | 1.79h | .04 | 1.75h | 1.75h | | .00 |
| 11 | 1.98 | 1.68 | .17 | 1.81 | .07 | 1.74 | 1.74 | | .00 |
| 8 | 2.01 | 1.97 | .19 | 1.81 | .08 | 1.73 | 1.74 | | −.02 |
| 6 | 2.07 | 2.21 | .22 | 1.85 | .09 | 1.76 | 1.73 | | .03 |
| 4 | 2.08 | 2.53 | .25 | 1.83 | .10 | 1.73 | 1.73 | | .00 |
| 3 | 2.12 | 2.75 | .28 | 1.83 | .11 | 1.72 | 1.73 | | −.01 |
| 2 | 2.20 | 3.05 | .30 | 1.90 | .12 | 1.78 | 1.76 | | −.02 |
| 1 | 2.26 | 3.57 | .36 | 1.90 | .14 | 1.76 | 1.76 | | .00 |

Notes: "Diff" contains unsmoothed "pushed-back" values, which are plotted in exhibit 20.

### P) PROBLEMS

19a) Repeat for the raw areas.

19a2) Plot the result.

19b) Repeat for the root areas.

19b2) Plot the result.

### S) SOURCE
Exhibit 5 of chapter 3.

**exhibit 20/19: Shapes of distribution**

### 19H. How far have we come?

This chapter has focussed on comparing counted fractions with a standard shape of distribution, doing this in various terms:

◊ all individual values.

◊ some binning of the values.

◊ letter values for the batch.

◊ half-octave-depth values for the batch.

As a reference standard from which to measure the deviations of the real world, we have introduced the Gaussian distribution, and we have been careful to avoid thinking of real data as in any sense "exactly Gaussian".

We are now ready to:

◊ approach the description of distributions by first finding flogs (or froots) for each of several cut locations and then describing the relationship by the combination of a fit (flog to location) and residuals (of flogs).

◊ do this using two or more different fits in different sections.

Moreover we are also ready to use letter values--or half-octave-depth values--to study shapes of distribution in a somewhat different way:

◊ by converting letter values into mids and 25% pseudospreads--by plot-

**exhibit 20 of chapter 19: county areas**

Plot of pushed back values (from "diff" column of exhibit 19)

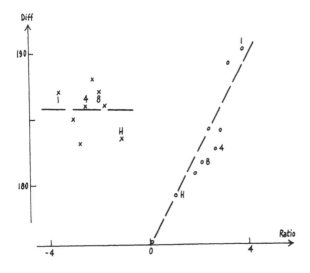

ting the result against--and then fitting to--the square of the divisor used to find the corresponding pseudospread.

◇ by alternatively using simple pushback on either letter values or half-octave-depth values.

And we have begun to understand the proper nature and relevance of the Gaussian distribution (sometimes misleadingly called the "normal" distribution), in particular understanding that:

◇ the Gaussian is a reference standard, *not* an example of what in fact is-- that it is something against which to compare the actual behavior of real data. (This can be by residuals--including pushed-back values--or by the behavior of mids and 25% pseudospreads.)

◇ we often can, and even should, use it as an initial approximation to actual behavior.

◇ in dealing with distributions of the real world, we are very lucky if

◇ **we know APPROXIMATELY how values are distributed.**
◇ **this approximation is itself not TOO far from being simple.**

# index for chapter 20

review questions  647

**20A. Binnings vs. distributions  648**
continuity and discreteness  648
fractions, densities, cumulatives  648
*(mathematical) density*  648
review questions  651

**20B. Densities for distributions vs. densities for binnings  651**
correction for curvature  652
*average density*  652
*(mathematical) density*  652
review questions  654

**20C. Tables and pictures comparing two sets of shapes of distribution  654**
identification  656
further comment  660
review questions  661

| EXHIBIT | PAGE |
|---|---|
| **20A** | |
| 1★ | 649 |
| **20B** | |
| 2 | 652 |
| 3 | 653 |
| **20C** | |
| 4★ | 655 |
| 5★ | 657 |
| 6★ | 658 |
| 7★ | 659 |
| 8 | 660 |

# Mathematical distributions   **20**

Suppose we have a binning involving many, many appearances--say a binning of U.S. taxpayers according to reported income (as in exhibit 5 of chapter 19). We could think of subdividing our bins more and more finely--if the incomes were in thousand-dollar bins, we could surely go to hundred-dollar bins, and probably much further, especially if we could have more individuals to bin.

It is natural for a mathematician, once started on the idea of narrower bins and more individuals, to push on until the bins are infinitely narrow and the individuals are infinitely many. The result is a mathematical distribution, a concept that has:

⋄ many uses in the theories of statistics and data analysis.

⋄ real convenience as a background against which to interpret and understand actual binnings.

It will clearly be convenient to try to represent the distributions we work with by simple formulas--but we must remember that this is only a hope--indeed that it is more reasonable to hope for only reasonable approximation by simple formulas. (Even this is asking a lot.)

This chapter uses the idea of mathematical distribution to tie together what we have been doing in the last two chapters. In particular, it looks more closely into what a binning suggests about a corresponding mathematical distribution, and into how the various mathematical distributions suggested by the last two chapters compare with one another.

### review questions

In principle, how small bins could we use? In practice? What is a mathematical density? A mathematical distribution? How does a mathematician's mind get to them? What are mathematical distributions good for? Can we expect real distributions to be given by simple formulas? Exactly or approximately? What is our aim in this chapter? What are two important parts of what we will do?

## 20A. Binnings vs. distributions

### continuity and discreteness

If we are going to think about arbitrarily narrow bins, we have to think of observed values, each of which is given to arbitrarily many decimals. To make arbitrarily narrow bins useful for any real data--assuming we could get as much of it as was asked for--each of these decimals would have to have entries not all the same. (Numbers, all of which were like 1063.2130000..., 2039.1170000..., and 1557.6420000..., might possibly use bins of width 0.001, but bins of width 0.0001 would tell us nothing more.) Many mathematical distributions assume that the "observations" are given to arbitrarily many decimals; an observation matching the given value to that number of decimals will come along if we wait long enough.

To the man with data, such unrealism hardly looks helpful, but to the worker with theory such an assumption turns out to make many things simpler. Since all the man with data can hope for is an approximation, and since an approximation to a few decimal places is often more than he deserves, he soon learns not to worry very often about this particular sort of unrealism. (But he may have to worry about theoretical consequences that others have drawn from it.)

Other kinds of mathematical distributions assume that only isolated values are possible. These values are usually, as a practical convenience, taken to be integers (which may, of course, have to be reexpressed).

### fractions, densities, cumulatives

When we think of dealing with more and more--and then many more--"observations", we soon start to find counts unhandy. Instead, it becomes natural to think about fractions of the form

$$\frac{\text{count}}{\text{total count}} = \text{fraction}.$$

(Let us not worry about "+1/6" here--we need not, since all counts are getting larger and larger) and, once our bins get very narrow, we tend to work with the results of dividing by bin width

$$\frac{\text{fraction}}{\text{bin width}} = \text{(average) density},$$

in close analogy to what we did when we dealt with counts for bins of unequal width.

One reasonable way to describe a mathematical distribution is to say how the

### (mathematical) density

depends on the value concerned. This will usually be a density for an "infinitely narrow" bin, meaning a bin so narrow that making it narrower will not change the density--to the precision we are concerned with. (Hidden in this definition is a more detailed assumption about the smoothness of distribution of the "observations" that we need to know is there, but which we can lump in with the other idealizations as part of "all is approximation anyway".) Panel A of exhibit 1 shows a few values of the mathematical density of a particular distribution.

Another way to describe a mathematical distribution is in terms of a

$$\text{cumulative fraction} = \frac{\left(\begin{array}{c}\text{count of all}\\ \text{values} < \text{cut}\end{array}\right) \text{plus} \left(\begin{array}{c}\text{half of count of}\\ \text{those equal to the cut}\end{array}\right)}{\text{total count}}$$

defined for any way of cutting the values into lower (to the left of the cut) and higher (to the right of the cut). We can do this by a formula--or at least for all practical purposes, by a numerical table--from either of which we can find the cumulative fraction, given the cut. Panel B (of exhibit 1) shows a few lines of such a table for the same distribution. Panel C gives a rather longer list of values from such a table for the same distribution. We can do equally well by a formula--or table--from which we can find the cut, given the cumulative fraction.

exhibit **1** of chapter 20: sample description

**A few values of the mathematical density, the cumulative distribution, and the representing function (% point function) for the standard distribution used in chapter 19 (with hinges at ±1)**

| A) A few VALUES of the MATHEMATICAL DENSITY | | B) A few VALUES of the CUMULATIVE DISTRIBUTION | |
|---|---|---|---|
| cut = Value of what is distributed | Density* | cut = Value of what is distributed | Fraction below = Cumulative distribution |
| −5 | .0009 | −5 | .0004 |
| −4 | .0071 | −4 | .0035 |
| −3 | .035 | −3 | .0215 |
| −2 | .108 | −2 | .0887 |
| −1 | .214 | −1 | .2500 |
| 0 | .269 | 0 | .5000 |
| 1 | .214 | 1 | .7500 |
| 2 | .108 | 2 | .9113 |
| 3 | .035 | 3 | .9785 |
| 4 | .0071 | 4 | .9965 |
| 5 | .0009 | 5 | .9996 |

* Mathematical density.

Note that, just as panel B leads from −1 to 0.25, so panel C leads from 0.25 back to −1. (And that, as panel C leads from 0.75 to +1, so panel B leads from +1 back to 0.75.)

Such formulas, if they exist, are said to give the "cumulative distribution" (or the "cumulative") when they give cumulative fraction in terms of cut, or to give "the representing function" or "the percent-point function" when they give the cut in terms of the cumulative fraction. (A good numerical table can, of course, be used either way, giving whichever we want in terms of whichever we have. Such a table thus gives both cumulative and representing function--to an indicated approximation.) We are, by now, familiar with tables of representing functions for a few distributions--disguised as "re-expressions of fractions" in the case of froots and flogs, or as "ratios" in the Gaussian case.

exhibit **1** of chapter 20(continued)

**C)** Selected VALUES of the REPRESENTING FUNCTION (or PERCENT POINT FUNCTION)

| Fraction below = Cumulative fraction | Cumulative percent | cut = Value of what is distributed |
|---|---|---|
| .001 | (0.1%) | −4.58 |
| .01 | (1%) | −3.45 |
| .05 | (5%) | −2.44 |
| .10 | (10%) | −1.90 |
| .15 | (15%) | −1.54 |
| .20 | (20%) | −1.25 |
| .25 | (25%) | −1.00 |
| .30 | (30%) | −.78 |
| .40 | (40%) | −.38 |
| .50 | (50%) | .00 |
| .60 | (60%) | .38 |
| .70 | (70%) | .78 |
| .75 | (75%) | 1.00 |
| .80 | (80%) | 1.25 |
| .85 | (85%) | 1.54 |
| .90 | (90%) | 1.90 |
| .95 | (95%) | 2.44 |
| .99 | (99%) | 3.45 |
| .999 | (99.9%) | 4.58 |

**P) PROBLEMS**

1a/b/c/d) Below what value do about 9%/45%/98%/99.9% of the values lie?

1e/f/g/h/i) At what value does the density amount to .1/.2/.01/.03/.001?

1j) How do panels A and C interrelate?

A few kinds of mathematical distributions--two, three, or a half-dozen, depending on your definition of simple--have simple formulas for all three of density, cumulative, and representing function. Rather more have simple formulas for one or two of the three. Others have not very simple formulas-- some of these are quite important--or even quite complex formulas for one or two or three of the triad: representing function, density, or cumulative. This will not bother us here because--at least in analyzing data--we will always do one of two things:

⋄ rely on some quite simple formula.
⋄ rely on a table (which gets rid of any need for a formula).

### review questions

Do most real distributions fit simple formulas? Could arbitrarily narrow bins be useful? When? In theory? In practice? Is a "continuity" assumption realistic? Why/why not? Does lack of realism here really matter in practice? Why/why not? What is a (mathematical) density? How is it related to any actual batch of numbers (pick the closest one)? To any mathematical distribution (pick the closest one)? What is a cumulative fraction? How can we give a cumulative fraction for all or many cuts? What is a cumulative distribution? A representing function? A percent-point function? How many of these are related? Which, how, and how closely? What are the three ways to specify a mathematical distribution? Do many mathematical distributions have simple formulas for all three ways? About how many? Dare we expect to stick to mathematical distributions with at least one simple formula? Why/why not? What do we do here (in this book)?

## 20B. Densities for distributions vs. densities for binnings

We learned, in chapter 17, to fit many binnings. We routinely fitted

$$\text{some simple expression of } \sqrt{\frac{\text{count}}{\text{bin width}}}$$

with a simple function of bin location--usually one constant plus another constant times (bin location − peak)$^2$. (When the bins were of equal width, we just took all their widths to be 1.0 and talked about an expression of $\sqrt{\text{count}}$.)

The first change we need to make is to replace "count" by fraction, and start with

$$\text{some simple expression of } \sqrt{\frac{\text{fraction}}{\text{bin width}}},$$

which is, of course,

$$\text{some simple expression of } \sqrt{\text{density}}.$$

### correction for curvature

The next thing we have to worry about is the difference between

### average density

for some not too narrow bin width, which is the only thing we can calculate from data, and

### (mathematical) density

mathematically defined, which is what appears in the mathematical distribution. Fortunately, a simple approximate correction is good enough for essentially every case that comes up in practice. So let us learn what it is.

The simplest kinds of mathematical distributions are shown in exhibit **2**;

exhibit **2** of chapter 20: various densities

**Densities for some very simple mathematical distributions**

**A)** Rectangular = constant density    **B)** Triangular = linear density

**C)** Quadratic density (enlarged scale)

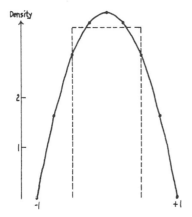

constant density, linear density, quadratic density. The dashed rectangles on the last two show what average densities we would get for particular bins. We see that, for the linear density, the average density is the same as the mathematical density at the center of the bin. Apparently we will need no correction when the density is linear--presumably also when the average density is linear.

For the quadratic density, and the rather wide, centrally placed bin shown, the average density is appreciably below the mathematical density at the center of the bin. We do need an adjustment. Exhibit 3 shows the same situation in more detail, using a convenient vertical scale. For three adjacent bins of equal size, the average densities are 0.7292, 0.9792, and 0.7292. If these were linear--and how clearly they are not!--the skip mean involving the two outer

exhibit **3** of chapter 20: quadratic density

**Some average and mathematical densities for a particular example**

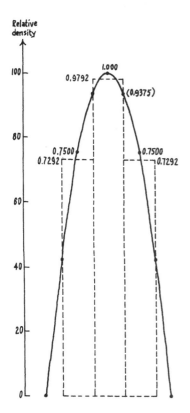

values, here .7292, would equal the central value, here .9792. Thus the difference

average density MINUS skip mean of average densities,

here equal to

$$.9792 - .7292 = .25,$$

is a natural measure of bend.

The correction needed here is

$$1.0000 - .9792 = .0208,$$

which is exactly 1/12 of .25. So it is reasonable to take, as a plausible correction,

$$\frac{\text{average density MINUS skip mean of average densities}}{12},$$

which means that we use, as the adjusted average density,

$$\text{average density PLUS } \frac{\left(\begin{array}{c}\text{average density MINUS skip}\\ \text{mean of average density}\end{array}\right)}{12}.$$

Note that this adjustment is only an approximation. If $\sqrt{\text{adjusted count}}$ differs from $\sqrt{\text{count}}$ by more than a few units, we should not trust the approximation too far--though the adjusted value will usually be much better than the unadjusted one. (Better approximations are more trouble than we care to face in this volume.)

Exhibit 4 shows how this works for an age-of-death example, one for deaths by coronary disease. The main feature of this exhibit is the small size of the adjustments to $\sqrt{\text{count}}$, only two of which are larger than ±1.1. (The adjustment of −2.6 applies at ages 95–99, where expert judgment tells us that the accuracy of ages is very dubious.) The adjustments are small, even though there are a total of nearly half a million carefully tabulated deaths.

### review questions

What is the aim of this section? How did we begin? What is an average density? A (mathematical) density? What are some very simple mathematical distributions? What mathematical distribution was a simple leading case? What formula did this suggest? How firmly do we believe in the corresponding answers? What example did we try? How big were the adjustments? What happened when we used wider bins?

### 20C. Tables and pictures comparing two sets of shapes of distribution

In chapter 17 we introduced seven choices of expression for $\sqrt{\text{count}}/\text{width}$ that might behave simply when plotted against $(B - \text{peak})^2$ or $(L - \text{peak})^2$.

exhibit 4 of chapter 20: coronary deaths

**Age binning and adjustment for 406,499 deaths assigned to cause 420.1--heart disease specified as involving coronary arteries--in the U.S. during 1965**

A) DATA and CALCULATIONS

| Bin* | Count | $\sqrt{\text{count}}$ | >† | Diff | Diff/12 | Adj count | Adj $\sqrt{\text{count}}$ | Adj in $\sqrt{\text{count}}$ |
|---|---|---|---|---|---|---|---|---|
| 0–4     | 32    | (5.7)   | 5       | 27     | 2.2    | 34    | 5.8   | (+.2)  |
| 5–9     | 10    | (3.2)   | 22      | −12    | −1.0   | 9     | 3.0   | (−.2)  |
| 10–14   | 12    | (3.5)   | 27h     | −15h   | −1.3   | 11    | 3.3   | (−.2)  |
| 15–19   | 45    | (6.7)   | 68      | −23    | −1.9   | 43    | 6.6   | (−.1)  |
| 20–24   | 124   | (11.1)  | 193h    | −69h   | −5.8   | 118   | 10.9  | (−.2)  |
| 25–29   | 342   | (18.5)  | 637     | −295   | −24.6  | 317   | 17.8  | (−.7)  |
| 30–34   | 1150  | (33.9)  | 1940    | −790   | −65.8  | 1084  | 32.9  | (−1.0) |
| 35–39   | 3538  | (59.5)  | 4900    | −1362  | −113.5 | 3424  | 58.5  | (−1.0) |
| 40–44   | 8650  | (93.0)  | 9666    | −1016  | −84.7  | 8565  | 92.5  | (−.5)  |
| 45–49   | 15794 | (125.7) | 17414   | −1620  | −135.0 | 15659 | 125.1 | (−.6)  |
| 50–54   | 26178 | (161.8) | 26252h  | −75h   | −6.2   | 26172 | 161.8 | (0)    |
| 55–59   | 36711 | (191.6) | 36955   | −244   | −20.3  | 36691 | 191.5 | (−.1)  |
| 60–64   | 47732 | (218.5) | 47441   | 291    | 24.2   | 47756 | 218.5 | (0)    |
| 65–69   | 58171 | (241.2) | 56200   | 1971   | 164.2  | 58335 | 241.5 | (+.3)  |
| 70–74   | 64668 | (254.3) | 59577   | 5091   | 424.2  | 65092 | 255.1 | (+.8)  |
| 75–79   | 60983 | (246.9) | 55021   | 5962   | 496.8  | 61480 | 248.0 | (+1.1) |
| 80–84   | 45374 | (213.0) | 43137h  | 2236h  | 186.4  | 45560 | 213.4 | (+.4)  |
| 85–89   | 25292 | (159.0) | 27336h  | −2044h | −170.4 | 25122 | 158.5 | (−.5)  |
| 90–94   | 9299  | (96.4)  | 13654h  | −4355h | −363.0 | 8936  | 94.5  | (−.9)  |
| 95–99   | 2017  | (44.9)  | 4787h   | −2770h | −230.9 | 1786  | 42.3  | (−2.6) |
| 100–‡   | 276   | (16.6)  | 1008h   | −732h  | −61.0  | 215   | 14.7  | (−1.9) |

(Age not stated: 101)

* Age in years.
† Split mean of count (not of ($\sqrt{\text{count}}$)).
‡ All taken as 100–104.

P) PROBLEMS

4a) Combine the counts in 15-year bins (0–14, 15–29, etc.) and make a similar calculation. Are the adjustments larger?

4a2) Divide the adjusted root counts thus obtained by $\sqrt{3}$, and compare the results with those for the 5-year intervals at the center of every fifteen-year interval. How do they compare?

4b/4b2) Do the same for 15-year bins (5–19, 20–34, etc.).

4c/4c2) Do the same for 15-year bins (10–24, 25–39, etc.).

4d) Why did we choose to divide by $\sqrt{3}$ in (4a2)?

S) SOURCE

*Vital Statistics of the United States, 1965, Volume II-Mortality, Part A.* U.S. Dept. Health, Education and Welfare, 1967. Pages 1–132 and 1–133.

After the discussion of the last section, we are likely to be interested in the seven shapes of distribution whose $\sqrt{}$(mathematical) density, when correspondingly expressed, is a linear function of $(B - \text{peak})^2$.

In chapter 19 we introduced three shapes of distribution--two deriving from convenient re-expressions of fractions and the third being the reference standard (the Gaussian). We may reasonably want to compare these shapes, both with each other and with the seven just mentioned.

It is now time to say that the Gaussian shape is exactly the shape for which $\log \sqrt{\text{density}}$ is linear in $(x - \text{peak})^2$--the reference standard of chapter 19 is exactly the middle shape of the seven of chapter 17. Thus, seven and three give nine, not ten.

Exhibit **5** compares the expressions of fractions corresponding to all nine distribution shapes, each adjusted to hinges = +1 for easy comparison. We know that this table can also be used as a skeleton table of representing functions. With 75% forced to give 1.000, we should not be surprised that the expressions of 70% are a lot alike (all but one between .789 and .764 (one at .726), as are those at 80% (1.221 to 1.279 and 1.376) and 60% (.388 to .362 and .325). Even at 90% things do not vary tremendously (1.727 to 2.141 and 3.077). At 99% things are more different (2.45 to 5.94 and 31.82). And so more and more, the farther we go into the tails.

Exhibit **6** gives the corresponding table of cumulatives, again matched to have hinges at ±1.0. Again, behavior between −1.5 and +1.5 is reasonably steady.

Exhibit **7** gives the corresponding relative densities. For larger deviations, the influence of the increasingly stretched tails of the distributions to the right of exhibit 7--and the increasingly squeezed-in tails of the distributions to the left of the same exhibit--is clearly evident. (The distributions are quite different--but smoothly and reasonably so.) Exhibit **8** pictures the spacings of the letter values. For the nine distributions, we see that the tails stretch more and more in the order $r_4$ (very squeezed), $r_6$, $r_{10}$, Gauss, $t_7$, $t_3$, $t_1$ (very very stretched), with:

◇ the shape corresponding to froots close to, but more squeezed than, $r_4$.

◇ the shape corresponding to flogs close to $t_7$, extreme tails being less stretched.

◇ the $t_1$ shape very much more tail-stretched than the others.

Those who use flogs or froots clearly straddle the Gaussian neatly (as would one who used $t_7$ and $r_4$--thus taking either $(\sqrt{\text{density}^2})$ or $1/(\sqrt{\text{density}^{1/2}})$ as linear in $(\text{Shift})^2$).

### identification

Some of the readers who reach this, the last section of the main body of this book, will have previously been introduced to classical statistics. They may recognize such terms as "Student's $t$", "Pearson's $r$", or "product-moment correlation". These paragraphs are for them--others are encouraged to skip.

exhibit **5** of chapter 20: tabular comparison

**Nine expressions of fractions, adjusted to "hinges = ±1".**

A) TABLE

| + | "froot" | $r_4$ | $r_6$ | $r_{10}$ | "Gauss" | "flog" | $t_7$ | $t_3$ | $t_1$ | − |
|---|---|---|---|---|---|---|---|---|---|---|
| 50% | .000 | .000 | .000 | .000 | .000 | .000 |  | .000 | .000 | 50% |
| 60% | .388 | .386 | .382 | .380 | .376 | .369 | .370 | .362 | .325 | 40% |
| 70% | .789 | .787 | .783 | .781 | .777 | .771 | .772 | .764 | .726 | 30% |
| 80% | 1.221 | 1.226 | 1.234 | 1.235 | 1.248 | 1.261 | 1.260 | 1.279 | 1.376 | 20% |
| 90% | 1.727 | 1.752 | 1.802 | 1.842 | 1.90 | 2.000 | 1.990 | 2.141 | 3.077 | 10% |
| 92% | 1.85 | 1.88 | 1.998 | 2.00 | 2.38 | 2.22 | 2.21 | 2.43 | 3.89 | 8% |
| 95% | 2.05 | 2.10 | 2.21 | 2.30 | 2.44 | 2.68 | 2.68 | 3.08 | 6.31 | 5% |
| 98% | 2.32 | 2.40 | 2.605 | 2.77 | 3.04 | 3.54 | 3.54 | 4.55 | 15.89 | 2% |
| 99% | 2.45 | 2.54 | 2.81 | 3.05 | 3.45 | 4.18 | 4.22 | 5.94 | 31.82 | 1% |
| 99.2% | 2.48 | 2.58 | 2.86 | 3.12 | 3.57 | 4.39 | 4.44 | 6.45 | 39.7 | .8% |
| 99.5% | 2.53 | 2.64 | 2.97 | 3.28 | 3.82 | 4.82 | 4.92 | 7.64 | 63.7 | .5% |
| 99.8% | 2.61 | 2.73 | 3.13 | 3.52 | 4.27 | 5.65 | 5.92 | 10.59 | 159.1 | .2% |
| 99.9% | 2.64 | 2.77 | 3.22 | 3.68 | 4.58 | 6.29 | 6.73 | 13.35 | 318.3 | .1% |
| 99.92% | 2.65 | 2.78 | 3.24 | 3.72 | 4.68 | 6.49 | 7.00 | 14.4 | 398. | .08% |
| 99.95% | 2.68 | 2.80 | 3.29 | 3.81 | 4.88 | 6.92 | 7.60 | 16.9 | 636. | .05% |
| 99.98% | 2.69 | 2.83 | 3.36 | 3.96 | 5.25 | 7.75 | 8.85 | 23.0 | 1591. | .02% |
| 99.99% | 2.70 | 2.85 | 3.40 | 4.05 | 5.51 | 8.38 | 9.95 | 29.0 | 3180. | .01% |
| (100%) | (2.73) | (2.88) | (3.56) | (4.34) |  |  |  |  |  |  |

Notes

| | |
|---|---|
| "froot" | marks the shape of distribution associated with froots |
| $r_4$ | means $(\sqrt{\text{density}})^2$ is linear in $(x - \text{peak})^2$ |
| $r_6$ | means $(\sqrt{\text{density}})$ is linear in $(x - \text{peak})^2$ |
| $r_{10}$ | means $(\sqrt{\text{density}})^{1/2}$ is linear in $(x - \text{peak})^2$ |
| "Gauss" | marks the Gaussian shape, for which log density is linear in $(x - \text{peak})^2$ |
| "flog" | marks the shape of distribution associated with flogs |
| $t_7$ | means $1/(\sqrt{\text{density}})^{1/2}$ is linear in $(x - \text{peak})^2$ |
| $t_3$ | means $1/\sqrt{\text{density}}$ is linear in $(x - \text{peak})^2$ |
| $t_1$ | means $1/(\sqrt{\text{density}})^2$ is linear in $(x - \text{peak})^2$ |

P) PROBLEMS

5a) Plot all eight representing functions (on the same plot) against the one associated with flogs. Discuss.

5b/c/d/e) Use these % values the way the letter values were used in section 19C to find (Gaussian) 25% pseudospreads and mids for the 1st/2nd/3rd/4th column. Plot against (Gaussian representing function)$^2$.

5f/g/h/i) Do the same for the 6th/7th/8th/9th column.

exhibit **6** of chapter 20: tabular comparison

**Nine cumulative distributions, each adjusted to "hinges = ±1" for easy comparison**

A) OUT to ±5

| Cut | Cumulative for appropriate multiple of | | | | | | | | |
|---|---|---|---|---|---|---|---|---|---|
|  | "froot" | $r_4$ | $r_6$ | $r_{10}$ | "Gauss" | "flog" | $t_7$ | $t_3$ | $t_1$ |
| −5 | X | X | X | X | .0004 | .0041 | .0046 | .0157 | .0526 |
| −4.5 | X | X | X | X | .0012 | .0071 | .0075 | .0206 | .0696 |
| −4 | X | X | X | .0001 | .0035 | .0122 | .0124 | .0274 | .0780 |
| −3.5 | X | X | .0000 | .0022 | .0091 | .0209 | .0208 | .0378 | .0886 |
| −3 | X | X | .0043 | .0113 | .0215 | .0357 | .0352 | .0528 | .1024 |
| −2.5 | .0033 | .0124 | .0259 | .0349 | .0459 | .0663 | .0593 | .0759 | .1711 |
| −2 | .0571 | .0678 | .0734 | .0803 | .0887 | .1000 | .0990 | .1118 | .1476 |
| −1.5 | .1422 | .1446 | .1490 | .1620 | .1558 | .1614 | .1608 | .1672 | .1872 |
| −1 | .2500 | .2500 | .2500 | .2500 | .2500 | .2500 | .2500 | .2500 | .2500 |
| −.5 | .3716 | .3711 | .3700 | .3691 | .3680 | .3660 | .3663 | .3638 | .3524 |
| 0 | .5000 | .5000 | .5000 | .5000 | .5000 | .5000 | .5000 | .5000 | .5000 |
| 1 | .7500 | .7500 | .7500 | .7500 | .7500 | .7500 | .7500 | .7500 | .7500 |
| 2 | .9429 | .9372 | .9266 | .9197 | .9113 | .9000 | .9010 | .8882 | .8524 |
| 3 | X | X | .9957 | .9887 | .9785 | .9643 | .9648 | .9472 | .8976 |
| 4 | X | X | X | .9999 | .9965 | .9878 | .9876 | .9725 | .9304 |
| 5 | X | X | X | X | .9995 | .9959 | .9954 | .9843 | .9474 |

(For column labels, see note on exhibit 5.)

B) FARTHER OUT

| Cut | Cumulative for appropriate multiple of | | | | |
|---|---|---|---|---|---|
|  | Gauss | "flog" | $t_7$ | $t_3$ | $t_1$ |
| −1000 |  |  |  | .0005 | .00032 |
| −300 |  |  |  | .0007 | .00106 |
| −100 |  |  |  | .0011 | .00317 |
| −30 |  | .00000 | .00000 | .0016 | .0106 |
| −10 |  | .00002 | .00010 | .00232 | .0317 |
| −10 |  | .00002 | .00010 | .00232 | .0303 |
| −9 |  | .00005 | .00018 | .00314 | .03522 |
| −8 |  | .00015 | .00037 | .00439 | .03968 |
| −7 | .00000 | .00046 | .00080 | .00637 | .04517 |
| −6 | .00003 | .00137 | .00186 | .00972 | .05287 |
| −5 | .00037 | .00410 | .00464 | .01574 | .06283 |

(For column labels, see note on exhibit 5.)

P) PROBLEMS

6a) For what cut are most cumulative distributions close to 37%? To 63%?

6b) Estimate, by eye and mind, the nine cumulative fractions corresponding to a cut at −1.8.

6c) The same, as nearly as may be, for a cut near −2.8.

6d) Or near −3.8.

exhibit **7** of chapter 20: tabular comparison

**Nine mathematical densities, each for a distribution adjusted to "hinges = ±1" for easy comparison**

A) The TABLE

| Cut | Density for appropriate multiple of ||||||||| 
| | "froot" | $r_4$ | $r_6$ | $r_{10}$ | "Gauss" | "flog" | $t_7$ | $t_3$ | $t_1$ |
|---|---|---|---|---|---|---|---|---|---|
| 0    | .2588 | .2605 | .2636 | .2651 | .2689 | .2747 | .2737 | .2811 | .3183 |
| ±.5  | .2523 | .2526 | .2532 | .2537 | .2540 | .2549 | .2549 | .2556 | .2546 |
| ±1.  | .2320 | .2291 | .2235 | .2196 | .2142 | .2050 | .2071 | .1969 | .1591 |
| ±1.5 | .1962 | .1898 | .1782 | .1706 | .1613 | .1487 | .1499 | .1358 | .0979 |
| ±2.  | .1404 | .1346 | .1233 | .1163 | .1084 | .0989 | .0992 | .0887 | .0836 |
| ±2.5 | .0398 | .0641 | .0675 | .0669 | .0650 | .0622 | .0617 | .0571 | .0439 |
| ±3.  | X     | X     | .0220 | .0300 | .0348 | .0378 | .0369 | .0370 | .0318 |
| ±3.5 |       |       | .0003 | .0089 | .0166 | .0225 | .0217 | .0245 | .0340 |
| ±4.  |       |       | X     | .0011 | .0071 | .0132 | .0127 | .0166 | .0187 |
| ±4.5 |       |       |       |       | .0027 | .0077 | .0074 | .0115 | .0150 |
| ±5.  |       |       |       |       | .0009 | .0045 | .0044 | .0081 | .0122 |
| ±5.5 |       |       |       |       | .0003 | .0028 | .0027 | .0059 | .0102 |
| ±6.  |       |       |       |       | .00010| .0015 | .0016 | .0044 | .0086 |
| ±6.5 |       |       |       |       | .0000 | .0009 | .0010 | .0033 | .0074 |
| ±7.  |       |       |       |       | .0000 | .0005 | .0006 |       |       |
| ±8.  |       |       |       |       |       | .0002 | .0003 | .0016 | .0049 |
| ±9   |       |       |       |       |       | .0001 | .0001 | .0010 | .0039 |
| ±10  |       |       |       |       |       | .0000 | .0001 | .0007 | .0032 |
| ±11  |       |       |       |       |       |       | .0000 | .0005 | .0026 |
| ±12  |       |       |       |       |       |       |       | .0003 | .0022 |

P) PROBLEMS

7a) What is the mathematical density for $t_3$ at −2.5? at −4.5? at +4.5?

7b) Same for $r_4$ at each of the three cuts in turn.

7c/d/e/f/g) Make up some problems of your own.

**660** /exhibit 8/20: Mathematical distributions

The seven distributions of chapter 17, where simple re-expressions of the square root of the density are linear in the square of the shift from the peak, all become conventional mathematical distributions, once we replace average density (as used in chapter 17) by mathematical density. Some will like to know that the correspondence is:

| | |
|---|---|
| $(\sqrt{\text{density}})^2$ | distribution of (product-moment) $r$ on 4 df. |
| $(\sqrt{\text{density}})$ | distribution of $r$ on 6 df. |
| $(\sqrt{\text{density}})^{1/2}$ | distribution of $r$ on 10 df. |
| $\log \sqrt{\text{density}}$ | Gaussian distribution. |
| $(1/\sqrt{\text{density}})^{1/2}$ | distribution of Student's $t$ on 7 df. |
| $(1/\sqrt{\text{density}})$ | distribution of Student's $t$ on 3 df. |
| $(1/\sqrt{\text{density}})^2$ | distribution of Student's $t$ on 1 df (Cauchy). |

**further comment**

The distributions of $r$ (of $t$) are those usually called by those names and apply to the product-moment correlation coefficient or to Student's $t$, when the

exhibit **8** of chapter 20: mathematical distributions

**Behavior of the letter values for the nine shapes of distribution (Gaussian shape solid, shapes corresponding to froots and flogs heavier dashes, $r$ and $t$ shapes lighter dashes)**

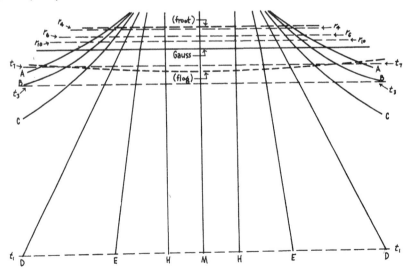

underlying values follow a Gaussian distribution EXACTLY and the population correlation (or difference) is IDENTICALLY zero. (When the underlying distribution, as always, is nonGaussian, the actual null distributions (those for zero population correlation or zero population difference) for observed $r$ and observed $t$ will be somewhat different--usually NOT catastrophically different but, particularly for $r$, often appreciably different.

**review questions**

What shapes of distribution from chapter 17 may interest us? What was tried out as behaving nicely in that chapter? What modification is now considered? What shapes of distribution did we introduce in chapter 19? Were they given in terms of observed distributions or mathematical ones? How many were there in chapter 17? In chapter 15? How many all together? Explain your answer! What have we given tables for in this section?

# index for postscript

21A. Our relationship to the computer  663
21B. What has been omitted?  664
21C. How should the past chapters look different?  665
21D. What have we been introduced to?  666

# Postscript    21

What ought to be said to those who have come through twenty chapters of diverse techniques? Where do they stand? In what ways should earlier parts be looked at differently in the light of later parts? What sorts of techniques have not been begun? Why? Brief answers to such questions are due all those who have worked their way to this point.

### 21A. Our relationship to the computer

Today the world is going through a process of computerization that is culturally rapid but humanly slow. The writer learned to calculate with a hand-crank desk calculator, and looked upon electric motor drive as a great advance. (Much better facilities than those that then cost rather more than half as much as a new car are now available for less than twenty dollars, and are being sold to housewives as a shopping aid.) A decade later, the writer helped a little with John von Neumann's computer project, which set the pattern for modern computing systems--large, medium, or small. After about a second decade, the first large array of statistical programs for computer use (Dixon's BIOMED, later BIMED, later BMD) began to appear. The two decades that followed have seen great diversification of approaches, techniques, and packages. Nothing in statistical computing a decade old looks at all up to date--and we would find it hard to say where we are to be a decade hence.

This book focusses on paper and pen(cil)--graph paper and tracing paper when you can get it, backs of envelopes if necessary--multicolor pen if you can get it, routine ballpoint or pencil if you cannot. Some would say that this is a step backward, but there is a simple reason why they cannot be right: Much of what we have learned to do to data can be done by hand--with or without the aid of a hand-held calculator--LONG BEFORE one can find a computing system--to say nothing of getting the data entered. It will be a long time until this fails to be so. Even when every household has access to a computing system, it is unlikely that "just what we would like to work with" will be easily enough available. Now--and in the directly foreseeable future--there will be a place for hand calculation.

This is not to undersell the virtues of good computing systems, large or small. Given only a little more data than one is willing to tackle by hand, they become necessary. Long before that they become a great convenience. Many of the things we have learned to do in this book are relatively novel. It will be a while before most computing systems are set up to furnish the facilities we all would like. Installation is never trivial; appreciable effort will be needed at every place where such facilities are made available.

But there is hope. Paul Velleman, with support from the National Science Foundation, has been putting together facilities adapted to the use of such techniques in research, a number of versions of SNAP-IEDA (stemming from Michael Godfrey's initial efforts) are in use here and there, and a variety of facilities have been provided locally to help with the teaching of this book's preliminary edition. Perhaps there will come to be supplementary accounts, complete with programs, meshing computer systems with the text of this book.

Actually, the attitudes that underlie the writing of this book, and many of the techniques used in it owe much to modern computers. Today's understanding of how well elementary techniques work owes much to both mathematical analysis and experimentation by computer. You couldn't see the computer in the 661 pages that have gone before, but its shadow lay heavily on many of them.

### 21B. What has been omitted?

If we were to ask a diverse group of well-informed users which statistical techniques were most used, we would get diverse answers. Most, I am sure, would mention regression and analysis of variance, though in special areas of application these words might be drowned out by contingency-table analysis, life tables, bioassay, or even item and factor analysis. How is it that we have seen only the barest of introductions to any of these?

We have fitted lines, but almost everyone would feel that "regression" is multiple regression. We have made row-PLUS-column fits--and their extensions to somewhat more complex models and to three factors--but most would agree that "real analysis of variance" begins when you must choose between error terms, something we have not done. Why have we not gone more deeply into the main highways of today's data analysis?

A number of reasons can be given--all plausible, and some, perhaps, correct. These include:

⋄ these main-highway methods have conventionally been thought of as confirmatory rather than exploratory (even though their most important uses are often in exploration).

⋄ these main-highway methods tend to involve more computing than anyone is prepared to face by hand.

⋄ understanding what is really going on in some main-highway methods is far from trivial.

⋄ first things should come first.

Any one or two of these would be enough reason for not covering these methods here.

When the work that led to this book began--in preparation for the teaching of a course in the spring of 1968--only a few of the techniques we have worked through were available. Many of them have been called forth first by the course and then by the book. Over these same years, we have learned a lot about the main-highway techniques, enough so that it may soon be time to start to develop an account of these techniques either as exploration, or as confirmation, or as both.

A quarter century ago, W. G. Cochran said that "Regression is the worst-taught part of statistics". The continuing truth of his statement has not kept (multiple) regression from being used perhaps more than any other technique. It has had many outstanding successes--and it has helped to mislead many. It will continue to be a main-highway technique--one you will and should learn if you become an analyzer of data--very few can do without it. Just do not expect it to be very easy to learn both just what it is about and just what cautions are needed in its use--another reason why we could not crowd it in here. (But see Mosteller and Tukey, 1977.)

### 21C. How should the past chapters look different?

A course of reading, like a course of instruction, has a beginning, a middle, and an end. It may well loop back and forth, but often only at the price of confusion.

We started with elementary techniques, including plotting scatters of points, and it was not till chapters 7 and 8 that we learned about smoothing and middling. Were someone to write a "second course in exploratory data analysis," one that, like a second course in high-school French, went over the same ground just a little more deeply, it would draw on chapters 7 and 8 when it started on scatter plots, for, once you know about smoothing, middling, and even delineation (chapter 9), it is both very natural and very useful to equip each and every scatter diagram either with one or more automatically-produced middles or with judgment-based smooths of same.

It was an early lesson that we gain a lot by plotting the points. It was a later lesson, not emphasized as hard as it should be, perhaps, that you often do not learn enough until you smooth or middle the points. Plotting the points is often not enough.

Both the scatter and its smooth or middle can help us. The scatter is still our best way to recognize the utterly unanticipated--the smooth or middle is a better way to recognize the more or less anticipated. (And the rough may be a way to recognize what neither scatter nor smooth shows clearly.) Having gone through our techniques once, we can move toward a more balanced attitude toward them, and toward wider uses for many.

In a more detailed way, the fitting techniques of chapters 10 and 11 can be used to organize and simplify the delineation of point scatters. Our results could be essentially the same, but the calculations would be neater, and one computer program, if available, would do for both purposes.

These are but examples. Every serious reader should look for additional ones, taking a personal responsibility for further organizing and usefully digesting the various techniques we have learned to use in isolation.

### 21D. What have we been introduced to?

When we try to structure what we have done, we find overlapping relationships, as the following list suggests (chapter numbers in parentheses):

Batches (1, 2, 3, 4, 15)

Lines and trends (5, 6, 7, 8)

Smooths and roughs (7, 8, 9, 16, 17)

Scatters (8, 9, 17)

Two-way analyses (10, 11, 12, 13)

Counted data (scattered throughout, 17, 18, 19)

It will help most readers to review, at least generally, what has been said under each of these broad topics as a whole. Some will then find it useful to consider how each one overlaps with or interrelates to each of the others.

### References

F. Mosteller and J. W. Tukey, 1977, Data Analysis and regression: A second course in statistics. Addison-Wesley, Reading, Mass.

The Holy Bible (King James Version). Revelation, Chapter 13, Verse 18.

# Glossary

(entries beginning with punctuation marks and digits will be found on page 676)

**adjacent value:** value nearest to, but still within, an inner fence.

**adjacent point:** point on the adjacent polygon.

**adjacent polygon:** the polygon exactly surrounding all points not outside.

**adjusted density:** in chapter 16, observed density adjusted for observed curvature.

**all:** abbreviation for value of common term; for something included (by addition) in all values of the fit.

**amount:** value that can never be less than zero and is not a count.

**average density:** (fraction of counts)/(bin width)

**balances:** variates that can be either + or − and have no obvious bounds.

**basic count:** a number of occurrences, as distinguished from the numbers of bins/cases/individuals/species for which a particular number of occurrences are observed.

**batch:** a set of values of similar meaning, obtained however. (By no means necessarily anything like a sample.)

**bin:** a cell or interval along a scale, into which individual values do or don't fall.

**blurring:** representing smoothed values as the centers of vertical bars (of carefully chosen length).

**box-and-whisker plots:** a box from hinge to hinge, barred at the median, with "whiskers" to (a) the extremes (b) the innermost identified values (c) the adjacent values.

**break table:** a no-interpolation table working through the "breaks"--the arguments at which the functional value changes. (Also called "critical table".)

**broken median:** horizontal bars at the median of each piece (as defined by letter cuts) and the portions of the letter cuts needed to connect them.

**clarity:** ratio of a measure of size of fit (usually highest value MINUS lowest value) to a measure of size of residual (usually a letter spread). (E.g. E-clarity, H-clarity.)

**coded residual:** display of the rough value of a residual as one of few (often ⊙, ○, °, ·, ×, X, ※, or ⊞).

**col:** frequently used to mean a value depending upon column alone.

**col-extra:** an additional quantity depending upon column.

**column effect:** see effect.

**common term:** something included, by addition, in all values of a fit.

**comparison values:** values of (row)(col)/(all) for a PLUS fit.

**completed fraction:** ratio of completed rank to total number of cases.

**completed rank:** *See* c'rank.

**condensed fit:** result of coalescing fits and medianing residuals when fitted "row" [or "col"] values come too close to one another.

**condensed residuals:** see condensed fit.

**coordinate:** a function of position, often defined by its level traces.

**count:** integer (positive or zero) obtained by counting.

**counted fraction:** a fraction obtained by counting both a total and the number that "do", and taking the ratio.

**counting in:** assigning an extreme value rank 1, the next value 2, and so on.

**c'rank:** the largest rank assigned to a given basic count (or size).

**cross median:** that point, often not among the given points, each of whose two coordinates is the median of the corresponding coordinates of the given points.

**cumulative:** a formula (or table) giving cumulative fractions in terms of cuts.

**cutting value:** one of a sequence of values of $x$ chosen to divide a batch of $(x, y)$ pairs into slices.

**D:** a letter value at depth $\frac{1}{2}(1 + e)$, where $e$ is the integer part of the depth of an E; crudely, a sixteenth.

**delineation:** a set of smoothed letter traces, crossed by the letter cuts used to begin their definition; a display of traces and cuts (often each D, E, H, M, H, E, D).

**density:** see average density, mathematical density.

**dependency ratio:** (actual MINUS median)/(most positive MINUS median), unless this is negative, when denominator becomes (median MINUS most negative).

**dependency trace:** a trace of dependency ratio (usually smoothed) against slice position.

**depth:** lesser of rank up and rank down.

**diagnostic plot:** After a PLUS analysis, a plot of residuals against comparison values. (The middle half, in terms of comparison values, is often replaced by its cross median.)

**double flog:** See fflog.

**double lines:** indication that the things they separate are to be added together.

**double root:** $\sqrt{2 + 4 \, (observed \, count)}$ (and sometimes $\sqrt{1 + 4(fitted \, count)}$).

**D-spread:** high D-value MINUS low D-value.

**D-trace:** a smooth curve indicating the behavior of D-values of slices, usually passing through points obtained by smoothing D-to-E or E-to-D differences and subtracting or adding the result to the appropriate E-trace.

**E:** a letter value at depth $\frac{1}{2}(1 + h)$, where $h$ is the integer part of the depth of a hinge; crudely an eighth, sometimes called one.

**E-box:** closed figure made up of E-traces and E-cuts. (*See* letter traces *and* letter cuts.)

**eff:** abbreviation for "effect".

**effect:** allocation of part of several observed values to a circumstance (more generally to a combination of circumstances).

**end-value rule:** using $y_1 = median(x_1, y_2, 3y_2 \dotminus 2y_3)$ as the smoothed end value ($x$ = data, $y$ = smooth).

**E-spread:** high E-value MINUS low E-value.

**E-trace:** a smooth curve indicating the behavior of E-values of slices, usually passing through points obtained by smoothing E-to-H or H-to-E differences and subtracting or adding the result to the appropriate hinge trace.

**E-value:** *See* E.

**extreme:** a highest or lowest value of a batch. (As of depth 1, often labelled "1".)

**factor:** a circumstance described in terms of specified versions.

**far out:** a value beyond the outer fences.

**fcent:** a folded percent, given by $p - (100 - p) = 2p - 100$.

**fence:** inner fences are one step outboard of the hinges; outer fences are two steps outboard.

**fence polygon:** in a plane, the inner polygon defined by all possible fence lines.

**fenced letter display:** a fairly complete letter display, including fence values and identification of outside points.

**fflog:** a doubly folded logarithm; a double flog, a sum and difference of four terms. (*See* definition of flog and Section 15E.)

**fit:** an incomplete description of the data, usually selected from a class of possible descriptions, sometimes by a specified procedure.

**flog:** a folded logarithm, specifically

$$\frac{1}{2}\log_e (\textit{fraction that "do"}) - \frac{1}{2}\log_e (\textit{fraction that "don't"}),$$

usually best taken as

$$\log_e \sqrt{\textit{number that "do"} + \frac{1}{6}} - \log_e \sqrt{\textit{number that "don't"} + \frac{1}{6}}.$$

**folded root:** see froot.

**folded log:** see flog.

**four steps (of median polish):** a frequently adequate stage of analysis reached by removing row and column medians (alternately) twice each. (Also called "two cycles of median polish".)

**froot:** a folded square root, specifically

$$\sqrt{2 (\textit{fraction that "do"})} - \sqrt{2 (\textit{fraction that "don't"})}.$$

**Gaussian reference:** the Gaussian shape of distribution regarded as something from which we assess deviations (almost certainly not zero) rather than as something that might really happen.

**Gaussian shape:** all distributions obtained from

$$\frac{1}{\sqrt{2\pi}} e^{-x^2/2} \, dx$$

by linear change of scale and location.

**grade:** any variate whose values are ordered, but which we do not care to consider numerical.

**h:** following a digit, means "and a half".

**H:** as a letter value, a hinge, a value whose depth is $\frac{1}{2}(1 + m)$, where $m$ is the integer part of the depth of the median; crudely, a quartile.

**H:** in smoothing, hanning (which *see*).

**half-octave values:** values proceeding by two steps to the octave--as 1, –, 2, 3, 4, 6, 8, 11, 16, ...

**hanning:** a smoothing process alternatively describable as the result of (i) two repetitions of running means of 2; (ii) taking the mean of each given value and its skip mean; (iii) a moving average with weights 1/4, 1/2, 1/4.

**hinge:** letter value at depth $\frac{1}{2}(1 + m)$, where $m$ is the integer part of the depth of the median; crudely, a quartile.

**hinge trace:** a smooth curve indicating the behavior of hinges of slices, usually passing through points obtained by smoothing H-to-M or M-to-H differences and subtracting or adding the result to a median trace.

**H-spread:** high hinge MINUS low hinge.

**H-value:** *See* H.

**integer part:** if $i$ is an integer, and $f$ a fraction with $0 < f < 1$, the integer part of $i + f$ is $i$.

**interpolated rank:** the value at interpolated rank $i + f$ ($i$ an integer, $f$ a fraction) is $(1 - f)x_i + fx_{i+1}$, where $x_i$ has rank $i$ and $x_{i+1}$ has rank $i + 1$.

**ladder of expressions:** the increasing sequence of powers (first negative powers, then the log, then positive powers).

**leaf:** in a stem-and-leaf display, the righthand part of one of the values displayed.

**letter cut:** a vertical cut defined by a slight outward displacement of a letter value.

**letter-value difference:** difference between adjacent letter values (e.g., H-to-M or M-to-H).

**letter display:** *See* letter-value display.

**letter trace:** a trace which tries to divide the points roughly like a letter value, doing this not only globally but somewhat locally.

**letter value:** one of the values tagged ..., B, C, D, E, H, M, H, E, D, C, B, A, Z, ..., where the depth of each letter is almost half-way from the depth of the letter value next nearer the median to 1 (the depth of an extreme). (Extremes and the like are sometimes included.)

**letter-value display:** a somewhat formalized display of the letter values. (It may include: fence positions, number outside, spreads, mids, identifications, and/or pseudospreads.)

**level curve:** curve along which some function, usually a possible coordinate, takes a given value (elsewhere also called *level line*).

**level line:** line along which some function takes a given value.

**level trace:** a level line or level curve.

**log:** a logarithm (probably to base 10).

**$\log_e$:** a natural or Napierian logarithm (to base $2.71828\ldots$, $\log_e u = 2.30259 \log_{10} u$).

**$\log_{10}$:** a common logarithm, a log to base 10.

**M:** as a letter value, the median.

**mathematical density:** limiting value of average density (as bin width $\to 0$ and bin count $\to \infty$).

**median:** value of depth $\frac{1}{2}(1 + n)$, where there are $n$ values in the batch; the middle value when all values are arranged in nondecreasing--or nonincreasing--order.

**median polish:** a process of (finding, adding elsewhere, and subtracting here) medians alternately. (In a two-way table, alternately those of rows and those of columns.)

**median removal:** *See* median polish.

**median trace:** a middle trace obtained by using medians (as most middle traces are).

**mid:** indicates the mean of two letter values (e.g. midhinge, mideighth, midD, midextreme).

**middle trace:** trace leaving roughly half of all points above and below itself, both semilocally and globally--in practice, usually the result of smoothing the cross medians of a set of pieces.

**mixed leaves:** leaves in a single stem-and-leaf display involving different numbers of leaf digits.

**modified (hinges, medians, etc.):** result of weighting exact zeroes $\frac{1}{2}$.

**negative reciprocal:** the result of dividing a chosen negative constant by the given number.

**octave:** values (perhaps counts) between A and 2A (for some A), or intervals of such form.

**outside point:** one outside the inner fence polygon.

**outside value:** a value outside the inner fences (and, if "far out" is also mentioned, inside the outer fences).

**peak:** inferred highest point of fit (will usually not fall at a bin center).

**piece:** those data points between two adjacent letter cuts.

**plurality:** difference between a fraction and its complement, $f - (1 - f) = 2f - 1$.

**PLUS fit:** a row-PLUS-column fit (or one of its simplest generalizations).

**PLUS-one fit:** a fit that can be put in either--and hence both--of the forms:

$$\text{all PLUS row PLUS col PLUS (row)(col)/constant}$$

$$\text{constant PLUS (row*)(col*)}$$

**PLUS-one plot:** a two-way plot appropriate to a PLUS-one fit (one set of parallel lines and one set of lines with a common intersection when raw response is represented).

**PLUS-one plot of residuals:** coded residuals placed where intersections in a PLUS-one plot would fall.

**PRODUCT:** in product–ratio analysis, basic count times c'rank (from above).

**product–ratio analysis:** calculation and use of product–ratio plots; study of the relation of root of PRODUCT to log of RATIO.

**product–ratio plots:** plots of $\sqrt{(basic\ count)(final\ rank\ from\ above)}$ against $\log((basic\ count)/(final\ rank\ from\ above))$, or of the result of changing "final rank" to "final fraction".

**(25%) pseudospread:** the spread between 25% points of a standard distribution shape (usually Gaussian) when the spread between $p\%$ points is that observed.

**R:** in "3R" means "3 repeated as many times as necessary to settle down."

**range:** high extreme MINUS low extreme.

**rank:** ordinal number of value when values are made nondecreasing (rank up) or nonincreasing (rank down).

**RATIO:** in product–ratio analysis, basic count divided by c'rank (from above).

**reciprocal:** the result of dividing 1 by the given number (and hence, loosely, the result of dividing any chose constant, instead, by the given number).

**reduced delineation:** an E-box in solid lines, M traces/cuts dashed within box, H-traces/cuts dashed outside it.

**re-expression:** expressing the same information by different numbers, as when $y$'s are replaced by log $y$'s.

**re-matching:** re-pairing of given $y$'s with given $x$'s, usually to make the re-matched relationship as strong (negative or positive) as possible.

**reroughing:** smoothing a rough and adding this smooth to the original smooth.

**residual:** result of subtracting fit from data; thus, what is not accounted for by incomplete description (fit).

**response:** variate, or value of variate, thought of, at least potentially and temporarily or schematically, as influenced by (or associated with) certain factors or circumstances.

**root:** square root.

**root rough:** the rough of the square roots of the given counts.

**root smooth:** the smooth of the square roots of the given counts.

**rough:** data MINUS smooth.

**row:** often means something depending upon row alone.

**row effect:** *See* effect.

**row-extra:** an additional quantity depending upon row alone (often appears multiplied by constant and col).

**row-PLUS-column fit:** a PLUS fit, a fit of the form "all PLUS row PLUS col" (for generalizations, see Sections 12A and 13D).

**row-TIMES-column fit:** a TIMES fit, a fit of the form "(row*) TIMES (col*)" (for generalizations, see 12A and 13D).

**running median of 3:** for a sequence, at each $i$ the median of $x_{i-1}$, $x_i$, and $x_{i+1}$.

**S:** in smoothing, treating peaks/valleys of length 2 by splitting the sequence between the 2 and applying first the end-value rule to each piece and then 3R to the result.

**schematic plot:** a schematic representation of a batch, including a hinge box, barred at the median with dashed whiskers extending to the adjacent values terminated by (dashed) bars, and with all outside values identified.

**schematic (x, y) plot:** a schematic representation of a batch, including an E-box, an adjacent polygon, and identified outside points.

**sequence:** an ordered set of $(x, y)$ data (where $x$ may have been suppressed), in which $x$ proceeds by steps of equal size.

**shape of distribution:** that which is common to the distributions of all trivial re-expressions of some quantity.

**shift:** displacement of value considered from the peak.

**skip mean:** in smoothing, $(x_{j-1} + x_{j+1})/2$ placed at $j$.

**slice:** those $(x, y)$ pairs whose $x$-values lie between chosen cutting values.

**sliced coordinate:** one used to determine letter cuts, and thus stress and various traces.

**smooth:** a sequence of smoother appearing values, calculated from the given values; the process of such calculation.

**smoother:** a process dividing a given sequence into smooth and rough (which sum, term by term, to the given sequence).

**split count:** a count of "those less" PLUS one-half of "those equal".

**squeezed stem-and-leaf display:** a display with 5 stems (usually labelled "*", "T", "F", "S", and "·") for each stem number.

**SS:** S applied once, and then applied once again.

**ss:** stands for "started split".

**ss-count:** the value of "(count below) $+ \frac{1}{2}$(count equal) $+ \frac{1}{6}$".

**ss-fraction:** the value of "$\left[(count\ below) + \frac{1}{2}(count\ equal) + \frac{1}{6}\right]$ divided by $\left(total\ count + \frac{1}{3}\right)$".

**standard cutting points:** for coding residuals, 1 or 2 H-spreads outboard of hinges. (Distinguish from fences. Nonstandard cutting points often used.)

**start:** a constant added before taking a log or root (or etc.). (Here 1/6 is popular. Elsewhere 1/10, 1/4, or 1 may appear.)

**started count:** for product–ratio analysis, the basic count increased by 1; (in general a count increased by a constant).

**stem:** a line in a stem-and-leaf display.

**stem-and-leaf display:** a generalized two-digit display, in which the lefthand portion of the values displayed is given by a stem value, while the righthand portion makes up a leaf (leaves follow stem juxtaposed if uniformally one digit, follow separated by commas if some or all involve 2 or more digits).

**stem label:** the stem value, followed (if there are 2 or more stems per value) by a character from "*", "T", "F", "S" and "·", the whole preceding a vertical line.

**stem-value:** that part of the label of a stem which consists of the common left-hand portion of the values recorded in that line.

**step:** 1.5 times the H-spread.

**stretched stem-and-leaf:** a stem-and-leaf display, with 2 stems per value, these being distinguished with "*" and "·".

**three-way analysis:** analysis of one response according to three kinds of circumstances and their combinations.

**tilt:** an $(x, y)$ data display is tilted if subtracting some $bx$ from $y$, with $b \neq 0$, is needed to make the display appear "untilted".

**TIMES fit:** a row-TIMES-column fit (or one of its simplest generalizations).

**traced coordinate:** one whose slice medians define the broken median, etc.

**trimean:** $\frac{1}{4}(low\ hinge) + \frac{1}{2}(median) + \frac{1}{4}(high\ hinge)$.

**trivial re-expression:** using $a + bx$ for some fixed $a$ and $b$, instead of $x$.

**twicing:** reroughing with the same smoother used to make the initial smooth and rough.

**two-way analysis:** analysis of one response according to two kinds of circumstances.

**two-way plot:** a representation of a fit by two families of lines--one family for each kind of circumstance, one line for each circumstance--such that the y-coordinate of each intersection gives the corresponding fitted value.

**two-way plot of residuals:** coded residuals, each plotted where the corresponding intersection of the two-way plot would fall.

**two-way schematic plot:** *See* schematic $(x, y)$ plot.

**untilted:** such that no subtraction of $bx$ from $y$ makes the $(x, y)$ display seem more nearly horizontal.

**untilted analysis:** in smoothing, schematic plotting, or delineation, an analysis of $(x, y - bx)$ values for a suitable $b$, whose results are then restored to $(x, y)$ form.

**version:** a version may be qualitative (when it is likely to be called a "category" or "class") or quantitative (when it is likely to be called a "level"). (Ordered versions may also be called "levels".)

**wandering schematic plot:** a description of an (x, y) batch involving hinge and median traces and the adjacent polygon (points outside that polygon will ordinarily be plotted and identified).

**",":** in smoothing, used to separate components related by reroughing or twicing.

**#:** a count, often a total count or batch size.

**\*:** may indicate a place that can be filled.

**\*:** when followed by a letter-value term or symbol (as in *E, *-fences, *H, *-hinge, *-letter-value, *M, etc.), indicates that exact zeroes have each been given a weight of 1/2.

**1:** in letter values, an extreme.

**3:** in smoothing, running medians of 3.

**5-number summary:** values of extremes, hinges, and median, values of 1HMH1.

**7-number summary:** values of 1EHMHE1.

**9-number summary:** values of 1DEHMHED1.

## Index to reference tables

break table for two-decimal logs, 62 *(also frontpaper 1)*
break table for (square) roots, 71 *(also frontpaper 2)*
break table for (negative) reciprocals, 72 *(also frontpaper 3)*
pluralities, folded roots, folded logarithms, 499 *(in part, rearpaper 4)*
values of $\log_e \sqrt{count + 1/6}$, 503 *(in part, rearpaper 5)*
values of $\sqrt{count + 1/6}$, 506 *(in part, rearpaper 6)*
double roots, useful numbers for, 578
the Gaussian shape as a re-expression of fractions, 624
pseudospreads from HED...spreads, 632, 633
Gaussian ratios at half-octave depths, 640

## alphabetical index

ABC Corp *(exhibit tag)*, 133
accelerated growth, 142
adjacent polygon, 285
adjacent values, 44, 272
adjusted gross income *(exhibit tag)*, 621, 622
adjustment, to mathematical density, 654
adjustments, rough and exact, 110
ages at marriage, 626
agreement of spread, 97
all, 335ff
alpha-particles, 579
alternative forms of display, 99
American colonies, population of, 167
amounts and counts, 57, 92
amounts, stretched-tailed patterns of, 589
Anderson, E. O., 606
Anderson, L. J., 166
anonymous, 123
Anscombe, F. J., 606
appearances per word, 590ff
arithmetic of two-way fitting, alternative organizations, 372ff
arithmetic, practical, 3ff
arithmetic, precision of, 3ff
Arizona temperatures, 332ff, 385
Arizona temperatures *(exhibit tag)*, 333, 336, 338, 340, 342, 349, 352, 405, 407
arrangement of three- and more-way analyses, 443ff
Ashby, E., 554

association constant, 26
Atherton, Pauline, 592, 599
athletic power, 517
atomic weights *(exhibit tag)*, 68
Auerbach, 598
aunt, 516
auto makes *(exhibit tag)*, 22
auto years *(exhibit tag)*, 18
% automobiles, 470
average wages, farm workers, 255

Babington, 611
background, 194ff
bacterial clumps, 605
BaDep, 287ff
balances, 57, 92
baloney, 481
Bancroft, T. A., 550, 551
bank deposits, 287ff, 313
banks, *see also* suspended deposits, 246
banks, suspended, 212
basic count, smallest, 604ff
basic count, zero, 605ff
basic counts, 550
basic counts, counts of, 550ff
basic counts, logs of, 552
basic counts, roots of, 552
basic counts, stretched-tailed patterns of, 589
basic counts, tied, 589
basic law, whether or not, 181
basketball players, 551
batch, 6

batches, 666
batches of points, 309ff
batches, summarizing, 626
Beadles, J. R., 371
Beall, Geoffrey, 122, 365
Beamish, F. E., 122
Beare, 610
beast, number of, 666
Beckman, A. O., 119
behavior, need to see, 128
Bennett, C. A., 160
Bentham, 611
binned data, 620ff
bins, 551ff
bins, added zeros, 545ff, 551
bins, going beyond, 589
bins, infinitely narrow, 647
bins of equal width, 551ff
bins of unequal width, 570ff
birds, breeding pairs of, 552
births and deaths *(exhibit tag)*, 149, 150, 152, 153
births to deaths, ratio of, 148
blind subjects, 445ff
Bliss, C. I., 606
blurring, 218
blurring the smooth, 223
boards, 605
bold numbers, use of, 6
bolding, use of, 343
Bolz, A., 202
box-and-whisker plots, 39
braking distances *(exhibit tag)*, 182, 183, 184, 185, 186, 187
break tables, 62
breaking a smooth, 237ff
Brindley, T. A., 550, 551
Brischle, H. A., 548
Bristol, Pennsylvania, 472
broken hinges, 276ff
broken median, 274ff
broken smooth, 239ff
bromine, atomic weights of, 68
Brown, Robert G., 203
Brownlee, K. A., 512, 513
Brownsville, Florida, 472
Bruner, J. S., 107
Buffalo, seasonal snowfall in, 117

Burrough, W., 371
Butler, 610
Buxton, P. A., 606

Cairo, seasonal snowfall in, 117
Cannamore, South India, 606
carbon monoxide, in air, 119
cardboard, pieces of, 563
Carroll County, N.H., 239ff
cars/advertisement *(exhibit tag)*, 7, 8
cement, shipments of, 255
cement, strengths of, 457
Census, Bureau of the, 148
*Census of Population, U.S.*, 301
Chavez, H., 572
check columns, 10
checking a two-way analysis, 366ff
checking arithmetic, 10ff
chemistry, authors, 597-8
Chevrolet prices *(exhibit tag)*, 9, 30, 32, 34, 44, 617-619
Chief of the Weather Bureau, 117
choice of expression, in smoothing, 247ff
choice of expression, reasons for, 97
Chu, C. C., 118
circle-negative rule, 364ff
circumstance, 126
circumstances, two kinds of, 331ff
city killings *(exhibit tag)*, 409, 410, 411, 412, 413, 414
Clarke, G. L., 103
Clausen, J. A., 512, 513
cleaning ability, 456
*Climatography of the United States*, 334, 382, 386
coal production *(exhibit tag)*, 212, 215, 216, 217, 219, 220, 224, 226, 228, 528, 532-534
Cochran, W. G., 371, 665
coded residuals, plot of, 384ff
coding residuals, 382ff
col extra, 437ff
Cole, L. C., 606
commercial banks, liabilities of, 255
common, 335ff
comparative display *(exhibit tag)*, 502
comparing expressions, 250

comparing several batches, 102
comparison, meaning of, 110
comparison of distributions *(exhibit tag)*, 657-660
comparison values, 353ff
comparison values, three-way, 458ff
comparisons of two batches, 64
complete rank, 590ff
completed fraction, 602
computer, our relationship to, 663ff
computing systems, good, 663
concurrence, near, 477ff
condensing fits and residuals, 384ff
confirmatory data analysis, 3
Connecticut, voting, 390ff
constant, fitting one more, 352ff
contamination, long-lived, 197
continuity, 648ff
Contreras G., M., 572
coordinates, 467ff
Coos County, N.H., 239ff
Coos-Carroll *(exhibit tag)*, 239, 241, 242, 243, 244
Cootner, P., 568
copying-on, 221
corn borer larvae, 551
corn borers, 561ff
corner totals, 518ff
corners, washed into, 59
coronary arteries, 655
coronary deaths *(exhibit tag)*, 655
coronary disease, deaths by, 654
Corps of Engineers, U.S., 63
counted data, 83ff, 250ff, 666
counted fractions, 92, 495ff, 648ff
counted fractions and percentages, 57
counted fractions, careful expression, 508ff
counted fractions, matched scales for, 498ff
counted fractions, parallel, 615ff
counted fractions, sequence of, 495
counting by tallying
counting-in, 29-31
counts, focus on individual, 589ff
counts in bin after bin, 551ff
counts, matched, 602ff, 607ff
counts of basic counts, 550ff

counts, re-expressed as double roots, 576ff
counts, re-expression of, 83
counts, residuals vs. roughs, 561ff
counts, started, 605
counts, two-way tables of, 510
*County and City Data Book*, 148, 149, 158, 159, 322, 470
county areas *(exhibit tag)*, 12, 35, 45, 48, 54, 65, 67, 635
cousin, 516
Crane, R. A., 119
c'rank, 590ff
criminal heads, two groups?, 610
criminal heights, 563ff
criminals, finger lengths, 563ff
cross-hinges, 276ff
cross-medians, 275ff
crowding, %, 470
Cummings, C. L., 118
cumulative distribution, 649ff
cumulative fraction 649
cumulatives, 649ff
curve, bulging side of, 173
curved traces, 476
cuts, 495
cuts, all possible, 616
cuts, sequence of, 495
cutting numbers, 4
cutting points, in coding residuals, 382ff
cutting values, 495

dams, reservoir capacity, 96
Daniel, C., 312
data and problems *(exhibit type)*, 25, 26, 94-96, 116-123, 158, 160-167, 199-203, 251, 302, 317, 323, 370, 381, 385, 389, 456, 512, 535, 554, 610, 626
data sets, how many needed, 327
data, tilted, 155
data, told how analyzed?, 397
data, untilted, 155
Davidson, N. R., 166
Davis, H. T., 597-8
Dean, J. A., 120
deaths by cause *(exhibit tag)*, 86
decay of radioactivity, 195

decimal points, 4
delay, 464
delineations, 309ff
delineations, what they miss, 319
densities, 648ff
densities, distributions vs. binnings, 651ff
density, average, 648
density, constant, 653
density, linear, 653
density, quadratic, 653
dependency ratios, 485ff
dependency trace, 486ff
deposits, U. S. Government, 255
depth, 30
design patents, 255
detective work, quantitative, 1ff
detergent combinations, 456
diagnostic plots, 355ff, 398
diagnostic plots, for three-way, 458ff
difference limen, 445ff
directed diameters, 492
discrepancy, 492ff
discreteness, 648ff
display, alternative forms of, 99
display, semischematic, 527, 530
distance to stop, 182ff
distance vision, unaided, 518
distances, coastwise, 415
distribution, Gaussian shape of, 623ff
distribution, reference standard, 615
distributions, mathematical, 647ff
distributions, shapes of, 615ff
distributions vs. binnings, 648ff
Dixon, W. J., 663
documented vessels, tonnage, 255
double flog, 516ff
double folding, 513ff
double lines, 332ff
double roots, 576ff
double-skip-mean, 534
drape measurements, consistency of, 118
D-traces, 309
ducks, weight of, 123

East Coast *(exhibit tag)*, 354, 356, 357, 401, 403, 423, 426-431
East Coast temperatures, 353, 381

easy re-expression, 57ff
easy summaries, numerical and graphical, 27ff
ecological counts *(exhibit tag)*, 605-609
econometric activity *(exhibit tag)*, 596. 597
education and research, 344ff
Edwards, 610
eff, 335ff
effects, 335ff
eighths, 53ff
electrical resistance measurements, 122
electricity, usage, 265ff
ElUse, 267ff
end values, 248
end-value smoothing, 221, 222
energy use, (*see also* Twin Rivers), 265ff
Engineers, U.S., Corps of, 63
England and Wales *(exhibit tag)*, 137, 138, 140
entomological field experiments, 122
establishment, size of, 510
E-traces, 309
exact adjustments, 110
exact zeros, 223, 344
exploratory data analysis, as foundation, 3
exploring numbers, 3
expression, cautionary examples, 582ff
extended two-way fits, simpler, 438ff
extremes, 29ff
eye smooth, 215ff
eyes, unhooking, 220
Ezekiel, Mordecai, 182

factor, 126
Fairfield County, Connecticut, 392
Fairley, J. L., Jr., 203
far out, 44
farm workers, average wages, 255
fathers, 516, 545
fcents, 498
feel what the data are like, 19
fenced letter display, 44
*-fences, 226
fences, 43ff
Feuell, A. J., 457
fflog, (*see also* flog), 516ff

field corn *(exhibit tag)*, 550, 551, 562
finger lengths of criminals, 563ff
finger limens *(exhibit tag)*, 446, 449, 451, 453-455, 459-463
finger pull, 445ff
fits, condensing, 384ff
fits symmetrical around a peak, 543, 555ff
fits, three-way, 443ff
fits, two different satisfactory, 582
fitting to smoothed roots, 555ff
five-figure logs may not be enough, 67
Flacker, S. B., 548
Flagstaff, Arizona, 332ff
flattening, gain by, 148
fleas, 605
flogs, easy, 520
flogs, *(see also* fflogs*)*, 499, 616
folded fraction, 500
folded logarithms, 499
folded logs, 501
folded percents, 498
folded roots, 499
food and tobacco, 344ff
"*" for place filler, 5
"#" for count, 5
forearm lengths *(exhibit tag)*, 545, 547, 556-560, 582, 585
fractions, parallel, 615
France, 510
Franklin, N. L., 160
froots, 499ff, 616
froots, easy, 520
function, percent point, 650ff
function, representing, 650ff

Gallup, George, 509
Garrett, A. E., 202
gas densities *(exhibit tag)*, 49-52
gas, usage, 265ff
GaUse, 267ff
Gauss, 623
Gaussian reference, 623ff
Gaussian shape, 623ff
Geiger, H., 579, 580
Germany, 510
Gerschenkron, 510
Gilpin, V., 203

ginnings vs. distributions, 648ff
GminB, defined, 291
Godfrey, Michael, 664
gold, 205ff
gold assays *(exhibit tag)*, 548, 549
gold concentrations, 547
gold production *(exhibit tag)*, 207
GoSal, 287ff
governor's salary, 287ff, 299, 313
governor's salary, changes, 298
governor's salary *(exhibit tag)*, 288, 289, 291-297, 312-316, 320
governor's salary, summary, 299
governor's salary, width, 298
grades and other ordered versions, 57, 92
grandfather, 516
graph paper, good and bad, 128
graph paper, kinds, 128
graphical subtraction, mechanical model, 135
graphs, enlarging, no substitute for, 52
graphs, friendly, 157
graphs, ticks and numbers along axes
Gray, W. A., 457

h for "and a half", 5
haemocytometer, 554
Hald, A., 457
half-octave calculation, 595
half-octaves, 615
handles, choice of a set of, 441
handles, two or more, 331ff
hanning, 231ff, 523ff, 571
Harban, A. A., 25
Hartford County, Connecticut, 392
Haviland, G. D., 123
headlice, 609
heart disease, 655
heats of solution *(exhibit tag)*, 312
Heckmann, O., 26
heights of criminals, 563ff
Herdan, G., 592
Hernandez D., T. P. E., 572
Hersh, S. P., 122
Hertwig, K., 202
high tide, 302
Hindman, J. C., 26

hinge traces, 281ff, 292-4
hinges, 32, 33ff
*Historical Statistics of the United States*, 168, 256, 541
hollow downward, 173
hollow upward, 173
horizontal scale above the plot, 129
H-spread, 44
H-to-M, defined, 280
Hull, Clark L., 78
hydroelectric plants, 14, 60, 63
hydrogen, atomic weights for, 68

illustrative *(exhibit type)*, 15, 17, 90, 91, 130, 132, 134, 155, 198, 211, 222, 232, 249, 280, 383, 415, 424, 432, 469, 477, 510, 627, 652, 653
income, adjusted gross, 620
income tax returns, individual, 620
incomplete descriptions, subtracting out, 143
index cards, 321ff
indications, 3
individuals, infinitely many, 647
initial pull, 445ff
inner fences, 44
insect counts *(exhibit tag)*, 364, 367-369, 372, 375, 377-381, 383-388
intelligence, 517
interpolated ranks, 31
isometric graph paper, 128
isopods, 605
Isserlis, L., 626

Jackson, D. S., 122
Johnson, Palmer O., 445ff, 447
Josselson, H. H., 592
journal names, 594
*Journal of Chemical Physics*, 600
journal, papers per, 590
judgment = choices, 215

Kailua-Lanikai, Hawaii, 472
Kare, M. R., 202
Katz, J. J., 26
Keenan, Stella, 592, 599
Kendall, M. G., 518-519, 566, 568

Kennedy, John F., 509
Kennedy, support, 508ff
kinds of grids, 128
Kingsley, A., 121
Kinsey, D., 121
Kjeldahl ultramicrodetermination of nitrogen, 121
Koch, G. S., Jr., 547, 548
Kuck, J. A., 121

ladder of powers, 89
ladder of simple expressions, 172, 196
lakes, largest, 94
Landes, D. S., 510, 511
large numbers, abnormal law of, 15
largest-over-smallest, 169ff
layers, 458ff
leaf, 8
Leavens, D. H., 596
Lee, A., 545ff, 546
*Leptinotarsa decemlineata*, 363ff
letter cuts, 270ff
letter segments, 482
*-letter values, 223
letter values, 54, 270ff, 615, 626ff
letter values for rough, 546ff
letter-value display, 33
level curves, 468
level lines, 468
level traces, 467ff
line mean, 534
lines and trends, 666
Link, R. F., 547, 548
logs, one-decimal, 61
logarithmic graph paper, 128
logarithms, 59ff
logs, 59ff
logs, doubly folded, 515
logs, how to think about, 93
logs, octaves for, 552
logs of amounts as balances, 57
logs, of negative numbers, 397
logs, of zero, 397
logs, quick, 61
logs, relation to powers, 86
looking at batches of points, in two or more ways, 265, 309, 467ff
Lord Rayleigh, 49

Loring, H. S., 203
Lübeck, K., 26

MacGillivray, 610
MacLeod, P., 606
MacRae, J. C., 457
magnifying glass !(before plotting), 145
main effects, 450ff
Manning, W. M., 26
marriage, ages at, 626
marriages, 317
matched, modes of expression, 498
mathematical distribution, uses, 647
mathematical distributions, 647ff
mathematics, authors, 597-8
McCrone, W., 203
McCullough, J. D., 119
McGlanery, R. M., 25
McGuire, J. U., 550, 551
mean removal, 405
median, 29ff
median age, 150ff
median age, urban and rural populations, 167
median polish, four steps of, 366ff
median traces, 275ff, 292-4
median traces, what matters, 481
medians of 3, 210
medians, partly graphic, 100
medians, ubiquity of, 492ff
medical care and health expenses, 347
Medway, W., 202
merchant vessels built, 255
mercury, vapor pressure of, 200
Methodist Church, membership, 255
Meyer, S., 194, 195, 200
Michigan counties *(exhibit tag)*, 636-639, 643, 644 (*see also* county areas)
Michigan, land areas of counties, 48, 65ff, 637, 639ff
microdetermination of CO in air, 119
microscope fields of milk, 605
midl (midextreme), 79
middle traces, 276ff, 473
middle traces, different for the same, 470ff
middle traces, where falling, 490
middling, 476ff

midE (mideighth), 79
mideighth (midE), 79
midextreme (midl), 79
midextreme removal, 404ff
midH (midhinge), 79
midhinge (midH), 79
midpoints and locating peaks, 556ff
mids, 630
midsummaries, 79
midsummaries, use to pick a re-expression, 80
Mississippi, land areas of counties, 11ff 65ff
Mitchell, H. H., 371
Mobile, Alabama, 606
modes of expression, matched, 498
modified hinges, 223
modified medians, 223
Montgomery, D. J., 122
Morgan, M. E., 606
Mosteller, F., 105, 107, 665
motor vehicle deaths, 317
moving medians, 210
M-to-H, defined, 280
murders, 409

natural law?, 586
Nebraska voting, 371, 389
negative reciprocals, 70
"never" cases, difficulty with, 77
New Bedford, Massachusetts, 6, 18
New England Coast, 255
New Hampshire, for president, 237
New Hanover, New Jersey, 472
New York City rainfall *(exhibit)*, 99, 100, 206, 209, 330
niece, 516
nitrogen, Kjeldahl det'n of, 121
nonnegligent manslaughters, 409
numbers, pairs of, 23
numbers, scratching down, 1ff

∅ as a tag, 448
objectives, must be limited, 27
oceans, areas of, 95
octave bins, 589
octaves for logs, 552
omitted, what?, 664

opinion polls, 509ff
origin, wise change of, 193
outer fences, 44
outside, 44
outside values, 43ff

packed tubes, 456
papers per journal, 590ff
parallel schematic plots, 265ff
patents issued, 255
peak, desired, 584ff
peak, finding the, 555ff, 573
peak, locating by midpoints, 556ff
peak, polishing, 584ff
peak, smooth symmetrical, 555
peak, too narrow for technology, 562
peak, too sharp for ..., 566
Pearson, E. S., 565
Pearson, Karl, 516, 517, 545ff
Pearson's $r$, 656ff
percent point function, 650ff
percentages, 92
personal consumption *(exhibit tag)*, 345, 359
perspective reversals, 106
Phoenix, Arizona, 332ff
physics, atomic & molecular, 590ff, 600ff
physics, authors, 597-8
physics papers *(exhibit tag)*, 599-601
plankton hauls *(exhibit tag)*, 102, 104, 111, 112, 113
plants, counts of, 554
platinum sulfide, precipitation of, 122
plot, how to, 126
plot, shape of, 129
plots, choice of, 52
plots of relationship, 125ff
plots, should follow aims, 51
plots, straightening, 169
plotting, basic ideas of, 42
plotting, choice of ruling, 127
plotting, choice of scale units, 127
plotting, made easy vs. made effective, 42
plotting without graph paper, 43
pluralities, 499
plurality, 498
PLUS-one fits, 421ff

PLUS-one fits, beyond, 401ff
PLUS-one fits, parallel line plots, 431ff
PLUS-one fits, pictures for, 424ff
PLUS-one plot, 425ff
PLUS-one plots of residuals, making, 428ff
plutonium (IV), 26
points, three representative, 171
polish, other kinds, 404ff
polonium, 579, 584ff
polonium scintillations *(exhibit tag)*, 579-581, 585, 586
polynomial ladder, 207
population density, 149-154
population of England and Wales, 136ff
population of the U.S.A., 141ff, 173, 176
porosities, 456
port distances *(exhibit tag)*, 417
Portland, Maine, 302
ports, tonnage shipped from, 94
Postman, L., 107
postscript, 663ff
powers and logs, *(exhibit tag)*, 90, 91
powers, relation to logs, 86, 90, 91
precipitation, annual, 206, 209, 255
presidential Connecticut *(exhibit tag)*, 390, 391, 393-395
presidential voting, 116, 232, 389, 390ff
Preston, F. W., 552ff
price changes *(exhibit tag)*, 567-569
price series, 566
prisoners, Hindu male, 605, 606
production, quality of, 512
product-moment correlation, 656ff
product-ratio analysis, 594ff
product/ratio plots, 591ff
products, root of, 595ff
proteins, biological value of, 371
pseudospreads, 630ff
psychological experiment, 445
public debt, per capita, 255
pushback technique, 637ff
Pushkin, 590ff
*Pyrausta nubilalis* (Hbn.), 551

Quaker Run Valley, New York, 552
quick logs, 61
quick reciprocal roots, 89

quick reciprocals, 70
quick roots, 69
quick squares, 89

radioactive decay, 194ff, 579
radioactivity *(exhibit tag)*, 195-197
rainfall, 205ff
ranges, 46
rank, 29ff
rank down, 30
rank up, 30
rat running *(exhibit tag)*, 78
rates of increase, 445ff
ratio, log of, 595ff
rats, 605
reciprocal graph paper, 129
reciprocal times, 76
reciprocals, negative, 70
reciprocals, quick, 70
recip-root, 582ff
Reclamation, Bureau of, 14, 60
reduced delineations, 313ff
re-expressing both variables, 191
re-expression, amounts and counts, 397ff
re-expression, balances, 398
re-expression, gain from, 69
re-expression, Gaussian, of fractions, 623ff
re-expression, guidance for, 396ff
re-expression, in fitting counts, 555
re-expression, in three-way analysis, 458ff
re-expression, in two-way tables, 358ff, 396ff, 408ff
re-expression, single most needed, 61
re-expression, trivial, 88, 619
reference, Gaussian, 623ff
relationship, neutral, 483
relationship, plots of, 125ff
relationship, strongest negative, 483
relationship, strongest positive, 483
relatives, quantitive relation, 516
rematching, 482ff
repeated medians, 212
representing function, 650ff
Republican presidential candidates, 116
reroughing, 523ff

residuals, 113ff, 125ff
residuals, as a basis of choice, 248
residuals, coding, 382ff
residuals, condensing, 384ff
residuals, examining, 143
residuals, looking at, 351ff
residuals, map of, 151ff
residuals, of two sizes, 259
residuals, PLUS-one plot of, 427ff
residuals, two-way, 332ff
resmoothing by eye, 214
response, 125
responses, reformulated, 464ff
*Ribes* (currents and gooseberries), 548
ridley turtle, eggs of the, 572
right form, exactly?, 586
rivers, lengths of important, 95
$H_2O$, vapor pressure of, 187
root rough, 551ff
root rough, reference values, 546ff
root smooth, 551ff
roots, quick, 69
roots, quick reciprocal, 89
Roth, W. A., 68
rough, 208
rough, supporting the smooth, 263
rounding numbers, 4
row extra, 437ff
row-PLUS-col-PLUS-one fit, 356ff
row-PLUS-column, AN analysis, 340
row-PLUS-column analysis, 332ff
row-PLUS-column fit, 337ff
row-PLUS-column fits, extended, 436ff
row-PLUS-column fits, looking at, 349ff
row-PLUS-column residuals, looking at, 349ff
rows-TIMES-column analysis, 344ff
row-TIMES-column fits, special response, 422ff
running medians, 210
Russian words, 590
Rutherford, E., 579, 580

*Salicornia europea*, 554
sample description *(exhibit tag)*, 649
Saunders, A. A., 552ff
scale values (on graphs), 43
scales, changing, 153

scales for looking, 130
scales for plotting, 130
Scammon, Richard M., 116, 371, 389, 391
scatter plots, 665
scatter plots with middles, 665
Scheel, K., 68
schematic delineations, 313ff
schematic plots, 47
schematic plots, what they miss, 319
schematic (x,y)-plots, 315ff
school boys *(exhibit tag)*, 517
Schroeder staircase *(exhibit tag)*, 106, 108
scintillations, 579, 584ff
scratching down numbers, 6ff
sea distances, 415
Seaborg, G. T., 26
seas, areas of, 95
seasonal snowfall in Buffalo, 117
seasonal snowfall in Cairo, 117
second coordinate, why less important, 475ff
second pass, 665ff
semilog graph paper, 128
separations, smoothing, 280
sequence, 208
shape of distribution, identification, 656ff
shapes of distribution, 615ff
shapes of distribution, comparison, 654ff
shapes of distribution, via letter values, 626ff
shear, rate of, 25
Sheehan, F., 121
Shift, I., 166
significant figures, about so many, 69
simulations *(exhibit tag)*, 564-566, 570-576
sixteenths, 53ff
sizes, tied, 589
skip-meaning, 533
slice hinges, 280
slice medians, 280
slices, 269ff
slicing coordinate, changing the, 476
slide rule, 520
smallest basic count, 604ff

Smedal, A., 203
smooth, 208
smooth, supported by the rough, 227
smooth, why?, 205ff
smoothed roots, vs. squared shift, 555
smoothed roots, fitting to, 555
smoothed roots, making the fit, 556
smoothers, heavier, 534
smoothing, better, 523ff
smoothing both coordinates, 307
smoothing, in two-way analysis, 390ff
smoothing sequences, 205ff
smooths, 523ff
smooths and roughs, 666
smooths, heavier, 234
Snedecor, G. W., 103, 371
Socolow, R. H., 268
soldiers, plans to return, 512
solids, electrical properties of, 590ff, 601
son, 516
South, 610
speed, 182ff
split counts, 496ff
splitting peaks and valleys, 227ff
spread, agreement of, 97
spread, symmetry of, 97
square-root graph paper, 128
squares, quick, 89
ss-count, 496ff
ss-fractions, 497ff, 628
Stainton, 610
stars, radial velocities of, 26
started counts, 496ff, 605
starting a count, 496ff
starting part, 9
State, data, 317
state heights *(exhibit tag)*, 20, 38-41, 45
state populations, 46
state populations *(exhibit tag)*, 46, 54, 85
States, *see also* governor's salary, 315
stem, 8
stem-and-leaf, displays, 8ff
stem-and-leaf, doing better, 7
stem-and-leaf *(exhibit type)*, 24
stem-and-leaf, for additional information, 23
stem-and-leaf, mixed leaves, 12ff
stem-and-leaf, squeezed, 12

stem-and-leaf, stretched, 11
stems, +0 and -0, 16
stems, right number of, 11
step, 44
stock market prices, 568
Stouffer, S. A., 513
straight lines, finding, 136
straight lines, subtracting different, 137
straight lines, sum of two, 139
straightening, gain by, 148
straightening out plots, 169
strength of relationship, 482, 484ff
Stuart, A., 518-519
Student, 554, 563ff
Student's $t$, 656ff
Student's $t$, simulation of, 536ff, 570, 571ff
subtracting straight lines, 135
subtraction, looking at, 131
sugar beet, yields of, 371
summaries, 5-number, 32, 33ff
summaries, don't give details, 27
summarizing batches, 626
*Sunday Standard-Times*, 6, 18, 23
supplementary line, 139
suspended banks, 212
suspended banks, counts, 256
suspended banks *(exhibit tag)*, 257, 258
suspended deposits, 241ff
suspended deposits *(exhibit tag)*, 212, 245-247, 259, 263,
Swigert, G. F., 121
*Swiss Statistical Abstract*, 203
symmetry of spread, 97

tables, three- and more-way, 443ff
tagging of three- and more-way analyses, 443ff
Tatoosh Island, Washington, 255
technique schematic *(exhibit tag)*, 525
Teixeira, N. A., 118
temperature, 332ff, 385
termites, head breadths of, 123
Texas data *(exhibit tag)*, 300
textile establishments, 510
*The Captain's Daughter*, 590ff
three- and more-way analyses, arrangement, 443ff

three participations *(exhibit tag)*, 597, 603, 604
three points, fitting lines to, 174
three points, looking at, 171
3R, 214, 248
3R', 222
3RSS, 230
3RSSH, twice, 527ff
three sets of occurrences *(exhibit tag)*, 591-593
three variables, 321ff
three-and more-way analyses, tagging, 443ff
three-way analyses, making, 452ff
three-way analysis, breakdown, 448
three-way fits, 443ff
three-way residuals, 450ff
*Tide Tables, 2975*, 305
tilt, 154ff
tilt, clear away, 156
tilt, defined by untilting, 154
time series, 210
times for things to happen, 76
times, reciprocal, 76
Tippett, L. H. C., 82, 319
tools, best(?), 1
townhouses, 265ff
trace, 468ff
traces, reconstituted, 488, 490
traces, steep, 299
*Trachelipus rathkei* (an isopod), 606
tracing paper, 42, 128, 131
transparent plastic, 42, 131
trilinear graph paper, 128
trimeans, 46
trivial re-expressions, 88, 619
Tsao, Fei, 445ff
Tukey, J. W., 665
twicing, defined, 526
Twin Rivers *(exhibit tag)*, 267-275, 277, 279, 281-286, 311, 482, 485-489
two batches, comparisons of, 64
two eyes *(exhibit tag)*, 518
two keys to smoothing, 208
two-decimal logs, 62
two-decimal logs, use first, 61
two-factor interactions, 450ff
two-way analyses, making, 363ff

two-way analyses, using, 331ff
two-way analysis, 666
two-way analysis, with smoothing, 390ff
two-way fits, advanced, 421ff
two-way fits, *see also* PLUS-one fits
two-way fits, unusual, 415ff
two-way plot, 349ff
two-way plot, making, 374ff
two-way plot of residuals, 351ff, 378ff
two-way plots with skew lines, 426ff, 438
two-way schematic plot, 315
two-way tables, 331ff
two-way tables, no panacea, 416
two-way tables, of counts, 510

ultimate powers *(exhibit tag)*, 13, 14, 45, 60, 63
unincorporated places *(exhibit tag)*, 322, 323, 471, 473, 474, 478-480
units, 4
untilted, 154-158
untilting, 134, 154
untilting, not a rigid motion, 135
U.S. exports, 255
U.S. Internal Revenue Service, 620, 621
U.S. Meteorological Yearbook, 117
U.S.A. population *(exhibit tag)*, 141, 143-147, 176, 179, 180

valley birds *(exhibit tag)*, 552, 553
values, how distributed, 543ff, 551ff, 589ff, 615ff, 647ff
vapor pressure *(exhibit tag)*, 188-190
vapor pressure of mercury, 200
vapor pressure of $H_2O$, 187
variable, expressed, 468
variable, measured, 468
variable, named, 468
Velleman, Paul, 664
viscosity, 25
*Vital Statistics of the United States*, 655
volcano heights *(exhibit tag)*, 40, 41, 73-76, 80, 629, 631
von Neumann, John, 663
von Schweidler, E., 194, 195, 200
voting, 116, 237, 371, 389ff

Wagg, R. E., 457
wandering schematic plots, 265ff, 283, 285
warp breaks *(exhibit tag)*, 82
Warren, Ernest, 123
washed into the corners, 59
waterfalls, heights of, 96
"water-level" scales, 376
wheat, 205ff
wheat drying, 370
wheat price changes, 566ff
wheat production *(exhibit tag)*, 206, 209
whisker, 40ff
Wiberg, E., 202
width, bins of equal, 551ff
width, bins of unequal, 570ff
Willard, H. H., 120
Williams, C. B., 610, 612
Williams, H. B., 202
Williams, V. R., 202
Willis, 611
Wine, R. L., 457
Winsor, C. P., 103
Wishart, John, 565
Witherby, 611
Witwatersrand, South Africa, 547
Wood, F. S., 312
word, appearances per, 590ff
work stoppages, 255
*World Almanac,*, 14, 20, 39, 60, 64, 66, 73, 85, 87, 94-96, 99, 191, 213ff, 257ff, 289, 319, 320, 347, 370, 409, 418, 434
*World's Telephones, 1961*, 434
world's telephones *(exhibit tag)*, 433, 435, 439.

x Mod 1, 634
x-spacing, equal, 208

yeast cells, counts of, 554
Yuma, 332ff

zeros, when they are counted, 605ff
Zipf, G. K., 590, 595
Zipf's law, 590

exhibit 2 of chapter 15: reference table

**Pluralities, folded roots, folded logarithms -- alternative expressions for counted fractions (take sign of answer from head of column giving %)**

A) MAIN TABLE

| + | \|Plur.\| | \|froot\| | \|flog\| | − | | + | \|Plur.\| | \|froot\| | \|flog\| | − |
|---|---|---|---|---|---|---|---|---|---|---|
| 50% | use | .00 | use | 50% | | 85% | .70 | .76 | .87 | 15% |
| 51 | → | .02 | ← | 49 | | 86 | .72 | .78 | .91 | 14 |
| 52 |   | .04 |   | 48 | | 87 | .74 | .81 | .95 | 13 |
| 53 |   | .06 |   | 47 | | 88 | .76 | .84 | 1.00 | 12 |
| 54 |   | .08 |   | 46 | | 89 | .78 | .87 | 1.05 | 11 |
| 55% | use | .10 | use | 45% | | 90.0% | .80 | .89 | 1.10 | 10.0% |
| 56 | → | .12 | ← | 44 | | 90.5 | .81 | .91 | 1.13 | 9.5* |
| 57 |   | .14 |   | 43 | | 91 | .82 | .92 | 1.16 | 9 |
| 58 |   | .16 |   | 42 | | 91.5 | .83 | .94 | 1.19 | 8.5 |
| 59 |   | .18 |   | 41 | | 92 | .84 | .96 | 1.22 | 8 |
| 60% | use | .20 | use | 40% | | 92.5% | .85 | .97 | 1.26 | 7.5% |
| 61 | → | .22 | ← | 39 | | 93 | .86 | .99 | 1.29 | 7 * |
| 62 |   | .24 |   | 38 | | 93.5 | .87 | 1.01 | 1.33 | 6.5 |
| 63 | .26 | .26 | .27 | 37 | | 94 | .88 | 1.02 | 1.37 | 6 |
| 64 | .28 | .28 | .29 | 36 | | 94.5 | .89 | 1.04 | 1.42 | 5.5 |
| 65% | .30 | .30 | .31 | 35% | | 95.0% | .90 | 1.06 | 1.47 | 5.0% |
| 66 | .32 | .32 | .33 | 34 | | 95.5 | .91 | 1.08 | 1.53 | 4.5 |
| 67 | .34 | .35 | .35 | 33 | | 96 | .92 | 1.10 | 1.59 | 4 |
| 68 | .36 | .37 | .38 | 32 | | 96.5 | .93 | 1.12 | 1.65 | 3.5 |
| 69 | .38 | .39 | .40 | 31 | | 97 | .94 | 1.15 | 1.74 | 3 |
| 70% | .40 | .41 | .42 | 30% | | 97.2% | .94 | 1.16 | 1.77 | 2.8% |
| 71 | .42 | .43 | .45 | 29 | | 97.4 | .95 | 1.17 | 1.81 | 2.6 |
| 72 | .44 | .45 | .47 | 28 | | 97.6 | .95 | 1.18 | 1.85 | 2.4 |
| 73 | .46 | .47 | .50 | 27 | | 97.8 | .96 | 1.19 | 1.90 | 2.2 |
| 74 | .48 | .50 | .52 | 26 | | 98.0 | .96 | 1.20 | 1.95 | 2.0 |
| 75% | .50 | .52 | .55 | 25% | | 98.2% | .96 | 1.21 | 2.00 | 1.8% |
| 76 | .52 | .54 | .58 | 24 | | 98.4 | .97 | 1.22 | 2.06 | 1.6 |
| 77 | .54 | .56 | .60 | 23 | | 98.6 | .97 | 1.24 | 2.13 | 1.4 |
| 78 | .56 | .59 | .63 | 22 | | 98.8 | .98 | 1.25 | 2.21 | 1.2 |
| 79 | .58 | .61 | .66 | 21 | | 99.0 | .98 | 1.27 | 2.30 | 1.0 |
| 80% | .60 | .63 | .69 | 20% | | 99.2% | .99 | 1.28 | 2.41 | 0.8% |
| 81 | .62 | .66 | .72 | 19 | | 99.4 | .99 | 1.30 | 2.55 | 0.6 |
| 82 | .64 | .68 | .76 | 18 | | 99.6 | .99 | 1.32 | 2.76 | 0.4 |
| 83 | .66 | .71 | .79 | 17 | | 99.8 | 1.00 | 1.35 | 3.11 | 0.2 |
| 84 | .68 | .73 | .83 | 16 | | 100.0% | 1.00 | 1.41 | * | 0.0 |

For more detail for flogs of fractions beyond 1% or 99%, see page 500.

exhibit 4 of chapter 15: reference table 503

**Values of $\log_e \sqrt{\text{count} + \frac{1}{8}}$ (a) for integer counts from 0 to 199, (b) for every 10th count from 200 to 3090, (c) for larger counts, (d) for half-integers $\leq 20$**

**A) TABLE for COUNTS up to 199--see panel D for half integers up to 20**

|     | 0    | 1    | 2    | 3    | 4    | 5    | 6    | 7    | 8    | 9    |
|-----|------|------|------|------|------|------|------|------|------|------|
| 00  | -.90 | .08  | .39  | .58  | .71  | .82  | .91  | .98  | 1.05 | 1.11 |
| 10  | 1.16 | 1.21 | 1.25 | 1.29 | 1.33 | 1.36 | 1.39 | 1.42 | 1.45 | 1.48 |
| 20  | 1.50 | 1.53 | 1.55 | 1.57 | 1.59 | 1.61 | 1.63 | 1.65 | 1.67 | 1.69 |
| 30  | 1.70 | 1.72 | 1.74 | 1.75 | 1.77 | 1.78 | 1.79 | 1.81 | 1.82 | 1.83 |
| 40  | 1.85 | 1.86 | 1.87 | 1.88 | 1.89 | 1.91 | 1.92 | 1.93 | 1.94 | 1.95 |
| 50  | 1.96 | 1.97 | 1.98 | 1.99 | 2.00 | 2.01 | 2.01 | 2.02 | 2.03 | 2.04 |
| 60  | 2.05 | 2.06 | 2.06 | 2.07 | 2.08 | 2.09 | 2.10 | 2.10 | 2.11 | 2.12 |
| 70  | 2.13 | 2.13 | 2.14 | 2.15 | 2.15 | 2.16 | 2.17 | 2.17 | 2.18 | 2.19 |
| 80  | 2.19 | 2.20 | 2.20 | 2.21 | 2.22 | 2.22 | 2.23 | 2.23 | 2.24 | 2.25 |
| 90  | 2.25 | 2.26 | 2.26 | 2.27 | 2.27 | 2.28 | 2.28 | 2.29 | 2.29 | 2.30 |
| 100 | 2.30 | 2.31 | 2.31 | 2.32 | 2.32 | 2.33 | 2.33 | 2.34 | 2.34 | 2.35 |
| 110 | 2.35 | 2.36 | 2.36 | 2.36 | 2.37 | 2.37 | 2.38 | 2.38 | 2.39 | 2.39 |
| 120 | 2.39 | 2.40 | 2.40 | 2.41 | 2.41 | 2.41 | 2.42 | 2.42 | 2.43 | 2.43 |
| 130 | 2.43 | 2.44 | 2.44 | 2.45 | 2.45 | 2.45 | 2.46 | 2.46 | 2.46 | 2.47 |
| 140 | 2.47 | 2.47 | 2.48 | 2.48 | 2.49 | 2.49 | 2.49 | 2.50 | 2.50 | 2.50 |
| 150 | 2.51 | 2.51 | 2.51 | 2.52 | 2.52 | 2.52 | 2.53 | 2.53 | 2.53 | 2.53 |
| 160 | 2.54 | 2.54 | 2.54 | 2.55 | 2.55 | 2.55 | 2.56 | 2.56 | 2.56 | 2.57 |
| 170 | 2.57 | 2.57 | 2.57 | 2.58 | 2.58 | 2.58 | 2.59 | 2.59 | 2.59 | 2.59 |
| 180 | 2.60 | 2.60 | 2.60 | 2.61 | 2.61 | 2.61 | 2.61 | 2.62 | 2.61 | 2.62 |
| 190 | 2.62 | 2.63 | 2.63 | 2.63 | 2.63 | 2.64 | 2.64 | 2.64 | 2.64 | 2.65 |

**B) TABLE for COUNTS from 200 to 3090--grab nearest; don't interpolate**

|      | 00    | 10    | 20    | 30    | 40    | 50    | 60    | 70    | 80    | 90    |
|------|-------|-------|-------|-------|-------|-------|-------|-------|-------|-------|
| 200  | 2.65  | 2.67  | 2.70  | 2.72  | 2.74  | 2.76  | 2.78  | 2.80  | 2.82  | 2.84  |
| 300  | 2.85  | 2.87  | 2.88  | 2.90  | 2.91  | 2.93  | 2.94  | 2.96  | 2.97  | 2.98  |
| 400  | 3.00  | 3.01  | 3.02  | 3.03  | 3.04  | 3.05  | 3.07  | 3.08  | 3.09  | 3.10  |
| 500  | 3.11  | 3.12  | 3.13  | 3.14  | 3.15  | 3.16  | 3.16  | 3.17  | 3.18  | 3.19  |
| 600  | 3.20  | 3.21  | 3.21  | 3.22  | 3.23  | 3.24  | 3.25  | 3.25  | 3.26  | 3.27  |
| 700  | 3.28  | 3.28  | 3.29  | 3.30  | 3.30  | 3.31  | 3.32  | 3.32  | 3.33  | 3.34  |
| 800  | 3.34  | 3.35  | 3.35  | 3.36  | 3.37  | 3.37  | 3.38  | 3.38  | 3.39  | 3.40  |
| 900  | 3.40  | 3.41  | 3.41  | 3.42  | 3.42  | 3.43  | 3.43  | 3.44  | 3.44  | 3.45  |
| 1000 | 3.45  | 3.46  | 3.46  | 3.47  | 3.47  | 3.48  | 3.48  | 3.49  | 3.49  | 3.50  |
| 1100 | 3.50  | 3.51  | 3.51  | 3.52  | 3.52  | 3.52  | 3.53  | 3.53  | 3.54  | 3.54  |
| 1200 | 3.545 | 3.549 | 3.553 | 3.557 | 3.562 | 3.566 | 3.569 | 3.573 | 3.577 | 3.581 |
| 1300 | 3.585 | 3.589 | 3.593 | 3.597 | 3.600 | 3.604 | 3.608 | 3.611 | 3.615 | 3.619 |
| 1400 | 3.622 | 3.626 | 3.629 | 3.633 | 3.636 | 3.640 | 3.643 | 3.647 | 3.650 | 3.653 |
| 1500 | 3.657 | 3.660 | 3.663 | 3.667 | 3.670 | 3.673 | 3.676 | 3.679 | 3.683 | 3.686 |
| 1600 | 3.689 | 3.692 | 3.695 | 3.698 | 3.701 | 3.704 | 3.707 | 3.710 | 3.713 | 3.716 |
| 1700 | 3.719 | 3.722 | 3.725 | 3.728 | 3.731 | 3.734 | 3.737 | 3.739 | 3.742 | 3.745 |
| 1800 | 3.748 | 3.751 | 3.753 | 3.756 | 3.759 | 3.762 | 3.764 | 3.767 | 3.770 | 3.772 |
| 1900 | 3.775 | 3.777 | 3.780 | 3.783 | 3.785 | 3.788 | 3.790 | 3.793 | 3.795 | 3.798 |

For continuation in text, for counts above 1900, see pages 504–505.